T0174288

Handbook
of
Synthetic
Antioxidants

ANTIOXIDANTS IN HEALTH AND DISEASE

Series Editors

LESTER PACKER, PH.D.
University of California
Berkeley, California

JÜRGEN FUCHS, PH.D., M.D.
Johann Wolfgang Goethe University
Frankfurt, Germany

Handbook of Synthetic Antioxidants

edited by
Lester Packer

University of California
Berkeley, California

Enrique Cadenas

University of Southern California
School of Pharmacy
Los Angeles, California

CRC Press
Taylor & Francis Group
Boca Raton London New York

CRC Press is an imprint of the
Taylor & Francis Group, an **informa** business

CRC Press
Taylor & Francis Group
6000 Broken Sound Parkway NW, Suite 300
Boca Raton, FL 33487-2742

© 1997 by Taylor & Francis Group, LLC
CRC Press is an imprint of Taylor & Francis Group, an Informa business

First issued in paperback 2019

No claim to original U.S. Government works

ISBN 13: 978-0-367-45585-9 (pbk)
ISBN 13: 978-0-8247-9810-9 (hbk)

This book contains information obtained from authentic and highly regarded sources. Reasonable efforts have been made to publish reliable data and information, but the author and publisher cannot assume responsibility for the validity of all materials or the consequences of their use. The authors and publishers have attempted to trace the copyright holders of all material reproduced in this publication and apologize to copyright holders if permission to publish in this form has not been obtained. If any copyright material has not been acknowledged please write and let us know so we may rectify in any future reprint.

Except as permitted under U.S. Copyright Law, no part of this book may be reprinted, reproduced, transmitted, or utilized in any form by any electronic, mechanical, or other means, now known or hereafter invented, including photocopying, microfilming, and recording, or in any information storage or retrieval system, without written permission from the publishers.

For permission to photocopy or use material electronically from this work, please access www.copyright.com (http://www.copyright.com/) or contact the Copyright Clearance Center, Inc. (CCC), 222 Rosewood Drive, Danvers, MA 01923, 978-750-8400. CCC is a not-for-profit organization that provides licenses and registration for a variety of users. For organizations that have been granted a photocopy license by the CCC, a separate system of payment has been arranged.

Trademark Notice: Product or corporate names may be trademarks or registered trademarks, and are used only for identification and explanation without intent to infringe.

Visit the Taylor & Francis Web site at
http://www.taylorandfrancis.com

and the CRC Press Web site at
http://www.crcpress.com

Series Introduction

In June of 1992, 17 international researchers in the field of free radical and antioxidant biology and preventive medicine met at the village of Saas Fee, Switzerland, and drew up the Saas Fee Declaration to recognize the importance of prevention in medicine and health. Since then, hundreds of researchers from around the world have signed the declaration:

Saas Fee Declaration
On the significance of antioxidants in preventive medicine.
1. The intensive research on free radicals of the past 15 years by scientists worldwide has led to the statement in 1992 that antioxidant nutrients may have major significance in the prevention of a number of diseases. These include cardiovascular and cerebrovascular disease, some forms of cancer and several other disorders, many of which may be age-related.
2. There is now general agreement that there is a need for further work at the fundamental scientific level, as well as in large-scale randomized trials and in clinical medicine, which can be expected to lead to more precise information being made available.
3. The major objective of this work is the prevention of disease. This may be achieved by use of antioxidants which are natural physiological substances. The strategy should be to achieve optimal intakes of these antioxidant nutrients as part of preventive medicine.

4. It is quite clear that many environmental sources of free radicals exist, such as ozone, sunlight, and other forms of radiation, smog, dust, and other atmospheric pollutants. The optimal intake of antioxidants provides a preventive measure against these hazards.
5. There is a great need for improvement in public awareness of the potential preventive benefits of antioxidant nutrient intake. There is overwhelming evidence that the antioxidant nutrients such as vitamin E, vitamin C, carotenoids, alphalipoic acid and others are safe even at very high levels of intake.
6. Moreover, there is now substantial agreement that governmental agencies, health professionals and the media should promote information transfer to the general public, particularly when evidence exists that benefits for human health and public expenditure are overwhelming.

This declaration arose from the overwhelming evidence now available indicating that antioxidants play a critical role in wellness, health maintenance, and the prevention of chronic and degenerative diseases. Antioxidants neutralize free radicals that are generated during normal metabolism and during exposure to environmental insult. Free radicals play a role in most major health problems of the industrialized world, including cardiovascular disease, cancer, and disorders of aging.

Some antioxidants are quite familiar as vitamins or vitamin-forming compounds: vitamin E, vitamin C, and the carotenoids, including beta-carotene. These antioxidants must be constantly replenished through the diet. Others, such as ubiquinols and the thiol antioxidants, including glutathione and lipoic acid, are manufactured by the body, but the levels of many of these can be bolstered through dietary supplementation. Until recently, it was thought that each antioxidant played its role in isolation from the others. But work in several laboratories indicates that there is a dynamic interplay among the systems. For example, when vitamin E neutralizes a free radical in a membrane, it becomes itself a relatively harmless free radical, which decomposes. However, vitamin C can regenerate vitamin E from the vitamin E radical, in effect "recycling" vitamin E. Vitamin E becomes a radical in the process, but it, too, can be recycled by interacting with other antioxidant systems. It has been shown that these interactions occur in the test tube, and nutritional supplementation studies support this idea for the whole organism. Thus, a picture is emerging of a complex interplay among the defense systems, with the various antioxidant cycles acting to prevent cell damage and disease. Out knowledge is far from complete but these findings already have implications in terms of recommendations for supplementation.

Hence, it seems particularly appropriate to offer this series at the present time. Never has the demand for knowledge about antioxidants been greater, and never has their potential for treating disease and improving health been clearer. The series highlights natural antioxidants and artificial antioxidants that mimic natural systems.

Synthetic antioxidants, either of natural sources or designed to mimic biological antioxidant systems, are proving to be of great value. Some are drugs that are effective in therapy. Others may be useful in bolstering the body's natural antioxidant systems by carefully designing biological mimics of natural antioxidants.

Synthetic antioxidants are an area of research actively being pursued by pharmacologists, physiologists, biochemists, and cell biologists seeking to find new and better means to enhance antioxidant defenses, particularly under conditions of environmental and/or oxidative stress diseases. This remains an activity that will undergo many new developments in the future, as more and more we turn to the use of such substances in therapy for human disorders and for biotechnological purposes.

Lester Packer
Jürgen Fuchs

Preface

The recognition of the involvement of oxygen radicals in several pathologies has led to the implementation of antioxidant therapy. Dietary nutrients with antioxidant properties are assuming great significance in the context of certain pathologies, such as atherosclerosis. In a classical sense, vitamin E is the prototype antioxidant with a probable role in therapeutic regimes directed toward oxidative stress–mediated diseases.

Synthetic antioxidants are of potential use in chemistry, the food industry, and medicine. Some of these compounds retain a functional group chemistry analogous to that of "natural" antioxidants and introduce new chemical groups that enhance their range of cellular action or make them available to cell sites hitherto restricted. Alternatively, some synthetic antioxidants bear no structural analogy with natural antioxidants and exhibit a high reactivity toward reactive oxygen species and/or display selective protection in certain tissues.

Despite the overwhelming number of reports on synthetic antioxidants, no treatise exists that deals with the chemical, biological, medical, and industrial aspects of these compounds. Likewise, a comprehensive and systematic classification of synthetic antioxidants is missing. This book, which is a sequel to *Handbook of Antioxidants* (edited by Enrique Cadenas and Lester Packer), covers these aspects. The chapters provide a rationale for a classification of synthetic antioxidants, and each group is surveyed exhaustively from the physicochemical properties of these compounds to their applications in clinical medicine and the food industry.

Lester Packer
Enrique Cadenas

Contents

Contents

Contributors

Mary E. Anderson   Department of Biochemistry, Cornell University Medical College, New York, New York

Frank N. Bolkenius, Ph.D.   Department of Enzymology, Marion Merrell Research Institute, Strasbourg, France

Clifford S. Collis, Ph.D., C. Biol.   International Antioxidant Research Centre, UMDS Guy's Hospital, London, England

Ian A. Cotgreave, Ph.D.   Biochemical Toxicology Unit, Institute of Environmental Medicine, Karolinska Institute, Stockholm, Sweden

James Dow, Ph.D.   Clinical Pharmacokinetics, Glaxo Wellcome Research and Development, Greenford, Middlesex, England

Lars Engman, Ph.D.   Department of Organic Chemistry, Uppsala University, Uppsala, Sweden

Dennis E. Epps, Ph.D.   Chemical and Biological Screening, Pharmacia & Upjohn, Inc., Kalamazoo, Michigan

Marc W. Fariss, Ph.D.   Pharmaceutical Sciences, College of Pharmacy, Washington State University, Pullman, Washington

Kevin M. Faulkner, Ph.D.   Texas Wesleyan University School of Law, Irving, Texas

Robert A. Floyd, Ph.D.   Free Radical Biology and Aging Research Program, Oklahoma Medical Research Foundation, Oklahoma City, Oklahoma

Irwin Fridovich, Ph.D.   Department of Biochemistry, Duke University Medical Center, Durham, North Carolina

Edward D. Hall, Ph.D.   CNS Diseases Research, Pharmacia & Upjohn, Inc., Kalamazoo, Michigan

Nobuya Haramaki, M.D., Ph.D.   Third Department of Internal Medicine, Kurume University School of Medicine, Kurume, Japan

Regine Kahl, M.D.   Institute of Toxicology, University of Düsseldorf, Düsseldorf, Germany

Teruyuki Kawabata   Department of Pathology, Okayama University Medical School, Okayama, Japan

Murali C. Krishna, Ph.D.   Radiation Biology Branch, National Cancer Institute, National Institutes of Health, Bethesda, Maryland

Guang-Jun Liu   Free Radical Biology and Aging Research Program, Oklahoma Medical Research Foundation, Oklahoma City, Oklahoma

Gilbert Marciniak, Ph.D.   Department of Chemistry, Marion Merrell Research Institute, Strasbourg, France

Lucia Marcocci   Department of Molecular and Cell Biology, University of California, Berkeley, California, and University of Rome, Rome, Italy

John M. McCall, Ph.D.   Discovery Research, Pharmacia & Upjohn, Inc., Kalamazoo, Michigan

Etsuo Niki, Ph.D. Research Center for Advanced Science and Technology, University of Tokyo, Tokyo, Japan

Noriko Noguchi, M.D. Research Center for Advanced Science and Technology, University of Tokyo, Tokyo, Japan

Lester Packer, Ph.D. Department of Molecular and Cell Biology, University of California, Berkeley, California

Margaret A. Petty Department of Pharmacology, Marion Merrell Research Institute, Strasbourg, France

Catherine A. Rice-Evans, Ph.D. International Antioxidant Research Centre, UMDS Guy's Hospital, London, England

Amram Samuni, Ph.D. Department of Molecular Biology, Hebrew University Medical School, Jerusalem, Israel

Craig E. Thomas, Ph.D. Biochemical Technology, Lilly Research Laboratories, Eli Lilly and Company, Indianapolis, Indiana

Peter K. Wong Free Radical Biology and Aging Research Program, Oklahoma Medical Research Foundation, Oklahoma City, Oklahoma

Handbook
of
Synthetic
Antioxidants

1

Approaches and Rationale for the Design of Synthetic Antioxidants as Therapeutic Agents

Craig E. Thomas
Eli Lilly and Company, Indianapolis, Indiana

I. INTRODUCTION

Over the eons, numerous physiological processes and biological organelles have developed that allow existence in a precarious aerobic environment. Even though molecular oxygen is absolutely required to provide the fuel [e.g., adenosine triphosphate (ATP), reduced form of nicotinamide-adenine dinucleotide phosphate (NADPH)] which sustains cell viability, it also poses an enormous threat. Dioxygen, containing two unpaired electrons with parallel spin in its outermost shell ($2p\pi^*$ antibonding orbital), readily undergoes one electron redox reactions to generate reactive intermediates. Under ideal aerobic situations, oxygen is fully reduced by four electrons to H_2O in tightly coupled mitochondria. It is apparent, however, that partially reduced species of dioxygen, generated via one-electron reduction processes, do exist in biological systems. Consequently, cells have evolved a series of enzymatic and nonenzymatic means to minimize the accumulation of potentially damaging radicals. Figure 1 shows the sequential one-electron reduction products of dioxygen and the enzymes that either interconvert or reduce the active oxygen species to nonradical products.

Situations do arise in which there is an imbalance in free-radical formation and removal. Increasing evidence suggests that such an imbalance may contribute to the pathophysiological abnormalities associated with disease, exposure to environment pollutants, and aging. A list of some such conditions is given in

1

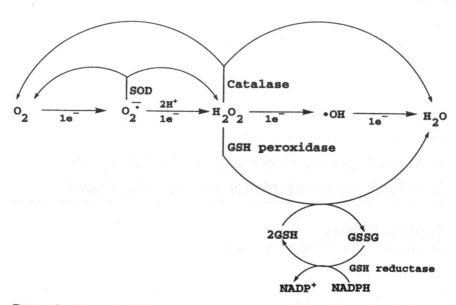

Figure 1 Oxygen radical formation and antioxidant enzymes. Scheme shows sequential univalent reduction of molecular oxygen and enzymes that act on partially reduced species of dioxygen. Only major products and substrates are depicted; thus, equations are not balanced. The major oxygen radicals observed are superoxide anion ($O_2^{\cdot-}$), hydrogen peroxide (H_2O_2), and hydroxyl radical ($\cdot OH$). Superoxide dismutase (SOD) converts $O_2^{\cdot-}$ to H_2O_2, which is reduced to H_2O by the enzymes catalase and GSH peroxidase (GSH-Px).

Table 1. For example, it was recently proposed that sleep represents a means by which free-radical accumulation is limited (1). It was suggested that radicals are generated during the awake period as a result of metabolism. Thus, a slower metabolic rate during sleep, coupled with a proposed increased efficiency of endogenous antioxidant mechanisms, would minimize radical-dependent damage. Although little evidence exists to support this suggestion, it does accent the wide-ranging opinions as to the formation, control, and consequences of reactive free radicals in biology.

Radicals are implicated in a wide range of diseases, but it is difficult to ascertain whether they represent a cause or an effect. Discerning between the two is hampered by (1) the lack of specific and sensitive assays for monitoring oxidative events in vivo and (2) in most cases, manifestation of the abnormality at a late stage, making determination of causality problematic. Nonetheless, for certain conditions, there is reasonable evidence to suggest that increased active oxygen formation contributes to disease initiation or progression. Accordingly, antioxidants represent a potentially valuable therapeutic modality. With

Table 1   Free-Radical Involvement (?) in
Pathophysiological Conditions

Adriamycin cardiomyopathy
AIDS
Adult respiratory distress syndrome (ARDS)
Aging
Alcoholism
Alzheimer's disease
Amyotrophic lateral sclerosis (ALS)
Anemia
Atherosclerosis
Autoimmune disorders
Batten disease
Cancer
Cataracts
Cystic fibrosis
Diabetes
Emphysema
Exercise
Favism
Iron overload
Huntington's disease
Kwashiorkor
Malaria
Multiple sclerosis
Muscular dystrophy
Myocardial infarction
Organ transplantation
Oxygen toxicity
Pancreatitis
Parkinson's disease
Porphyria
Radiation injury
Septic shock
Stroke
Trauma
Ulcerative colitis

regard to pharmaceuticals, there is only one synthetic "designer antioxidant"
available (Freedox, approved for use in three European countries). There are
a number of retrospective and prospective clinical trials that have or are evaluating the impact of natural antioxidants such as β-carotene, α-tocopherol, and
ascorbic acid (2,3) on, for example, cardiovascular events such as atheroscle-

rosis and heart attacks. It is reasonable to assume that, as new information emerges, there will be considerable impetus to develop antioxidants for acute and chronic prophylactic treatment. This chapter overviews some of the approaches that can be used to control oxidative stress and gives examples of synthetic organic compounds that have been demonstrated to be antioxidants in chemical systems or show activity in animal models of free-radical-mediated diseases or in human clinical trials.

II. APPROACHES TO THE CONTROL OF OXIDATIVE DAMAGE

When attempting to design a synthetic antioxidant, one must consider variables such as the site of radical production, the type of radicals produced, and the critical cellular targets, which, when oxidized, lead to a loss of cell function or viability (Table 2). Within any pathophysiological condition there may be more than one "player" for each category, thus making the number of possible destructive combinations virtually endless. Furthermore, it is often difficult to determine some of the variables. Thus, in the design of antioxidants, it is informative to consider three events common to virtually all free-radical-mediated events: initiation, propagation, and termination (4). This is best illustrated by the well-studied phenomenon of lipid peroxidation (Fig. 2). Assuming that similar principles apply to other biomolecules, control of initiation and propagation reactions is a reasonable general approach to designing a synthetic antioxidant (Table 3). Refining this approach, such as targeting specific tissues or subcellular organelles, is discussed later.

A. Prevention of Initiation

Ideally, preventing the formation of an initiating radical species or the initiation event is the best approach to controlling oxidative damage. After initiation, the explosive nature of free-radical chain propagation makes it inherently difficult to control. Preventing initiation is challenging. Often, one does not know what radical is being formed, and its reactivity with biological molecules is often at a rate constant approaching diffusion control. In biological systems the iron-catalyzed Haber-Weiss reaction is often invoked as being involved in initiation via formation of $\cdot OH$:

$$O_2^{\cdot-} + Fe^{3+} \rightarrow O_2 + Fe^{2+}$$

$$O_2^{\cdot-} + O_2^{\cdot-} \xrightarrow{2H^+} H_2O_2 + O_2$$

$$H_2O_2 + Fe^{2+} \rightarrow \cdot OH + {}^-OH + Fe^{3+}$$

Table 2 Free Radicals in Biological Systems: Sources and Targets

Radical/radical precursors	Radical sources	Targets
$O_2^{\cdot-}$	Xanthine oxidase	Cellular
H_2O_2	Electron transport chains	Plasma membrane (lipids and proteins)
$\cdot OH$	Hemoglobin	Mitochondria
ROOH	Phagocytic cells (NADPH oxidase)	Endoplasmic reticulum
RO$\cdot$	Xenobiotic metabolism	Cytosolic proteins
ROO$\cdot$	Ionizing radiation	DNA (mitochondrial and nuclear)
$^1O_2$	Autoxidation of reduced catechols	Extracellular
NO$\cdot$	Photochemical reactions	Lipoproteins
ONOO$^-$		Connective tissue

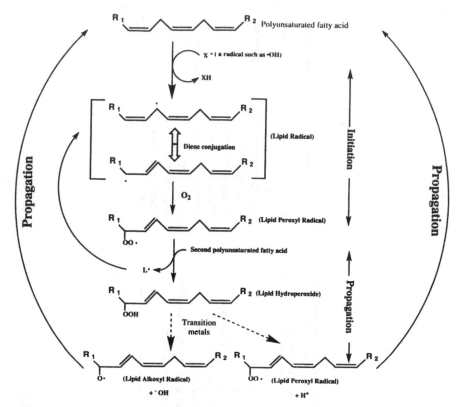

Figure 2 Schematic representation of lipid peroxidation. Lipid peroxidation is initiated via H atom abstraction by a radical (X·). The resultant lipid radical forms a conjugated diene and adds dioxygen to produce a lipid peroxyl radical. This species can abstract another H atom to produce a second lipid radical (L·) and a lipid hydroperoxide (LOOH) in the propagation phase. Transition metals (M) can cleave LOOH to alkoxyl and peroxyl radicals, which can abstract H atoms also. Thus, the process is cyclic in nature and proceeds rapidly following a single initiation event. (From Ref. 37; courtesy of The New York Academy of Sciences.)

If this is true, then minimizing O_2^{-} formation, or facilitating its removal, is an attractive option because it is required for metal reduction and H_2O_2 production. This approach has been tried by several investigators. It is difficult to prevent the formation of O_2^{-} due to generation via multiple pathways. For example, during ischemia or hypoxia the lack of sufficient oxygen to accept electrons from the mitochondrial electron transport system may lead to an accumulation of reduced components of the chain such as $FADH_2$ and coenzyme Q. Upon reperfusion, these molecules can directly transfer electrons to dioxygen to produce O_2^{-}, H_2O_2 or both. Prevention of this reaction would require the use

Table 3 Mechanisms and Examples for Preventing Oxidative Damage

Inhibition of initiation reactions
  Prevention of $O_2^{\cdot-}$ formation
    Allopurinol, oxypurinol, methylene blue, OPC-15161, OPC-6335
  Removal or dismutation of $O_2^{\cdot-}$
      SOD mimetics (CuDIPS, SC-52608)
      Modified SOD (PEG-SOD, HB-SOD, IgA-SOD)
      Nitroxides (OXANO)
      Hydroxylamines (OXANOH)
  Trapping of $\cdot$OH
    Mannitol, aspirin
    Spin traps (DMPO, PBN, MDL 101,002)
  Iron chelators
    Desferrioxamine, L-1
Prevention or limitation of propagation reactions
  Reactions with carbon-centered radicals
    Aminoxyls (indolinonic and quinolinic structures related to
TEMPO)
    Nitroxides (PBN, MD101,002)
  Reactions with oxygen-centered radicals
    α-Tocopherol, Trolox, U-74006F (Tirilazad), MDL 74,405
    6-Hydroxychroman-2-carbonitriles, WIN 62079, U-78517F
    IRFI-016, probucol
  Removal of peroxides
    Ebselen, selenosubtilisin, PEG-catalase

of alternative electron acceptors, which would not likely be feasible because of the efficiency with which dioxygen reacts with such molecules.

The primary approach to control of $O_2^{\cdot-}$ formation relies on its generation via the action of xanthine oxidase. Allopurinol (4-hydroxypyrazolol [3,4,-d] pyrimidine) is a potent inhibitor of xanthine oxidase, which has been shown to prevent oxidative damage to cardiac tissue (5) and the eye lens (6). It must be considered, however, that allopurinol works in tissues devoid of xanthine oxidase activity, suggesting that it may have other unrecognized properties and that its metabolic product, oxypurinol, is a scavenger of $\cdot$OH and hypochlorous acid. Therefore, it cannot be concluded with certainty that inhibition of $O_2^{\cdot-}$ formation is a reasonable approach, although the approach is conceptually appealing. Recently, Salaris and colleagues suggested a new approach (7). These authors capitalized on the ability of methylene blue to act as an electron acceptor from xanthine oxidase and proposed that it be considered for use in attenuation of ischemia and reperfusion injury. Indeed, methylene blue prevented the formation of lipid peroxidation products in rat kidney and liver slice models of reperfusion injury. In practice, the usefulness of this approach is unclear because the compound was administered before reperfusion and because the reduced methylene blue gives rise to H_2O_2, which could still participate in oxidative reactions.

Many disease states include some inflammatory component that evokes the neutrophil "oxidative burst" response. Inhibition of this response is an alternative method to inhibit $O_2^{\cdot-}$ formation. Otsuka Pharmaceutical Co. has isolated a pyrazine-based compound from a fungal broth and demonstrated that it (OPC-15161) and its major degradation product (OPC-15160) could inhibit $O_2^{\cdot-}$ generation from guinea pig macrophages (8). This group has recently reported on the synthesis of thiazole derivatives, particularly OPC-6335, which inhibited formylmethionylleucyl phenylalanine (fMLP)-stimulated $O_2^{\cdot-}$ formation with IC_{50} values in the 1-μM range (9). The usefulness of this approach in vivo has yet to be tested.

An alternative to prevention of formation of $O_2^{\cdot-}$ is to remove it from the system by enhancing dismutation or by "scavenging." Fridovich and colleagues originally described in vitro studies with superoxide dismutase (SOD) mimetics, which were complexes of MnO_2 and desferrioxamine (10,11). Similarly, copper complexed with diisopropyl salicylate catalyzes $O_2^{\cdot-}$ dismutation; however, free copper can be released intracellularly, resulting in toxicity (C. Thomas, unpublished data). This approach has been adopted by Monsanto Corp., which has prepared a number of low-molecular-weight manganese macrocyclic ligand complexes as SOD mimics (12,13). The lead compound SC-52608 has shown protection in cell culture studies and in vivo in a rabbit model of myocardial ischemia and reperfusion injury (14). Metal-independent SOD mimics such as OXANO have been well characterized by Russo et al. (15–18). (Fig. 3). These stable nitroxides undergo sequential reduction and oxidation by $O_2^{\cdot-}$, resulting, overall, in formation of H_2O_2 and O_2.

Although studies with SOD mimetics have been limited to in vitro systems and animal models, clinical trials with various forms of SOD are ongoing. As with any peptide-based approach, major issues concerning half-life and cell penetration exist. To increase circulating lifetime, Eastman Kodak has developed PEG-SOD, which is in phase 3 clinical trials for central nervous system (CNS) trauma. Cetus and Chiron are jointly developing a recombinant SOD, which is in phase 2 trials in Japan for the treatment of osteoarthritis; other companies are also pursuing the recombinant enzyme or other forms, such as extracellular or erythrocyte-derived, of SOD. As discussed with methylene blue, the product of these enzymes is H_2O_2, which may still contribute to oxidative stress. Also, the cellular formation of H_2O_2 by direct two-electron reduction would bypass $O_2^{\cdot-}$ as an intermediate and render SOD and its mimetic useless. An alternative approach is to remove hydrogen peroxide. It has been shown that poly-L-histidine-catalase complexes could protect endothelial cells from H_2O_2

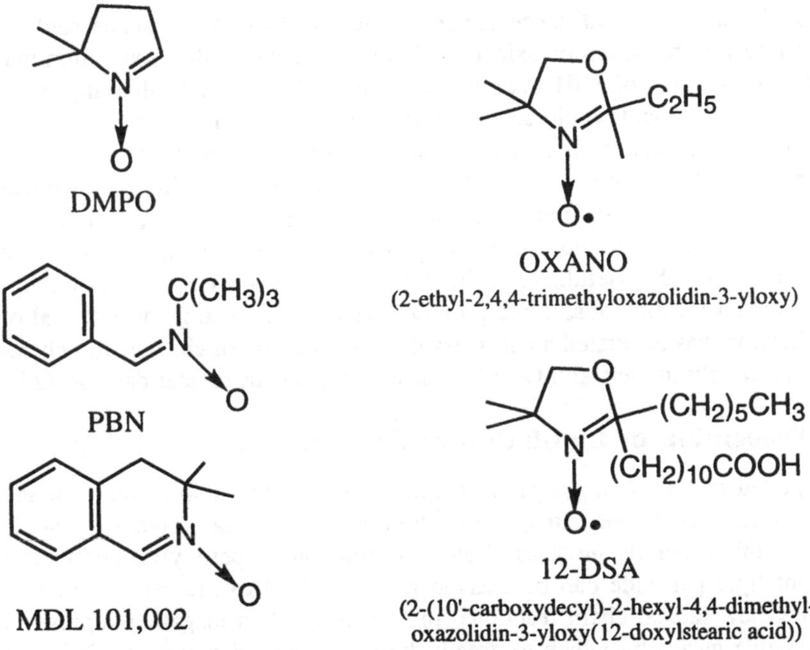

Figure 3 Nitroxides and nitrone spin traps. These compounds have been shown to be effective inhibitors of oxidative stress in model systems by virtue of SOD-like activity or radical scavenging.

mediated injury (19), and it has been reported that PEG-catalase is being studied preclinically.

The deleterious effect of O_2^{-} generation can also be negated by direct scavenging of the radical. This approach is unlikely because O_2^{-} is a relatively "passive" radical that functions primarily as a reductant. To work it would require the scavenger to be present in high concentration. Even so, it has been shown that the spin trap dimethyl pyrroline-*N*-oxide (DMPO) (Fig. 3) can protect cerebellar granule cells in culture from excitatory amino acid–induced toxicity and that the trapped radical was O_2^{-} (20). Furthermore, nitrone spin traps that react with O_2^{-}, albeit poorly, have shown efficacy in animal models of cerebral ischemia and reperfusion (Sec. III).

With regard to preventing initiation, there remain at least two additional possibilities for intervention. First, direct scavenging of ·OH could be attempted. This approach is less likely to succeed than scavenging O_2^{-} because the reactivity of ·OH with virtually all biomolecules is 10^8–10^{10} M^{-1} S^{-1}, thus requiring unrealistically large concentrations of a competing scavenger. Agents such as mannitol have shown protection in ischemia and reperfusion models, yet the ability of mannitol to influence oxygen radical formation by means such as modulating iron reduction or oxidation (21) makes tenuous the assumption that it protects by virtue of ·OH trapping. Therefore, it is unlikely that any pharmaceutical antioxidant would act solely by scavenging ·OH.

Finally, minimization of initiation can be theoretically achieved by control of iron redox chemistry if ·OH is involved. Numerous studies have shown that agents such as desferrioxamine (22) minimize ·OH formation by preventing reduction of ferric iron and, therefore, provide protection against reperfusion injury; however, the usefulness of the drug is limited by lack of oral activity, a requirement for high doses, and poor cell penetration. The great potential of iron chelators has generated a considerable amount of research, but iron chelators vary widely in their ability to inhibit iron-dependent radical damage (23).

B. Prevention or Limitation of Propagation

As depicted in Fig. 2, propagation reactions lead to the amplification of the initial radical-dependent damage. In the case of polyunsaturated fatty acids (PUFA), this generally involves H atom abstraction by peroxyl radicals. The resultant lipid peroxide can be cleaved by metal ions to radicals capable of initiating new radical chain reactions; thus, control of propagation represents a reasonable means by which to greatly limit oxidative damage. Again, even though PUFA are used for purposes of illustration, proteins also undergo radical-dependent reactions to generate carbon-centered radicals, peroxyl radicals, and protein peroxides (24,25).

1. Reactions with Carbon-Centered Radicals

The complexity of propagation reactions opens many avenues from which intervention can be attempted and, accordingly, numerous approaches have been attempted. An initial abstraction of an allylic hydrogen from the methylene carbon of a polyunsaturated fatty acid generates a carbon-centered radical (Fig. 2). The subsequent addition of molecular oxygen proceeds rapidly so that attempting to intercede at this level is akin to trapping ·OH. Nevertheless, a few studies suggest that this approach may be viable, even though evidence for the trapping of carbon-centered radicals as a sole or primary mechanism is lacking. The usefulness of carbon-centered radical traps as antioxidants is suggested by studies with nitroxides and nitrone spin traps (Fig. 3). As discussed, nitroxides can function as metal-independent SOD mimics (15–18). In addition, they can readily react with carbon-centered radicals as demonstrated for a number of indolinonic and quinolinic aminoxyls (26). Highlighted in this work was the demonstration of the ability of selected aminoxyls to abstract allylic hydrogens from linoleate, thereby initiating lipid peroxidation. Comparison of various aminoxyls indicated that the overall "antioxidant" activity of the compound reflected a balance between the ability to abstract hydrogen atoms and the propensity to trap carbon-centered radicals to stop propagation reactions. A detailed study of the effects of nitroxides and their corresponding hydroxylamines on lipid peroxidation was conducted by Nilsson et al. (27). These authors demonstrated that nitroxides could inhibit peroxidation at three levels: (1) blocking of peroxide-independent initiation, (2) prevention of peroxide-dependent initiation, and (3) scavenging of lipid-derived radicals. The efficiency of their antioxidant activity was related to numerous factors, including lipophilicity, cyclic oxidation, and reduction between nitroxides and the hydroxylamines, and the ability to interact with hydrophilic sites in the membrane environment.

In recent years, several groups have examined nitrone spin traps, particularly α-phenyl-*N-tert*-butyl nitrone (PBN), as protective agents in animal models of oxidative stress. PBN reacts readily with ·OH and carbon-centered radicals to form relatively stable nitroxide spin adducts, which can be detected by ESR spectroscopy (28). Oliver and colleagues originally reported that PBN provided protection against cell loss and neurological deficits associated with transient global ischemia in a gerbil model of stroke (29). This work has since been confirmed (30,31). PBN has also been shown to provide protection in models of transient (32) and permanent (33) focal ischemia. This latter finding could suggest protection afforded by a mechanism other than radical trapping or that collateral flow in the permanent ischemia model provides sufficient oxygen to generate active oxygen species. PBN has also been extensively studied in various models of reperfusion-induced myocardial injury, with conflicting results (34).

The results obtained with PBN have prompted persons in the pharmaceutical industry to consider nitrones as therapeutic agents. SmithKline Beecham has examined PBN in animal models of oxidative stress (31), but no reports on novel, structurally related compounds have appeared. Hoechst Marion Roussel has synthesized a series of cyclic variants of PBN, which are much more potent inhibitors of lipid peroxidation than PBN. Against iron-dependent oxidation of liposomes, the most potent compound (MDL 100,630) had an IC_{50} of 28 μM compared with 14.3 mM for PBN (35). The cyclic nitrones were similarly active against cupric-ion-induced oxidation of low-density lipoprotein (LDL) (36). In vivo, the cyclic nitrone MDL 101,002 afforded marked protection against ischemia-induced injury in the gerbil (37) and against endotoxin-mediated oxidative stress and mortality in the rat (38). These studies and those with PBN are intriguing, yet the mechanism of action of nitrones in vivo is unclear. Even under ideal in vitro conditions, only 5–10% of the radicals generated are trapped by the nitrones. Considering that those studies use as much as 100 mM of spin trap, it is likely that, in vivo, less than 1% of the radicals is trapped. Thus, it is highly unlikely that the remarkable protection these agents provide against neurodegeneration can be accounted for solely by radical trapping. Certainly, whether they function in vivo as carbon-centered radical traps is deserving of additional study.

2. Reaction with Alkoxyl and Peroxyl (Oxygen-Centered) Radicals

By far, the greatest body of literature concerning antioxidants encompasses agents that function as terminators of the radical chain (Table 3). Although the previously described nitrones are capable of such action, their activity requires formation of a covalent bond that is a slow reaction in comparison to an H atom transfer. Not surprisingly then, α-tocopherol has evolved as the major chain-breaking antioxidant in biological systems. It is postulated to react with peroxyl or alkoxyl radicals to generate hydroperoxides and alcohols, respectively. It is somewhat controversial as to whether this is achieved via H atom transfer or as separate electron transfer and proton equilibration steps (39). In either case, the oxygen-centered radical derived from a biological substrate such as a PUFA is reduced to a nonradical species, which leads to chain termination. However, the number of radicals is conserved because α-tocopherol is present as a phenoxyl radial. The stability of the phenoxyl radical is such that it does not generally affect hydrogen abstraction reactions but undergoes a second reaction with a peroxyl radical to produce two nonradical species, thus providing a stoichiometric factor of 2 for radical scavenging (40).

Agents that act via rapid H atom transfer and have stoichiometric factors of 2 represent ideal antioxidants. As discussed, α-tocopherol is being examined for prevention of cardiovascular disease (2). However, its lipophilicity and carefully controlled absorption and distribution eliminate its use in acute care settings. For this reason, considerable activity has, and continues to be, devoted to the design of synthetic phenolic antioxidants. Much of the current design on new compounds is based on the classical studies of α-tocopherol by Burton and Ingold (41,42). Those studies pinpointed the steric and stereoelectronic effects that stabilized the phenoxyl radical formed by reaction of α-tocopherol with peroxyl radicals. In particular, overlap between a $2p$ lone electron pair on the para ether oxygen and the aromatic π system is crucial for stabilization. Importantly, this overlap is maximized upon condensation of the six-membered heterocyclic ring of α-tocopherol to a 2,3-dihydrobenzofuran (42). More recently, Barclay et al. (43) have undertaken a systematic and thorough investigation of four classes of phenolic antioxidants. This work substantiated the contribution of the stereoelectronic effect to antioxidant efficacy. It was also shown that, in model membrane systems, hydrogen bonding by water with the para ether oxygen minimized the stereoelectronic effect compared with results obtained in homogeneous solution. Thus, in the study of synthetic antioxidants, the biological milieu in which the compound is expected to act may significantly influence activity, and data obtained in in vitro systems are not readily extrapolated to the in vivo situation.

Considerable resources have been expended in the preparation and study of synthetic dimethyl-*ortho*-phenols. Tocopherol analogs are discussed in detail elsewhere in this volume. The original member of this class of compounds was Trolox, in which the phytyl side chain of α-tocopherol is replaced with a carboxylic acid moiety to increase water solubility (Fig. 4). Although it is still used in the laboratory, it has not been developed clinically and is available commercially from Aldrich Chemical Co. (Milwaukee, WI). Other companies have continued to synthesize 6-hydroxy polyalkyl chromans. Janero et al. at Hoffmann-LaRoche reported on a large series of 6-hydroxychroman-2-carbonitriles, which demonstrated potent antioxidant activity in myocardial membrane phospholipids (44). More recently, an analog of Trolox synthesized by Sterling Winthrop (WIN 62079) was shown to be much more active than Trolox in cell culture and animal models of reperfusion injury (45). Similarly, the second-generation "Lazaroids" from Upjohn (U-78517F) combine the chromanol portion of α-tocopherol with the amine of the original 21-aminosteroids such as U-74006F (discussed in a subsequent chapter) (46). Others have prepared a series of novel, ethanaminium analogs of α-tocopherol (47–49), which will be referred to in Sec. III and discussed elsewhere in this volume. A variation on the chromanol structure was recently reported by Konica Co., which is investigating benzodioxin derivatives as antioxidants (50).

The work of Burton et al. (42), which demonstrated the improvement in radical trapping with the benzofuran ring, has led to considerable activity in the synthesis of similar phenolic compounds. The patent literature reports on at least two groups that have claims on dihydrobenzofuran derivatives as antioxidants (51,52). A similar approach has recently been reported for agents targeted toward neuroprotection (53). Investigators at the Institute of International Pharmacological Research (Italy) have amassed considerable data on IRFI-016 [2(2,3-dihydro-5-acetoxy-4,6,7-trimethylbenzofuranyl)acetic acid]. This compound has shown efficacy in a rat model of myocardial injury (54,55) and against reperfusion damage in the gerbil model of stroke (56). It also increased mucus production and exhibited anti-inflammatory activity (57). Battioni et al. (58) have synthesized a number of benzofurans that were tested as inhibitors of peroxidation of rat liver microsomes. Overall, the compounds were more active than α-tocopherol and Trolox, with a 3,4-dihydro-6-hydroxy-2*H*-1-naphthopyran being particularly active.

Stabilization of the phenoxyl radical to increase antioxidant activity can be achieved by substitution of *tert*-butyl groups for the methyls *ortho* to the hydroxyl. In a homogeneous solution with relatively small radicals, the presence of these bulky groups tends to stabilize the phenoxyl and thus favor H atom transfer reactions. For this reason, such compounds are in widespread use in commercial applications to prevent undesirable oxidation. In membranous systems, it is conceivable that the *tert*-butyl groups may actually deter reaction between the antioxidant and large radical species such as lipid peroxyl radicals. Nonetheless, such compounds are being investigated. Perhaps one of the most well studied in this class is probucol (Fig. 4), which is discussed in Chapter 6. This compound, marketed as Lorelco, was originally prescribed for treatment of atherosclerosis based on its lipid lowering activity. Recently it has been suggested that its efficacy may also be attributable to prevention of LDL oxidation (59,60). Unfortunately, it lowers high-density lipoprotein (HDL) cholesterol. Recent studies have attempted to identify analogs with potent antioxidant activity but devoid of HDL lowering action (61). One analog (MDL 29,311) has shown efficacy in two models of insulin-dependent diabetes mellitus (62). Probucol may still hold promise for treatment of other conditions of oxidative stress. It has been shown to prevent restenosis following angioplasty (63) and is being examined in phase 2 clinical trials in HIV-positive patients by Vyrex Corporation under the tradename PANAVIR.

The proven antioxidant efficacy of di-*tert*-butyl phenolics has continued to foster activity on this class of compounds. Most of these compounds are ex-

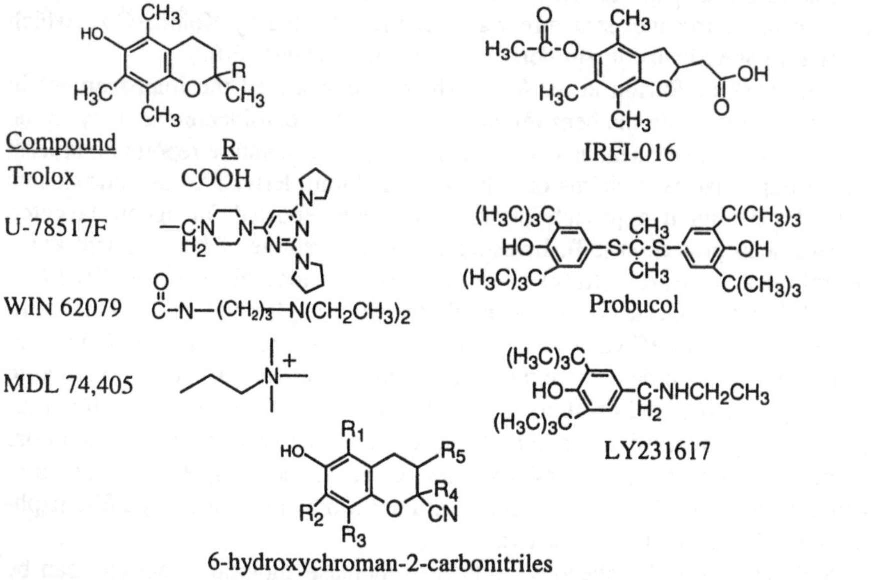

Figure 4 Phenolic antioxidants. Hindered phenols are generally potent inhibitors of lipid peroxidation and ·OH scavengers. For this reason, many compounds have been examined for activity in model systems of oxidative damage.

tremely hydrophobic and thus have limitations with regard to formulation, bioavailability, and distribution. Their lipophilicity also favors partitioning into membranes and may leave additional critical cellular targets, such as protein thiols (64–66), susceptible to oxidative damage. Therefore, whereas probucol-like compounds are ideal LDL-protective antioxidants, they are limited in their utility, and other related compounds have been prepared. A diglutaric acid ester of probucol was shown to have greater stability, cell penetration, and protective capacity than probucol (67). The compound S12340 (8[3,5-di-*tert*-butyl-4-hydroxyphenyl-thio}propyl]-1-oxa-2-oxo-3,8-diazaspiro[4.5]decane) was reported to be approximately threefold less hydrophobic than probucol and demonstrated good antioxidant activity (68). Kita et al. have studied a number of phenolic ethers, particularly (3,5-di-*tert*-butyl-4-hydroxyphenylthio) alkanoic acids as antioxidants and antiatherosclerotic agents (69). A series of di-*tert*-butylhydroxyphenylthio-substituted hydroxamic acids were prepared, and a three-carbon tail-containing compound demonstrated activity greater than that of probucol against rat brain homogenate peroxidation (70). This compound is unique in that the hydroxamic acid functionality contributes iron chelation ability or radical trapping activity distinct from the phenolic end of the compound.

A series of compounds in this chemical class has been synthesized and studied by investigators at Eli Lilly (71). LY 231617 (2,6-bis(1,1-dimethyl ethyl)-4-[[1-ethyl amino]methyl]phenol hydrochloride) demonstrated neuroprotection in rat models of both global (72) and focal ischemia (73). Significantly, the compound had efficacy when administered 30 min after occlusion and showed activity following oral administration. This suggests that the compound may be useful in providing prophylactic protection against second strokes. LY 213829 (5-(3,5-di-*tert*-butyl-4-hydroxybenzyl)thiazolidin-4-one) was shown to protect against acute inflammation in acetic acid–induced colitis in rats and may be used in treatment of inflammatory bowel disease (74).

A number of nonphenolic compounds inhibit lipid peroxidation and prevent reperfusion-induced injury (Fig. 5). Perhaps one of the most well studied is MCI-186 (Norphenazone, 3-methyl-1-phenyl-pyrazolin-5-one) from Mitsubishi Kasei (Fig. 5). MCI-186 was first reported to inhibit hydroperoxide-induced injury to endothelial cells (75). More recently, it has been shown to inhibit iron-dependent peroxidation of rat brain homogenate and trap ·OH as assessed by the salicylate hydroxylation assay (76). Furthermore, MCI-186 appears efficacious in animal models of stroke (77,78) and prevents edema formation and leukotiene biosynthesis following transient ischemia in the gerbil (79). All of these activities suggest that MCI-186 may be useful in treatment of cerebral ischemia and reperfusion injury, and phase 2 clinical trials are underway in stroke and Alzheimer's patients. From a chemical perspective, the phenyl ring

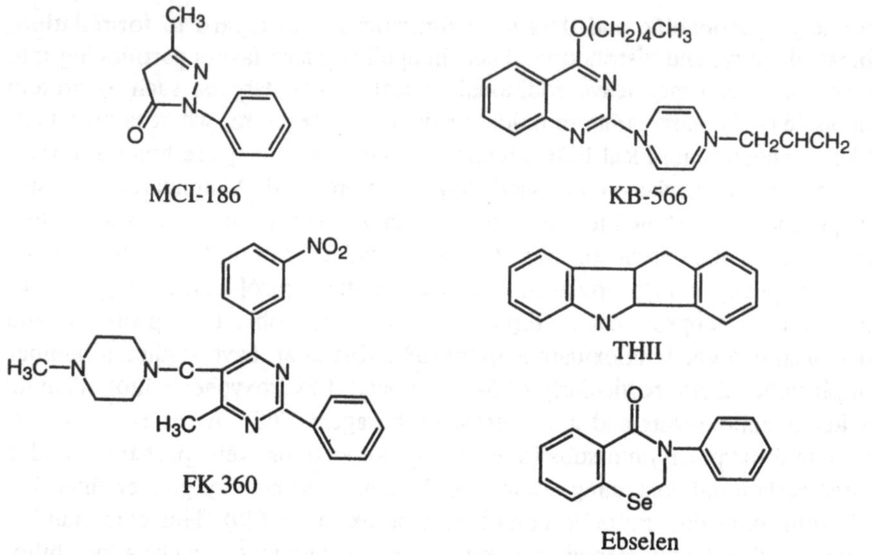

Figure 5 Nonphenolic compounds with "antioxidant" activity. The compounds shown have been reported to be inhibitors of lipid peroxidation or to prevent anoxia-induced injury. Most can function as radical traps, whereas Ebselen limits oxidative stress by virtue of its peroxidase activity.

of MCI-186 lends itself to ·OH trapping; a mechanistic understanding of its antilipid peroxidation activity is not yet available.

Over the past several years, Fujisawa Pharmaceutical Co. has synthesized a number of 4-aryl-1,4-dihydropyrimidines, which have been tested for anti-anoxic (enhanced survival of mice to 100% N_2) and antilipid peroxidation activity (80). As with researchers working with MCI-186, this group was interested in identifying compounds that could prevent arachidonate-induced edema. It was noted that agents that have antianoxic activity, such as nimodipine and idebenone, contain a basic nitrogen moiety; thus, synthetic efforts were focused on this type of modification at the C-5 position because substitution at that position did not affect antilipid peroxidation activity. This effort produced FK 360 (6-methyl-5-(4-methylpiperazin-1-yl-carbonyl)-4-(3-nitrophenyl)-2-phenyl-pyrimidine), which had potent activity and relatively low toxicity (Fig. 5) (81). Recently, additional studies have produced FR 72707, which replaces the carbonyl of FK 360 with a methyl (82). Molecular modeling indicated similar electrostatic potentials around the nitrogenous basic moiety, which correlated well with the observed similar efficacy of the two compounds. The toxicity of FR 72707, however, was much less than that of FK 360. These series of papers provide an excellent example of the use of structure activity relationships in the design of inhibitors of lipid peroxidation. What is lacking, however, is an understanding of the mechanism by which these arylpyrimidines exert their in vivo effects.

A class of compounds that has received more attention recently as antioxidants is the indoles. As reported by Sainsbury and Schertzer, these compounds function as potent antioxidants by virtue of their hydrophobicity and ability to undergo oxidation to stable cation species (83). Thus, these dihydroindenoindoles protect against lipid peroxidation and cytotoxicity in a number of model

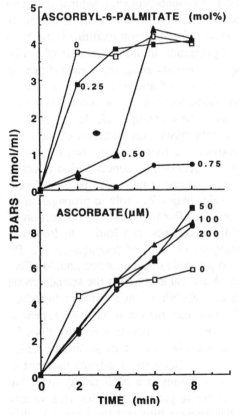

Figure 6 Effect of ascorbyl-6-palmitate (top) or ascorbate (bottom) on lipid peroxidation. Liposomes were prepared from soybean phosphatidylcholine by ethanol injection. Peroxidation was initiated by the addition of Fe^{2+} and histidine-Fe^{3+} and peroxidation determined by the formation of thiobarbituric acid–reactive substances (TBARS). Ascorbate was added at 50–200 μM, and ascorbyl-6-palmitate was incorporated into the liposomes at 0.25, 0.50, and 0.75 mol%. (From Ref. 120; with permission of AOCS Press.)

systems (84–86). Several years ago, a patent was filed on a series of indole derivatives that protected against both NADPH and Fe^{2+}-induced lipid peroxidation and provided protection against brain injury elicited by trauma and ischemia and reperfusion (87). A similar approach is being pursued by Otsuka, which has recently reported on 1-amino-2,3-dihydro-7-hydroxy-1H-indenes, which are predicted to have antioxidant activity and protect against hypoxia (88).

Another nitro-containing compound that has been reported to protect against ischemic insult is KB-5666 (2-(allyl-1-piperazinyl)-4-n-amyloxyquinazoline fumarate) from Kanebo Ltd. (Fig. 5) (89). KB-5666 protected against both cellular and functional alterations induced by transient ischemia in the gerbil. In review, a diverse array of chemical structures has been examined and been claimed to be antioxidants that provide protection in animal models of oxidative stress; however, virtually anything can function as an antioxidant if present at sufficiently high concentration. The true test of an antioxidant is its ability to prevent oxidation of other oxidative substances when it is present at a substantially lower concentration (90). Thus, for each compound, the in vivo data must be scrutinized to ascertain whether the tissue concentration of the compound would be sufficient to exert antioxidant activity. If not, then it must be speculated that its subcellular localization increases its concentration at cellular targets; otherwise, an alternative mode of action must also be considered.

Figure 2 shows that lipid hydroperoxides play a key role in propagation by virtue of transition-metal-dependent cleavage. Furthermore, the production of lipid hydroperoxides, for example by lipoxygenases, can lead to hydroperoxide-dependent initiation reactions analogous to those of propagation (4). By inference, then, prevention of lipid hydroperoxide decomposition should effectively control lipid or protein oxidation. A natural example is the selenoenzyme glutathione peroxidase (91–93), which uses glutathione as a cofactor to reduce hydroperoxides to less reactive alcohols. Considerable effort has been expended in the design of synthetic seleno compounds with peroxidase activity (94,95). Early efforts were hampered by the intracellular release of selenium, which resulted in toxicity. Finally, Ebselen, or PZ51 (2-phenyl-1,2-benzisoselenazol-3-(2H)-one), was shown to not provide selenium in a biologically available form. Accordingly, Ebselen has been shown to prevent oxidation in a variety of model systems, including isolated hepatocytes (96) and liver mitochondria (97). Ebselen was also shown to prevent Cu^{2+}-dependent oxidation of LDL, presumably by preventing decomposition of preformed lipid hydroperoxides (98). Recently, using purified phospholipid hydroperoxide glutathione peroxidase and a mercury drop electrochemical detector to quantitate hydroperoxides, it was confirmed that cleavage of preexisting lipid hydroperoxides in LDL is an important determinant in LDL oxidation (99).

Important in the catalytic activity of Ebselen is the presence of a reducing cofactor. There is little reduced glutathione present in plasma, suggesting that Ebselen would not be an effective agent for minimizing LDL oxidation *in vivo*. However, other thiols such as dihydrolipoate can support the peroxidase activity of Ebselen (100). Results of a recent study have shown that blood cells release diffusable, low-molecular-weight thiols, which will support Ebselen-dependent reduction of cholesteryl ester hydroperoxides in LDL (101). The varied activities of Ebselen are discussed in Chapter 9.

There are multiple and varied approaches to the control of oxidative events at the level of propagation. Not surprisingly, most compounds tested and being developed are phenolics, which are well characterized as antioxidants. Most of these compounds tend to be hydrophobic; therefore, they are ideal as protectants of PUFA in biological membranes. It is becoming increasingly apparent, however, that a loss of cell function or viability may involve merely a subtle oxidation of cellular macromolecules rather than rampant lipid peroxidation, which leads to cell membrane blebbing and leakiness. This may explain, in part, why the most potent inhibitors of peroxidation in vitro are not always the most active compounds in vivo. Perhaps antioxidants that are more amphipathic, or slightly less hydrophilic, may be desirable. Thus, thorough characterization of an antioxidant should take into account its activity in various milieu, such as a homogeneous solution versus a bilayer or two-phase system. It is also prudent to characterize the ability of the antioxidant to prevent protein oxidation if possible.

The search for agents that act at the level of propagation should be fruitful. Prevention of initiation requires an exogenous antioxidant to be present at exceedingly high concentrations. Also, tight control over initiation reactions necessitates knowledge of the initiating species. This is not agreed upon even in simple model systems of lipid peroxidation (102). The development of agents that reduce hydroperoxides is a particularly attractive approach because, if propagation reactions are controlled, then the impact of initiation on cellular membranes should be minimal. In this regard, only Ebselen has been extensively reported on, although a semisynthetic selenium-containing enzyme, selenosubtilisin, has been shown to reduce *tert*-butyl hydroperoxide with high efficiency (103). It is also reported that researchers at Oxis International (Bonnevil-sur-Marne, France) have identified promising selenoorganic compounds with good peroxidase activity. Future developments in this regard are awaited.

III. TARGETING AND DELIVERY OF THERAPEUTIC ANTIOXIDANTS

One of the major pitfalls encountered in randomly introducing antioxidants into biological systems is that much is left to chance or random action. By necessity, high concentrations of the compound must be achieved with the hope that it accesses either the site of radical production or their target. The ability to deliver an antioxidant to a relevant site of action would decrease the amount of drug required, thereby decreasing cost to the patient and minimizing the potential for toxic side effects. Ideally, for each pathophysiological condition described in Table 1, the pathology would be understood to the degree at which an initiating radical could be identified. At the next level the sources of this radical could be pinpointed. Taken together, it would then be possible to "target" radical specific scavengers to this site. For example, there is considerable evidence that endothelial cell xanthine oxidase contributes to ischemia and reperfusion injury to the intestines (104,105) and renal injury during transplantation (106). One approach would be to use an inhibitor of xanthine oxidase, such as allopurinol, which has been shown to reduce intestinal reperfusion injury (105) and decrease renal injury following transplantation (107). Similar protection was also afforded by SOD; therefore, it is likely that allopurinol acted via prevention of O_2^- formation. It could be argued that SOD is unlikely to penetrate the endothelial cell to any great extent. However, Arroyo et al. (108) demonstrated by EPR spin trapping that endothelial cells generated oxygen radical exogenously. It is these radicals that may contribute to upregulation of adhesion molecules, promoting neutrophil adherence and additional oxidative damage (106).

A more sophisticated approach would be to target the antioxidant or radical scavenger to the endothelial cell. This approach has been taken by Inoue and colleagues (109), who initially coupled styrene-(co-maleic acid butyl ester) to Cys^{111} or lysyl residues of human SOD. This ligand binds to albumin and accumulates in tissues with lowered pH (i.e., ischemic tissue) (109). This novel SOD derivative was subsequently improved by constructing a fusion gene consisting of cDNA encoding for human Cu/Zn-SOD and a C-terminal heparin binding peptide (110). Immunocytochemical studies demonstrated not only that the complex rapidly bound to heparin sulfate on the cell surface but that it also could undergo translocation in certain tissues. This is an attractive approach in that it facilitates a rapid accumulation of antioxidant to the site of radical formation, thus requiring less protein than administration of a nontargeted compound such as PEG-SOD.

In most instances the locale of the oxidant source or the critical cellular target remains enigmatic. This does not, however, rule out targeting for antioxidant intervention. Rather, intervention must be designed to occur at the tissue level rather than the cellular or subcellular level. This approach has been taken with a series of hydrophilic α-tocopherol analogs, which are discussed in Chapter 2. Replacement of the phytyl tail of α-tocopherol with a quaternary amino functionality led to accumulation of the compound MDL 74,720 in cardiac tissue (111). Subsequently, MDL 74,405 (the deesterified $R(+)$ enantiomer of

MDL 74,720) was shown to decrease reperfusion-induced myocardial stunning (112,113) and infarction (114–116). The same approach was taken for piperazinobenzofuran derivatives, which have been recently reported to be potent antioxidants with excellent brain penetration that may provide neuroprotection after trauma or stroke (53).

One obvious disease state for antioxidant targeting is atherosclerosis (117, 118). Oxidation of PUFA in LDL and subsequent modification of apolipoprotein B impart to the LDL biological activities that can contribute to atherogenesis, including cytotoxicity, induction of cytokine release, cholesterol loading in macrophages, and smooth muscle cell proliferation. Consequently, compounds preventing LDL oxidation should be effective antiatherosclerotic agents. A chronic disease such as atherosclerosis is multifactorial, and LDL oxidation may be only one, and perhaps not the primary, contributor. Nonetheless, much effort has gone into the study of LDL protective agents from 1986, to now. As discussed, probucol has been well studied, and many analogs have been prepared and studied as well. The efficacy of probucol and its analogs in preventing LDL oxidation relates, in part, to their accumulation in the LDL particle. Viewing the LDL particle as a homogeneous system is misleading, however; it contains oxidizable PUFA in both a phospholipid outer monolayer and as cholesteryl esters in a hydrophobic core region. Our studies have suggested that probucol may reside primarily in the cholesteryl ester phase and that α-tocopherol may favor the outer monolayer (119). Studies with probucol analogs in liposomes also indicate that phenolic antioxidants can orient differently within membranes, thus affecting their activity (120). Targeting of LDL-protective antioxidants must take into account the physical makeup of LDL. For example, β-carotene is associated with LDL and provides good antioxidant protection (121). Part of this may stem from its location within LDL. Accordingly, a flavonoid derivative containing a hydrocarbon chain similar to β-carotene was synthesized and shown to be 15 times more potent than α-tocopherol at preventing LDL oxidation in vitro (122). A recent study has claimed that thiocholesterol was more effective at preventing Cu^{2+}-induced LDL oxidation than other antioxidants, including probucol and thiopalmitic acid (123). This may stem from the ability of the compound to access the core of the LDL. These examples of "targeted" antioxidants are summarized in Table 4.

Altering the cellular distribution of natural antioxidants is another means by which to selectively protect certain tissues or organelles. This has been achieved primarily by modifications in lipophilicity. This is best exemplified by derivatives of ascorbic acid. Although ascorbate often functions as a potent antioxidant, its hydrophilicity and redox potential can render it prooxidant in the presence of transition metal ions. Figure 6 shows ascorbate enhancing liposomal oxidation promoted by iron salts. Conversely, the inclusion of low amounts of the hydrophobic derivative ascorbyl-6-palmitate provided significant antioxidant

Table 4 Examples of "Targeted" Antioxidants

Name[a]	Chemical class/type	Target
HB-SOD	Enzyme	Vascular endothelium
Aerosolized SOD	Enzyme	Lungs
Probucol	Di-*tert*-butyl phenol	Lipoproteins
Thiocholesterol	Steroid	Lipoproteins
MDL 74,405	Quaternary amine phenol	Heart
MDL 74,722	Phenylpiperazine phenol	Brain
EPC-KI	Tocopherol/ascorbate ester	Dermis

[a]The listed compounds represent means by which antioxidants can be targeted to specific organs, tissues, or cells. Those listed represent chemically modified structures, modified proteins, and delivery systems that can all serve to increase the likelihood of the agent reaching the desired site of action.

activity. This work is in agreement with that of Nagao and Terao, who demonstrated that ascorbyl-6-palmitate was an effective peroxyl radical trap and prevented peroxyl radical-dependent liposomal peroxidation, whereas ascorbic acid was only weakly active (124). In comparing a series of 3-*O*-alkyl ascorbic acids, Nihro et al. concluded that the ability to inhibit lipid peroxidation highly correlated with hydrophobicity (125). Takeda Chemical Industries has prepared 2-octadecyl ascorbic acid (CV-3611) and demonstrated its efficacy in reducing endotoxin-induced mortality (126) and development of acute pancreatitis (127).

A variant on this theme has been introduced by Senju Pharmaceutical Co., which has prepared phosphoric diesters with the 2 position of ascorbic acid and the hydroxyl of various tocopherols (128). This not only provides both ascorbic acid and α-tocopherol upon hydrolysis and deesterification but also prevents spurious oxidation of the tocopherol. The diester containing α-tocopherol was the most active isomer against rat brain homogenate peroxidation and was more potent than free α-tocopherol. In the same system, free ascorbic acid provided no protection. The potassium salt of this diester (EPC-KI) has shown efficacy in a number of model systems of oxidative stress and is currently in phase 2 clinical trials for treatment of stroke.

Also of interest is the work by Fariss and colleagues with the succinate ester of α-tocopherol (129–131). This compound has proved to be much more cytoprotective than α-tocopherol in various model systems. The ester does not appear to increase accumulation of free α-tocopherol, which would account for its efficacy. It has not been reported whether the succinate ester leads to a unique cellular or subcellular distribution, which could be considered a form of antioxidant targeting. Refer to Chapter 4 for further information.

While we have given a few examples of antioxidant targeting, this area deserves further attention. For example, mitochondrial oxygen radical production is hypothesized to contribute to a number of free-radical pathologies such as aging (132,133). The ability of the electron transport chain to bleed off O_2^- and H_2O_2 (134,135) leads to damage of mitochondrial DNA (136) and most likely other subcellular organelles. Furthermore, in hepatocytes and perhaps other cells, a loss of mitochondrial glutathione cell (GSH) is a critical link to ultimate loss of cell viability (137–140). It is not surprising then that mitochondria contain antioxidants such as Mn-SOD and GSH peroxidase. Nonetheless, it is reasonable to speculate that these mitochondrial antioxidant defenses may be impaired or overwhelmed, precipitating loss of cell function; therefore, targeting of antioxidants to the mitochondria may represent an untapped means by which to increase efficiency of therapeutic antioxidants.

An alternative to directly targeting a putative antioxidant to a particular site is to use a delivery system that introduces the compound in a manner likely to increase its penetration to the desired site of action. For many classes of drugs, liposomes are being used more frequently as drug delivery systems (141). Entrapment of SOD in liposomes was shown to protect aortic endothelial cells against oxygen-induced injury (142) and prevent acetaminophen-induced liver necrosis (143). Recently, it was reported that liposomes containing antioxidants such as α-tocopherol, ascorbic acid, and SOD are being investigated for prevention of virus induction by environmental stress (144). Other methods of delivery are also amenable to antioxidants. For example, numerous pulmonary diseases may be a manifestation of oxidative stress; therefore, delivery of agents to the lungs would be desirable. It has been shown that aerosolization of recombinant SOD could produce a sustained increase in SOD activity in respiratory epithelial lining fluid (145). Numerous drugs used in treatment of asthmatic patients are delivered by aerosolization and inhalation, and the potential for delivery of other antioxidants in this fashion should be tested.

A facile means of delivering drugs is via topical application. An obvious use of this technology would be in skin care lotions because of the widespread incidence of dermatological disorders. For example, linoleic acid–derived hydroperoxides have been reported in psoriatic scales (146), although this may be a secondary epiphenomenon and not causal in lesion formation. The sodium salt of the Senju phosphate diester of ascorbic acid and α-tocopherol was shown to be readily deesterified in the skin of the hairless mouse (147). Ultraviolet-(UV) light–blocking agents such as *para*-aminobenzoic acid (PABA) are in widespread use in suntan lotions. Because photochemical reactions can lead to formation of reactive oxygen products and skin damage, there is an upswing in the number of formulations containing antioxidants such as *p*-hydroxy-benzaldehyde (148).

In other situations, topical application of antioxidants is practical. It has been reported that topical application of an antioxidant provided significant radioprotection of the rat bowel mucosa (149). The addition of allopurinol and GSH to tissue preservation solutions (Viaspan, Dupont Pharmaceuticals) is a form of topical application that has dramatically improved the success rate for renal transplantation (150). Taken as a whole, the evidence suggests that the ability to deliver an antioxidant to the critical site translates to efficacy. Therefore, some of the negative results in clinical trials with antioxidants may reflect an inability to access the proper tissue, cell, or subcellular organelle. Given this evidence, there should be more efforts devoted to developing pharmacological and physical means to target antioxidants to the relevant site of radical formation or damage.

IV. SECONDARY MECHANISMS CONTRIBUTING TO THE EFFICACY OF ANTIOXIDANTS

Over the past several years, a new appreciation for the diverse role that oxygen-derived radicals play in biological systems has emerged. No longer are they viewed as merely destructive, but it is now recognized that they mediate critical processes such as signal transduction and gene expression. Much of this awareness has developed through the use of antioxidants as pharmacological tools. Perhaps the best example of this is the elegant work of Bauerle and colleagues (151–153). These investigators and others (154) have shown that NF-κB, a ubiquitous transcription factor, can be activated by oxidants such as H_2O_2. Furthermore, antioxidants such as N-acetylcysteine (NAC) and pyrrolidine dithiocarbamate (PDTC) prevented such activation, leading to the proposal that oxygen radicals contribute to NF-κB activation via a redox-sensitive mechanism. Since NF-κB activation is required for expression of a host of gene products, modulation of its activation can have profound consequences.

Many mediators involved in inflammatory disorders require transcriptional activation by NF-κB. Therefore, many studies have examined antioxidants, particularly sulfhydryl compounds such as NAC, in various immune disorder models. It is not clear whether NAC functions as an antioxidant by replenishing GSH (155,156) or by other mechanisms not fully understood (157). Although GSH repletion may explain the efficacy of NAC in septic shock and adult respiratory distress syndrome (ARDS) (158), the number of inflammatory mediators upregulated during sepsis would suggest that direct thiol-dependent prevention of NF-κB regulation could also be therapeutically effective. Atherosclerosis has a significant inflammatory component, and Marui and colleagues (159) have shown that NAC and PDTC could prevent interleukin-1 beta (IL-1β) activated vascular cell adhesion molecule 1 (VCAM-1) gene expression.

Presumably, this was the result of modifying redox-sensitive regulatory mechanisms for NF-κB-driven expression of VCAM-1. The significance of these findings is that VCAM-1 expression on vascular endothelial cells is a noted feature in atherogenesis and is necessary for recruitment of mononuclear leukocytes to the endothelium (160).

Perhaps nowhere is the potential for antioxidants to modulate disease progression via secondary effects such as cytokine secretion greater than in treatment of acquired immunodeficiency syndrome (AIDS). Transcription of integrated latent HIV DNA appears to require a signal or second messenger. When this signal is received, NF-κB is released from its inhibitor protein I-κB and translocates to the nucleus to regulate viral replication. Based on work in isolated cells, production of oxidants could conceivably contribute to NF-κB activation and replication of the virus (161,162). Treatment of immune cells with NAC was found to increase GSH levels and decrease viral replication (163–165); furthermore, NAC prevented upregulation of cytokines in response to oxidants and inhibited tumor necrosis factor-alpha (TNF-α)–induced apoptosis of infected cells (166,167). These multiple, potential beneficial effects argue strongly for further investigations into the use of NAC in AIDS patients (168).

The potential impact of antioxidants on disease progression by multiple mechanisms is not limited to thiol reagents. Suzuki and Packer (169) have shown that the catechol derivatives nitecapone and OR-1246 inhibited TNF-α-induced activation of NF-κB in Jurkat T cells. The 21-aminosteroid U-74389F has shown beneficial effects in animal models of oxidative stress, which may be related to radical scavenging. In a mouse model of hemorrhage and resuscitation, U-74389F reduced cytokine mRNA levels in intraparenchymal, but not peripheral, pulmonary mononuclear cells (170). This same compound significantly improved skin flap survival after secondary ischemia (171), and protection correlated with both an attenuation of neutrophil infiltration and a decrease in malondialdehyde (MDA) formation.

An excellent example of an antioxidant drug that may be efficacious by virtue of multiple activities is probucol (Table 5). As described, probucol exhibits powerful antioxidant activity against in vitro or ex vivo oxidation of LDL. This alone may be sufficient to account for its antiatherosclerotic properties; however, probucol has shown other interesting activities. One factor examined is the physical effect of probucol on LDL and cells. It is probably underappreciated how many hydrophobic antioxidants derive a portion of their activity through physical effects on membranes. Alterations in fluidity of a bilayer can have a profound effect on the rate of chain propagation (172,173), and it has been suggested to contribute to the efficacy of the 21-aminosteroids (174,175) (Chap. 3). Similar to α-tocopherol, probucol has profound physical effects on membranes, as is shown by its effect on the liquid-crystalline-phase transition of dipalmitoylphosphatidylcholine using differential scanning calorimetry (120).

Table 5 Activities of Antioxidants that May Be Derived from Nonantioxidant Properties

Compound[a]	Process affected	Mechanism	Reference
N-Acetylcysteine	Gene expression	⇓ NF–κB	159
Pyrrolidine dithiocarbamate	Gene expression	⇓ NF–κB	159
U-74389F	Cytokine release	—	170
Probucol	IL-1β release	Antioxidant/physiological effect?	179
	Promotion of cholesterol efflux	Physical effect	177
α-Tocopherol	Antiproliferation	—	187

[a]These compounds are known to be antioxidants and radical scavengers. In addition, they have been reported to modulate additional physiological processes such as gene expression and cytokine release. These additional activities may, in some cases, not be related to either radical scavenging or ability to inhibit lipid peroxidation.

Probucol has also prevented lipid oxidation–dependent modifications of the lipoprotein surface lysine residues (176). These properties of probucol led to further investigations wherein it was found that probucol alters the phase state of cholesteryl ester droplets in cells to a more fluid state, which facilitates cholesteryl ester mobilization (177). Studies with closely related analogs confirmed that these hydrophobic antioxidants disrupt the packing of cellular cholesteryl esters (178). This physical effect could increase the turnover of cholesteryl esters and contribute to the observed reduction in arterial lesions in cholesterol-fed rabbits.

Numerous studies have also examined the effect of probucol on release of macrophage cytokines, particularly IL-1β. Work by Ku et al. (179,180) demonstrated that macrophages isolated from probucol-treated mice secreted much less IL-1β in response to endotoxin. Subsequent work demonstrated that both probucol and α-tocopherol could inhibit phorbol ester–induced release of IL-1β from THP-1 cells (181). Although it is still not well understood, both an antioxidant (182) and a physical effect (183) may account for these results. Whether probucol exerts its effect on IL-1β release via an NF-κB-dependent pathway is not known. It has also been shown that lipid oxidation products such as 9-hydroxyoctadecadienoic acid (9-HODE) (184) and unsaturated aldehydes (185) could elicit the release of IL-1β from macrophages. These are both components of oxidized LDL; thus, probucol could also ameliorate IL-1β release in vivo by virtue of its antioxidant activity. This approach can also be extended to other classes of LDL-protective antioxidants. For example, the spin trap PBN prevented both LDL oxidation and IL-1β release from macrophages (186). Although PBN was not very potent, the newly described cyclic nitrones (36) may offer more promise as antiatherosclerotic drugs.

Thus, the efficacy of an antioxidant in a disease model may reflect multiple pharmacological activities. In certain instances the measured activity influenced by the antioxidant may have nothing to do with antioxidant activity. As an example, Azzi and colleagues studied a series of tocopherols and a series of 6-hydroxy-chroman-2-carbonitriles for their effect on smooth muscle cell proliferation (187). For both series a poor correlation between antioxidant and antiproliferative properties was observed. Therefore, if these agents were proposed for use as antiatherosclerotic agents, they may impact on the disease by an antioxidant mechanism (inhibition of LDL oxidation) and an antiproliferative mechanism (inhibition of smooth muscle cell proliferation). The challenge faced by the pharmaceutical researcher is to balance these two activities to obtain the most efficacious drug.

Many of these additional activities may be epiphenomena associated with a reduction in oxidative events. The determination of primary versus secondary effects is often difficult in in vivo models; however, from the viewpoint of pharmaceutical research, the most important fact is that these additional effects, that is, inhibition of cytokine production or neutrophil infiltration and activation, could all contribute to improved efficacy of drug treatment, which is the ultimate goal. Only when additional pharmacological properties lead to deleterious consequences, such as toxicity, is this of major concern.

V. ANTIOXIDANT ACTIVITY OF PHARMACEUTICALS WITH ALTERNATE PRIMARY MECHANISMS OF ACTION

With the increasing awareness of a possible contribution of oxidative stress to a variety of disease states, there has been a retrospective examination of the ability of various drugs, which were designed with one mechanism of action in mind, to potentially exert activity via an antioxidant mechanism (Table 6). For example, the products of lipoxygenases, such as leukotrienes, are potent physiological mediators that may influence inflammatory conditions. Because lipoxygenases contain nonheme iron and catalyze the oxygenation of PUFA (188), it is expected that antioxidants or iron chelators act as enzyme inhibitors. Antioxidants that function as lipoxygenase inhibitors are typified by the di-*tert*-butyl phenolic compounds R-830, E-5110, and BI-L-93BS (Fig. 7), which act as dual cyclooxygenase and lipoxygenase inhibitors. Many analogs of BI-L-93BS have been prepared, and modification of the *ortho-tert*-butyl groups was found to highly influence the relative potencies for lipoxygenase versus cyclooxygenase inhibition (189). It was demonstrated that alkylation of the phenol resulted in loss of activity, which further strengthens the argument that the compounds inhibit the enzymes via a redox mechanism. α-Tocopherol and related hydroxy chromans can inhibit soybean lipoxygenase in vitro, al-

Table 6 Drugs[a] Demonstrating Additional Activity as Antioxidants

Lipoxygenase inhibitors	Cardiovascular drugs	Others
Zileuton	Nifedipine	Tamoxifen
BW755c	Verapamil	Diethylstilbesterol
AA-86	Diltiazem	Chlorpromazine
R-830	Captopril	Trifluoperazine
KME-4	Bepridil	Naproxen
E-5110	Propranolol	
BI-L-93BS	Dipyridamole	
NDGA	U-92032	
Quercetin	Carvedilol	

[a]These drugs are used as pharmaceuticals for treatment of a variety of disorders. They have also been shown to possess antioxidant activity that could contribute to their efficacy profile.

though the phytyl tail of α-tocopherol diminished its activity relative to the other analogs (190). This work was confirmed by Maccarrone et al. (191), who also reported that ascorbic acid and ascorbyl-6-palmitate could competitively inhibit the enzyme.

In accordance with the potency of benzofuran analogs of α-tocopherol, 5-hydroxy-2,3-dihydrobenzofurans have been synthesized as leukotriene biosynthesis inhibitors (192). Furthermore, nonphenolic compounds such as BW755c are thought to inhibit lipoxygenase via an antioxidant mechanism. It is not surprising, then, that the ability of antioxidants to also inhibit lipoxygenase has spawned considerable debate concerning lipoxygenases versus nonenzymatic mechanisms in oxidative modification of LDL in vivo (193,194).

Some polyhydroxylated natural compounds such as quercetin and nordihydreguaiaretic acid (NDGA) are known to inhibit 5-lipoxygenase (195). Phenolic compounds, in general, however, lack oral activity that would be essential for treatment of chronic inflammatory disorders such as arthritis and allergic rhinitis. The potential for compounds to inhibit lipoxygenase by virtue of iron chelation or reduction led to the replacement of catechols with hydroxamic acids in the hope of achieving oral activity. Much effort has been devoted to this approach, with the first major breakthrough being the realization that, although some degree of lipophilicity was required to bind to the enzyme active site, the original eicosanoid backbone could be replaced with phenyl, benzyl, naphthyl, or benzothiophene groups (196). Subsequently, it was determined that replacement of the alkyl group of the hydroxamic acid with an amine group to generate an N-hydroxyurea derivative increased in vivo stability and oral efficacy. This finding led to the development of Zileuton (N-(1-benzo[b]thien-2-yl-ethyl)-N-hydroxyurea) by Abbott Laboratories, which has demonstrated inhibition of leukotriene biosynthesis in humans (Fig. 7). Other companies have continued to search for even more potent compounds with longer duration of action. CIBA-GEIGY recently reported on a series of N-hydroxyureas based on a benzodioxan template (197). This work identified CGS 25997 as a potent inhibitor and provided evidence that stereochemistry was an important determinant because it was twice as active as its R enantiomer. Other researchers have examined bishydroxamic acids that required the second hydroxamic acid group (198). These compounds are among the most potent lipoxygenase inhibitors known, although some of the best compounds lack in vivo activity.

The potency of the phenolic compounds, the hydroxamic acids, and the N-hydroxyureas as lipoxygenase inhibitors suggests that they would likely be effective at preventing generalized membrane lipid oxidation. For acute care situations such as heart attack, stroke, and neurotrauma, it would be of interest to study even those agents lacking oral activity because the intravenous route would be preferred. Several lipoxygenase inhibitors, including BW 755c, AA-861, and dipyridamole, were shown to inhibit xanthine oxidase-dependent oxi-

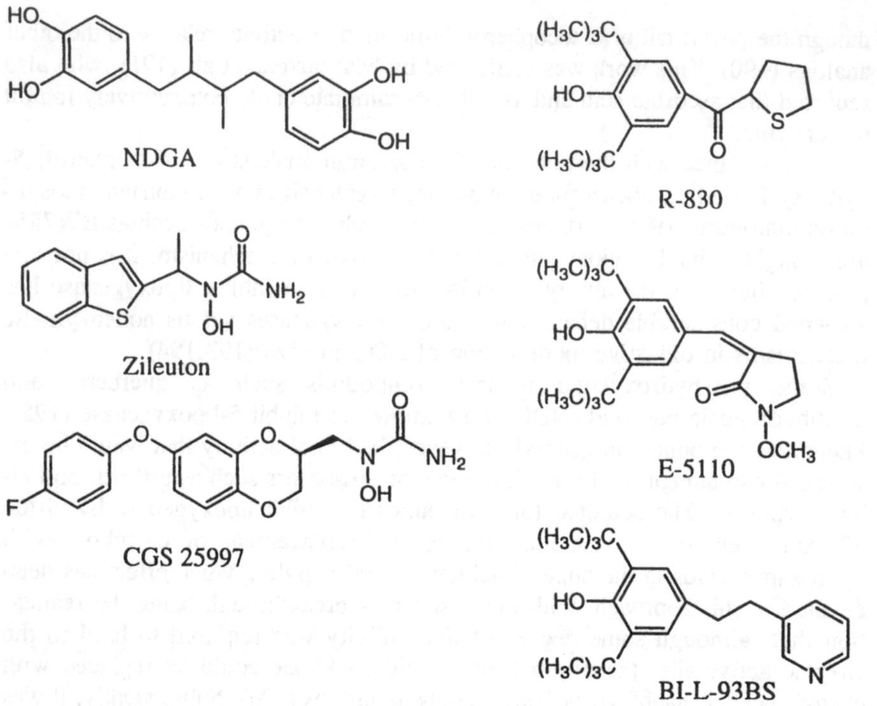

Figure 7 Lipoxygenase inhibitors with known antioxidant activity. Many inhibitors of lipoxygenase are hindered phenols or hydroxamic acids and, therefore, are potent antioxidants or iron chelators.

dation of myocardial membrane phospholipids (199). The cyclooxygenase inhibitors aspirin, indomethacin, and naproxen were inactive. The active compounds did not scavenge O_2^{-} and the kinetics of the inhibition suggested that they functioned as peroxyl radical traps. Further study of lipoxygenase inhibitors in animal models of acute oxidative injury may uncover potent antioxidants with desirable therapeutic profiles.

The potential for lipoxygenase inhibitors to act as antioxidants is apparent; less obvious is the antioxidant activity displayed by some cardiovascular drugs. Weglicki et al. (200) have tested a number of β-blockers and calcium channel blockers for antioxidant activity (Fig. 8). Among β-blockers, inhibition of lipid peroxidation correlated with hydrophobicity (200). In sarcolemmal membranes, nifedipine was the most potent agent, followed by verapamil and diltiazem (201). Based on structural considerations, it was proposed that the antioxidant activity was a reflection of interaction with lipid peroxyl radicals via an electron transfer process with a highly significant correlation between activity and lipophilicity. Weglicki and colleagues have also shown that thiol-containing

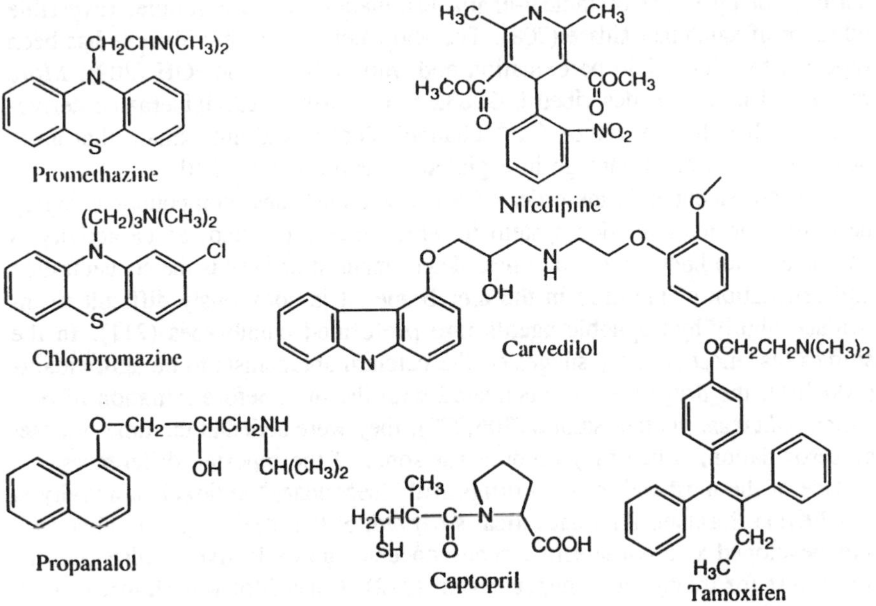

Figure 8 Pharmaceutical compounds shown to possess antioxidant activity. The compounds shown are used for a variety of medicinal purposes, most for treatment of cardiovascular disorders. They have also been shown to function as antioxidants, and it is speculated that this may also contribute to their efficacy.

angiotensin-converting enzyme (ACE) inhibitors such as captopril and *epi*-captopril can scavenge $O_2^{\cdot-}$ (202) and inhibit lipid peroxidation in endothelial and smooth muscle cells (203,204). These authors presented an interesting hypothesis focused on the ability of many of these vasoactive drugs to bind metal, which led to the proposal that transition-metal interactions in the plasmalemmal membranes of smooth muscle cells may be an important mediator of vasodilation (205).

Inhibition of lipid peroxidation by calcium antagonists has been studied by others with mixed results. Janero et al. (206) reported that nifedipine was inactive, diltiazem weakly active (IC_{50} = 510 μM), and bepridil very active (IC_{50} = 55 μM) against xanthine oxidase-dependent injury to isolated cardiac phospholipids. No inhibition of the enzyme was observed and the results were interpreted as inhibition via interception of lipid-derived radical intermediates. In rat heart and liver microsomes, nifedipine inhibited NADPH or ascorbate and ADP-iron driven peroxidation, although it was relatively weak (207). The β-adrenergic antagonist propranolol was shown to inhibit phospholipid per-oxidation, not by virtue of radical-chain termination but as a simple, reversible inhibitor of xanthine oxidase (208). The antiplatelet drug dipyridamole has been reported to inhibit lipid peroxidation and trap both O_2^- and $\cdot OH$ (209). More recently, Upjohn has described U-92032 as a diphenylmethylpiperazine derivative, which is both a T-type Ca^{2+} channel blocker and antioxidant that ameliorated hippocampal damage in a global ischemia model (210).

The assessment and rank order of potency among these compounds is highly dependent on the oxidation system in which they are tested. Since activity is correlated with lipophilicity, a critical determinant of activity is the concentration and orientation of the drug in the membrane. It is notoriously difficult to introduce highly hydrophobic agents into preformed membranes (211). In the studies that most strongly suggested the calcium antagonists to be antioxidants (200,201), the drugs were preincubated with the lipid before initiation of oxidation, whereas, in other studies (206,207), they were added at the time of onset of peroxidation, which may account for some of the reported differences.

One of the most well-studied drugs with "secondary" antioxidant activity is SmithKline Beecham Pharmaceutical's carvedilol (Kredex) (Fig. 8). This drug was developed as a nonselective β-adrenoceptor and selective α-adrenoceptor antagonist for treatment of hypertension (212). Carvedilol was shown to be a potent inhibitor of rat brain homogenate peroxidation and to compete with the spin trap DMPO for $\cdot OH$ (213). It was a much more active peroxidation inhibitor than other β-blockers such as propranolol and atenolol. The major structural feature distinguishing carvedilol from the other drugs is the carbazole moiety, and, indeed, studies with carvedilol analogs indicated that oxidation of the carbazole was responsible for its antioxidant activity (214). Carvedilol is known to be extensively metabolized; thus, its major metabolites were identified and studied. Two major ring hydroxylated metabolites, the *para*-hydroxylation product (SB 209995) and the *meta*-hydroxylated species (SB 211475), were further investigated. Although not directly compared, both agents exhibited similar activity in a wide variety of antioxidant tests, including inhibition of LDL oxidation and prevention of $\cdot OH$-induced death of neurons in culture (215,216). The redox potential of carvedilol was 1.1; it is 0.5 for SB 209995 and 0.2 for SB 211475, which is expected based on the hydroxyl substitution. The potency of these compounds suggests that their antioxidant activity may partially contribute to their therapeutic effects.

As previously discussed, the physical effects imparted on membranes by hydrophobic compounds should not be overlooked as a contributor to antioxidant activity. This physical effect has been attributed to account for the observed antioxidant action of the antiestrogen tamoxifen, which is used in the treatment of breast cancer. Both tamoxifen and its metabolite 4-hydroxytamoxifen have been shown to inhibit metal ion-dependent oxidation in various systems

(217,218). The drugs did not induce a lag phase typical of chain-breaking antioxidants, although this has been reported (219). Rather it was proposed that the drugs stabilize membranes analogous to cholesterol and that the increased rigidity decreases the rate of lipid peroxidation (220). The same phenomenon has been described for related synthetic estrogens such as diethylstilbesterol (221). These effects have led Wiseman (220) to propose that the membrane effects of tamoxifen contribute to both its anticancer activity and reversal of multidrug resistance, and that it should be considered for use as an antiviral agent.

The contribution of physical effects to the antioxidant activity of drugs has also been described for the surface-active anesthetics cepharanthine, chlorpromazine, and trifluoperazine (222). These drugs did not scavenge O_2^- nor act similar to chain-breaking antioxidants; therefore, it was concluded that their membrane-stabilizing ability led to a reduction in peroxidation. Some drugs, then, may possess antioxidant activity that could contribute to their observed clinical effects. The relevance of these studies must be considered in the context of whether the in vivo concentration of the drug approaches that at which in vitro antioxidant activity is observed. Furthermore, multiple mechanisms, such as radical trapping and physical effects, can contribute to the antioxidant effect. The activity of these drugs is often difficult to evaluate because of the variety of systems in which they are tested. For example, the nonsteroidal antiinflammatory drug naproxen was reported to act as an antioxidant in preventing cataracts (223). Conversely, in isolated rat hepatocytes, naproxen metabolism was shown to promote oxidative stress (224).

VI. CONCLUSIONS AND FUTURE DIRECTIONS

Without a doubt, interest in the study of oxygen radicals in causality of disease and the potential for antioxidants to be effective therapies is on the rise. Established investigators in this area are keenly aware, however, that, at a high enough concentration, nearly any agent can be an antioxidant. Thus, the challenge lies in clearly establishing mechanisms and relevance in a number of in vitro and in vivo studies.

It is only within the past few years that rational drug design has entered the antioxidant arena. In this class of drugs, it is often difficult to know the molecular target or the radical "bad guy" that must be captured. Advances in technology such as x-ray crystallography and molecular modeling have made it possible to envision interaction of enzyme inhibitors and receptor antagonists with their targets and allowed conceptualization of synthetic molecules before their actual synthesis. One technique that has been introduced in the study of phenols is molecular mechanic (225) and molecular orbital (226) calculations.

These calculations do provide some knowledge regarding antioxidant activity, from a strictly chemical point of view, for a relatively simple class of molecules that function as H atom donors. Molecular orbital calculations have also been used to aid in the study of nitrone spin traps (227).

Large numbers of antioxidants can be synthesized rapidly when aided by computer design and applied to combinatorial chemistry approaches, although the technology is limited for nonpeptidic compounds. Full use of this technology necessitates the requirement for assays that are capable of high throughput. Some attempts have been made to develop assays that can handle relatively large numbers of samples (228), and recently a method was described for water-soluble antioxidants that couple analogous organic synthesis and a resin-free deconvolutive assay in a 96-well format (229). Although these assays are of some usefulness in rank-ordering compounds, they are somewhat limited because the assays generally need to be conducted in a homogeneous solution. In other test systems and in vivo, antioxidants (at least inhibitors of lipid peroxidation) exert their activity in lipid bilayers or more complex structures such as LDL. Several determinants of activity in nonhomogeneous systems have not been identified by assays conducted in a single phase, as discussed for localization of hydrophobic antioxidants in LDL.

Another consideration that could be overlooked in the design of phenolic antioxidants is the potential for prooxidant activity. This has been highlighted by studies of α-tocopherol in LDL. Generally, α-tocopherol is a highly efficient inhibitor of membrane oxidation. In LDL, however, it can act as a chain transfer agent to propagate peroxidation (230). Thus, Ingold et al. (230) suggested that antioxidants targeted for LDL should be evaluated for their ability to export their phenoxyl radical from the LDL particle. Kagan and colleagues (231) have also recently described the reactivity of phenoxyl radicals with intracellular thiols. It was determined that the radical of the antitumor drug etoposide (VP-16) was reduced by GSH and protein thiols. A similar ability to oxidize cellular thiols could comprise the antioxidant activity of putative phenolic antioxidant drugs and would not be predicted by high-throughput assays or molecular-modeling studies.

For the reasons described, there may be limitations or toxicities imposed by the resultant phenoxyl radicals that preclude development of such compounds because of lack of a sufficient therapeutic index. Therefore, it is expected that there will be study of new chemical classes as antioxidants. Such a trend is already emerging. As discussed the 21-aminosteroids and other compounds such as MCI-186 represent nonphenolic antioxidants that have demonstrated clinical efficacy. Continued examination of other chemical structures is warranted. An excellent treatise by Roberfroid et al. (232) focused on the design of captodative molecules. This chemistry is based on stabilization of electron-deficient carbocations by electron-donating substituents (dative) and electron-rich carbanions by electron-withdrawing substituents (captor). Molecules possessing captor and dative substituents are stabilized after donating or accepting electrons and thus function as antioxidants. Examples of naturally occurring captodative structures are semiquinone and flavin radicals whose stability is essential for carrying out biological electron transfer reactions. Many of the lipoxygenase inhibitors such as NDGA and BW755c can be oxidized with captodative stabilization of the resultant radical. The advantages of this approach are that first, the substituents can be varied considerably to aid in "targeting" the antioxidant and, second, that the stabilized radical is unlikely to be a prooxidant.

Throughout the pharmaceutical industry there is interest in natural products as drugs. In many cultures, plant extracts have been used for centuries as medicinal agents and many are being carefully analyzed for composition. A recent review highlights the status of research and clinical use of Chinese medicinal plants (233). Many extracts contain flavonoids, which are clearly potent antioxidants (Fig. 9). Flavonoids isolated from Indian medicinal plants were shown to be scavengers of oxygen radicals and inhibitors of lipid peroxidation (234). Many phenolic compounds, mostly flavonoids, were shown to inhibit carbon tetrachloride–induced peroxidation in rat liver microsomes, with some comparable in activity to the well-studied compound (+)-catechin (235). How-

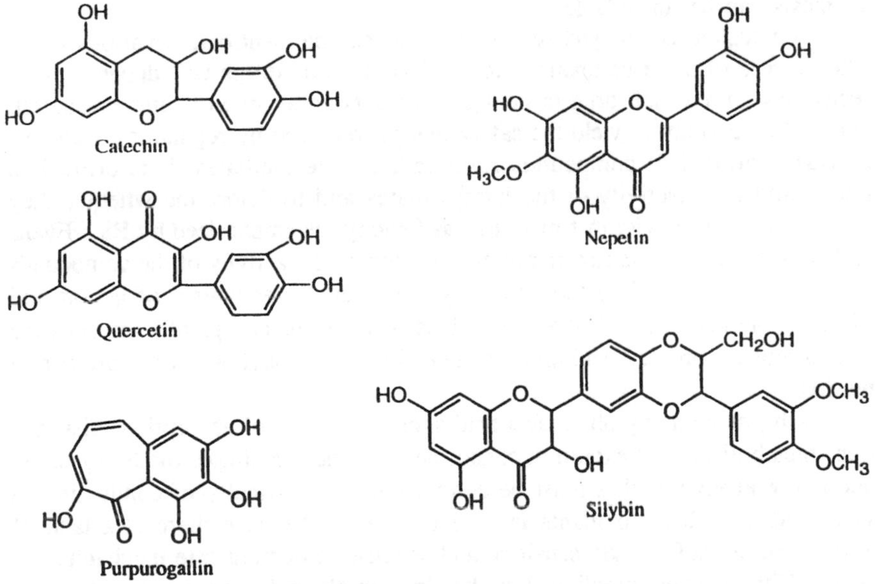

Figure 9 Natural products which are antioxidants and radical scavengers. Many of the plant flavonoids are polyhydroxylated compounds and are therefore potent antioxidants. Only a few of those studied are shown.

ever, these compounds can also exhibit prooxidant activity, as has been described for the catechins, epicatechin, and ferulic acid (236). The flavonoid purpurogallin was shown to protect against myocardial ischemia and reperfusion injury in vivo (237). Not surprisingly, the work with natural compounds has given rise to medicinal chemistry efforts based on flavone structures, and active molecules have been identified (238).

In addition to plant extracts, a relatively unexplored source of antioxidants is microbial broths, which are widely used as a source of novel anti-infectives. Kirin Brewery in Japan has identified several agents in their search for cytoprotective agents. Rumbrin, which was isolated from the fungus *Auxarthron umbrinum* (239), and pyrrolostatin, from *Streptomyces chrestomyceticus* EC40 (240), were both found to be active antioxidants with little acute toxicity in mice (>200 mg/kg intraperitoneal). The culture broth of *Thielavia minor* OFR-1561, isolated from soil in Okinawa-ken, Japan, yielded a novel pyrazinoxide compound, NF-1616-904, which decreased macrophase release of O_2^{--} in response to fMLP (241). Although it is capable of yielding novel molecules, this approach can be laborious. First, it requires isolation and structural identification of the active ingredient. Second, once the ingredient is identified, it is necessary to be able to synthesize the molecule by conventional synthetic organic methods. Because these molecules are often complex, structure elucidation and synthesis can be formidable.

In conclusion, novel synthetic antioxidants can be identified in various ways. The success of these compounds to be developed as drugs will depend on effectiveness, for which drug targeting represents an avenue of future research, and on low toxicity to yield the safety margin required by regulatory agencies. Because antioxidant compounds form radical intermediates, it is critical to understand the reactivity of the intermediates and to determine whether they have deleterious activity that manifests as toxicity, as emphasized by Rice-Evans and Diplock (242). It is also important to examine the activity of the compounds in a variety of model systems to fully understand mechanism of action and identify potential prooxidant activity. This is laborious and greatly impedes the drug discovery process, and improvements in high-throughput assays are sorely needed.

Finally, as with any pharmaceutical agent, the roles of absorption, distribution, metabolism, and excretion cannot be discerned in simple model systems, and many in vivo studies must be conducted. The difficulty of conducting in vivo studies with antioxidants is especially acute because there is a lack of assays that are sufficiently sensitive and specific to demonstrate mechanism of action. This hampers preclinical studies in animals and is even more of a barrier to the conduct of human clinical trials. Recently, better methods have been reported. The use of breath ethane as a marker of lipid peroxidation represents a sensitive, noninvasive marker that can be monitored (243). Similarly, measurement of peroxidation-derived F_2-isoprostanes has proved effective in animal studies (244,245) and has been shown to be useful in a study of human smokers (245). Ames and colleagues have also developed a gas chromatography–mass spectrometry method to detect lipid-derived aldehydes (247,248). The method can detect endogenous baseline levels of many species, which are increased in response to oxidative stress associated with conditions such as a cold virus. Only small amounts of plasma are required, and perhaps even tissue biopsy samples can be studied. These advances in methodology will aid greatly in the design and development of synthetic antioxidants as pharmaceuticals.

ACKNOWLEDGMENTS

I thank Jean Thomas for typing the manuscript and Ann Marie Ogden for critical review and discussion.

REFERENCES

1. Reimund E. The free radical flux theory of sleep. Med Hypotheses 1994; 43: 231–233.
2. Gey KF. Prospects for the prevention of free radical disease, regarding cancer and cardiovascular disease. Br Med Bull 1993; 49:679–699.
3. Hennekens CH. Antioxidant vitamins and cancer. Am J Med 1994; 97:2S–4S.
4. Pryor WA. The role of free radical reactions in biological systems. In: Pryor WA, ed. Free Radicals in Biology. Vol. 1. New York: Academic Press, 1976:1–49.
5. Coghlan JG. Allopurinol pretreatment improves postoperative recovery and reduces lipid peroxidation in patients undergoing coronary artery bypass grafting. J Thorac Cardiovasc Surg 1994; 107:248–256.
6. Augustin AJ, Boker J, Blumenroder SH, Lutz J, Spitznas M. Free radical scavenging and antioxidant activity of allopurinol and oxypurinol in experimental lens-induced uveitis. Invest Ophthalmol Vis Sci 1994; 35:3897–3904.
7. Salaris SC, Babbs CF, Voorhees WD III, Methylene blue as an inhibitor of superoxide generation by xanthine oxidase. Biochem Pharmacol 1991; 42:499–506.
8. Nakano Y, Kawaguchi T, Sumitomo J, Takizawa T, Uetsuki S, Sugawara M, Kido M. Novel inhibitors of superoxide anion generation, OPC-15160 and OPC-15161. Taxonomy, fermentation, isolation, physico-chemical properties, biological characteristics and structure determination. J Antibiot (Tokyo) 1991; 44:52–53.
9. Chihiro M, Nagamoto H, Takemura I, Kitano J, Komatsu H, Sekiguchi K, Tabusa F, Mori T, Tominaga M, Yabuuchi Y. Novel thiazole derivatives as inhibitors of superoxide production by human neutrophils: synthesis and structure-activity relationships. J Med Chem 1995; 38:353–358.
10. Beyer W Jr, Fridovich I. Characterization of a superoxide dismutase mimic prepared from desferrioxamine and MnO_2. Arch Biochem Biophys 1989; 271:149–156.

11. Darr D, Zarilla JA, Fridovich I. A mimic of superoxide dismutase activity based upon desferrioxamine B and manganese (IV). Arch Biochem Biophys 1987; 258:351–355.

12. Riley DP, Weiss RH. Manganese macrocyclic ligand complexes as mimics of superoxide dismutase. J Am Chem Soc 1994; 116:387–388.

13. Weiss RH, Flickinger AG, Rivers WJ, Hardy MM, Aston KW, Ryan US, Riley DP. Evaluation of activity of putative superoxide dismutase mimics: direct analysis by stopped-flow kinetics. J Biol Chem 1993; 268:23049–23054.

14. Kilgore KS, Friedrichs GS, Johnson CR, Schasteen CS, Riley DP, Weiss RH, Ryan U, Lucchesi BR. Protective effects of the SOD-mimetic SC-52608 against ischemia/reperfusion damage in the rabbit isolated heart. J Mol Cell Cardiol 1994; 26:995–1006.

15. Samuni A, Krishna CM, Riesz P, Finkelstein E, Russo A. A novel metal-free low molecular weight superoxide dismutase mimic. J Biol Chem 1988; 263:17921–17924.

16. Mitchell JB, Samuni A, Krishna MC, DeGraff WG, Ahn MS, Samuni U, Russo A. Biologically active metal-independent superoxide dismutase mimics. Biochemistry 1990; 29:2802–2807.

17. Krishna SA, Riesz P, Russo A. Superoxide reaction with nitroxide spin-adducts. Free Radic Biol Med 1989; 6:141–148.

18. Samuni A, Ahn MC, Krishna MC, Mitchell JB, Russo A. SOD-like activity of 5-membered ring nitroxide spin labels. Adv Exp Med Biol 1990; 264:85–92.

19. Gibbs D, Varani J, Ginsburg I. Formation and use of poly-L-histidine–catalase complexes: protection of cells from hydrogen peroxide-mediated injury. Inflammation 1989; 13: 465–474.

20. Lafon-Cazal M, Pietri S, Culcasi M, Bockaert J. NMDA-dependent peroxide production and neurotoxicity. Nature 1993; 364:535–537.

21. Minotti G, Aust SD. The requirement for iron (III) in the initiation of lipid peroxidation by iron (II) and hydrogen peroxide. J Biol Chem 1987; 262:1098–1104.

22. Giulivi C, Boveris A, Cadenas E. Hydroxyl radical generation during mitochondrial electron transfer and the formation of 8-hydroxydeoxyguanosine in mitochondrial DNA. Arch Biochem Biophys 1995; 316:909–916.

23. Dean RT, Nicholson P. The action of nine chelators on iron-dependent radical damage. Free Radic Res Commun 1994; 20:83–101.

24. Simpson JA, Narita S, Gieseg S, Gebicki S, Gebicki JM, Dean RT. Long-lived reactive species on free-radical–damaged proteins. Biochem J 1992; 282:621–624.

25. Davies MJ, Fu S, Dean RT. Protein hydroperoxides can give rise to reactive free radicals. Biochem J 1995; 305:643–649.

26. Antosiewicz J, Bertoli E, Damiani E, Greci L, Popinigis J, Pryzybylski S, Tanfani F, Wozniak M. Indolinonic and quinolinic aminoxyls as protectant against oxidative stress. Free Radic Biol Med 1993; 15:203–208.

27. Nilsson UA, Olsson L-I, Carlin G, Bylund-Fellenius A-C. Inhibition of lipid peroxidation by spin labels. J Biol Chem 1989; 264:11131–11135.

28. Janzen EG. A critical review of spin trapping in biological systems. In: Pryor

WA, ed. Free Radicals in Biology. Vol. 4. New York: Academic Press, 1980:115–154.

29. Oliver CN, Starke-Reed PE, Stadtman ER, Liu GJ, Carney JM, Floyd RA. Oxidative damage to brain proteins, loss of glutamine synthetase activity, and production of free radicals during ischemia/reperfusion-induced injury to gerbil brain. Proc Natl Acad Sci U S A 1990; 87:5144–5147.

30. Phillis JW, Clough-Helfman C. Free radicals and ischaemic injury: protection by the spin trap PBN. Med Sci Res 1990; 18:403–404.

31. Yue T-L, Gu J-L, Lysko PG, Cheng H-Y, Barone FC, Feurstein G. Neuroprotective effects of phenyl-t-butyl nitrone in gerbil global brain ischemia and in cultured rat cerebellar neurons. Brain Res 1992; 574:193–197.

32. Zhao Q, Pahlmark K, Smith M-L, Siesjo BK. Delayed treatment with the spin trap α-phenyl-N-tert-butyl nitrone (PBN) reduces infarct size following transient middle cerebral artery occlusion in rats. Acta Physiol Scand 1994; 152:349–350.

33. Cao X, Phillis JW. Alpha-phenyl-tert-butyl nitrone reduces cortical infarct and edema in rats subjected to focal ischemia. Brain Res 1994; 644:267–272.

34. Kalyanaraman B, Konorev EA, Joseph J, Baker JE. A critical review of myocardial protection by spin traps. In: Free Radicals in the Environment, Medicine and Toxicology. London: Richelieu Press, 1994: 313–326.

35. Thomas CE, Ohlweiler DF, Schmidt CJ, Cheng HC, Bernotas RC, Nieduzak TR, Carr AA. Radical trapping and antioxidant activity of cyclic nitrone spin traps (abstr.) Second Aging Symposium—Molecular Mechanisms of Degenerative Disease, Kansas City, MO, May 6–8, 1994.

36. Thomas CE, Ohlweiler DF, Kalyanaraman B. Multiple mechanisms for inhibition of low density lipoprotein oxidation by novel cyclic nitrone spin traps. J Biol Chem 1994; 269:28055–28061.

37. Thomas CE, Carney JM, Bernotas RC, Hay DA, Carr AA. In vitro and in vivo activity of a novel series of radical trapping agents in model systems of CNS oxidative. Ann NY Acad Sci 1994; 738:243–249.

38. French JF, Thomas CE, Downs TR, Ohlweiler DF, Carr AA, Dage RC. Protective effects of a cyclic nitrone antioxidant in animal models of endotoxic shock and chronic bacteremia. Circ Shock 1994; 43:130–136.

39. Njus D, Kelley PM. Vitamins C and E donate single hydrogen atoms *in vivo*. FEBS Lett 1991; 284:147–151.

40. Liebler DC. The role of metabolism in the antioxidant function of vitamin E. Crit Rev Toxicol 1993; 23:147–169.

41. Burton GW, Ingold KU. Autoxidation of biological molecules. 1. The antioxidant activity of vitamin E and related chain-breaking phenolic antioxidants in vitro. J Am Chem Soc 1981; 103:6472–6477.

42. Burton GW, Hughes L, Ingold KU. Antioxidant activity of phenols related to vitamin E. Are these chain-breaking antioxidants better than α-tocopherol? J Am Chem Soc 1983; 105:5950–5951.

43. Barclay LR, Vinqvist MR, Mukai K, Itoh S, Morimoti H. Chain-breaking phenolic antioxidants: steric and electronic effects in polyalkylchromanols, tocopherol analogs, hydroquinones, and superior antioxidants of the polyalkylbenzochromanol and naphthofuran class. J Am Chem Soc 1993; 58:7416–7420.

44. Janero DA, Cohen N, Burghardt B, Schaer BH. Novel 6-hydroxychroman-2-carbonitrile inhibitors of membrane peroxidative injury. Biochem Pharm 1990; 40:551–558.
45. Silver PJ, Gordon RJ, Horan PJ, Bushover CR, Gorczyca WP, Etzler JR, Buchholz RA, Schlegel D, Ellames GJ, Smith DI, Ezrin AM. Low molecular weight analogs of Trolox with potent antioxidant activity in vitro and in vivo. Drug Dev Res 1992; 27:45–52.
46. Hall ED, Braughler JM, Yonkers PA, Smith SL, Linseman KL, Means ED, Scherch HM, Von Voigtlander PF, Lahti RA, Jacobsen EJ. U-78517F: A potent inhibitor of lipid peroxidation with activity in experimental brain injury and ischemia. J Pharmacol Exp Ther 1991; 258:688–694.
47. Grisar JM, Petty MA, Bolkenius FN, Dow J, Wagner J, Wagner E, Haegele KD, De Jong W. A cardioselective, hydrophilic N, N, N-trimethylethanaminium α-tocopherol analogue that reduces myocardial infarct size. J Med Chem 1991; 34:257–260.
48. Petty MA, Dow J, Grisar M, De Jong W. Effect of a cardioselective α-tocopherol analogue on reperfusion injury in rats induced by myocardial ischaemia. Eur J Pharmacol 1991; 192:383–388.
49. Zughaib ME, Tang XL, Schleman M, Jeroudi MO, Bolli R. Beneficial effects of MDL 74,405, a cardioselective water soluble α-tocopherol analogue, on the recovery of function of stunned myocardium in intact dogs. Cardiovasc Res 1994; 28:235–241.
50. Kaneko Y, Kita H, Mukai K, Kawano K. Synthesis and antioxidant activity of 4H-1,3-benzodioxin-6-ol derivatives: new vitamin E analogs. Bull Chem Soc Jpn 1994; 67:1371–1379.
51. Hirose N, Hamamura K, Banba T, Kijima S. 2,2-Dimethyl-5-hydroxy-2,3-dihydrobenzofuran derivatives as antioxidants and radical scavengers. Japanese patent JP 02,121,975 [90,121,975] 1990.
52. Tamura K, Kato Y, Yoshida M, Cynshi O, Ohba Y. Preparation of 4-alkoxy-2,6-di-t-butylphenol derivatives as antiarteriosclerotic agents. PCT Int Appl WO 94 08,930, 1994.
53. Grisar JM, Bolkenius FN, Petty MA, Verne J. 2,3-Dihydro-1-benzofuran-5-ols as analogues of α-tocopherol that inhibit in vitro and ex vivo lipid autoxidation and protect mice against central nervous system trauma. J Med Chem 1995; 38:453–458.
54. Campo GM, Squadrito F, Loculano M, Avenoso A, Zingarelli B, Calandra S, Scuri R, Saitta A, Caputi AP. IRFI-016, new radical scavenger, limits ischemic damage following coronary artery occlusion in rats. Res Commun Chem Pathol Pharmacol 1992; 76:287–303.
55. Campo GM, Squadrito F, Ioculano M, Altavilla D, Zingarelli B, Pollicinok AM, Rizzo A, Calapai G, Calandra S, Scuri R. Protective effects of IRFI-016, a new antioxidant agent, in myocardial damage, following coronary artery occlusion and reperfusion in the rat. Pharmacology 1994; 48:157–166.
56. Calapai G, Squadrito F, Rizzo A, Crisafulli CA, Campo GM, Marciano MC, Mazzaglia G, Scuri R. A new antioxidant drug limits brain damage induced by transient cerebral ischaemia. Drugs Exp Clin Res 1993; 19:159–164.

57. Scuri R, Giannetti P, Paesano A. 2-(2,3-Dihydro-5-acetoxy-4,6,7-tribenzo-furanyl)acetic acid (IRFI 016): a new antioxidant mucoactive drug. Drugs Exp Clin Res 1990; 16:649–656.

58. Battioni JP, Fontecave M, Jaouen M, Mansuy D. Vitamin E derivatives as new potent inhibitors of microsomal lipid peroxidation. Biochem Biophys Res Commun 1991; 174:1103–1108.

59. Barnhart RL, Busch SJ, Jackson RL. Concentration-dependent antioxidant activity of probucol in low density lipoproteins in vitro: probucol degradation precedes lipoprotein oxidation. J Lipid Res 1989; 30:1703–1710.

60. Bittolo-Bon B, Cazzolato G, Avogaro P. Probucol protects low-density lipoproteins from in vitro and in vivo oxidation. Pharmacol Res 1991; 29:337–344.

61. Mao SJT, Yates MT, Parker RA, Chi EM, Jackson RL. Attenuation of athero-sclerosis in a modified strain of hypercholesterolemic watanabe rabbits with use of a probucol analogue (MDL 29,311) that does not lower serum cholesterol. Arterioscler, Thromb 1991; 11:1266–1275.

62. Heineke EW, Johnson MB, Dillberger JE, Robinson KM. Antioxidant MDL 29311 prevents diabetes in nonobese diabetic and multiple low-dose STZ-injected mice. Diabetes 1993; 42:1721–1730.

63. Schnedier JE, Berk BC, Gravanis MB, Santoian EC, Cipolla GD, Tarazona N, Lassegue B, King SB. Probucol decreases neointimal formation in a swine model of coronary artery balloon injury. A possible role for antioxidants in restenosis. Circulation 1993; 88:628–637.

64. Reed DJ, Pascoe GA, Olafsdottir K. Some aspects of cell defense mechanisms of glutathione and vitamin E during cell injury. Arch Toxicol Suppl. 1987; 11:34–38.

65. Brodie AE, Reed DJ. Reversible oxidation of glyceraldehyde 3-phosphate dehy-drogenase thiols in human lung carcinoma cells by hydrogen peroxide. Biochem Biophys Res Commun 1987; 148:120–125.

66. Thomas JA, Poland B, Honzatko R. Protein sulfhydryls and their role in the antioxidant function of protein S-thiolation. Arch Biochem Biophys 1995; 319:1–9.

67. Parthasarathy S. Evidence for an addtional intracellular site of action of probucol in the prevention of oxidative modification of low density lipoprotein. J Clin Invest 1992; 89:1618–1621.

68. Iliou JP, Thollon C, Robin F, Cambarrat C, Guillonneau C, Regineri G, Lenaers A, Vilaine JP. Protective effect of S12340 on cardiac cells exposed to oxidative stress. Eur J Pharmacol 1993; 248:263–272.

69. Kita T, Narumiya S, Narisada M, Watanabe F, Matsumoto S, Doteuchi M, Mizui T. Preparation of phenolic thioethers, specifically (3,5-di-*tert*-butyl-4-hydroxy-phenylthio) alkanoic acids and related compounds, useful as antiarteriosclerosis agents. Eur Pat Appl 148,203.

70. Matsumoto S, Mizui T, Doteuchi M. Preparation of di-*tert*-butylhydroxyphenyl-thio-substituted hydroxamic acid derivatives as drugs. Eur Pat Appl 405,788.

71. Panetta JA, Clemens JA. Novel antioxidant therapy for cerebral ischemia-reperfusion injury. Ann NY Acad Sci 1994; 723:239–245.

72. Clemens JA, Saunders RD, Ho PP, Phebus LA, Panetta JA. The antioxidant

LY231617 reduces global ischemic neuronal injury in rats. Stroke 1993; 24:716–723.

73. Clemens JA, Panetta JA. Neuroprotection by antioxidants in models of global and focal ischemia. Ann NY Acad Sci 1995; 738:250–256.
74. Patent application: EP434394.
75. Watanabe T, Morita I, Nishi H, Murota S. Preventive effect of MCI-186 on 15-HPETE induced vascular endothelial cell injury in vitro. Prostaglandins Leukot Essent Fatty Acids 1988; 33:81–87.
76. Watanabe T, Yuki S, Egawa M, Nishi H. Protective effects of MCI-186 on cerebral ischemia: possible involvement of free radical scavenging and antioxidant actions. J Pharmacol Exp Ther 1993; 268:1597–1604.
77. Abe K, Yuki S, Kogure K. Strong attenuation of ischemic and post-ischemic brain edema in rats by a novel free radical scavenger. Stroke 1988; 19:480–485.
78. Nishi H, Watanabe T, Sakurai H, Yuki S, Ishibashi A. Effect of MCI-186 on brain edema in rats. Stroke 1989; 20:1236–1240.
79. Watanabe T, Egawa M. Effects of an antistroke agent MCI-186 on cerebral arachidonate cascade. J Pharmacol Exp Ther 1994; 271:1624–1629.
80. Kuno A, Sugiyama Y, Katsuta K, Kamitani T, Takasugi H. Studies on cerebral protective agents: I. Novel 4-arylpyrimidine derivatives with anti-anoxic and anti-lipid peroxidation activities. Chem Pharmacol Bull 1992; 40:1452–1461.
81. Kuno A, Sugiyama Y, Katsuta K, Sakai H, Takasugi H. Studies on cerebral protective agents: II. Novel 4-arylpyrimidine derivatives with anti-anoxic and anti-lipid peroxidation activities. Chem Pharmacol Bull 1992; 40:2423–2431.
82. Kuno A, Katsuta K, Sakai H, Ohkubo M, Sugiyama Y, Takasugi H. Studies on cerebral protective agents: III. Novel 4-arylpyrimidine derivatives with anti-anoxic and anti-lipid peroxidation activities. Chem Pharmacol Bull 1992; 41:139–147.
83. Sainsbury M, Shertzer HG. US Patent: 5,185,360, 1993.
84. Shertzer HG. Retardation of benzo[a]pyrene-induced epidermal tumor formation by the potent antioxidant 4b,5,9b,10-tetrahydroindeno[1,2-b]indole. Cancer Lett 1994; 86:209–214.
85. Liu RM, Vasiliou V, Zhu H, Duh JL, Tabor MW, Puga A, Nebert DW, Sainsbury M, Shertzer HG. Regulation of [Ah] gene battery enzymes and glutathione levels by 5,10-dihydroindeno[1,2-b]indole in mouse hepatoma cell lines. Carcinogenesis 1994; 15:2347–2352.
86. Liu RM, Sainsbury M, Tabor MW, Shertzer HG. Mechanisms of protection from menadione toxicity by 5, 10-dihydroindeno[1,2-b]indole in a sensitive and resistant mouse hepatocyte line. Biochem Pharmacol 1993; 46:1491–1499.
87. Szporny L, Kiss B, Gere A, Szantay C, Bihari M, Pellionisz Paroczai M, Hegedus B, Karpati E, Domany G, et al. Antioxidant 1-[phenyl-1α,2,3,4,6,7,12,ii bα-octahydropyrimido[1'6':1,2]pyrido[3,4-b]indole derivatives and pharmaceutical composition containing them, and process for their production. Hungarian Patent HU 66,017, 1994.
88. Tsujimori H, Kamigaki N, Taniguchi Y, Tafusa F, Namikawa J. Preparation of 1-amino-2,3-dihydro-7-hydroxy-1H-indenes as pharmaceuticals and antioxidants. Japanese Patent JP 06,239,812 [94,239,812], 1994.

89. Hara H, Kogure K. Prevention of hippocampus neuronal damage in ischemic gerbils by a novel lipid peroxidation inhibitor (quinazoline derivative). J Pharmacol Exp Ther 1990; 255: 906–913.

90. Halliwell B. How to characterize a biological antioxidant. Free Radic Res Commun 1990; 9:1–32.

91. Flohe L. Determination of glutathione peroxidase. In: Handbook of Free Radicals and Antioxidants in Biomedicine. Vol. 3. Miquel J, Quintanilha AT, Weber H, eds. Boca Raton, FL: CRC Press, 1989: 281–286.

92. Mirault ME, Tremblay A, Furling D, Trepanier B, Puymirat J, Pothier F. Transgenic glutathione peroxidase mouse models for neuroprotection studies. Ann NY Acad Sci 1994; 738:104–115.

93. Chambers SJ, Lambert N, Williamson G. Purification of a cytosolic enzyme from human liver with phospholipid hydroperoxide glutathione peroxidase activity. Int J Biochem 1994; 26:1279–1286.

94. Parnham MJ, Leyck S, Graf E, Dowling EJ, Blake DR. The pharmacology of Ebselen. Agents Actions 1991; 32:4–9.

95. Parnham MJ, Graf E. Seleno-organic compounds and the therapy of hydroperoxide-linked pathological conditions. Biochem Pharmacol 1987; 36:3095–3102.

96. Cotgreave IA, Sandy MS, Berggren M, Moldeus PW, Smith MT. N-Acetylcysteine and glutathione-dependent protective effect of PZ51 (Ebselen) against diquat-induced cytotoxicity in isolated hepatocytes. Biochem Pharmacol 1987; 36:2899–2904.

97. Vasanthy N, Sies H. Oxidative damage to mitochondria and protection by Ebselen and other antioxidants. Biochem Pharmacol 1990; 40:1623–1629.

98. Thomas CE, Jackson RL. Lipid hydroperoxide involvement in copper-dependent and independent oxidation of low density lipoproteins. J Pharmacol Exp Ther 1991; 256:1182–1188.

99. Thomas JP, Kalyanaraman B, Girotti AW. Involvement of preexisting lipid hydroperoxide in Cu^{2+}-stimulated oxidation of low-density lipoprotein. Arch Biochem Biophys 1994; 315:244–254.

100. Haenen GRMM, De Rooij BM, Vermeulen NPE, Bast A. Mechanism of the reaction of Ebselen with endogenous thiols: dihydrolipoate is a better cofactor than glutathione in the peroxidase activity of Ebselen. Mol Pharmacol 1990; 37:412–422.

101. Christison J, Seis H, Stocker R. Human blood cells support the reduction of low-density-lipoprotein-associated cholesteryl ester hydroperoxides by albumin-bound Ebselen. Biochem J, 1994; 304:341–345.

102. Aust SD, Morehouse LA, Thomas CE. Role of metals in oxygen radical reactions. Free Radic Biol Med 1985; 1:3–25.

103. Wu Z-P, Hilvert D. Selenosubtilisin as a glutathione peroxidase mimic. J Am Chem Soc 1990; 112:5647–5648.

104. Granger DN, Rutilli G, McCord JM. Superoxide radicals in feline intestinal ischaemia. Gastroenterology 1981; 81:22–29.

105. Parks DA, Bulkley GB, Granger DN, Hamilton SR, McCord JM. Ischemic injury in the cat small intestine: role of superoxide radicals. Gastroenterology 1982; 82:9–15.,

106. Rangan U, Bulkley GB. Prospects for treatment of free-radical–mediated tissue injury. Br Med Bull 1993; 49:700–718.

107. Hoshino T, Maley WR, Bulkley GB, Williams GM. Ablation of free radical–mediated reperfusion injury for the salvage of kidneys taken from nonheartbeating donors: a quantitative evaluation of the proportion of injury caused by reperfusion following periods of warm and cold ischemia. Transplantation 1988; 45:284–289.

108. Arroyo CM, Carmichael AJ, Bouscarel B, Liang JH, Weglicki WB. Endothelial cells as a source of oxygen-free radicals. An ESR study. Free Radic Res Commun 1990; 9:287–296.

109. Inoue M, Ebashi I, Watanabe N, Morino Y. Synthesis of a SOD derivative that circulates bound to albumin and accumulates in tissues whose pH is decreased. Biochemistry 1989; 28:6619–6624.

110. Inoue M, Watanabe N, Morino Y, Sasaki J, Tanaka Y, Amachi T. Inhibition of oxygen toxicity by targeting SOD to endothelial cell surface. FEBS Lett 1990; 269:89–92.

111. Dow J, Petty Ma, Grisar JM. Wagner ER, Haegele KD. Cardioselectivity of alpha-tocopherol analogues, with free radical scavenger activity, in the rat. Drug Metab Dispos 1991; 19:1040–1045.

112. Zughaib ME, Tang XL, Schleman M. Jeroudi MO, Bolli R. Beneficial effects of MDL 74,405, a cardioselective water soluble alpha tocopherol analogue, on the recovery of function of stunned myocardium in intact dogs. Cardiovasc Res 1994; 28:235–241.

113. Tang XL, McCay PB, Sun JZ, Hartley CJ, Schleman M, Bolli R. Inhibitory effect of a hydrophilic α-tocopherol analogue, MDL 74,405, on generation of free radicals in stunned myocardium in dogs. Free Radic Res Commun 1995; 2:293–302.

114. Petty M, Grisar JM, Dow J, De Jong W. Effects of a α-tocopherol analogue on myocardial ischaemia and reperfusion injury in rats. Eur J Pharmacol 1990; 179:241–242.

115. Petty MA. MDL 73,404: a cardioselective antioxidant that protects against myocardial reperfusion injury. Cardiovasc Drug Rev 1992; 10:413–424.

116. Lukovic L, Petty MA, Bolkenius FN, Grisar JM, Dow J, De Jong W. Protection of infarcted chronically reperfused hearts by an α-tocopherol analogue. Eur J Pharmacol 1993; 233:63–70.

117. Jackson RL, Ku G, Mao SJT, Sheetz MJ, Robinson KM, Thomas CE. Natural and synthetic antioxidants: action and treatment implications for atherosclerosis and diabetes. In: New Horizons in Coronary Heart Disease. London: Science Press, 1993:20.1–20.13.

118. Jackson RL, Ku G, Thomas CE. Antioxidants: a biological defense mechanism for the prevention of atherosclerosis. Med Res Rev 1993; 13:161–182.

119. Thomas CE. Development of antioxidant pharmaceuticals. In: KJA Davies, ed. The Oxygen Paradox. Padova, Italy. Cleup Press, 1995:615–625.

120. Thomas CE, McLean LR, Parker RA, Ohlweiler D. Ascorbate and phenolic antioxidant interactions in prevention of liposomal oxidation. Lipids 1992; 27:543–550.

121. Esterbauer H, Gebicki J, Puhl H. Jurgens G. The role of lipid peroxidation and antioxidants in oxidative modification of LDL. Free Radic Biol Med 1992; 13:341–390.

122. Beck G, Bergmann A, KeBeler K, Wess G. Synthesis of a new flavonoid-antioxidant. Tetrahedron Lett 1990; 50:7293–7296.

123. Tanaka M, Nakagawa M. Antioxidant activity of thiocholesterol on copper induced oxidation of low-density lipoprotein. Lipids 1995; 30:321–325.

124. Nagao A, Terao J. Antioxidant activity of 6-phosphatidyl-L-ascorbic acid. Biochem Biophys Res Commun 1990; 172: 385–389.

125. Nihro Y, Miyataka H, Sudo T, Matsumoto H, Satoh T. 3-O-alkylascorbic acids as free-radical quenchers: synthesis and inhibitory effect on lipid peroxidation. J Med Chem 1991; 34:2152–2157.

126. Nonaka A, Manabe T, Tobe T. Effect of a new synthetic free radical scavenger, 2-octadecyl ascorbic acid, on the mortality in mouse endotoxemia. Life Sci 1990; 47:1933–1939.

127. Nonaka A, Manabe T, Tobe T. Effect of a new synthetic ascorbic acid derivative as a free radical scavenger on the development of acute pancreatitis in mice. Gut 1991; 32:528–532.

128. Shimamoto N, Ogata K. European Patent Application 89100205.7, 1989.

129. Fariss MW. Cadmium toxicity: unique cytoprotective properties of alpha tocopheryl succinate in hepatocytes. Toxicology 1991; 69:63–77.

130. Fariss MW, Fortuna MB, Everett CK, Smith JD, Trent DF, Djuric Z. The selective antiproliferative effects of alpha-tocopheryl hemisuccinate and cholesteryl hemisuccinate on murine leukemia cells result from the action of the intact compounds. Cancer Res 1994; 54:3346–3351.

131. Ray SD, Fariss MW. Role of cellular energy status in tocopheryl hemisuccinate cytoprotection against ethyl methansulfonate-induced toxicity. Arch Biochem Biophys 1994; 311:180–190.

132. Yu BP. Cellular defenses against damage from reactive oxygen species. Physiol Rev 1994; 74:139–162.

133. Shigenaga MK, Hagen TM, Ames BN. Oxidative damage and mitochondrial decay in aging. Proc Natl Acad Sci U S A 1994; 91:10771–10778.

134. Turrens JF, Freeman BA, Crapo JD. Hyperoxia increases H_2O_2 release by lung mitochondria and microsomes. Arch Biochem Biophys 1982; 217:411–421.

135. Turrens JF, Alexandre A, Lehninger AL. Ubisemiquinone is the electron donor for superoxide formation by complex III of heart mitochondria. Arch Biochem Biophys 1985; 237:408–414.

136. Ames BN, Shigenaga MK, Hagen TM. Oxidants, antioxidants, and the degenerative diseases of aging. Proc Natl Acad Sci U S A 1993; 90:7915–7922.

137. Meredith MJ, Reed DJ. Status of the mitochondrial pool of glutathione in the isolated hepatocyte. J Biol Chem 1982; 257:3747–3753.

138. Olafsdottir K, Pascoe GA, Reed DJ. Mitochondrial glutathione status during Ca^{2+} ionophore-induced injury to isolated hepatocytes. Arch Biochem Biophys 1988; 263:226–235.

139. Thomas CE, Reed DJ. Effect of extracellular CA^{++} omission on isolated hepatocytes: I. Induction of oxidative stress and cell injury. J Pharmacol Exp Ther 1988; 245:493–500.

140. Shan X, Jones DP, Hashmi M, Anders MW. Selective depletion of mitochondrial glutathione concentrations by (R,S)-3-hydroxy-4-pentenoate potentiates oxidative cell death. Chem Res Toxicol 1993; 6:75–81.

141. Shek PN, Suntres ZE, Brooks JI. Liposomes in pulmonary applications: physicochemical considerations, pulmonary distribution and antioxidant delivery. J Drug Targeting 1994; 2:431–442.

142. Freeman BA, Young SL, Crapo JD. Liposome-mediated augmentation of superoxide dismutase in endothelial cells prevents oxygen injury. J Biol Chem 1983; 258:12534–12542.

143. Nakae D, Yamamoto K, Yoshiji H, Kinugasa T, Maruyama H, Farber JL, Konishi Y. Liposome-encapsulated superoxide dismutase prevents liver necrosis induced by acetaminophen. Am J Pathol 1990; 136:787–795.

144. In: Scrip (1994), No. 1935, p 27.

145. Gillissen A, Roum JH, Hoyt RF, Crystal RG. Aerosolization of superoxide dismutase. Augmentation of respiratory epithelial lining fluid antioxidant screen by aerosolization of recombinant human Cu^{++}/Zn^{++} superoxide dismutase. Chest 1993; 104:811–815.

146. Baer AN, Costello PB, Green FA. Free and esterified 13 (R,S)-hydroxyoctadecadienoic acids: principal oxygenase products in psoriatic skin scales. J Lipid Res 1990; 31:125–130.

147. Tojo K, Lee AC. Bioconversion of a provitamin to vitamins C and E in skin. J Soc Cosmet Chem 1987; 38:333–339.

148. Kato T, Murakami Y, Mimura M, Takahara Y. UV-induced damage preventing topical preparations containing p-hydroxybenzaldehyde. Japanese Patent JP 07, 02,640 [95 02,640].

149. Bulkley GB. Free radicals and other reactive oxygen metabolites: clinical relevance and the therapeutic efficacy and antioxidant therapy. Surgery 1993; 113:479–483.

150. Harris ML, Schiller HJ, Reilly PM, Donowitz M, Grisham MG, Bulkley GB. Free radicals and other reactive oxygen metabolites in inflammatory bowel disease: cause, consequence or epiphenomenon? Pharmacol Ther 1992; 53:375–408.

151. Schreck R, Rieber P, Baeuerle PA. Reactive oxygen intermediates as apparently widely used messengers in the activation of the NF-κB transcription factor and HIV-1. EMBO J 1991; 10:2247–2258.

152. Schreck R, Meier B, Mannel DN, Droge W, Baeuerle PA. Dithiocarbamates as potent inhibitors of nuclear factor κB activation in intact cells. J Exp Med 1992; 175:1181–1194.

153. Schreck R, Baeuerle P. A role for oxygen radicals as second messengers. Trends Cell Biol 1991; 1:39–42.

154. Anderson MT, Staal FJT, Gitler C, Herzenberg LA, Herzenberg LA. Separation of oxidant-initiated and redox-regulated steps in the NF-κB signal transduction pathway. Proc Natl Acad Sci U S A 1994; 91:11527–11531.

155. Christman BW, Bernard GR. Antilipid mediator and antioxidant therapy in adult respiratory distress syndrome. New Horiz 1993; 1:623–630.

156. Ferrari R, Ceconi C, Curello S, Cargnoni A, Alfieri O, Pardini A, Marzollo P, Visioli O. Oxygen free radicals and myocardial damage: protective role of thiol-containing agents. Am J Med 1991; 91:95S–105S.

157. Langley SC, Kelly FJ. N-acetylcysteine ameliorates hyperoxic lung injury in the preterm guinea pig. Biochem Pharmacol 1993; 45:841–846.

158. Bernard GR. N-Acetylcysteine in experimental and clinical acute lung injury. Am J Med 1991; 91(suppl C):3C54S–3C59S.

159. Marui N, Offermann MK, Swerlick R, Kunsch C, Rosen CA, Ahmad M, Alexander RW, Medford RM. Vascular cell adhesion molecule-1 (VCAM-1) gene transcription and expression are regulated through an antioxidant-sensitive mechanism in human vascular endothelial cells. J Clin Invest 1993; 92:1866–1874.

160. Cybulsky MI, Gimbrone MAJ. Endothelial expression of a mononuclear leukocyte adhesion molecule during atherogenesis. Science 1991; 251:788–791.

161. Piette J, Legrands-Poels A. HIV-1 reactivation after an oxidative stress mediated by different reactive oxygen species. Chem Biol Interact 1994; 91:79–89.

162. Simon G, Moog C, Obert G. Valpoic acid reduces the intracellular level of glutathione and stimulates human immunodeficiency virus. Chem Biol Interact 1994; 91:111–121.

163. Simon G, Moog C, Obert G. Effects of glutathione precursors on human immunodeficiency virus replication. Chem Biol Interact 1994; 91:217–224.

164. Roederer M, Staal FJT, Raju PA, Ela WS, Herzenberg LA. Cytokine HIV replication is inhibited by N-acetyl-L-cysteine. Proc Natl Acad Sci U S A 1990; 87:4884–4888.

165. Klebic T, Kinter A, Poli G, Anderson ME, Meister A, Fauci AS. Suppression of HIV expression in chronically infected monocytes by glutathione, glutathione-esters and N-acetylcysteine. Proc Natl Acad Sci U S A 1991; 88:986–990.

166. Grimble RF. Nutritional antioxidants and the modulation of inflammation: theory and practice. New Horizons 1994; 2:175–185.

167. Malorni R, Rivabene MT, Santini MT, Donelli G. N-Acetyl cysteine inhibits apoptosis and decreases viral particles in HIV chronically infected U937 cells. FEBS Lett 1993; 1327:75–78.

168. Favier A, Sappey C, Leclerc P, Faure P, Micoud M. Antioxidant status and lipid peroxidation in patients infected with HIV. Chem Biol Interact 1994; 91:165–180.

169. Suzuki YJ, Packer L. Inhibition of NF-kappa B transcription factor by catechol derivatives. Biochem Mol Interact 1994; 32:299–305.

170. Shenkar R, Abraham E. Effects of treatment with the 21-aminosteroid, U74389F, on pulmonary cytokine expression following hemorrhage and resuscitation. Crit Care Med 1995; 23:132–139.

171. Shin MS, Angel MF, Im MJ, Manson PN. Effects of 21-aminosteroid U74389F on skin-flap survival after secondary ischemia. Plast Reconstr Surg 1994; 94:661–666.

172. Nakazawa T, Nagatsuka S, Yukawa O. Effect of membrane stabilizing agents and radiation on liposomal membranes. Drugs Exp Clin Res 1986; 12:831–835.

173. McLean LR, Hagaman KA. Effect of lipid physical state on the rate of peroxidation of liposomes. Free Radic Biol Med 1992; 12:113–119.

174. Hinzmann JS, KcKenna RL, Pierson TS, Han F, Kezdy FJ, Epps DE. Interaction of antioxidants with depth-dependent fluorescence quenchers and energy transfer probes in lipid bilayers. Chem Phys Lipids 1992; 62:123–138.

175. Horan KL, Lutzke BS, Cazers AR, McCall JM, Epps DE. Kinetic evaluation of lipophilic inhibitors of lipid peroxidation in DLPC liposomes. Free Radic Biol Med 1994; 17:587–596.

176. McLean LR, Hagaman KA. Effect of probucol on the physical properties of low-density lipoproteins oxidized by copper. Biochemistry 1989; 28:321.

177. McLean LR, Thomas CE, Weintraub B, Hagaman KA. Modulation of the physical state of cellular cholesteryl esters by 4,4′-(isopropylidenedithio)bis(2,6-di-t-butylphenol) (probucol). J Biol Chem 1992; 267:12291–12298.

178. McLean LR, Brake N, Hagaman KA. Interactions of MDL 29,311 and probucol metabolites with cholesteryl esters. Lipids 1994; 29:819–823.

179. Ku G, Doherty NS, Wolos JA, Jackson RL. Inhibition by probucol of interleukin 1 secretion and its implication in atherosclerosis. Am J Cardiol 1988; 62:77B–81B.

180. Ku G, Doherty NS, Schmidt LR, Jackson RL, Dinerstein RJ. Ex vivo lipopolysaccharide-induced interleukin-1 secretion from murine peritoneal macrophages inhibited by probucol, a hypocholesterolemic agent with antioxidant properties. FASEB J 1990; 4:1645–1653.

181. Akeson AL, Woods CW, Mosher LB, Thomas CE, Jackson RL. Inhibition of IL-1β expression in THP-1 cells by probucol and tocopherol. Atherosclerosis 1991; 86:261–270.

182. Kasama T, Kobayashk K, Fukushima T, Tabata M, Ohno L, Negishi M, Ide H, Takahashi T, Niwa Y. Production of interleukin 1–like factor from human peripheral blood monocytes and polymorphonuclear leukocytes by superoxide anion: the role of interleukin 1 and reactive oxygen species in inflamed sites. Clin Immunol Immunopathol 1989; 53:439–448.

183. Mahoney CV, Azzi A. Vitamin E inhibits protein kinase C activity. Biochem Biophys Res Commun 1988, 154:694–697.

184. Ku G, Thomas CE, Akeson AL, Jackson RL. Induction of interleukin 1β expression from human peripherol blood monocyte-derived macrophages by 9-hydroxy-octadecadienoic acid. J Biol Chem 1992; 267:14183–14188.

185. Thomas CE, Jackson RL, Ohlweiler DF, Ku G. Multiple lipid oxidation products in low density lipoproteins induce interleukin-1 beta release from human blood mononuclear cells. J Lipid Res 1994; 35:417–427.

186. Thomas CE, Ku G, Kalyanaraman B. Nitrone spin trap lipophilicity as a determinant for inhibition of low density lipoprotein oxidation and activation of interleukin-1β release from human monocytes. J Lipid Res 1994; 35:610–619.

187. Boscoboinik D, Ozer NK, Moser U, Azzi A. Tocopherols and 6-hydroxy-chroman-2-carbonitrile derivatives inhibit vascular smooth muscle cell proliferation by a nonantioxidant mechanism. Arch Biochem Biophys 1995; 318:241–246.

188. Yamamoto S. Mammalian lipoxygenases: molecular structures and functions. Biochem Biophys Acta 1992; 1128:117–131.

189. Lazer ES, Wong HC, Wegner CD, Graham AG, Farina PR. Effect of structure on potency and selectivity in 2,6-disubstituted 4-(2-arylethenyl) phenol lipoxygenase inhibitors. J Med Chem 1990; 33:1892–1898.

190. Arai H, Nagao A, Terao J, Suzuki T, Takama K. Effect of d-α-tocopherol analogues on lipoxygenase-dependent peroxidation of phospholipid-bile salt micelles. Lipids 1995; 30:135–140.

191. Maccarrone M, Veldink GA, Vliegenthart JFG, Agro AF. Inhibition of soybean lipoxygenase-1 by chain-breaking antioxidants. Lipids 1995; 30:51–54.

192. Belanger PC, Lau CK, Dufresne C, Rokach J, Guindon Y, Schiegetz J. Therien M, Young RN. Preparation of 5-hydroxy-2,3-dihydrobenzofuran analogs as leukotriene biosynthesis inhibitors. European Patent Appl EP 447,189, 1991.

193. Rankin SM, Parthasarathy S, Steinberg D. Evidence for a dominant role of lipoxygenase(s) in the oxidation of LDL by mouse peritoneal macrophages. J Lipid Res 1991; 32:449–456.

194. Sparrow CP, Olszewski J. Cellular oxidative modification of low density lipoprotein does not require lipoxygenases. Proc Natl Acad Sci U S A 1992; 89:128–131.

195. Furukawa M, Yoshimoto T, Ochi K, Yamamoto S. Studies on arachidonate 5-lipoxygenase of rat basophilic leukemia cells. Biochem Biophys Acta 1984; 795:458–465.

196. Bell RL, Young PR, Lanni AC, Summers JB, Brooks DW, Rubin P, Carter GW. The discovery and development of zileuton: an orally active 5-lipoxygenase inhibitor. Int J Immunopharmacol 1992; 14:505–510.

197. Satoh Y, Powers C, Toledo LM, Kowalski TJ, Peters PA, Kimble EF. Derivatives of 2-[[N-(aminocarbonyl)-N-hydroxyamino]methyl]-1,4-benzodioxan as orally active 5-lipoxygenase inhibitors. J Med Chem 1995; 38:68–75.

198. Ohemeng KA, Nguyen VN, Schwender CF, Singer M, Steber M, Ansell J, Hageman W. Novel bishydroxamic acids as 5-lipoxygenase inhibitors. Bioorg Med Chem Lett 1994; 2:187–193.

199. Janero DR, Burghardt B, Lopez R, Cardell M. Influence of cardioprotective cyclooxygenase and lipoxygenase inhibitors on peroxidative injury to myocardial-membrane phospholipid. Biochem Pharmacol 1989; 38:4381–4387.

200. Weglicki WB, Mak IT, Simic MG. Mechanisms of cardiovascular drugs as antioxidants. J Mol Cell Cardiol 1990; 22:1199–1208.

201. Mak IT, Weglicki WB. Comparative antioxidant activities of propranolol, nifedipine, verapamil and diltiazem against sarcolemmal membrane lipid peroxidation. Circ Res 1990; 66:1449–1452.

202. Westlin W, Mullane K. Does captopril attenuate reperfusion-induced myocardial dysfunction by scavenging free radicals. Circulation 1988; 77:130–139.

203. Weglicki WB, Dickens BF, Pflug BR, Mak IT. Protective effects of sulfhydryl-containing ACE inhibitors against lipid peroxidation in endothelial and smooth muscle cells. FASEB J 1989; 3:A593.

204. Mak IT, Dickens BF, Weglicki WB. Hydroxyl radical scavenging and attenuation of free radical injury endothelial cells by SH-containing ACE inhibitors. Circulation 1989; 80:II-242.

205. Weglicki WB and Mak IT. Antioxidant drug mechanisms: transition metal-binding and vasodilation. Mol Cell Biochem 1992; 118:105–111.

206. Janero DR, Burghardt B, Lopez R. Protection of cardiac membrane phospholipid against oxidative injury by calcium antagonists. Biochem Pharmacol 1988; 37:4197–4203.

207. Engineer F, Sridhar R. Inhibition of rat heart and liver microsomal lipid peroxidation by nifedipine. Biochem Pharmacol 1989; 38:1279–1285.

208. Janero DR, Lopez R, Pittman J, Burghardt B. Propanolol as xanthine oxidase inhibitor: implications for antioxidant activity. Life Sci 1989; 44:1579–1588.

209. Iuliano L, Violi F, Ghiselli A, Alessandri C, Balsano F. Dipyridamole inhibits lipid peroxidation and scavenges oxygen radicals. Lipids 1989; 24:430–433.

210. Ito C, Im WB, Takagi H, Takahashi M, Tsuzuki K, Liou SY, Kunihara M. U-92032, a T-type Ca^{2+} channel blocker and antioxidant, reduces neuronal ischemic injuries. Eur J Pharmacol 1994; 257:203–210.

211. Carini R, Poli G, Dianzani MV, Maddix SP, Slater TF, Cheeseman KH. Comparative evaluation of the antioxidant activity of alpha-tocopherol, alpha-tocopherol polyethylene glycol 1600 succinate and alpha tocopherol succinate. Biochem Pharmacol 1990; 39:1597–1601.

212. Ruffolo RR Jr, Boyle D, Venuit RP, Lukas M. Carvedilol (Kredex): a novel multiple action cardiovascular agent. Drugs Today 1991; 27:465–492.

213. Yue TL, Lysko PG, Barone FC, Gu JL, Ruffolo RR Jr, Feuerstein GZ. Carvedilol, a new antihypertensive drug with unique antioxidant activity: potential role in cerebroprotection. Ann NY Acad Sci 1994; 738:230–242.

214. Yue TL, Cheng HY, Lysko PG, McKenna PJ, Feuerstein R, Gu JL, Lysko KA, Davis LL, Feuerstein G. Carvedilol, a new vasodilator and beta adrenoceptor antagonist, is an antioxidant and free radical scavenger. J Pharmacol Exp Ther 1992; 263:92–98.

215. Feuerstein R, Yue TL. A potent antioxidant, SB209995, inhibits oxygen-radical-mediated lipid peroxidation and cytotoxicity. Pharmacology 1994; 48:385–391.

216. Yue TL, McKenna PJ, Lysko PG, Gu JL, Lysko KA, Ruffolo RR Jr, Feuerstein GZ. SB 211475, a metabolite of carvedilol, a novel antihypertensive agent, is a potent antioxidant. Eur J Pharmacol 1994; 251:237–243.

217. Wiseman H, Cannon M, Arnstein RV, Halliwell B. Mechanism of inhibition of lipid peroxidation by tamoxifen and 4-hydroxytamoxifen introduced into liposomes. FEBS Lett 1990; 274:107–110.

218. Wiseman H. The antioxidant action of a pure antiestrogen: ability to inhibit lipid peroxidation compared to tamoxifen and 17β-oestradiol and relevance to its anticancer potential. Biochem Pharmacol 1994; 47:493–498.

219. Custodio JB, Denis TC, Almeida LM, Medeira VM. Tamoxifen and hydroxytamoxifen as intramembraneous inhibitors of lipid peroxidation. Evidence for peroxyl radical scavenging activity. Biochem Pharmacol 1998; 47:1989–1998.

220. Wiseman H. Tamoxifen: new membrane-mediated mechanisms of action and therapeutic advances. Trends Pharmacol Sci 1994; 15:83–89.

221. Wiseman H, Quinn P. The antioxidant action of synthetic oestrogens involves decreased membrane fluidity: relevance to their potential use as anticancer and

cardioprotective agents compared to tamoxifen? Free Radic Res Commun 1994; 21:187–194.

222. Janero DR, Burghardt B. Prevention of oxidative injury to cardiac phospholipid by membrane-active "stabilizing agents." Res Commun Chem Pathol Pharmacol 1989; 63:163–173.

223. Gupta SK, Joshi S. Role of naproxen as anti-oxidant in selenite cataract. Ophthalmic Res 1994; 26:226–231.

224. Yokoyama H, Horie T, Awazu R. Oxidative stress in isolated rat hepatocytes during naproxen metabolism. Biochem Pharmacol 1995; 49:991–996.

225. Drew MGB, Hopkins WA, Mitchell PCH. Molecular mechanics study of hindered phenols used as antioxidants. J Chem Soc Faraday Trans 1990; 86:47–52.

226. Tanaka K, Sakai S, Tomiyama S, Nishiyama T, Yamada F. Molecular orbital approach to antioxidant mechanisms of phenols by an ab initio study. Chem Soc Jpn 1991; 64:2677–2680.

227. Thomas CE, Ohlweiler DF, Carr AA, Neiduzak TR, Hay DA, Adams G, Vaz R, Bernotas R. Characterization of the radical trapping activity of a novel series of cyclic nitrone spin traps. J Biol Chem 1996; 271:3097–3104.

228. Pryor WA, Cornicelli JA, Devall LJ, Tait B, Trivedi BK, Witiak DT, Wu M. A rapid screening test to determine the antioxidant potencies of natural and synthetic antioxidants. J Org Chem 1993; 58:3521–3532.

229. Kurth MJ, Randall LAA, Chen C, Melander C, Millere RB. Library-based lead compound discovery: antioxidants by an analogous synthesis/deconvolutive assay strategy. J Org Chem 1994; 59:5862–5864.

230. Ingold KU, Bowry VW, Stocker R, Walling C. Autoxidation of lipids and antioxidation by alpha-tocopherol and ubiquinol in homogeneous solution and in aqueous dispersions of lipids: unrecognized consequences of lipid particle size as exemplified by oxidation of human low density lipoprotein. Proc Natl Acad Sci U S A 1993; 90:45–49.

231. Tyurina YY, Tyurin VA, Yalowich JC, Quinn PJ, Claycamp HG, Schor NF, Pitt BR, Kagan VE. Phenoxyl radicals of etoposide (VP-16) can directly oxidize intracellular thiols: protective versus damaging effects of phenolic antioxidants. Toxicol Appl Pharmacol 1995; 131:277–288.

232. Roberfroid MB, Viehe HG, Remacle J. Free radicals in drug research. Adv Drug Res 1987; 16:1–84.

233. Liu J, Xiao PG. Recent advances in the study of antioxidative effects of Chinese medicinal plants. Phytother Res 1994; 8:445–451.

234. Sanz MJ, Ferrandiz ML, Cejudo M, Terencio MC, Gil B, Bustos G, Ubeda A, Gunasegaran R, Alcarax MJ. Influence of a series of natural flavonoids on free radical generating systems and oxidative stress. Xenobiotica 1994; 24:689–699.

235. Cholbi MR, Paya M, Alcaraz MJ. Inhibitory effects of phenolic compounds on CCl_4-induced microsomal lipid peroxidation. Experientia 1991; 47:195–199.

236. Schott BC, Butler J, Halliwell B, Aruoma OI. Evaluation of the antioxidant actions of ferulic acid and catechins. Free Radic Res Commun 1993; 19:241–153.

237. Wu TW, Wu J, Zeng LH, Au JX, Carey D, Fung KP. Purpurogallin: in vivo evidence of a novel and effective cardioprotector. Life Sci 1994; 54:PL23–28.

238. Cotelle N, Bernier JL, Henichardt JP, Catteau JP, Gaydou E, Wallet JC. Scavenger and antioxidant properties of ten synthetic flavones. Free Radic Biol Med 1992; 13:211–219.

239. Yamgishi Y, Matsuoka M, Odagawa A, Kato S, Shindo K. Rumbrin, a new cytoprotective substance produced by *Auxarthron umbrium*: I. Taxonomy, production, isolation and biological activities. J Antibiot (Tokyo) 1993; 46:884–887.

240. Kato S, Shindo K, Kawai H, Odagawa A, Matsuoka M. Pyrrolostatin, a novel lipid peroxidation inhibitor from *Streptomyces chrestomyceticus*. Taxonomy, fermentation, isolation, structure elucidation and biological properties. J Antibiot 1993; 46:892–899.

241. Nakano Y, Sugawara M, Uetsuki S, Izawa T, Kawaguchi T, Wada A. US Patent 5,021,419, 1991.

242. Rice-Evans CA, Diplock AT. Current status of antioxidant therapy. Free Radic Biol Med 1993; 15:77–96.

243. Kuzui M, Andreoni KA, Norris EJ, Klein AS, Burdick JF, Beattie C, Sehnert SS, Bell WR. Breath ethane: a specific indicator of free-radical–mediated lipid peroxidation following reperfusion of the ischemic liver. Free Radic Biol Med 1992; 13:509–515.

244. Roberts LJ II, Morrow JD. Isoprostanes. Novel markers of endogenous lipid peroxidation and potential mediators of oxidant injury. Ann NY Acad Sci 1994; 744:237–242.

245. Morrow JD, Roberts LJ II. Mass spectrometry of prostanoids: F_2-isoprostanes produced by noncyclooxygenase free radical–catalyzed mechanism. Methods Enzymol 1994; 233:163–174.

246. Morrow JD, Frei B, Longmire AW, Gaziano JM, Lynch SM, Shyr Y, Strauss WE, Oates JA, Roberts LJ II. Increase in circulating products of lipid peroxidation (F2-isoprostanes) in smokers—smoking as a cause of oxidative damage. N Engl J Med 1995; 332:1198–1203.

247. Yeo HC, Helbock HJ, Chyu DW, Ames BN. Assay of malondialdehyde in biological fluids by gas chromatography–mass spectrometry. Anal Biochem 1994; 220:391–396.

248. Yeo HC, Wagner JR, Helbock HJ, Chyu DW, Ames BN. Detection of aldehyde profile in biological systems using gas chromatography–mass spectrometry (abstr.) Oxygen Club of California, 1995: 130.

2

Design and Biological Evaluation of Tissue-Directed α-Tocopherol Analogs

Margaret A. Petty, Frank N. Bolkenius, and Gilbert Marciniak
Marion Merrell Research Institute, Strasbourg, France

James Dow
Glaxo Wellcome Research and Development, Greenford, Middlesex, England

I. INTRODUCTION

There is considerable evidence to suggest that free radicals are a major component of many pathophysiological disorders such as ischemia/reperfusion, inflammation, and neurodegenerative diseases (Alzheimer's and Parkinson's diseases). In fact, oxygen-derived free radicals have been implicated in the pathology of more than 100 diseases (1). Depending on their chemical structure, free radicals can be very reactive and toxic, leading to cellular damage and even death if their production is not well controlled (2).

Cells have evolved antioxidant defense systems that serve to control the level of these reactive oxygen species. Under physiological conditions, protection is ensured by enzymes such as superoxide dismutase, catalase or gluthathione peroxidase, or by nonenzymatic mechanisms (such as α-tocopherol, ascorbic acid, glutathione, or uric acid) (3). In a situation of oxidative stress when the production of free radicals exceeds the defense systems the use of an antioxidant therapy such as synthetic antioxidants (radical scavengers) would be effective.

Vitamin E, which consists mainly of α-tocopherol, is the principal chain-breaking, lipid-soluble antioxidant in mammalian tissue (4), and in particular is capable of quenching the propagation of free-radical reactions within cell membranes. Although many other chemical structures have been used to de-

53

sign novel radical scavengers, the structure of α-tocopherol has often served as a starting point for the design of potent synthetic free-radical scavengers (5). These α-tocopherol analogs belong to a class of antioxidant that can react with superoxide (O_2^-), hydroxyl, and lipoperoxyl radicals to form less reactive products (5).

This chapter describes the design and biological evaluation of two different classes of α-tocopherol analogs that have been targeted for specific tissues. The first class of compounds are water-soluble, quaternary ammonium tocopherol analogs that are able to accumulate in heart tissue and hence have the potential for cardioprotective use in conditions of acute reperfusion. One other interesting property of these compounds is their lack of oral availability, suggesting a possible use in the treatment of intestinal diseases. The second class of compounds are 2,3-dihydrobenzofuran-5-ol derivatives which penetrate brain tissue and have potential use for cerebrovascular diseases such as stroke, head trauma, or subarachnoid hemorrhage.

II. POTENTIAL USE OF QUATERNARY AMMONIUM α-TOCOPHEROL ANALOGS IN MYOCARDIAL REPERFUSION INJURY

A. Introduction

It is now generally accepted that reperfusion of the acutely ischemic myocardium can be beneficial, both reducing tissue damage and mortality (6–8). However, reperfusion can also be detrimental by causing further damage to jeopardized reversibly injured cells, thereby extending necrosis. Explosive myocardial swelling, calcium overload, and release of toxic oxygen-derived free radicals are the mechanisms most frequently implicated in the genesis of reperfusion damage (9–12).

Together with their related toxic oxygen species free radicals are thought to be involved in postischemic myocardial dysfunction in humans. Free-radical production has been measured directly during angioplasty reperfusion by means of electron paramagnetic resonance spectroscopy (13). An increase in markers of free-radical activity associated with successful thrombolysis for acute myocardial infarction has been demonstrated, (14,15) which correlated with reduced left ventricular function. Increased lipid peroxidation, apparent as an increase in malondialdehyde, was evident in the plasma of patients with unstable angina (16).

Vitamin E levels are reduced in heart during cardiopulmonary bypass surgery (17–19). Pretreatment with vitamin E, for 14 days, prior to coronary artery bypass surgery reduced free-radical-mediated metabolic dysfunction (20,21). Pretreatment with vitamins E and C prior to cardiopulmonary bypass resulted in lower levels of malondialdehyde and attenuated the decrease in red blood cell catalase (22). Therefore, in circumstances of ischemia/reperfusion and free-radical-induced injury in vivo, endogenous α-tocopherol levels are reduced, but unfortunately the acute administration of the vitamin is not feasible due to its high lipophilicity which results in slow tissue incorporation. Thus, more hydrophilic compounds with greater affinity for cardiac tissue would be of potential therapeutic use for reperfusion injury.

B. Chemistry

A series of α-tocopherol analogs was prepared in which the benzopyran-5-ol moiety of α-tocopherol, which is responsible for its antioxidant properties, was retained while the lipophilic phytyl chain of α-tocopherol, which is responsible for its intercalation into lipid bilayers, was replaced by more hydrophilic side chains (22,23). Among the compounds of the series, 6-acetoxy-3,4-dihydro-N,N,N,-2,5,7,8-heptamethyl-2H-1-benzopyran-2-ethanaminium-4-methylbenzene sulfonate, MDL74,270, was found to have these properties (22; Chart I). MDL74,270 can be considered to be a prodrug, as hydrolysis of the acetate moiety on the 6-position of the benzopyran ring liberates the active free-radical scavenging phenol, MDL73,404 (24). In order to determine whether the cardioselectivity is (1) dependent on the permanent cationic character of the molecule, (2) dependent on the nature of the counterion, or (3) related to the molecular geometry at the chiral center of the molecule, the corresponding tertiary amino analog MDL74,366, the corresponding bromide salt MDL73,954, and the two enantiomers of MDL73,404 were respectively synthesized. X-ray analysis of one of the chiral intermediates of the synthesis allowed, by extrapolation, the determination of the absolute configuration of the chiral center at position 2. Thus, the R configuration was assigned to the (+)-enantiomer (MDL74,405) and the S configuration was assigned to the (−)-enantiomer (MDL74,406).

C. Biochemistry

Stimulated neutrophilic polymorphonuclear leukocytes (neutrophils), undergoing the "oxidative burst" reaction (25), are one source of excessive production of superoxide radicals ($O_2^{.-}$) and other pathogenic oxygen-derived reactive species. Xanthine oxidase, metabolizing the xanthine derived from energy-storing purines used up during ischemia, as well as mitochondria with an uncoupled respiratory chain, are possible other sources of enhanced $O_2^{.-}$ production (26).

1. Superoxide Radical ($O_2^{.-}$) Scavenging

Superoxide dismutase catalyses the formation of hydrogen peroxide and molecular oxygen from $O_2^{.-}$ with $k_2 = 10^9$ M^{-1} s^{-1}. In contrast, α-tocopherol analogs,

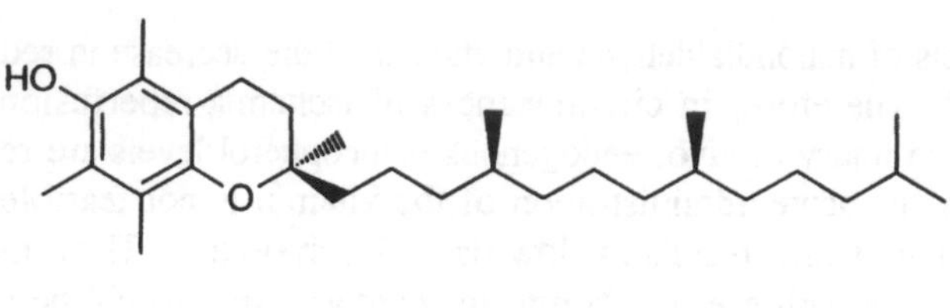

α-Tocopherol

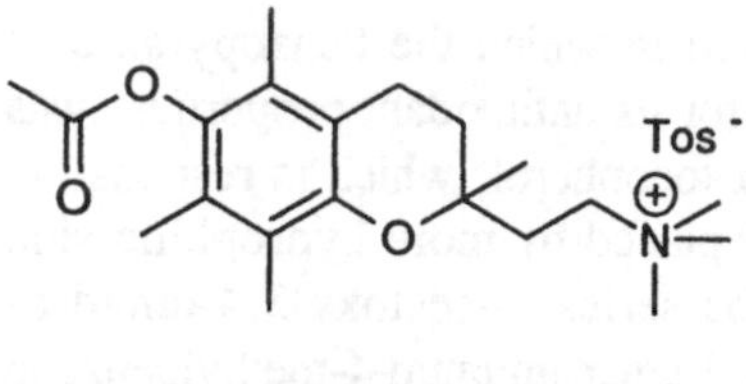

MDL74,270

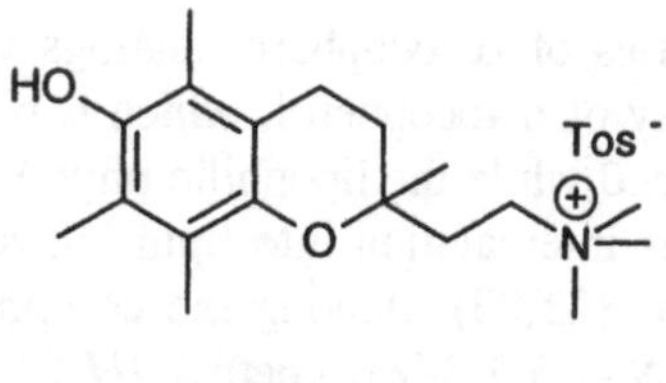

MDL73,404

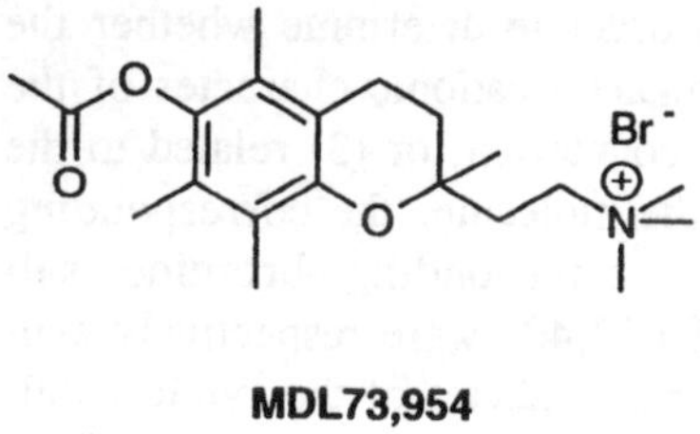

MDL73,954

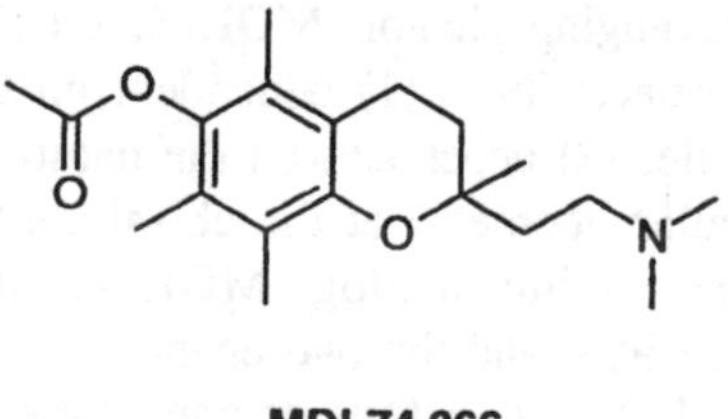

MDL74,366

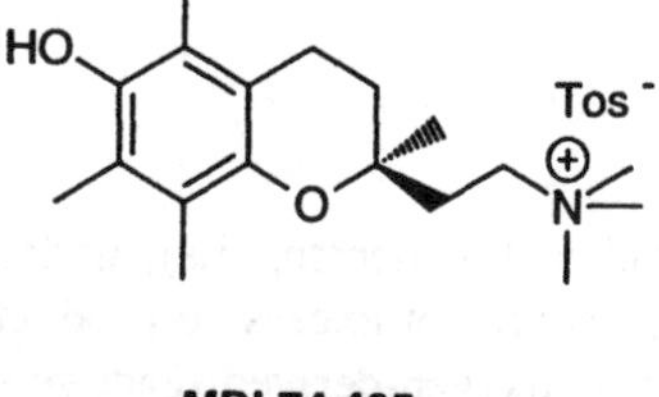

MDL74,405

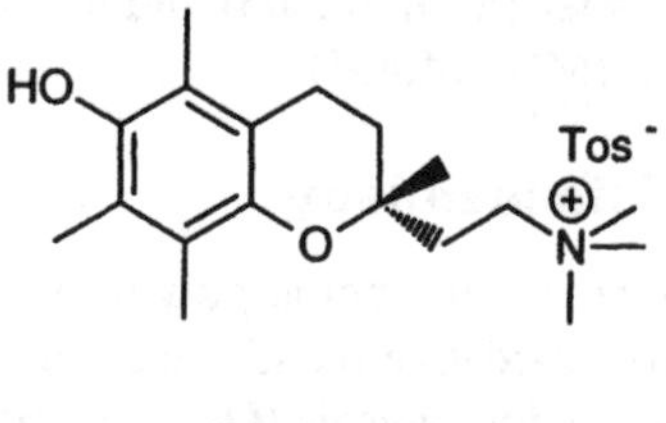

MDL74,406

Chart 1 Chemical structure of the α-tocopherol analogs cited in this section.

although slower with k_2 around 2×10^4 M^{-1} s^{-1} (27,28), have the potential to reduce hydrogen peroxide formation, due to covalent binding of $O_2^{\cdot-}$ (29–31), which in turn leads to a decreased generation of other reactive species such as hydroxyl radicals, lipoperoxyl radicals, hypochlorite or reduced metal ions. To demonstrate this, xanthine oxidase/lumazine for $O_2^{.-}$ generation was used. The catalase-induced release of molecular oxygen from the hydrogen peroxide, produced as a result of spontaneous $O_2^{.-}$ dismutation, was measured using a Clark oxygen electrode as previously described (32). By applying this method, we could demonstrate that the addition of MDL73,404 to the reaction mixture decreased the formation of hydrogen peroxide (Table 1). It is believed that this effect is responsible for the protection afforded by MDL73,404 against oxidative inactivation of α_1-proteinase inhibitor (33), an endogenous elastase inhibitor (34). Oxidative inactivation of endogenous protease inhibitors, e.g., by activated neutrophils, is a mechanism that entails high amplification of the oxidant damage during pathogenic processes (35). Constituting an important part of the immune system, the excessive $O_2^{.-}$-derived formation of hypochlorite by stimulated neutrophils can overwhelm the host's defenses. These cells have thus been taken to be responsible for part of the myocardial ischemia/reperfusion damage (36).

2. In Vitro Lipid Peroxidation

Lipid peroxidation is an important consequence of primary excessive $O_2^{.-}$ production. Barclay and Vinqvist, using a pressure transducer technique to measure oxygen consumption, have shown (37) that the quaternary ammonium derivative, MDL73,404, reacts faster with lipoperoxyl radicals in negatively charged liposomes than in zwitterionic environments (Table 2). Trolox, a negatively charged α-tocopherol analog, showed the opposite behavior. Thus, using a liposome preparation from rat brain homogenate, naturally containing a mixture of negatively charged and neutral lipids, MDL73,404 was superior to trolox to inhibit spontaneous lipid peroxidation, based on this charge dependence

Table 1 Reduced Hydrogen Peroxide Formation from Superoxide Radical in the Presence of MDL73,404

Incubation mixture	Hydrogen peroxide formed/Oxygen consumed[a]
Xanthine oxidase[b]	1.0 ± 0.1 ($n = 12$)
Xanthine oxidase +0.1 mM MDL73,404	0.7 ± 0.2 ($n = 10$)[c]
Xanthine oxidase + 0.1 mM MDL73,404 +50 U superoxide dismutase	0.9 ± 0.2 ($n = 5$)[d]

[a]Measured by using a Clark oxygen electrode and addition of 200 U of catalase after 20% oxygen has been consumed, 10 mM potassium phosphate at pH 7.4 and 37°C.
[b]27 mU xanthine oxidase in the presence of 0.1 mM lumazine.
[c]($p < 0.001$; ANOVA) versus xanthine oxidase.
[d]($p < 0.05$; ANOVA) versus xanthine oxidase + 0.1 mM MDL73,404.

Table 2 Charge Dependence of Inhibition of Lipid Peroxidation

	Inhibition of lipid peroxidation in		
	Single zwitterionic lipid vesicles[a]	Single anionic lipid vesicles[b]	Rat brain homogenate[c]
	k_{inh} (M^{-1} s^{-1})		IC$_{50}$ (µM)
Trolox	2.88×10^3	No inhibition	12
MDL73,404	0.93×10^3	1.62×10^3	1.7

[a]Dilinoleylphosphatidyl choline vesicles at pH 7; measurement of oxygen consumption (data taken from Ref. 37).
[b]Dilinoleylphosphatidyl glycerol vesicles at pH 7; measurement of oxygen consumption (data taken from Ref. 37).
[c]Rat brain homogenate 1000 × g supernatant at pH 7.4; measurement of formation of thiobarbituric acid reactive substances (data taken from Ref. 27).

(27; Table 2). A similar advantage for the positively charged analog has been observed in the reaction between ascorbic acid and the tocopheroxyl radicals, formed on laser flash photolysis from MDL73,404 and trolox which reacted at pH 7 with $k_2 = 4 \times 10^7$ M^{-1} s^{-1} and $k_2 = 1.44 \times 10^7$ M^{-1} s^{-1}, respectively (38). Thus, regeneration by ascorbic acid of the tocopheroxyl radical to the reduced form, after having served as a free-radical scavenger, can be expected to occur faster with MDL73,404 than with trolox.

3. Ex Vivo Lipid Peroxidation

In addition to a strong radical scavenging effect in vitro, high tissue penetration or rather tissue specificity would be needed for an antioxidant to be considered for therapeutic applications. Measuring the inhibition of ex vivo lipid peroxidation allows an estimate of the relative tissue penetration of compounds as active antioxidants. As an example the effect of subcutaneous administration of MDL73,404 in mice is shown (Fig. 1). Using this method the equipotency of the enantiomers of MDL73,404 was confirmed and cardioselective α-tocopherol analogs with a sulfonium or phosphonium instead of the ammonium side chain (23) were identified according to their relative ex vivo antioxidant efficiencies.

D. Drug Metabolism

1. Tissue Distribution of Tertiary and Quaternary Ammonium Analogs

A free-radical scavenger which is also cardiac selective could be of potential therapeutic use for reperfusion injury. Tissue distribution of MDL74,270 was

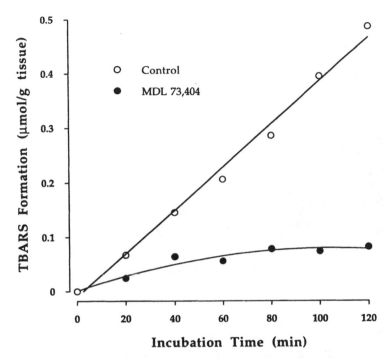

Figure 1 Effect of subcutaneous administration of MDL73,404 on the ex vivo lipid peroxidation induced in mouse heart homogenate. Male CD1 mice (5 in each group) were administered either vehicle or 10 μmol MDL73,404/kg s.c. and the hearts were taken 1 h later. The hearts of each group were pooled and homogenized with potassium phosphate at pH 7.3, containing 140 mM potassium chloride. The homogenates, overall diluted by a factor of 200, were incubated in duplicate for the indicated time intervals at 37°C under an oxygen atmosphere, followed by the measurement of thiobarbituric acid reactive substances (TBARS) as reported (82) (points representing means of duplicate experiments).

compared with its tertiary amine analog, MDL74,366, by following the heart, skeletal muscle, and brain tissue/blood ratio of total radioactivity in rats administered ^{14}C-labeled drug (24).

   Concentrations of total radioactivity in heart, skeletal muscle, brain, and blood 0.25–6 h after i.v. administration of MDL74,366 and MDL74,270 are shown in Fig. 2a–b, respectively. Although blood concentrations of radioactivity were similar for both compounds at all time intervals studied, much higher concentrations of radioactivity were found in heart tissue after MDL74,270 (8 times at 1 h and 20 times at 3 h). These were also the highest concentrations of radioactivity found in the evaluated tissues. Skeletal muscle uptake of radio-

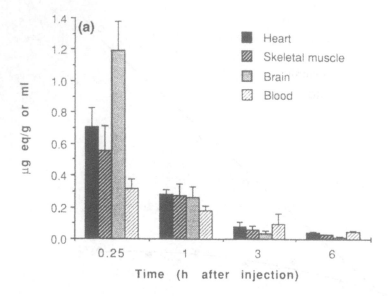

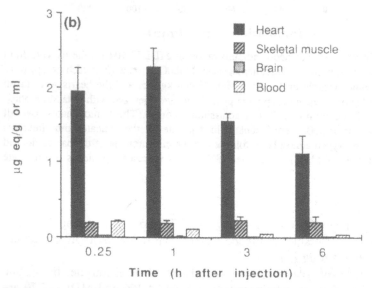

Figure 2 Comparison of the tissue distribution of total radioactivity after i.v. administration (1 mg/kg) of [¹⁴C]MDL74,366 (a) and [¹⁴C]MDL74,270 (b) in rat. Animals (n = 3) were killed 0.25, 1, 3, and 6 h after injection. Results (mean $\pm$ S.D.) are expressed as μg/g tissue or ml blood.

activity for both compounds was always lower than that found in heart. Concentrations were initially higher after MDL74,366 but decreased rapidly, and at 3 h were higher in rats treated with MDL74,270. Concentrations of radioactivity in brain, 0.25 h after administration of MDL74,366, were the highest of all tissues studied, but decreased rapidly. As would be expected for a quaternary amine, brain concentrations of radioactivity were very low after MDL74,270 administration.

Tissue/blood ratios (T/B) of MDL74,366 and MDL74,270 are shown in Fig. 3a–b, respectively. The heart T/B was greater than 20, 1–6 h after administration of MDL74,270 and less than 2 after MDL74,366. Skeletal muscle T/B varied from 1.8–5 after MDL74,270 and 1.5–0.6 after MDL74,366, over the 1–6 h period. These values were much lower than heart T/B after MDL74,270 administration. Brain T/B was initially much higher after administration of the tertiary amine, MDL74,366. Fifteen minutes after drug administration, ratios were around 4, but decreased rapidly. Results demonstrated a marked cardioselectivity of total radioactivity after MDL74,270 administration.

2. Cellular Distribution

Cellular distribution of radioactivity after administration of [^{14}C]MDL74,270 showed that radioactivity was equally distributed between the 900-g pellet (fraction containing nuclei and cell walls) and the 9000-g supernatant (fraction containing microsomes and cytosol). Only 5% of radioactivity in heart tissue was associated with mitochondria (24; Table 3).

3. HPLC of Rat Heart Homogenates

Rat heart homogenates, from tissue distribution studies, were analyzed by HPLC (24) for the presence of the phenol derivative of MDL74,270 (MDL 73,404), the active free-radical scavenger. Even as early as 15 min postadministration no [^{14}C]MDL74,270 was detectable, and the major radioactivity (88% of radioactivity injected) was present as MDL73,404. Thus, in vivo, hydrolysis of MDL74,270 is rapid and the major metabolite found in rat heart is the

Table 3 Cellular Fractionation of Heart Homogenates Prepared from Rats 1 h after i.v. Administration (1 mg/kg) of [^{14}C]MDL74,270[a]

Fraction	% radioactivity in heart
900-g pellet—cell walls and nuclei	33.5 ± 3.7
9000-g supernatant—microsomes, cytosol	45.5 ± 8.5
9000-g pellet—mitochondria	5.3 ± 1.3

[a]Results are mean ± SD of four rats, expressed as % total radioactivity in whole heart.

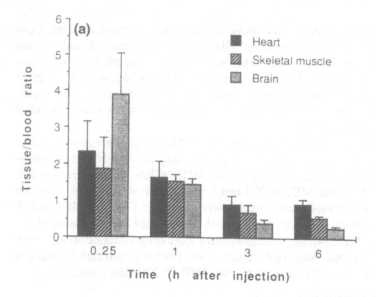

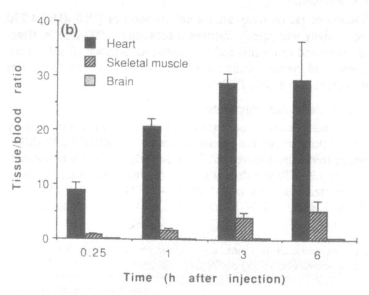

Figure 3 Comparison of tissue/blood ratios (mean ± S.D.) of total radioactivity after i.v. administration (1 mg/kg) of [^{14}C]MDL74,366 (a) and [^{14}C]MDL74,270 (b). Animals (n = 3) were killed 0.25, 1,3, and 6 h after injection.

phenol analog, MDL73,404. No unchanged drug was detected at any time
interval. A minor metabolite, accounting for 9–12% of total radioactivity, was
identified as the quinone analog. The proposed metabolic pathway of
MDL74,270 in rat is shown in Fig. 4.

E. Pharmacology

1. *Reperfused Infarcted Myocardium*

Short-Term Reperfusion In a pentobarbitone anaesthetized rat model of coronary artery ligation for 60 min followed by 30 min of reperfusion, intravenous (i.v.) infusion of the α-tocopherol analogs MDL74,270, MDL73,404, and MDL74,405 (the 2S-(−)-enantiomer of MDL73,404), in the dose range of 0.03–3.0 (mg/kg)/h beginning 10 min before occlusion until the end of the 30-min reperfusion period, produced very similar results—that is, a reduction in infarct size compared to saline-treated controls (Fig. 5) which was not associated with any alteration in systemic pressure and heart rate (39–42). Infarct and occluded areas were determined by means of the triphenyl tetrazolium chloride-Evans blue technique, which demonstrated that the area at risk was not different between control and treated groups, indicating that any reduction in infarct size could not be attributed to differences in the size of the ischemic areas. The maximum decrease was obtained with a dose of 0.3 (mg/kg)/h. Doses below 0.1 (mg/kg)/h had no effect on infarct size, and doses higher than 0.3 (mg/kg)/h induced no greater decrease, indicating the existence of an extremely steep dose-response relationship. Due to the similarity of the effects of the acetylated prodrug MDL74,270 and MDL73,404 there appeared to be no advantage to the use of MDL74,270 and the use of the active metabolite was preferred.

In this model MDL74,270 (tosylate salt) and MDL73,954 (bromide salt) were equipotent in reducing infarct size, demonstrating that the effectiveness of the compounds was not salt-dependent (39).

MDL74,366, the tertiary amine analog of MDL73,404 was found to be at least as effective as an oxygen radical scavenger as MDL74,270 in a variety

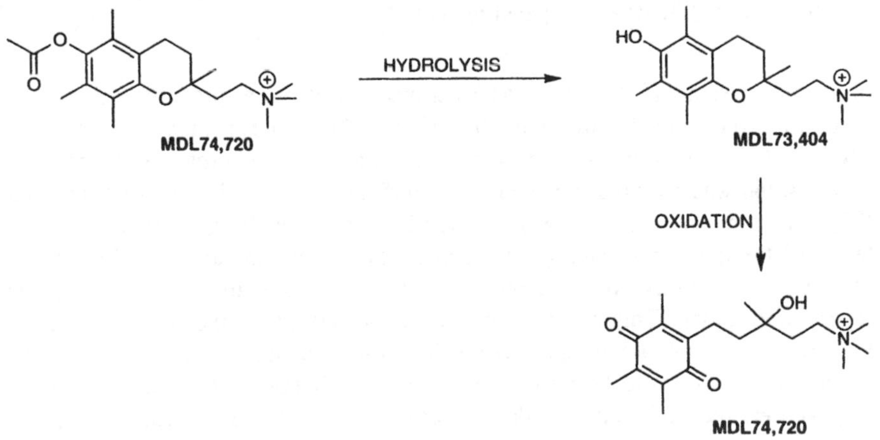

Figure 4 Proposed metabolic pathway of MDL74,270 in the rat.

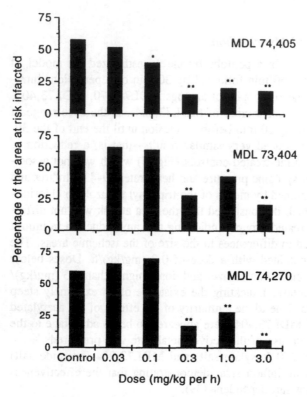

Figure 5 Effect of increasing doses of the α-tocopherol analogs on the percentage of the area at risk infarcted. The left descending coronary artery was occluded for a 60-min period followed by 30-min reperfusion. Infusion (2.3 ml/h) of a particular compound or saline (control) commenced 10 min before occlusion until the end of reperfusion. *p < 0.05, **p < 0.01 when compared to control.

of in vitro systems (27), but it was 30 times less potent at protecting the myocardium from reperfusion damage. Using the same protocol as above, MDL74,270 was active at a dose of 0.3 (mg/kg)/h, whereas 10 (mg/kg)/h MDL74,366 was required to produce a significant decrease in the infarcted area (39). Presumably, the myocardial selectivity and resulting high myocardial levels of the quaternary analog (24) can explain why comparatively low doses of these compounds are required for myocardial salvage in conditions of reperfusion injury. This relative myocardial selectivity of the quaternary analogs represents an advantage over other antioxidants, such as superoxide dismutase which has to be introduced into the ischemic myocardium by coronary venous retroinfusion in order to achieve adequate concentrations (43).

Infusion of MDL73,404 (3 (mg/kg)/h) to rats commencing 30 min before reperfusion (i.e., 30 min after the onset of occlusion) until the end of reperfusion still induced a significant ($p < 0.01$) decrease in infarct size compared to saline-treated controls. The infarct size was 57 ± 16 mg with MDL73,404 infusion compared to 127 ± 3 mg after saline. This protocol more closely resembles the clinical situation where the drug is given after the onset of ischemia and presumable enters the occluded area on reperfusion (41). Further evidence that MDL73,404 actually enters the occluded tissue has been obtained by measuring the incorporation of [^{14}C]MDL73,404 into rat heart tissue (44).

MDL73,404 is rapidly distributed from a subcutaneous (s.c.) injection site reaching a maximum in the heart 1 h after injection, where the levels remain constant for up to 3 h. These findings correlate well with the decrease in infarct size following MDL73,404 injection (3 mg/kg s.c.) injected 1 h before reperfusion (45).

Reperfusion following myocardial ischemia leads to an increase in circulating lipid peroxide levels, which probably reflects lipid peroxidation by free radicals (46). Membrane lipid peroxidation results in an alteration of membrane integrity and an increase in fluidity and permeability. In left ventricular blood samples obtained from rats 2 min after the onset of reperfusion, following 60-min coronary artery occlusion, a significant increase in thiobarbituric acid adducts (TBARS), used as an indication of lipid peroxides, was apparent which was significantly reduced in those rats receiving an infusion of MDL73,404 (44). The MDL73,404-induced decrease in TBARS was evident 10 min after the start of reperfusion when the concentration of TBARS was already declining, being 35 ± 9 pmol/100 μl and 47 ± 12 pmol/100 μl ($p < 0.05$) in MDL73,404 and saline-treated animals respectively. These findings suggest that MDL73,404 is inhibiting free-radical generation and the resulting lipid peroxidation at the time of reperfusion.

Long-Term Reperfusion The beneficial effects of the α-tocopherol analogs, described so far, have been demonstrated in studies using short periods of reperfusion. To show that the reductions in infarct size are not merely due to a delay in tissue necrosis but that the results are permanent, the infarct size was determined after 60 min of occlusion and 8 days of reperfusion. In MDL73,404-treated rats infarct size was significantly reduced compared to saline-treated animals after 8 days of reperfusion, and cardiac output, used as an indication of left ventricular function, was significantly greater in the antioxidant-treated rats compared to controls. In fact, cardiac output was not different from that determined in sham-treated controls (44). Similarly, after 8 days of reperfusion following MDL74,405 treatment, an increase in cardiac output was associated with the reduction in infarct size (Fig. 6; 42). The increase in cardiac output may be due to an increase in contractility, since in Langendorff-perfused iso-

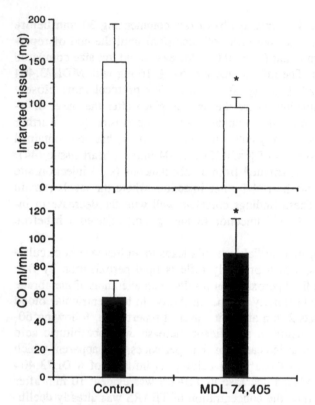

Figure 6 The effects of MDL74,405 on infarct size and cardiac output. Rats were infused with MDL74,405 (0.3 (mg/kg)/h) or saline beginning 10 min before coronary artery occlusion (1 h) and continued for 30 min into the 8-day period of reperfusion. After 8 days aortic blood flow was measured as an indication of cardiac output and infarct size by means of tetrazolium. Such a long period of reperfusion was used to ascertain that the beneficial effects of these quaternary ammonium compounds are not merely due to a delay in tissue necrosis but that the reduction in infarct size and improvement in left ventricular function are permanent results of their administration. *$p < 0.05$ compared to control.

lated hearts obtained from both MDL73,404- and MDL74,405-treated rats after 8 days of reperfusion all indices of contractility including maximal pressure development $[+(dP/dt)_{max}]$ and relaxation $[-(dP/dt)_{max}]$ and left ventricular systolic pressure were elevated compared to saline-treated controls (42,44). Recent experiments have demonstrated that in MDL73,404-treated rats subjected to the ischemia and the prolonged reperfusion protocol an increase in β_2-adrenoceptor density was apparent in the infarcted tissue, noninfarcted septum, and right ventricle tissue. This increased density of β_2-receptors in the noninfarcted tissue may contribute to the MDL73,404-induced improvement of cardiac performance (47). The reduction in infarct size induced by MDL73,404 or its S-(–)-enantiomer MDL74,405 could be due to a free-radical scavenger-induced decrease in lipid peroxidation since both inhibited TBARs formation (42,44). Thus the timely administration of either antioxidant reduces infarct size, which is confirmed by a reduction in the circulating levels of creatine phosphokinase, and as a result left ventricular function is maintained in vitro and in vivo.

It has been reported (48) that MDL74,405 attenuates the contractile dysfunction caused by myocardial stunning in dogs following 15 min of coronary artery ligation. Regional myocardial function (assessed as systolic wall thickening) was significantly improved in MDL74,405-treated dogs 2 h after the restoration of flow and was sustained throughout the rest of the reperfusion period. The enhanced recovery in MDL74,405-treated dogs could not be attributed to nonspecific factors such as differences in collateral blood flow during occlusion, coronary flow after reperfusion, systemic pressure, heart rate, or other hemodynamic variables.

2. *Global Ischemia and Reperfusion*

The protective effects of MDL73,404 and MDL74,405 are apparent in in vivo studies in animals subjected to coronary artery ligation and various periods of reperfusion. They are also very evident in the isolated Langendorff-perfused rat heart, subjected to global ischemia and reperfusion. After a 30-min period of global ischemia the presence of either antioxidant in the perfusion buffer increased left ventricular systolic pressure and reduced the elevated left ventricular diastolic pressure, compared to controls, to values not significantly different from baseline, preischemic values. Maximal pressure development and relaxation and heart rate were also significantly elevated compared to controls (41,42) After the heart was stopped for 2 min with a cardioplegic solution before 30-min global ischemia, MDL73,404 further augmented contractility and left ventricular pressure as compared to control hearts treated with cardioplegic solution alone (41).

F. Conclusion

The studies described in this part have demonstrated the marked cardioselectivity of quaternary amine compounds when tissue distributions of the quaternary and tertiary analogs of MDL74,270 were compared. Bretylium (49), N,N-dimethylpropranolol (50), and clofilium (51) are long-acting antiarrhythmic compounds that contain a quaternary amine group. Their long duration of action may be due to cardiac tissue uptake of drug, as all compounds are highly concentrated in myocardium, and myocardial tissue concentrations 20–40 times greater than those found in plasma have been measured. These findings are similar to results found in the present work and may suggest that a specific transport process exists for quaternary amine compounds into heart muscle. Interestingly, MDL74,270, bretylium (49), and clofilium (51) also show specific uptake for heart when compared to skeletal muscle.

The ammonium α-tocopherol analogs are cardioselective free-radical scavengers that enter the myocardial area at risk during ischemia, where they increase the endogenous antioxidant defense. Postischemic dysfunction is markedly attenuated by quaternary ammonium α-tocopherol analog therapy, implemented before the restoration of flow. They appear to maximize myocardial salvage achieved by reperfusion and, hence, preserve ventricular function. The rationale for the design of MDL73,404 and its prodrug combining a short-chain hydrophilic analog of α-tocopherol for good mobility in the membrane, with a quaternary ammonium group for cardioselectivity, appears to be correct. Hence, these compounds may represent a potentially useful therapy during reperfusion of cardiac tissue, after thrombolysis, angioplasty, and coronary bypass surgery in man.

III. SCAVENGING OF FREE RADICALS IN INTESTINAL DISEASE

A. Introduction

Free-radical-mediated ischemia/reperfusion injury appears also in the intestine (52,53). Also in this condition oxidant-producing neutrophils are involved (53). Using a rat model of colonic ischemia/reperfusion, created by occlusion of the superior mesenteric artery, it was shown that infusion of MDL73,404 at a rate of 0.3 (mg/kg)/h inhibited Evans blue extravasation as a measure of intestinal damage (54).

Another free-radical-mediated disease of the intestine is chronic inflammation. Chronic inflammatory bowel disease (IBD) consists of two histologically different entities called ulcerative colitis (UC) and Crohn's disease (CD). In IBD the recurrent inflammation of the intestine gives rise to periods of diarrhea, weight loss, fever, and anorexia. Moreover, a growing body of clinical and experimental data suggests that severe long-standing inflammation of the colon is associated with an increased risk of colorectal cancer (55,56).

Neutrophils are attracted to the site of inflammation and infiltrate the intestinal wall during active episodes of IBD (57,58). The attracted neutrophils are activated by different factors such as complement, aggregated immunoglobulin, endotoxin, or cytokines (59). Neutrophils isolated from IBD patients during acute attacks are in a state of increased free-radical production (60,61). The cell-membrane-associated "oxidative burst" oxidase is stimulated under these conditions to produce high amounts of $O_2^{\cdot-}$. Through $O_2^{\cdot-}$-dependent formation of a lipid-derived, albumin-bound chemotactic factor in the blood plasma, further neutrophils are attracted (59). This process of self-amplification is inhibited by superoxide dismutase.

However, the hydrogen peroxide formed from $O_2^{.-}$ by spontaneous or enzyme-catalyzed dismutation can in turn be the substrate of myeloperoxidase. This enzyme, secreted by activated neutrophils, produces the potent oxidant hypochlorite and the N-chloramines derived from it (35). Oxidants, in concert with the proteases secreted by activated neutrophils, normally are supposed to act as microbicidal agents; however, they also have the potential to mediate endothelial cell death and further neutrophil invasion into the surrounding tissue. Moreover, by increasing the intracellular Ca^{2+} level (62), the oxidant attack can lead to phospholipase A2 activation (63). As a result, enhanced arachidonic acid metabolism may occur, producing further inflammatory mediators, such as prostaglandins and leukotrienes. It may therefore be assumed that $O_2^{.-}$ plays a fundamental role in this cascade of events, and that its scavenging has a beneficial effect in IBD.

At present, long-term drug therapy of IBD is mainly limited to sulfasalazine, used in the maintenance of UC. Sulfasalazine acts as a prodrug which is poorly absorbed from the small intestine after oral administration. Reaching the colon, the active metabolite 5-aminosalicylic acid is liberated (64), where it is assumed to act locally. Inhibition of 5-lipoxygenase and the scavenging of reactive oxygen species are its proposed mechanisms of action (65). Due to important side effects of sulfasalazine or 5-aminosalicylic acid (e.g., nephrotoxicity), and the inefficiency of sulfasalazine in maintaining remission or preventing recurrence of CD, a novel therapy is highly desirable. A free-radical scavenger with low gastrointestinal absorption, which is not appreciably transformed in the gastrointestinal tract, would be of potential therapeutic use in IBD.

B. Biochemical Properties of MDL73,404

MDL73,404 scavenges $O_2^{.-}$ with $k_2 = 1.8 \pm 0.2$ [10^4 M^{-1} s^{-1}] (mean $\pm$ SD, $n = 3$) with no difference between its two enantiometers (Fig. 7). Due to its $O_2^{.-}$ scavenging property, MDL73,404 and its enantiomers attenuate the potential of activated human leukocytes to oxidatively inactive α_1-proteinase inhibitor (α_1PI) (33). α_1PI is an endogenous inhibitor of neutrophil elastase, being able to protect the inflamed host tissue from proteolytic digestion, however, it is sensitive to hypochlorite-mediated oxidative inactivation (34). As a probable indication of its involvement in the inflammatory process, enhanced serum concentration and fecal excretion of α_1PI is observed in patients with IBD (66,67). IBD patients also show increased serum levels of neutrophil elastase (68).

The concentration of MDL73,404 protecting 50% of α_1PI in vitro (relative PC_{50}) is about 0.6 nmol/μg leukocyte protein (Fig. 8). Again, there was no significant difference in the potencies of the racemate and the two enantiomers.

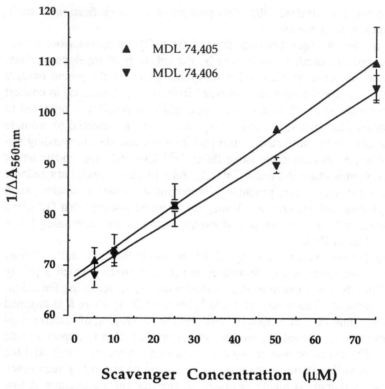

Scavenger Concentration (µM)

Figure 7 Scavenging of superoxide radicals by MDL74,405 and MDL74,406, the R-(+)- and S-(−)-enantiomers, respectively, of MDL73,404. Formation of $O_2^{\cdot-}$ by xanthine oxidase/xanthine, was followed spectroscopically via its reducing effect on nitro-blue tetrazolium (NBT), which results in the generation of a blue chromophore. Due to competition between the test compounds and NBT for reaction with $O_2^{\cdot-}$, the color formation was slowed down. The relative rate constants k_2 were calculated based on $k_{NBT} = 6 \times 10^4$ M^{-1} s^{-1} for the reaction of NBT with $O_2^{\cdot-}$. This was done by plotting the inverse absorbance increases versus the corresponding scavenger concentrations. The slope of the resulting line equals $k_2/k_{NBT}[NBT]A_0)$ (A_0 = absorbance increase in the absence of scavenger; [NBT] = 40 µM) (points representing means of three experiments ± S.D.).

MDL73,404 and its enantiomers are also inhibitors of 5-lipoxygenase, producing 50% inhibition at 1.6 µM, irrespective of the geometry at the chiral center (not shown). Inhibition of 5-lipoxygenase by MDL73,404 may be explained by the scavenging of lipoperoxyl radicals, which are intermediates in this reaction. The IC_{50} was 1000 times lower as compared with sulfasalazine (IC_{50} = 0.9 to 1.5 mM) (69) or 5-aminosalicylic acid (IC_{50} = 5 mM) (70). Leukotriene B_4, produced by leukocyte 5-lipoxygenase, is a chemotactic substance that at-

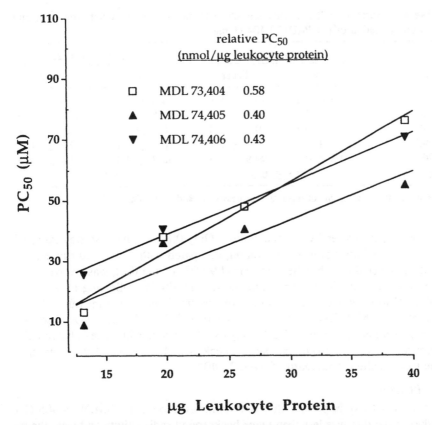

Figure 8 Dependence of the concentration of MDL73,404 required for protection of 50% of the α_1-proteinase inhibitor (αPI) on the amount of activated leukocytes. Freshly prepared, zymosan-activated human leukocytes in a total volume of 250 μl were used at four different dilutions to oxidatively inactivate human αPI. Protection was achieved at five different scavenger concentrations. The points represent the PC_{50} values obtained by plotting the scavenger concentrations versus the residual capacity of αPI to inhibit porcine pancreatic elastase (33). Incubation of leukocytes with αPI in the absence of scavenger furnished control values for 100% inactivation (points representing means of duplicate experiments).

tracts further leukocytes toward inflamed tissues in a self-stimulatory process. Increased formation of leukotriene plays a role in UC (71).

C. Drug Absorption

1. *In Rats*

Over 90% of the administered dose (1 mg/kg) was found in 0–24 h feces of rats after oral administration of [^{14}C]MDL73,404 in the rat (Table 4), compared

Table 4 Excretion of Total Radioactivity in Urine and Feces of Rat After p.o. and i.v. Administration of [^{14}C]MDL73,404 (1 mg/kg)[a]

Time (h)	Oral dose		Intravenous dose	
	Urine	Feces	Urine	Feces
0–24	0.5 ± 0.1	90.2 ± 5.1	34.2 ± 0.4	17.8 ± 14.7
24–48	0.1 ± 0.1	9.0 ± 2.8	1.8 ± 0.4	21.1 ± 16.8
48–72	0.1 ± 0.1	0.7 ± 0.3	0.5 ± 0.1	1.6 ± 0.4
72–96	0.1 ± 0.1	0.1 ± 0.1	0.3 ± 0.1	0.5 ± 0.1
0–96	0.8 ± 0.1	99.9 ± 6.5	36.7 ± 0.2	40.9 ± 2.6
Recovery	100.8 ± 6.6		77.7 ± 2.7	

[a]Results are the mean ± SD of three rats, expressed as % dose administered.

with only 17.8% after i.v. administration. Only 0.8% of the dose was excreted in 0–96 h urine after oral administration, compared with 36.7% after iv administration. Over the 96 h collection period 99.9% of the drug was recovered in feces after oral administration compared to only 40.9% following intravenous dosing. If urinary excretion of total radioactivity is used as a measure of oral absorption of MDL73,404, then no more than 2% of the dose was absorbed after oral administration. HPLC analysis of extracted rat feces showed the presence of only one major radioactive peak, which corresponded in retention time to an authentic standard of MDL73,404.

2. In Dogs

Total radioactivity in plasma after oral administration of [^{14}C]MDL74,405 (10 mg/kg) in the dog was less than twice background radioactivity and was, therefore, not quantified. These results thus indicate that the compounds was very poorly absorbed. Urinary excretion of radioactivity confirmed the low absorption of the compound, as 4.8% of the dose was found in 0–96 h urine after oral administration and 42.5% after intravenous infusion (Table 5). Over 80% of the administered dose was found in 0–96 h feces of dogs after oral administration, compared with around 28% after i.v. administration. HPLC analysis of 0–24 h feces after p.o. administration showed that parent drug was the major compound in dog feces and represented over 90% of radioactivity injected on-column.

3. Mouse Acute Colitis Model

In a model of acute colitis male mice were given 5% of dextran sulfate sodium in their drinking water over a period of 5 days. This pretreatment caused weight loss, loose stool, and intestinal bleeding. A disease activity index has been defined (72,73). After 5 days treatment, the disease activity index was 1.8 ±

Table 5 Excretion of Total Radioactivity in Urine and Feces of Dog After 30-Min Intravenous Infusion and Oral Administration of [^{14}C]MDL74,405 (10 mg/kg)[a]

Time (h)	Oral dose		Intravenous dose	
	Urine	Feces	Urine	Feces
0–24	35.1 ± 0.9	19.4 ± 1.3	3.7 ± 0.6	75.9 ± 21.5
0–48	39.0 ± 1.4	24.8 ± 0.7	4.4 ± 0.6	78.5 ± 22.2
0–72	41.1 ± 1.7	27.2 ± 0.9	4.7 ± 0.7	80.1 ± 21.4
0–96	42.5 ± 1.8	28.2 ± 0.9	4.8 ± 0.7	80.7 ± 21.5
Recovery	70.7 ± 2.4		85.3 ± 21.3	

[a]Results are the mean ± SD of three dogs, expressed as % dose administered.

0.4 (mean ± SD; $n = 5$), similar to values which have been reported (74). On day 6, the mice, along with a control group, were given an oral dose of [^{14}C]MDL73,404 (50 mg/kg). This dose is five times the therapeutic dose in this disease model (75). Urine and feces were collected separately on each of the subsequent 4 days.

Over 80% of the dose was excreted in feces of control and treated mice during the first 24 h (Table 6). The amount of radioactivity excreted into urine was used as a measure of drug absorption. In control mice, 1.9 ± 0.9% of the dose was excreted in 0–96 h urine compared with 2.8 ± 2.0% in mice with acute colitis (Table 6). Thus, in spite of the inflammatory reaction produced by the dextran sulfate sodium, no increase in the urinary excretion of radioactivity compared to normal mice was noted. Using urinary excretion as a measure of drug absorption the amount of MDL73,404 entering the circulation was

Table 6 Excretion of Total Radioactivity in Urine and Feces of Control Mice and Mice with Acute Colitis After Oral Administration of [^{14}C]MDL73,404 (50 mg/kg)[a]

Time (h)	Control ($n = 6$)		Acute colitis ($n = 5$)	
	Urine	Feces	Urine	Feces
0–24	1.3 ± 0.6	83.1 ± 11.0	2.3 ± 1.6	98.0 ± 5.5
24–48	0.4 ± 0.6	1.0 ± 1.2	0.4 ± 0.4	0.8 ± 0.8
48–72	0.1 ± 0.1	1.6 ± 3.5	0.1 ± 0.1	0.2 ± 0.1
72–96	0.1 ± 0.1	0.7 ± 1.4	ND[b]	0.2 ± 0.1
0–96	1.7 ± 0.9	86.4 ± 7.5	2.8 ± 2.0	99.2 ± 5.0
Recovery	88.3 ± 7.6		102.0 ± 3.6	

[a]Results are the mean ± SD, expressed as % administered dose.
[b]ND = not detectable (radioactivity less than twice background radioactivity).

very low for both groups of animals. MDL73,404 was the major compound in feces and represented 55 ± 8% of the dose in feces of control mice compared with 74 ± 5% of the dose in feces of mice with acute colitis. The quinone metabolite, MDL73,876, was excreted in similar quantities in both groups.

4. Mouse Chronic Colitis Model

In a model of chronic colitis mice were used at the age of 17 weeks, when they were given 5% of dextran sulfate sodium in their drinking water for 7 days, followed by a recovery period of 14 days (74). This regimen has been reported to induce chronic inflammation (74), starting around day 21. The disease activity was measured periodically. On day 21 the mice, along with a control group, were orally administered [^{14}C]MDL73,404 (50 mg/kg). Urine and feces were collected separately on each of the subsequent 4 days. Total radioactivity in urine and feces was expressed as a percentage of the administered dose.

The severity of the disease in these animals was less pronounced than in the acute phase, loose stool consistency being the most prominent symptom. Over 90% of the dose was excreted in feces of control and treated mice during the first 24 h (Table 7). In control mice 2.4 ± 1.7% of the dose was excreted in 0–96 h urine compared with 2.1 ± 0.9% in mice with chronic colitis (Table 7). Thus, using urinary excretion as a measure of drug absorption, no difference in drug absorption was noted between the two groups of mice. As for the acute experiment, drug absorption was very low. The amounts of total radioactivity, MDL73,404, and MDL73,876 excreted in 0.24 h feces were very similar for both control mice and mice with chronic colitis. In both groups, parent drug was the major compound and represented over 50% of the administered dose excreted in 0–24 h feces. Again, the quinone metabolite MDL73,876 represented around 10% of the administered dose.

Table 7 Excretion of Total Radioactivity in Urine and Feces of Control Mice and Mice with Chronic Colitis After Oral Administration of [^{14}C]MDL73,404 (50 mg/kg)[a]

Time (h)	Control ($n = 6$)		Acute colitis ($n = 5$)	
	Urine	Feces	Urine	Feces
0–24	1.3 ± 0.3	91.5 ± 6.0	1.3 ± 0.6	90.4 ± 12.8
24–48	0.6 ± 0.9	0.9 ± 0.4	0.4 ± 0.2	6.0 ± 8.8
48–72	0.3 ± 0.7	0.2 ± 0.3	0.3 ± 0.5	0.6 ± 1.1
72–96	0.2 ± 0.3	0.2 ± 0.3	0.1 ± 0.1	0.2 ± 0.2
0–96	2.4 ± 1.7	92.8 ± 5.5	2.1 ± 0.9	97.2 ± 8.8
Recovery	95.2 ± 5.0		99.3 ± 8.2	

[a]Results are the mean ± SD, expressed as % administered dose.

5. Transport Across Caco-2 Cell Monolayers

The intestinal absorption of drugs can be studied in vitro using monolayers of Caco-2 cells, which are human intestinal epithelial cells, to study passive drug absorption (76,77). MDL74,405 demonstrated very low permeability across Caco-2 cell monolayers, whereas the tertiary amine MDL74,366 showed around 30% transport after 2 h incubation (Fig. 9). These results confirm the low absorption of MDL74,405.

MDL73,404 and MDL74,405, like the prodrug MDL74,270 (24), are poorly absorbed in the rat. Results in dog with the S-(-)-enantiomer, MDL74,405, also showed that, after oral administration, plasma concentrations of radioactivity were less than twice background and could not be quantified. Thus, in a second species, drug absorption was shown to be very low. The study in a third species, mice, has also demonstrated low oral absorption and further demonstrated that the presence of acute or chronic colitis did not enhance drug absorption. In addition, studies with the human intestinal epithelial cell line Caco-2 also demonstrated that Caco-2 cell monolayers showed very low permeability of MDL74,405. Thus, as well as in the three animal species studied, a human intestinal epithelial cell line also demonstrated low permeability of MDL74,405.

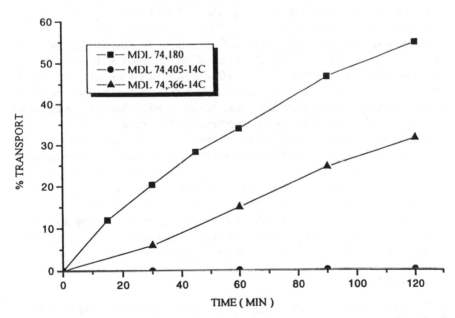

Figure 9 Transport of MDL74,405, MDL74,366, and MDL74,180 across Caco-2 cell monolayers. Results are plotted as the cumulative amount of drug transported, expressed as a percentage of the initial concentration of drug.

MDL73,404 and MDL74,405 are not only poorly absorbed, but they are excreted, in majority, unchanged in feces of rat, mouse, and dog. These results are encouraging for the potential therapeutic use of MDL73,404 or its enantiomers in IBD.

D. Therapeutic Effect of MDL73,404 in Dextran Sulfate Sodium–Induced Colitis in Mice (75)

In female Swiss Webster mice 5% DSS in drinking water for 5 consecutive days induced acute colitis on day 5. The chronic phase, following DSS treatment for 7 days, started on day 21. Quantitative evaluation was done by determining the disease activity index (DAI).

The effect of MDL73,404 was tested at 1,3,10,25, and 50 mg per kg body weight per day p.o. To examine its prophylactic effect, MDL73,404 was administered orally during the 5 days of colitis induction. There was a significant decrease of the DAI at the dose of 10 (mg/kg)/day. The overall effective dose for 50% inhibition (ED_{50}) was calculated to be 1.5 (mg/kg)/day. The acute-phase colitis was established after 5 days DSS, and the DAI was recorded over the subsequent 8 days. In this experiment MDL73,404, administered from day 5 on, inhibited the colitis with an ED_{50} = 2.1 (mg/kg)/day. The maximal inhibition observed was about 90% at 10 (mg/kg)/day, and the minimal effective dose was 1 (mg/kg)/day, giving 16% inhibition. Treatment of the advanced (chronic) phase with oral administration of MDL73,404 started on day 21. Maximal inhibition of the disease was 80% to 90% at 10 mg/kg, and the ED_{50} = 1.7. By comparison, the ED_{50} for olsalazine, a 5-aminosalicylic acid preparation currently used to treat IBD, was > 150 (mg/kg)/day p.o. in the acute phase, and it had no effect on the induction of DSS-mediated UC.

The histology and the analysis of inflammatory mediators corroborate the DAI data. MDL73,404 reestablished normal crypt, as well as villous architecture to the damaged intestinal mucosa. It significantly reduced the levels of 6-keto prostaglandin F_{1a}, prostaglandin E_2, thromboxane B_2, and leukotriene B_4, which were all increased as a consequence of the inflammation. The latter is in accordance with the in vitro inhibition of 5-lipoxygenase as mentioned, which is therefore an additional potential mechanism by which MDL73,404 can interfere during active episodes of intestinal inflammation.

E. Conclusion

IBD is mediated by activated neutrophils, secreting reactive oxygen species along with proteases and other inflammatory mediators. MDL73,404, by scavenging the free radicals formed, is capable of attenuating the oxidative potential of these cells, thereby maintaining the protease inhibitor/protease balance and inhibiting formation of chemotactic substances. This antioxidant property is further amplified, due to the permanent cationic character of MDL73,404, by its apparent lack of gastrointestinal absorption, leading to enhanced local activity in the intestine. In a mouse colitis model MDL73,404 was thus found to have a significant therapeutic effect at oral doses as low as 10 (mg/kg)/day. Further studies are needed to evaluate its therapeutic potential for the prevention of acute attacks and maintenance of remission in human IBD.

IV. THE USE OF FREE RADICAL SCAVENGERS IN CNS DISORDERS

A. Introduction

The brain is particularly sensitive to oxidative damage because of its high concentration of polyunsaturated fatty acids and its high rate of oxygen consumption. Furthermore, certain brain areas are rich in iron, a potent catalyst of free-radical formation and lipid peroxidation. The brain has low concentrations of natural antioxidants such as α-tocopherol, glutathione, and vitamin C. Excitatory amino acids can induce oxygen radical formation, and oxidative stress can further increase the sensitivity of neurons to glutamate. Hence oxygen-derived free radicals are linked to many of the processes involved in cerebral damage and necrosis, and particularly related to cerebrovascular diseases such as ischemia, trauma, and subarachnoid hemorrhage (78–80).

B. Chemistry

Starting from our successful targeting of a hydrophilic α-tocopherol analog to accumulate in heart tissue (23,24), a new series of α-tocopherol derivatives with CNS activity was designed (81). The 3,4-dihydro-benzopyran-6-ol moiety of α-tocopherol was replaced by the more potent free-radical scavenger five-membered ring system 2,3-dihydrobenzofuran-5-ol (82,83). Different tertiary amino functions were attached at position 3 of the benzofuranol ring to confer CNS penetration. These amino functions which are positively charged at physiological pH should promote the interaction with negatively charged brain lipids and therefore enhance inhibition of lipid peroxidation (37).

MDL74,180 (2,3-dihydro-2,2,4,6,7-pentamethyl-3-[(4-methylpiperazino)-methyl]benzofuran-5-ol; Chart 2), was selected from this series of analogs since this compound possesses strong antioxidant properties and penetrates the brain.

C. Biochemistry

The method of ex vivo inhibition of lipid peroxidation was used to select a radical scavenger which penetrates brain tissue (81). MDL74,180 was selected

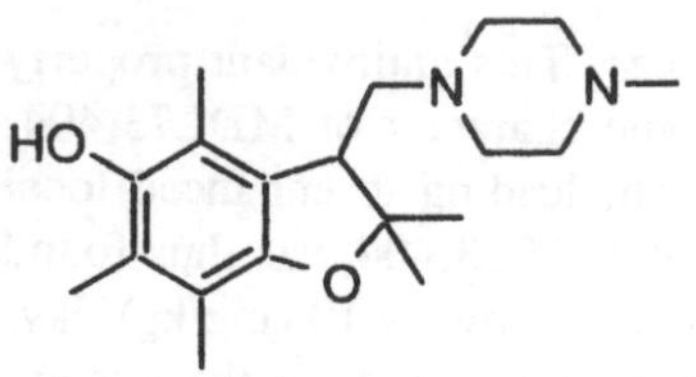

Chart 2 Chemical structure of MDL74,180.

from a series of α-tocopherol analogs because it showed better antioxidant properties and brain penetration than MDL73,335 as well as a number of related five-membered ring derivatives. Moreover, MDL74,180 was shown to be protective against $O_2^{\cdot-}$- and hydroxyl-radical-mediated, oxidative inactivation of protease nexin-1 (glia-derived nexin), a neurotrophic and neuroprotective endogenous thrombin inhibitor possibly involved in Alzheimer's disease (84).

MDL74,180 has also a vitamin E–sparing effect. The formation of thiobarbituric-acid-reactive substances, as a measure of lipid peroxidation in brain homogenate during in vitro incubation (27), was closely paralleled by a time-dependent formation of malondialdehyde, as followed by HPLC (Fig. 10a). A concomitant decrease of the endogenous α-tocopherol content became evident under these conditions. When MDL74,180 was added to the incubation mixture, the production of malondialdehyde was suppressed while the endogenous content of α-tocopherol was preserved (Fig. 10a). In the same experiment at 30-min incubation time, the presence of MDL74,180 enabled a concentration-dependent attenuation of both parameters (Fig. 10b). It appears that MDL74,180 prevented the loss of α-tocopherol at equimolar concentration. This is due to the amphiphilic rather than lipophilic character of the α-tocopherol analogs which prevents them from sticking to membranes, as well as to the above-mentioned charge effect (37). Therefore, MDL74,180 is capable of sparing tissue α-tocopherol content in vivo in an oxidative stress situation.

D. Drug Metabolism

1. *Permeability Across the Blood Brain Barrier (BBB)*

For CNS activity an obvious requirement is that a drug can cross the blood brain barrier (BBB). The penetration of drugs into the CNS can be studied using in vitro models of the BBB. Bovine brain microvessel endothelial cells (BME) grown in culture with astrocyte-conditioned medium have been reported to be a suitable in vitro cell culture model of the BBB (86,87).

MDL74,180, was around 40% transported across BME cell monolayers after 2 h incubation (Fig. 11), which was similar to propranolol, a drug known to

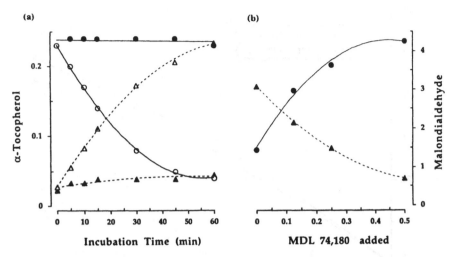

Figure 10 Consumption of endogenous α-tocopherol and formation of malondialdehyde during lipid peroxidation in rat brain homogenate and protective effect of MDL74,180. Lipid peroxidation in diluted rat brain homogenate was induced by incubation at 37°C and pH 7.4 as described (27). The changes, as measured by HPLC with fluorescence detection, of endogenous α-tocopherol (O) and the formed malondialdehyde (MDA (Δ), measured as its thiobarbiuric acid adduct (85)) are shown in (a) as a function of incubation time. The presence of 0.125-nmol MDL74,180/0.5-ml sample (closed symbols) preserved normal α-tocopherol and MDA content. In (b) are shown the protective effects of three concentrations of MDL74,180 at 30 min incubation time. Note that MDL74,180 replaces an equal amount of α-tocopherol (concentration scales are in nmol/0.5 ml sample; points representing means of duplicate experiments).

cross the BBB. These results suggest that the compound would cross the BBB in vivo. MDL74,180 also had good penetration across the Caco-2 cell monolayer, with over 50% of the compound transported after 2 h incubation (Fig. 9) demonstrating that the compound crosses cell membranes and has good oral absorption.

2. In Vitro Drug Metabolism

MDL74,180 was incubated at a concentration of 100 μM with rat and human liver microsomes in the presence of an NADPH generating system. The major metabolite formed was the quinone analog of parent drug. This metabolic pathway was similar to that of the cardioselective α-tocopherol analog MDL74,720, which also gave a quinone analog as a major metabolite in rat urine (data not shown).

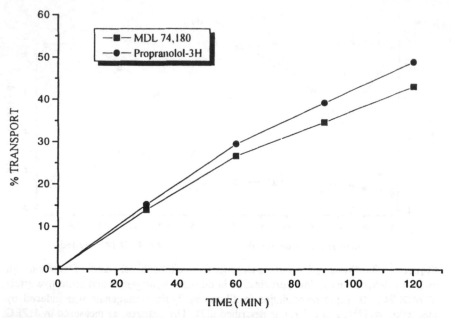

Figure 11 Transport of MDL74,180 and propranolol across bovine brain microvessel endothelial cell monolayers. Results are plotted as the cumulative amount of drug transported, expressed as a percentage of the initial concentration of drug.

E. Pharmacology

1. *Experimental Head Injury*

Blunt head trauma in mice resulted in severe sensorimotor function disturbances determined by means of the grip test (88), 1 h after the injury. MDL74,180 injected intravenously 3–5 min after the injury induced a bell-shaped dose-response relationship, the maximum prolongation of the "grip time" occurring with a dose of 10 µg/kg. One hour after head injury rota rod performance was also significantly improved by MDL74,180. Saline-treated injured mice fell off the rota rod, with only 10% remaining after 20 s. In contrast, sham-injured mice produced a very shallow curve, with more than 40% remaining on the rota rod for the full 60-s period. After MDL74,180 treatment (10 µg/kg injected i.v. 3–5 min after injury) 40% of the mice were still on the rota rod at 20 s and 30% remained after 60 s (Fig. 12). A similar improvement of neurological status was achieved after injection of MDL74,180 s.c. 30 min. before the injury (81). In this study prophylactic administration of 12 mg/kg MDL74,180 s.c. 30 min before head injury induced a significant reduction in the percentage of paraparetic mice measured 1 h after head injury compared to saline-treated injured controls.

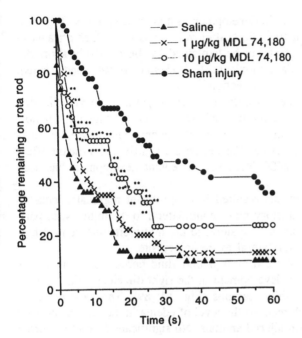

Figure 12 Rota rod performance of head-injured mice. The rota rod performance of mice 1 h after concussive head injury or sham injury (●). Injured mice were treated with MDL74,180 or saline i.v. 3–5 min after the injury. The results are expressed as the percentage number of mice remaining on the rota rod at a given time after the injury. **$p < 0.01$ when MDL74,180 treated animals are compared to saline-treated injured mice or sham-injured animals.

Using magnetic resonance imaging (MR) the development of brain edema caused as a result of blunt head injury in mice has been followed, and the effects MDL74,180 on the edema development have been studied. In this particular mouse model the brain water content reaches a maximum at about 24 h after the injury, most of the edema being present in the area of the olfactory bulb and surrounding the lateral ventricle. However, as the percentage edema in the vicinity of the olfactory bulb continues to increase up to and including 28 h, the periventricular edema is at maximum between 4 and 20 h and is rapidly resorbed over the subsequent hours (89).

The integrity of the blood brain barrier can be assessed by means of gadolinium and measuring its penetration on T1 weighted images. In fact, the blood brain barrier was ruptured in very few mice 4 h after head injury; therefore, it seems likely that in the remaining animals any rupture of the blood brain barrier may have already sealed (89).

MDL74,180 reduced the edema development 4 h after head injury (Fig. 13), but was without significant activity 20 h after the trauma (89). Such a transient reduction in edema development can be explained by the relatively short duration of action of this particular antioxidant (F. Bolkenius, unpublished results), when administered as a bolus i.v. injection.

In MR imaging a technique has been used which permits the visualization and eventual measurement of edema in small areas of brain, which under normal circumstances cannot be measured since the edematous change becomes diluted by the bulk of normal tissue. Furthermore a positive, significant effect of the free-radical scavenger MDL74,180 on the edema development has been demonstrated.

Fluid percussion injury in rats resulted in a significant regional edema 48 h after injury, in the right olfactory bulb, contralateral to the injury site, parietal cortex, both ipsilateral and contralateral to the injury site and ventral hippocampus, ipsilateral and contralateral to the injury site.

Infusion of MDL74,180 for 2 h after lateral fluid percussion head injury significantly attenuated edema development in the right olfactory bulb, parietal cortex, striatum, and ventral hippocampus (Fig. 14; 89). The most effective dose was 1 (mg/kg)/h, which reduced the level of edema in the aforementioned areas to that apparent in sham-injured animals. No significant decrease in brain

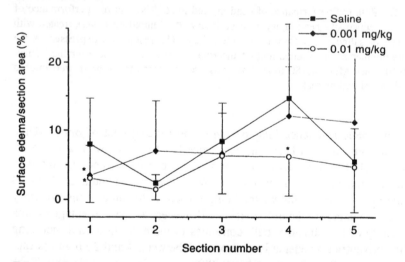

Figure 13 The effects of MDL74,180 on edema development 4 h after head injury in mice. Using MR imaging cerebral edema development was monitored in the same mice. When the edema present on each section is expressed as a percentage of the section area a protective effect of MDL74,180 could be visualized at the level of the olfactory bulb (section 1) and at the level of the lateral ventricle (section 4) when treated animals are compared to saline-injured controls.

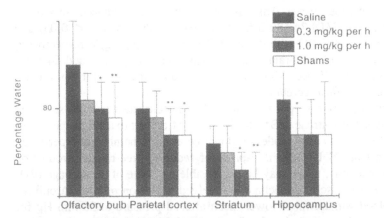

Figure 14 The effect of MDL74,180 infusion (2 h) on brain edema development following head injury in rats. Brain regional edema development 48 h after lateral fluid percussion head injury in the rat. The edema was measured by the wet weight/dry weight method and is expressed as percentage water. MDL74,180 or saline infusion began 3–5 min after head injury or sham injury. $*p < 0.05$, $**p < 0.01$ when compared to saline-treated controls.

edema after infusion of MDL74,180 was evident in areas on the left, injured side of the brain. This may be due to greater irreversible tissue damage which occurs on the impact side, since in these areas greater hemorrhage and a more prolonged decrease in regional cerebral blood flow occurs.

Subsequent to head injury, there is a secondary cerebral arteriolar vasodilation together with a loss of reactivity to vasoactive agents. Concomitantly a disruption of the blood brain barrier occurs, resulting in sodium and protein accumulation and osmotic fluid expansion of the brain extracellular space. Suggestive evidence from experimental edema models and the effects of exogenously applied antioxidants support an involvement of oxygen-derived free radicals and lipid peroxidation in the development of posttraumatic brain edema. Therefore, it seems likely that MDL74,180 is inhibiting the superoxyl radicals (81) produced as a result of the initial injury (90) and, hence, the ensuing hydroxyl radical formation and lipid peroxidation, resulting in an interruption of the cascade of events.

2. Cerebral Ischemia/Reperfusion

In male Wistar rats, subjected to 2 h of middle cerebral artery occlusion (MCA) and 8 days of reperfusion, the area of necrotized tissue was measured by means of an image analysis processing system after staining with triphenyl tetrazolium chloride. Intravenous infusion of saline, beginning 15 min before the onset of reperfusion and continuing for 2 h into the reperfusion period, resulted in 158

± 40 mm³ of necrotic tissue. This area was significantly reduced, in a dose-related fashion, by MDL74,180 infusion to 109 ± 48 mm³ ($p < 0.05$) and 73 ± 35 mm³ ($p < 0.01$) with doses of 0.1 and 1.0 (mg/kg)/h respectively; i.e., the percentage of forebrain infarcted was reduced by 31% and 54% respectively (Fig. 15). Hence, MDL74,180 can reduce the cerebral tissue damage resulting from focal cerebral ischemia and reperfusion.

Application of electron spin resonance (ESR) techniques and the spin trap α-phenyl-*N-tert*-butylnitrone (PBN) to the study of the antioxidant properties of MDL74,180 in vivo in a rat model of global cerebral ischemia and reperfusion demonstrated that MDL74,180 is capable of reducing free-radical production under such conditions, which may be responsible for some of the damage (91). The global cerebral ischemia technique involved occlusion of both carotid arteries and blood withdrawal to lower mean arterial pressure to 30 mm Hg for 20 min followed by reperfusion for 5 min. MDL74,180 infusion commenced 15 min before the onset of ischemia until the end of reperfusion. In doses of

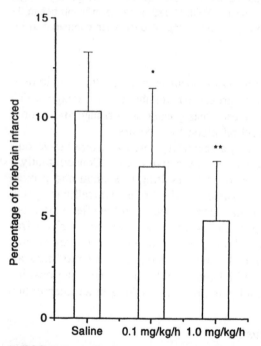

Figure 15 MDL74,180 and its effects on infarct size after MCA occlusion and reperfusion. The effect of MDL74,180 infusion beginning 15 min before the onset of reperfusion and continuing for 2 h into the reperfusion period on the percentage of forebrain infarcted after MCA occlusion (2 h) and 8 days of reperfusion. *$p < 0.05$, **$p < 0.01$ when compared to saline-treated controls.

1 and 3 (mg/kg)/h MDL74,180 reduced the level of trapped radicals in a dose-dependent fashion from 0.404 ± 0.005 after saline infusion to 0.377 ± 0.005 and 0.3 ± 0.002 after MDL74,180, respectively.

3. *Subarachnoid Hemorrhage*

Doses of MDL74,180 that were active at reducing the tissue damage in cerebral ischemia and the edema in the rat model of head trauma were also active at decreasing the acute vasospasm in a rat model of subarachnoid hemorrhage (SAH). This model involves injecting a small volume of autologous blood into the cisterna magna and measuring cerebral blood flow by means of radioactive microspheres. SAH produced a 20–30% fall in cerebral blood flow (CBF), which was at maximum 30 min after the injection of blood and then gradually returned toward baseline values over the subsequent 90 min. Infusion of MDL74,180 (1(mg/kg)/h) beginning 10 min before SAH reduced the fall in CBG, particularly in the brain stem region, where the decrease was significantly reduced 60 min after SAH (Fig. 16). One of the major complications of SAH in humans after rupture of an intracranial aneurysm is a delayed cerebral ischemia due to vasospasm which usually occurs 6–10 days after the SAH. Hence the rat model of SAH and acute vasospasm is not identical to the human situation, although the molecular mechanisms, which depend on the severity of bleeding and include a modification of vascular reactivity before the vasospasm, are likely to be similar (92). The cascade of events includes free-radical generation, catalyzed by oxyhemoglobin released from the clot (93), which would trigger secondary endothelial damage leading to a modification of the response to vasodilators and the resulting vasospasm. Hence MDL74,180 reduces the vasospasm by inhibiting one of the first events in the cascade.

F. Conclusion

MDL74,180 is a benzofuran derivative of α-tocopherol. It is a free-radical scavenger that can penetrate the brain. At present no clinically effective treatment exists for cerebral ischemia, hemorrhage, and head trauma, because the molecular events leading to the modification of vascular reactivity and neuronal damage have only recently been elucidated and are probably similar for the different brain insults. With MDL74,180 the possibility of inhibiting free-radical generation and, hence, interrupting the cascade of events leading to neuronal damage exists. Therefore, MDL74,180 may be a potential treatment for cerebrovascular disease.

V. GENERAL CONCLUSION

In this chapter studies have been reviewed that confirm that the structure of α-tocopherol is a good basis for the design of new potent free-radical scavengers.

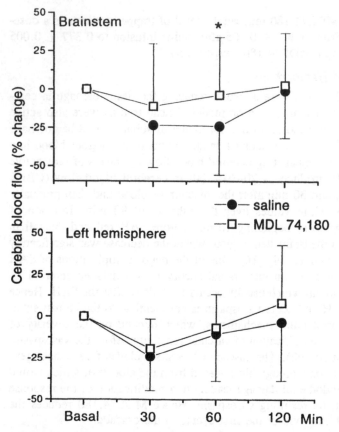

Figure 16 The effects of MDL74,180 on the reduction of cerebral blood flow following subarachnoid hemorrhage. Infusion of saline or MDL74,180 (1(mg/kg)/h) began 10 min before subarachnoid hemorrhage until the end of the experiment. MDL74,180 reduced the hypoperfusion, particularly in the brain stem area.

The water-soluble benzopyran-6-ol series together with the benzofuran-5-ol series displayed strong antioxidant properties. They inhibit lipid peroxidation and scavenge superoxide and hydroxyl radicals more effectively than α-tocopherol. By modification of the chemical structure of the compounds and as a consequence of their physicochemical properties, it has been possible to target the compounds for specific tissues.

The quaternary ammonium analog MDL73,404 was demonstrated to accumulate in heart tissue, reaching concentrations 10 to 30 times higher than in blood and other tissues after i.v. administration. This cardioselectivity contributes to its potency when being used for myocardial protection in situations of myocardial reperfusion injury. Due to its permanent cationic character, the ammonium derivative is poorly absorbed after oral administration, suggesting use in diseases such as IBD as demonstrated by its effect in a mouse colitis model.

The benzofuran-5-ol derivative, MDL74,180, entered the brain and may have therapeutic potential in cerebrovascular disease as demonstrated by its beneficial effects in a mouse head injury model and rat models of ischemia and reperfusion.

Although the α-tocopherol analogs presented here proved to be effective in animal models, further work is required to fully characterize these classes of compounds in other disease states and to fully assess their efficacy in the clinical situation.

ACKNOWLEDGMENT

The authors wish to acknowledge Dr. J. M. Grisar, who initiated this research program.

REFERENCES

1. Gutteridge JMC. Free radicals in disease processes: a compilation of cause and consequence. Free Radic Res Commun 1993; 19: 141–158.
2. Halliwell B, Gutteridge JMC. Free Radicals in Biology and Medicine. 2d ed. Oxford: Clarendon Press, 1989.
3. Halliwell B. Drug antioxidant effects: a basis for drug selection? Drugs 1991; 42: 569–605.
4. McCay PB, King MM. Vitamin E: its role as a biological free radical scavenger and its relationship to the microsomal mixed-function oxidase system. In: LJ Machlin, ed. Vitamin E, A Comprehensive Treatise, New York: Marcel Dekker, 1980: 289–317.
5. van Acker SABE, Koymans LMH, Bast A. Molecular pharmacology of vitamin E : structural aspects of antioxidant activity. Free Radic Biol Med 1993; 15: 311–328.
6. Reimer KA, Murry CE, Richard VJ. The role of neutrophils and free radicals in the ischemic-reperfused heart: Why the confusion and controversy? J Mol Cell Cardiol 1989; 21: 1225–1239.
7. Ritchie JL, Davies KB, Williams DL, Caidwell J, Kennedy JW. Global and regional left ventricular function and tomographic radionuclide perfusion: The western Washington intracoronary streptokinase in myocardial infarction trial. Circulation 1984; 70: 867–875.
8. White HD, Norris RM, Brown MA. Effect of intravenous streptokinase on left ventricular function and early survival after acute myocardial infarction. N Engl J Med 1987; 317: 850–855.

9. Hearse DJ. Reperfusion of the ischemic myocardium. J Mol Cell Cardiol 1977; 9: 605–616.

10. Becker LC, Ambrosio G. Myocardial consequences of reperfusion. Prog Cardiovasc Dis 1987; 30: 23–44.

11. Simpson PJ, Lucchesi BR. Free radicals and myocardial ischemia and reperfusion injury. J Lab Clin Med 1987; 110: 13–20.

12. Weisfeldt ML. Reperfusion and reperfusion injury. Clin Res 1987; 35: 13–20.

13. Grech ED, Dodd NJF, Bellamy CM, Perry RA, Morrison WL, Ramsdale DR. Free radical generation during angioplasty for acute myocardial infarction. Lancet 1993; 341: 990–991.

14. Davies SW, Ranjadayalan K, Wickens DG, Dormandy TL, Timmis AD. Lipid peroxidation associated with successful thrombolysis. Lancet 1990; 335: 741–743.

15. Davies SW, Ranjadayalan K, Wickens DG, Dormandy TL, Umachandran V, Timmis AD. Free radical activity and left ventricular function after thrombolysis for acute infarction. Br Heart J 1993; 69: 114–120.

16. McMurray J, Choprar M, Abdullah I, Smith WE, Dargie HJ. Evidence for oxidative stress in unstable angina. Br Heart J 1992; 68: 454–457.

17. Weisel RD, Mickle DAG, Finkle CD, Tumati LC, Madonik MM, Ivanov J, Burton GW, Ingold KU. Myocardial free-radical injury after cardioplegia. Circulation 1989; 80(suppl III): 14–18.

18. Mickle DAG, Weisel RD, Burton GW, Ingold KU. Effect of orally administered α-tocopheryl acetate on human myocardial α-tocopherol levels. Cardiovasc Drugs Therap 1991; 5: 309–312.

19. Baracchi R, Pelosi G, Maffei S, Baroni M, Salvatore L, Ursini F, Verunelli F, Biagini A. Myocardial vitamin E is consumed during cardiopulmonary bypass: indirect evidence of free radical generation in human ischaemic heart. Int J Cardiol 1992; 37: 339–343.

20. Yau T, Weisel RD, Mickle DAG, Ivnov J, Tumati LC, Ingold K, Burton G. Vitamin E improved myocardial protection. Circulation 1990; 82(suppl III): 111–146.

21. Barta E, Pechan L, Cornak V, Lukarova O, Rendekova V, Verchovodko P. Protective effect of alpha-tocopherol and ascorbic acid against the ischaemic-reperfusion injury in patients during open-heart surgery. Bratisl Lek Histy 1991; 92: 174.-183.

22. Grisar JM, Petty MA, Bolkenius FN, Dow J, Wagner J, Wagner ER, Haegele KD, De Jong W. A cardioselective, hydrophilic N,N,N-trimethylethanaminium α-tocopherol analogue that reduces myocardial infarction. J Med Chem 1991; 34: 257–260.

23. Grisar JM, Marciniak G, Bolkenius FN, Verne-Mismer J, Wagner ER. Cardioselective ammonium, phosphonium and sulfonium analogues of α-tocopherol and ascorbic acid that inhibit in vitro and ex vivo lipid peroxidation and scavenge superoxide radicals. J Med Chem 1995; 38: 2880–2886.

24. Dow J, Petty MA, Grisar JM, Wagner ER, Haegele KD. Cardioselectivity of α-tocopherol analogues, with free radical scavenger activity, in the rat. Drug Metab Dispos 1991; 19: 1040–1045.

25. Babior, BM. Oxidants from phagocytes: agents of defense and destruction. Blood 1984; 4: 959–966.

26. Fridovich L. Superoxide radical: an endogenous toxicant. Ann Rev Pharmacol Toxicol 1983; 23; 239–257.

27. Bolkenius FN, Grisar JM, De Jong W. A water-soluble quaternary analogue of α-tocopherol, that scavenges lipoperoxyl, superoxyl and hydroxyl radicals. Free Radic Res Commun 1991; 14: 363–372.

28. Gotoh N, Niki E. Rates of interactions of superoxide with vitamin E, vitamin C and related compounds as measured by chemiluminescence. Biochim Biophys Acta 1992; 1115: 201–207.

29. Dürckheimer W, Cohen LA. The chemistry of 9-hydroxy-α-tocopherone, a quinone hemiacetal. J Am Chem Soc 1964; 86: 4388–4393.

30. Nishikimi M, Machlin LJ. Oxidation of α-tocopherol model compound by superoxide anion. Arch Biochem Biophys 1975; 170: 684–689.

31. Matsuo M, Matsumoto S, Iitaka Y. Oxygenations of vitamin E (α-tocopherol) and its model compound 2,2,5,7,8-pentamethylchroman-6-ol in the presence of the superoxide radical solubilized in aprotic solvents: unique epoxidations and recyclizations. J Org Chem 1987; 52: 3514–3520.

32. Britigan BE, Roeder TL, Buettner GR. Spin traps inhibit formation of hydrogen peroxide via the dismutation of superoxide: implications for spin trapping the hydroxyl free radical. Biochim Biophys Acta 1991; 1075: 213–222.

33. Bolkenius FN. Leukocyte-mediated inactivation of α_1-proteinase inhibitor is inhibited by amino analogues of α-tocopherol. Biochim Biophys Acta 1991; 1095: 23–29.

34. Carrell R, Travis J. α_1-Antitrypsin and the serpins: variation and countervariation. Trends Biol Sci 1985; 10: 20–24.

35. Weiss SJ. Tissue destruction by neutrophils. New Engl J Med 1989; 320: 365–376.

36. Lucchesi BR, Mullane KM. Leukocytes and ischaemia-induced myocardial injury. Ann Rev Pharmacol Toxicol 1986; 26: 201–224.

37. Barclay LRC, Vinqvist MR. Membrane peroxidation: Inhibiting effects of water-soluble antioxidants on phospholipids of different charge types. Free Radic Biol Med 1994; 16: 779–788.

38. Bisby RH, Parker AW. Reaction of ascorbate with the α-tocopheroxyl radical in micellar and bilayer membrane systems. Arch Biochem Biophys 1995; 317: 170–178.

39. Petty MA, Dow J, Grisar JM, De Jong W. Effect of a cardioselective α-tocopherol analogue on reperfusion injury in rats induced by myocardial ischaemia. Eur J Pharmacol 1991; 192: 383–388.

40. Petty MA, Grisar JM, Dow J, De Jong W. Effect of an α-tocopherol analogue on myocardial ischemia and reperfusion injury in rats. Eur J Pharmacol 1990; 179: 241–242.

41. Petty MA, Grisar JM, De Jong W. Protective effects of an α-tocopherol analogue against myocardial reperfusion injury in rats. Eur J Pharmacol 1992; 210: 85–90.

42. Petty MA, Lukovic L, Grisar JM, Dow J, Bolkenius FN, De Jong W. Myocar-

dial protection by a cardioselective free radical scavenger. Eur J Pharmacol 1994; 255: 215–222.

43. Hatori N, Miyazaki A, Tadokoro H, et al. Beneficial effects of coronary retro-infusion of superoxide dismutase and catalase on reperfusion arrhythmias, myocardial function and infarct size in dogs. J Cardiovasc Pharmacol 1989; 14: 396–404.

44. Lukovic L, Petty MA, Bolkenius FN, Grisar JM, Dow J, De Jong W. Protection against myocardial reperfusion injury with an α-tocopherol analogue. Eur J Pharmacol 1993; 233: 63–70.

45. Petty M, Dow J. Subcutaneously administered MDL73,404: its distribution and effects on infarct size. Br J Pharmacol 1991; 104: 176P.

46. Ambrosio G, Flaherty JT, Duilio C, et al. Oxygen radicals generated at reflow induce peroxidation of membrane lipids in reperfused hearts. J Clin Invest 1991; 87: 2056–2066.

47. De Jong W, Jones CR, Petty MA, Lukovic L. Receptor density in rat heart after ischaemia and prolonged reperfusion. Clin Exp Pharmacol Physiol 1994; 22(Suppl 1): 5279.

48. Zughaib ME, Tang XL, Schleman M, Jeroudi MO, Bolli R. Beneficial effects of MDL74,405, a cardioselective, water-soluble α-tocopherol analogue, on the recovery of function of stunned myocardium in intact dogs. Cardiovasc Res 1994; 28: 235–241.

49. Namm DH, Wang CM, El-Sayad S, Copp FC, Maxwell RA. Effects of bretylium on rat cardiac muscle: the electrophysiological effects and its uptake and binding in normal and immunosympathectomized rat hearts. J Pharmacol Exp Ther 1995; 193: 194–208.

50. Patterson E, Stetson P, Lucchesi BR. Plasma and myocardial tissue concentrations of UM-272 (N,N-dimethylpropranolol) after oral administration in dogs. J Pharmacol Exp Ther 1980; 214: 449–453.

51. Lindstrøm TD, Murphy FF, Smallwood JK, Wiest SA, Steinberg MI. Correlation between the disposition of [14C] clofilium and its cardiac electrophysiological effect. J Pharmacol Exp Ther 1982; 221: 584–589.

52. Parks DA, Granger DN. Contributions of ischemia and reperfusion of mucosal lesion formation. Am J Physiol 1986; 250: G749–G753.

53. Arndt H, Kubes P, Granger DN. Involvement of neutrophils in ischemia-reperfusion injury in the small intestine. Klin Wochenschr 1991; 69: 1056–1060.

54. Sakai T, Qi Q.-H, Murthy S, Grisar M, Fondacaro JD. Ischemia/reperfusion in the rat colon. A discrete surgical model for biochemical and pharmacological studies. Gastroenterology 1993; 104 (suppl 4): A278.

55. Collins RH, Feldman M, Fordtran JS. Colon cancer, dysplasia and surveillance in patients with UC: a critical review. N Engl J Med 1987; 316: 1654–1658.

56. Korelitz BI. Carcinoma of the intestinal tract in Crohn's disease: Results of a survey conducted by the National Foundation for Ileitis and Colitis. Am J Gastroenterol 1983; 78: 44–46.

57. Yamada T, Grisham MB. Role of neutrophil-derived oxidants in the pathogenesis of intestinal inflammation. Klin Wochenschr 1991; 69: 988–994.

58. Saverymuttu SH, Peters AM, Hodgson HJF, Chadwick VS. Assessment of disease

activity in UC using [111]Indium labelled faecal leukocyte excretion. Scand J Gastroenterol 1983; 18: 907–912.

59. Petrone WF, English DK, Wong K, McCord JM. Free radicals and inflammation: Superoxide-dependent activation of a neutrophil chemotactic factor in plasma. Proc Natl Acad Sci 1980; 77: 1159–1163.

60. Shiratora Y, Aoki S, Takada H, Kiriyama H, Ohto K, Hai K, Teraoka H, Matano S, Matsumoto K, Kamii K. Oxygen-derived free radical generating capacity of polymorphonuclear cells in patients with UC. Digestion 1989; 44: 163–171.

61. Simmonds NJ, Allen RE, Stevens TRJ, v. Someren RNM, Blake DR, Rampton DS. Chemiluminescence assay of mucosal reactive oxygen metabolites in inflammatory bowel disease. Gastroenterology 1992; 103: 186–196.

62. Jones DP, Thor H, Smith MT, Jewell SA, Orrenius S. Inhibition of ATP-dependent microsomal Ca^{2+} sequestration during oxidative stress and its prevention by glutathione. J Biol Chem 1983; 258: 6390–6393.

63. Otamiri T. Oxygen radicals, lipid peroxidation and neutrophil accumulation after small-intestinal ischaemia and reperfusion. Surgery 1988; 105: 593–597.

64. Ireland A, Jewell DP. Mechanism of action of 5-aminosalicylic acid and its derivatives. Clin Sci 1990; 78: 119–125.

65. Nielsen OH, Ahnfelt-Rønne I. Involvement of oxygen-derived free radicals in the pathogenesis of chronic inflammatory bowel disease. Klin Wochenschr 1991; 69: 995–1000.

66. Weeke B, Jarnum S. Serum concentrations of 19 serum proteins in Crohn's disease and UC. Gut 1971; 12: 297–302.

67. Fischbach W, Deubel M, Boege F, Mössner J. Erste klinische Erfahrung mit einer vereinfachten Bestimmungsmethode des fäkalen alpha-1-Antitrypsins. Z Gastroenterol 1991; 29: 650–654.

68. Adeyemi EO, Neumann S, Chadwick VS, Hodgson HJF, Pepys MB. Circulating human leukocyte elastase in patients with inflammatory bowel disease. Gut 1985; 26: 1306–1311.

69. Sharon P, Stenson WF. Enhanced synthesis of leukotriene B_4 by colonic mucosa in inflammatory bowel disease. Gastroenterology 1984; 86: 453–460.

70. Allgayer H, Stenson WF. A comparison of effects of sulfasalazine and its metabolites on the metabolism of endogenous vs exogenous arachidonic acid. Immunopharmacology 1988; 15: 39–46.

71. Staerk-Laursen L, Naesdal J, Bukhave K, Lauritsen K, Rask-Madsen J. Selective 5-lipoxygenase inhibition in UC. Lancet 1990; 335: 683–685.

72. Okayasu I, Hatakeyama S, Yamada M, Ohkusa T, Inagaki Y, Nakaya R. A novel method in the induction of reliable experimental acute and chronic UC in mice. Gastroenterology 1990; 98: 694–702.

73. Murthy SNS, Cooper HS, Shim H, Shah RS, Ibrahim SA, Sedergran DJ. Treatment of dextran sulfate sodium-induced murine colitis by intracolonic cyclosporin. Dig Dis Sci 1993; 38: 1722–1734.

74. Cooper HS, Murthy SNS, Shah RS, Sedergran DJ. Clinicopathological study of dextran sulfate sodium experimental colitis. Lab Invest 1993; 69: 238–249.

75. Murthy SNS, Fondacaro JD, Murthy NS, Cooper HS, Bolkenius F. Beneficial

effect of MDL73,404 in dextran sulfate-mediated murine colitis. Agents Actions 1994; 41: C233–C234.

76. Artursson P, Karlsson J. Correlation between oral drug absorption in humans and apparent drug permeability coefficients in human intestinal epithelial (Caco-2) cells. Biochem Biophys Res Commun 1991; 175: 880–885.

77. Artursson P. Epithelial transport of drugs in cell culture. I: A model for studying the passive diffusion of drugs over intestinal absorbtive (Caco-2) cells. J Pharm Sci 1990; 79: 476–482.

78. Kontos HA. Oxygen radicals in cerebral vascular injury. Circ Res 1985; 57: 508–516.

79. Hall ED, Braughler JM. Central nervous system trauma and stroke. II: Physiological and pharmacological evidence for involvement of oxygen radicals and lipid peroxidation. Free Radic Biol Med 1989; 6: 303–313.

80. Siesjo BK, Ahardh CD, Bengtsson F. Free radicals and brain damage. Cerebrovasc Brain Metab Rev 1989; 1: 165–211.

81. Grisar JM, Bolkenius FN, Petty MA, Verne J. 2,3-Dihydro-1-benzofuran-5-ols as analogues of α-tocopherol that inhibit in vitro and ex vivo brain lipid autoxidation and protect mice against central nervous system trauma J Med Chem 1995; 38: 453–458.

82. Burton GW, Ingold KU. Vitamin E : Application of the principles of physical organic chemistry to the exploration of its structure and function. Acc Chem Rev 1986; 19: 194–201.

83. Ingold KU, Burton GW, Foster DO, Hughes L. Further studies of a new vitamin E analogue more active than α-tocopherol in the rat curative myopathy bioassay. FEBS 1990; 267: 63–65.

84. Bolkenius FN, Monard D. Inactivation of protease nexin-1 by xanthine oxidase-derived free radicals. Neurochem Int 1995; 26: 587–592.

85. Draper HH, Squires EJ, Mahmoodi H, Wu J, Agarwal S, Hadley M. A comparative evaluation of thiobarbituric acid methods for the determination of malondialdehyde in biological materials. Free Radic Biol Med 1993; 15: 353–363.

86. Rubin LL, Hall DE, Porter S, Barbu K, Cannon C, Horner HC, Janatpour M, Liaw CW, Manning K, Morales J, Tanner LI, Tomaselli KJ, Bard F. A cell culture model of the blood brain barrier. J Cell Biol 1991; 115: 1725–1735.

87. Miller DW, Audus KL, Borchards RT. Application of cultured endothelial cells of the brain microvasculature in the study of the blood brain barrier. J Tiss Cult Meth 1992; 14: 217–224.

88. Hall ED, Yonkers PA, McCall JM, Braughler JM. Effects of the 21-amino steroid U-74006F on experimental head injury in mice. J Neurosurg 1988; 68: 456–461.

89. Petty MA, Poulet P, Haas A, Namer I, Wagner J. Reduction of traumatic brain injury-induced cerebral oedema by a free radical scavenger. Eur J Pharmacol 1996; 307: 149–155.

90. Kukreja RC, Kontos HA, Hess ML, Ellis EF. PGH synthetase and lipoxygenase generate superoxide in the presence of NADH or NADPH. Circ Res 1986; 59: 612–619.

91. Cowley DJ, Lukovic L, Petty MA. MDL74,180 reduces cerebral infarction and free radical concentrations in rats subjected to ischaemia and reperfusion. Eur J Pharmacol 1996; 298: 227–233.

92. Hatake K, Wakabayashi I, Kakishita E, Hishida S. Impairment of endothelium dependent relaxation in human basilar artery after subarachnoid haemorrhage. Stroke 1992; 23: 1111–1117.

93. Sano K, Asano T, Tanishima T, Sasaki T. Lipid peroxidation as a cause of cerebral vasospasm. Neurol Res 1980; 2: 253–272.

3

Physical and Chemical Mechanisms of the Antioxidant Action of Tirilazad Mesylate

Dennis E. Epps and John M. McCall
Pharmacia & Upjohn, Inc., Kalamazoo, Michigan

I. INTRODUCTION

Reactive oxygen species play major roles in biological systems, for example, in the synthesis of prostaglandins, thromboxanes, and leukotrienes (1). Because of their high reactivity, these free radicals must be strictly confined within the compartments where they are needed, since in the wrong environment and with the wrong target molecules they may initiate lethal peroxidative chain reactions. Thus, the production and the use of reactive oxygen species is normally contained by a host of protective systems such as superoxide dismutase, catalase, the glutathione peroxidase system, vitamin E/ascorbic acid system, etc. Weakening or overwhelming the control systems can lead to oxidative damage of lipids, proteins, and nucleic acids and, ultimately, to cell death. Such toxic effects have been observed in, for instance, traumatic tissue injury (2), chronic inflammatory conditions (3), ischemia/reperfusion injury (4), and aging (5). Prevention of such oxidative tissue damage is, at the present time, the target of considerable effort in drug development; the types of antioxidant drugs currently under consideration run the spectrum from the enzyme superoxide dismutase to hydrophobic, membrane-bound compounds.

The first line of defense of a cell against extracellular toxic agents is the membrane. At this location, reactive oxygen species that contact the cell would react readily with the bisallylic groups of polyunsaturated fatty acyl chains to produce lipid peroxides and oxidized membrane proteins. The buildup of peroxy lipids can ultimately lead to disruption of the organization of membrane lipids, resulting in altered enzyme activities, increased membrane permeability, and interference with the function of ion channels (6). Vitamin E, together with its coantioxidants, is one of the major protectors against oxidative disruptions of membranes. However, because of its limited capacity, the vitamin E–based system is readily overwhelmed in a variety of pathology-induced oxidative stresses, and thus bolstering the antioxidative capacity of the cell membrane is an important target for therapeutic intervention. The drugs developed for this purpose should preferentially be lipophilic in order to be concentrated within the lipid bilayer, where they could directly protect the membranes from oxidative injury and, thus, indirectly preserve cell integrity.

The 21-aminosteroids (lazaroids), members of one such class of novel antioxidants, have been tested in a variety of animal injury and disease models. One compound in particular, tirilazad mesylate (U-74,006F), is protective in many of these disease/injury models and is in advanced clinical trials for head and spinal cord injury, subarachnoid hemorrhage, and ischemic stroke. In fact, significant efficacy on clinical vasospasm and improved 3-month outcome, as well as safety, have been reported after the completion of phase III trials. It is not our purpose to discuss the efficacy of lazaroids in animal and human injury models. Instead we focus on the mechanism of action of these novel lipophilic antioxidants, and the reader is referred to a review (7) for detailed information on the efficacy of tirilazad mesylate, hereinafter called U-74,006F, in disease and injury models.

This compound, discovered in part by serendipity, was found to be an antioxidant both in vitro and in vivo, but its complete mechanism of action was not immediately obvious. In particular, no direct experimental evidence was available to show whether the antioxidant effect in different models arose purely from the free-radical-scavenging pathway or other—perhaps purely physical— changes in membrane structure were also operative. Several experimental approaches were applied to answer these questions and this chapter reports on the results of these investigations. Specifically, using the examples of U-74,006F and related congeners, we wish to summarize the types of experiments that delineate the physical and chemical modes of action of lipophilic antioxidants and help in the design and optimization of membrane-targeted antioxidant drugs.

II. BASIC CONCEPTS

Biological lipid peroxidation, as all chain reactions, comprises the chemical reaction steps of initiation, propagation, and termination. With an externally generated initiator free radical, I·, the first step in the chain reaction is the withdrawal of a hydrogen atom from the central carbon of the bisallylic system, BH, according to the scheme

$$I^\bullet + BH \xrightarrow{\ k_i\ } RH + B^\bullet \qquad\qquad \text{initiation (1)}$$

The bisallylic radical reacts readily with any available molecular oxygen to give the oxidized bisallylic radical

$$B^\bullet + O_2 \rightarrow BO_2^\bullet \qquad\qquad \text{[propagation-1](2)}$$

The cycle of the chain reaction is then completed when this radical reacts with another bisallylic group to yield the lipid hydroperoxide and the bisallylic free radical which continues the chain reaction:

$$BO_2^\bullet + BH \rightarrow BOOH + B^\bullet \qquad\qquad \text{[propagation-2](3)}$$

Thus, the propagation step can be described formally as

$$B^\bullet + O_2 + BH \xrightarrow{\ kp\ } BOOH + B^\bullet \qquad\qquad \text{propagation (4)}$$

The chain reaction is terminated by withdrawing the free radical from the reacting pool of lipids, which may be accomplished by physical or chemical means. Physical termination of lipid peroxidation occurs when the free radical becomes hydrophilic and diffuses out of the lipid portion of the membrane. Chemically, free-radical chain reactions are only terminated by the recombination of two free radicals to form a product, X:

$$B^\bullet + B^\bullet \xrightarrow{\ k_t\ } X \qquad\qquad \text{termination (5)}$$

From the purely experimental point of view, the term "chain termination" is often used to signify that the lipid free radical has reacted with an antioxidant to form a radical that is stable on the time scale of the experiment. If the less reactive species is still reactive within this time frame, the propagation is inhibited but not terminated. The ultimate fate of the scavenger free radical may be recombination with itself, reductive recycling back to the parent, or diffusion out of the compartment of lipid peroxidation.

Inhibition of a lipid peroxidation chain reaction can a priori be accomplished by manipulating any one of the foregoing, that is, by decreasing the rates of initiation or propagation, or by increasing the rate of chemical, physical, or scavenging chain termination. There are no obvious chemical means by which to accelerate selectively the rate of chemical termination, but diffusional and scavenging terminations are possible targets of drug action. Inhibition may be more efficient, however, when the point of intervention of antioxidant drugs is the inhibition of the initiation and/or the propagation step. In a sense, it is fortunate that scavenging is not the only possible way of reducing the rate of lipid peroxidation and that other means are also available. For example, initiation can be inhibited by elimination of the reactive oxygen initiators of lipid peroxidation by SOD (superoxide dismutase), catalase, glutathione peroxidases, of free-radical scavengers located outside the lipid membrane. Also, chelators may complex the metal ions which catalyze the conversion of LOOH (lipid hydroperoxide) to LO˙ or LOO˙ and of H_2O_2 to HO˙. Recently, the importance of physical alterations of the lipid membrane and how they relate to normal membrane processes has become evident. We and others have shown that physical alterations of membranes can also have a profound influence on the rate of propagation and termination of lipid peroxidation. In this regard, one must consider the actual physical organization of the membrane and the effect that lipid peroxidation and/or the presence of a lipophilic antioxidant might exert on this organization.

A. Membrane Effects and Lipid Peroxidation

The hydrophilic headgroups of phospholipids delineate the boundaries of the lipid layer of biological membranes, whereas their hydrophobic fatty acyl chains serve as a backbone for the lipid core of the membrane bilayer. The organization of the lipid chains is dynamic and is governed by steric factors and hydrophobic and electrostatic forces. Compounds that have a high affinity for the lipid bilayer of membranes and have preferred orientations within it will often reorder the acyl chains. Amphiphiles with a hydrophilic cross section that is larger than the hydrophobic cross section (see structures in Fig. 1) will compress selectively the phospholipid headgroups. As a consequence, the fatty acyl chains become disordered and the hydrophobic core thins out. This should influence the rate of lipid peroxidation not only by limiting access of initiators and of oxygen to the hydrophobic core, but also by decreasing the rate of propagation through altered alignment (disordering) of the bisallylic groups on the lipid chains of the phospholipids. The latter mechanism changes the value of k_p; the same result can be achieved by other physical means: (1) A viscotropic agent, such as glycerol or sucrose, greatly slows the rate of diffusion of the initiator radical into the membrane and thereby lowers the local steady-state concentration of the initiator. (2) Lowering the temperature decreases k_i and k_p, due not only to the term of the activation energy but also to the strong temperature dependence of chain viscosity. Finally, as mentioned, lipid peroxidation may be terminated not only by the recombination of two lipid free radicals but also by withdrawing the free radical from the site of free-radical chain propagation. This may be achieved—besides scavenging, i.e., forming a stable free radical—by forming a free radical that diffuses out of the membrane into the aqueous environment, i.e., forming a hydrophilic free radical.

At this point, we wish to emphasize that one must define the phrases antioxidant/free-radical scavenger, on the one hand, and inhibitor of lipid per-

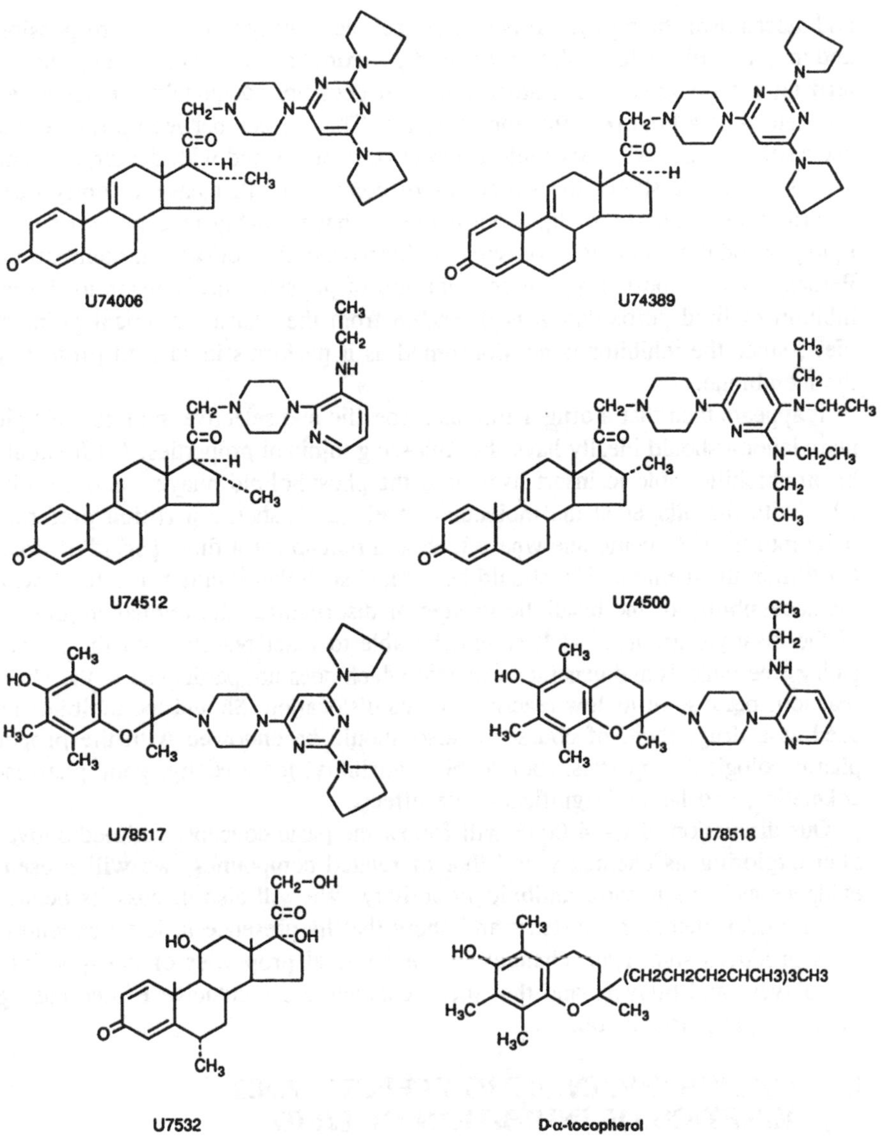

Figure 1 Structures of U-74,006F, congeners, and other antioxidants.

oxidation, on the other. The former refers to a chemical activity, whereas the latter can be applied to a compound that acts via chemical and/or physical mechanisms. The demonstration of chemical antioxidant/free-radical scavenging activities in vitro using a free-radical generating system cannot be extrapolated directly to activity in vivo. Depending on the cell type, locus of injury, and mechanism, both physical and chemical mechanisms of action are possible and may contribute to inhibition of lipid peroxidation. In other words, characterizing a compound's antioxidant potential does not adequately predict its effectiveness as a lipid peroxidation inhibitor. The balance between physical and chemical activities will vary with different compounds and with the target membranes. Figure 2 shows the structures of compounds that have recently been described as inhibitors of lipid peroxidation that probably owe much of their lipid peroxidation inhibitory activity to physicochemical effects on membranes. Perhaps more importantly, the contribution of physical mechanisms to the inhibition of lipid peroxidation is desirable from the pharmacological point of view, since the inhibitor is not consumed as it performs its task in protecting the membrane.

It appears then that a drug acting as a specific and selective inhibitor of lipid peroxidation should ideally have the following minimal properties: (1) It should be amphiphilic, able to insert itself into the phospholipid bilayer, and be miscible with the bilayer at the molecular level. (2) It should partition preferentially into the cell membrane where, it should remain for a finite period of time. (3) Within the membrane, it should be located such that it either interferes with the accessibility of the bisallylic groups or disorganizes the regular alignment of the bisallyic groups. (4) It should be able to react readily with the propagating free radicals and generate a radical which does not participate in the chain reaction, because of its low reactivity or its dislocation. Should the inhibitor be used as a drug, then, of course, it also should be endowed with the proper pharmacological properties, such as bioavailability, low toxicity, good pharmacokinetics, and lack of significant side effects.

Our discussion of U-74,006F will follow the basic concepts outlined above; after exploring its chemistry and that of related compounds, we will present evidence as to its in vitro antioxidant activity. We will also discuss its behavior in model membrane systems and show that its presence in low concentrations produces significant changes in the physical properties of phospholipid monolayers and bilayers and that these changes are conducive to decreasing rates of lipid peroxidation.

III. CHEMISTRY, IN VITRO EFFECTS, AND KINETICS OF INHIBITION OF LIPID PEROXIDATION BY AMINOSTEROID ANTIOXIDANTS

A. Structures and Chemistry of U-74,006F and Congeners

One of the most important features of a compound designed to be an inhibitor of lipid peroxidation is its chemical structure. As discussed, the compound should be sufficiently amphiphilic to localize at the proper depth in the mem-

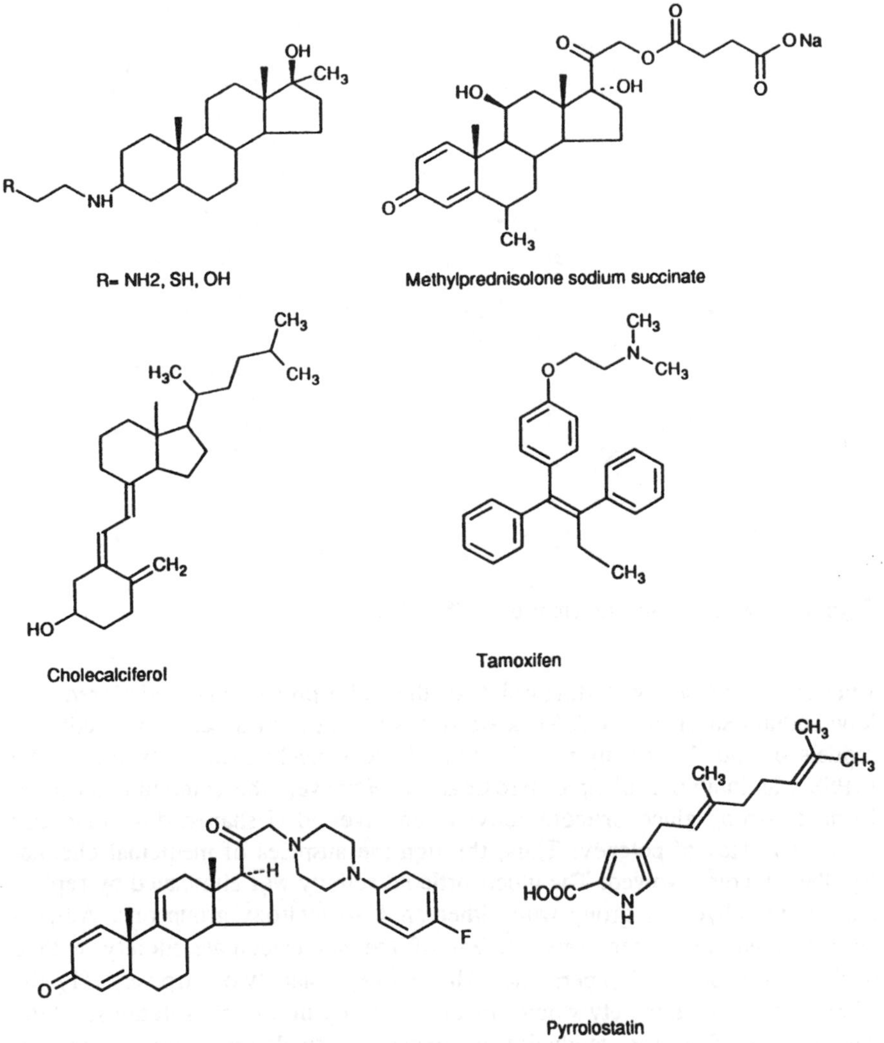

R= NH2, SH, OH

Methylprednisolone sodium succinate

Cholecalciferol

Tamoxifen

Pyrrolostatin

Figure 2 Structures of antioxidants that should have a significant physical component to their activity.

brane and have at least moderate radical scavenging capability. In this section we will discuss the rationale for the design, side effects, chemical reactivity, and ionization behavior of U-74,006F.

1. Structures

The structures of U-74,006F, congeners, and other antioxidants are shown in Fig. 1, and the x-ray crystal structure of U-78,517F is shown in Fig. 3. The conceptualization for the structure of U-74,006F actually evolved from two

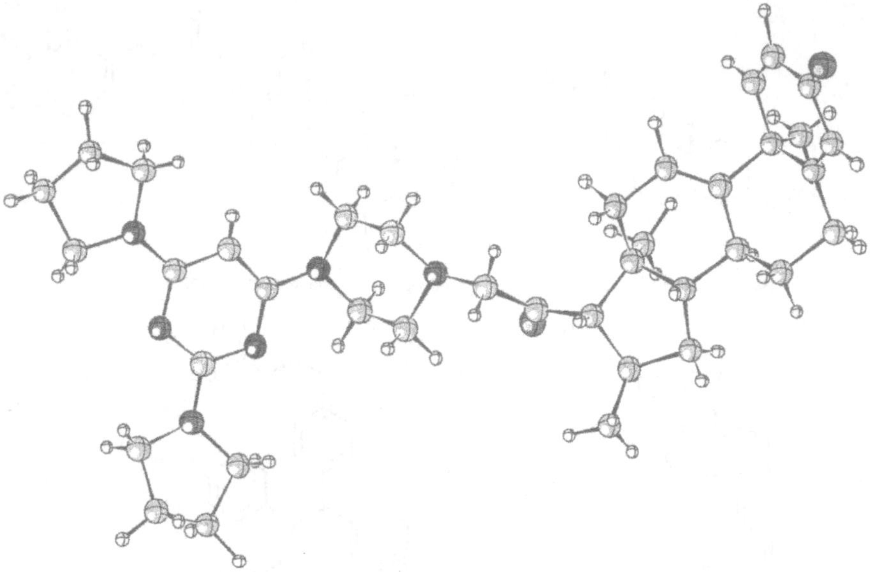

Figure 3 X-ray crystal structure of U-78,517F.

venues: The chemistry starting point for the entire project was methylpredniso-lone sodium succinate (MPSS), a steroid which at high doses is protective in models of spinal cord injury. The hypothetical mechanism of protection by MPSS was inhibition of lipid peroxidation. However, the compound suffered from its strong glucocorticoid activity, an inverted U-shaped dose response curve, and lack of potency. Thus, through the auspices of medicinal chemis-try, the structure evolved: The glucocorticoid activity was eliminated by replac-ing the 11-β-hydroxy group with either an 11-α-hydroxy group or a Δ-9,11-double bond, and later, replacement of the hemisuccinate moiety with a pyrimidine-substituted piperazine. This strategy had two purposes: (1) the triaminopyrimidine moiety would impart solubility in low pH solutions, which was desirable from the standpoint of injectable formulations, and (2) the tria-minopyrimidines were already demonstrated to be good antioxidants (8). Since the importance of lipid peroxidation in spinal cord and other severe CNS in-juries was well known, the introduction of the antioxidant moiety into the molecule was logical.

2. Protonation

Perhaps one of the more important properties of U-74,006F is its potential to be protonated; this property affects solubility, free-radical reactivity, and the possible formation of ionic complexes with negatively charged biological molecules. Protonation of U-74,006F was studied by potentiometric titration in an ethanol/water mixture which gave solution pK_a's of 4.2 and 6.5 upon extrapolation to 0% ethanol. Analysis of the pH-solubility profile gave a pK_{a1} of 4.6 with high uncertainty since the low solubility of the free base limited the study to a narrow pH range. By spectrophotometric methods, pK_{a2} was estimated to be 6.5, but pK_{a1} could not be measured. The pK_{a1} value was determined using ^{13}C NMR. Two ^{13}C signals showed shifts $> 1\delta$ over a pH range of 1.5 to 6.5, and these were assigned to the C-21 and C-2',6' position of the piperazine ring. These results, although not directly extrapolatable to surface pK_a's in a membrane, indicate that U-74,006F will be at least partially protonated and thus positively charged in vivo and, hence, capable of forming ionic complexes with negatively charged moieties, such as phospholipids, in membranes.

B. In Vitro Antioxidant Activity of U-74,006F and Congeners

1. Reactions of U-74,006F and Congeners with Radicals

As discussed, the quantitation of the antioxidant activity of a given compound depends, to a large extent, on the experimental system. U-74,006F was shown to inhibit lipid peroxidation in many models, although the quantitation of this inhibition by U-74,006F and its congeners is further complicated by extremely low solubility in water, which makes it difficult to assess the exact concentration at the site of action. In most experimental systems, U-74,006F shows a lower reactivity than electron-rich phenolic antioxidants such as vitamin E and U-78,517F, but higher than methylprednisolone. In organic solvents, U-74,006F can scavenge free radicals: It reacts with valeryl hydroperoxyl radical derived from azobis(2,4-dimethylvaleronitrile) to form an unusually stable oxyradical at the pyrimidine C-5 position. The hyperfine splitting constants of the ESR (electron spin resonance) spectrum of this pyrimidine peroxyl-like radical are fully consistent with the ones expected from the 5-pyrimidinoxyl radical (9). Also, in an aqueous environment, U-74,006F reacts with ferrous ammonium sulfate, hydrogen peroxide, or benzoyl peroxide to produce, respectively, 5-hydroxypyrimidine, or 5-O-benzoyloxypyrimidine derivatives. These findings establish the reactivity of U-74,006F's ring toward oxygen-derived radicals and delineate the preferred position of attack on the pyrimidine at ring carbon 5.

2. Antioxidant Activity of U-74,006F and Congeners in Cells and Animal Models

Inhibition of lipid peroxidation by 21-aminosteroids, including U-74,006F, is also prominent in physiologically relevant model systems. The inhibition was first demonstrated in rat brain homogenates with ferrous ammonium sulfate (FAS) as the free-radical generator. Using the decrease in the amount of TBARs (thiobarbituric acid reactive substances) formed for assessing inhibitory activity, the 21-aminosteroids were found to be inhibitors of iron-dependent lipid peroxidation in rat brain homogenates, with IC_{50}'s as low as 3 μM (10). As assessed by inhibition of TBAR formation, isolated liver microsomes were also protected from FAS (ferrous ammonium sulfate)-induced oxidative injury by U-74,006F delivered from ethanol. The apparent IC_{50} was 3.8 μM when apparent drug concentrations were considered as the analytical concentration. U-74,006F gave an apparent decreased IC_{50} of 0.68 μM when delivered in a lipid emulsion. The emulsion delivery probably enhanced the potency because of the formation of a colloidal microprecipitate upon addition from ethanol. Heat inactivation of the rat liver microsomes did not impair the activity of U-74,006F (11).

In a different experimental system, fetal mouse spinal cord neuronal cultures were treated with 200 μM FAS for 30 min to peroxidize lipids. Membrane damage and cell viability were assessed by measuring uptake of ^{3}H-AIBA (aminoisobutyric acid). Several compounds protected significantly the AIBA uptake, with the rank order of protection U-83,836E > U-74,500A > U-74,006F. U-83,836E is a single enantiomer of U-78,517F. U-74,006F at a nominal concentration of 5 μM also protected murine neocortical cell cultures against substantial neuronal degeneration and overt glial damage that was produced by 24-h treatment with 50 μM ferric and 50 μM ferrous iron (12). In the same cultures, combined oxygen and glucose deprivation for 45–60 min (ischemia), produced neuronal degeneration which was attenuated by U-74,006F. The combination of a 21-aminosteroid plus dextromethorphan, an NMDA (N-methyl-D-aspartate) antagonist, produced a greater benefit than either drug alone.

Iron-dependent lipid peroxidation was also inhibited in microsomes that were insulted with ADP:Fe (III) and NADPH by both U-74,500A and U-74,006F. This inhibition was dependent on both NADPH and Fe(III) and was enhanced by preincubation with U-74,006F (13). These results indicate that U-74,500A inhibits lipid peroxidation by directly affecting iron redox chemistry, whereas U-74,006F-mediated inhibition did not involve iron chelation or changes in iron redox chemistry. U-74,500A also inhibits, at concentrations as low as 1 μM, copper- and H_2O_2-induced erythrocyte and plasma lipid peroxidation. The inhibition of erythrocyte lipid peroxidation was accompanied by an inhibition of hemolysis (14).

U-74,006F inhibited iodoacetic acid–induced injury of cultured human astroglial cells (UC-11MG), as measured by changes in ^{3}H-AIBA uptake. Four-hour exposure to 50-μM IAA (iodoacetic acid) totally inhibited glycolysis and led to irreversible breakdown of cellular membranes and, ultimately, to cell death. At the beginning of the injury, IAA rapidly depleted cellular levels of ATP and decreased active uptake of ^{3}H-AIBA. Subsequent irreversible cellular injury was characterized by the release of large amount of free arachidonic acid into the extracellular medium, massive calcium influx, and leakage of cytoplasmic contents (^{51}Cr release). The appearance of 15-hydroxyeicosatetraenoic acid in membrane phospholipids and loss of cellular thiol groups indicated oxidative stress, which was alleviated by U-74,006F. The IAA-induced release of tritiated arachidonic acid was inhibited with an apparent IC50 of 6 μM. U-74,006F was effective even when it was administered 1 h after the onset of the metabolic insult (15).

The cytoprotective properties of U-74,006F were also evaluated in a system of cerebellar granular cells exposed to peroxynitrite. Cytoprotection by U-74,006F was assessed by measuring preservation of ^{3}H-AIBA uptake and decreases in lipid hydroperoxide and nitrotyrosine formation. U-74,006F was found to be an effective cytoprotectant ($EC_{50} = 100$ μM) even as a posttreatment following exposure of cells to peroxynitrite, while post-treatment with superoxide dismutase (50 units/ml) and allopurinol (100 μM) were not protective (16).

Lipid peroxidation in tissues can be followed by using F_2 isoprostanes that are produced by free-radical peroxidation of arachidonate as markers. Infusion of isoprostanes into the kidney reduces renal blood flow and glomerular filtration rates, suggesting that these metabolites may partly account for the hemodynamic alterations seen in free-radical-linked acute renal injury models. LLC-PK1 (renal tubular) cells in confluent cell layers were incubated with 1.5 mM H_2O_2 in the presence or absence of 20 μM U-74,389G for 3 h. This concentration of hydrogen peroxide increased TBARS from 0.07 to 0.42 nmol, and the presence of U-74,389 reduced TBARS back to 0.08 nmol. Similarly, treatment with U-74,006F reduced cell injury as evidenced by reduction of ^{51}Cr release. F2 isoprostanes were produced in a dose-dependent manner by H_2O_2 and their production was inhibited by U-74,006F (17).

3. Preservation of Stored Erythrocytes

The 21-aminosteroid U-74,500A contributes to the preservation of organs prior to transplant, probably by effects on the microvasculature (18,19). Oxidative damage, either to lipids or to proteins, is a major factor contributing to red cell senescence (20–22). This suggested that U-74,006F might prevent, or at least delay, the membrane-related changes which occur in stored red blood cells, the most "transplanted" biological material. Indeed, U-74,006F prevented membrane-associated damage of RBCs (red blood cells) resulting from prolonged storage under blood bank conditions (23), as assessed by several membrane-associated parameters, such as hemoglobin leakage, osmotic fragility, and markers of lipid peroxidation. Several metabolic parameters not expected to be sensitive to the presence of the drug did not show any change. As shown in Fig. 4, there was a dramatic change in the osmolytic profile from that of freshly collected RBCs to that in 60-day-old samples. Analysis of the lysis curves in terms of the error function (erf, integrated Gaussian distribution) showed no age-dependent differences in the midpoints of the hemolysis curves, but the degree of lysis measured at low hypoosmolality was much larger for the aged erythrocytes than that measured for the freshly collected erythrocytes. Thus, during storage, the osmotic fragility of changes in a discontinuous manner: Instead of all the erythrocytes becoming more susceptible to lysis, the only change was the progressive formation of a subpopulation of cells by a direct change from a low-susceptibility state to a high-susceptibility state. The aging cell population was thus composed of apparently healthy, "low NaCl-lysing" robust cells and damaged, "high-lysing" fragile cells. Addition of U-74,006F during storage decreases significantly the population of fragile cells.

The appearance of F_2 isoprostanes, a series of oxygen-radical-derived prostaglandins which are formed nonenzymatically from arachidonic acid located in the cell membranes was also measured for the erythrocytes after 60 days of storage. Two different species of F_2 isoprostanes were quantified: 8-*epi*-PGF$_{2\alpha}$ and PGF$_{2\alpha}$. We found that the aging of the erythrocyte is accompanied by the production of a large amount of these lipid peroxidation products and that U-74,006F mitigates this degradative process in a dose-dependent manner.

The foregoing experimental results show that U-74,006F and congeners display significant antioxidant activity in all the experimental biological systems

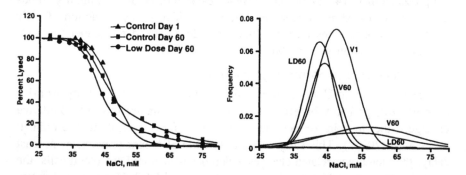

Figure 4   Hypoosmotic lysis curves for erythrocytes. (Left:) Vehicle control, day 1 (Δ---Δ); vehicle control, day 60 (□---□); low dose, day 60 (○---○). (Right:) Deconvolution of the theoretical curves from the left.

where they were examined. The quantitative expression of this activity depends critically on the system used. The principal cause of this apparent discrepancy may reside partly in the delivery system and the lack of information as to the concentration of the drug in the particular membrane compartment where the experimentally measured lipid peroxidation occurs. While all the results indicate that lipid peroxidation is the principal target of protection by U-74,006F, the experimental systems produce many different secondary effects which render these systems awkward for studying the mechanistic details of protection. Thus, simpler models which would be more amenable to gaining information concerning antioxidant mechanisms were sought.

C. Kinetics in Experimental Free-Radical Reaction Systems

The chemistry and kinetics of the minimum reaction pathway for free-radical scavenging by an antioxidant have been reviewed (24). Whether inhibition of liquid peroxidation is by chemical or physical means, the extent of inhibition is always measured by the decrease in the steady-state free-radical concentration. For monitoring free-radical concentrations, one of the most sensitive and simple experimental systems consists of a fluorescent indicator that reacts with the free radicals to yield a nonfluorescent compound at such a slow rate as not to deplete significantly the free radicals. Parinaric acid incorporated into phospholipid bilayers proved to be one such indicator (25). In this system, the steady-state concentration of the bisallylic free radicals, $B^{\cdot}$, can be calculated from the rate equation

$$\frac{dB^{\cdot}}{dt} = v_i - k_t B^{\cdot 2} - k_l L_o B^{\cdot} \tag{6}$$

where L_o is the concentration of the free-radical scavenger, k_l is the rate constant of free-radical scavenging, and v_i is the constant rate of initiation from an external source of free radicals, such as cumene hydroperoxide, hydrogen peroxide, t-butylhydroperoxide, etc.

The free-radical concentration can also be monitored in a system that is perhaps less than ideal from the theoretical point of view but very simple experimentally. This system consists of a colorimetric titrant for measuring the quantity of lipid peroxide formed as a function of time. This experimental method suffers from the fact that the lipid peroxides formed are ultimately unstable and that the concentration of oxidizable lipids influences the steady-state concentration of free radicals. For these reasons, the use of such systems is by and large limited to measuring initial rates of lipid peroxidation.

We wish to show that U-74,006F and other 21-aminosteroids are inhibitors of lipid peroxidation that, by virtue of their lipophilicity, partition preferentially and specifically into phospholipid bilayers. U-74,006F is a cone-shaped molecule; it compresses selectively the headgroups of phospholipids and disorders their acyl chains. In addition to these physical effects, the triaminopyrimidine portion of the molecule also acts as a free-radical scavenger. Thus, the antioxidant action of this compound is a judicious combination of the chemical and physical mechanisms by which lipid peroxidation can be inhibited. The free-radical-scavenging properties of U-74,006F and congeners were determined in two systems designed to measure the intrinsic ability of the compounds to react with the lipid free radicals.

1. Erythrocyte Ghosts

The intrinsic free-radical-scavenging properties of U-74,006F and other putative antioxidants were determined at the membrane level by examining the kinetics of oxidation of cis-parinaric acid in erythrocyte ghosts (24). The probe cis-parinaric acid (PnA), cis-trans-trans-cis-9,11,13,15-octadecatetraenoic acid, is fluorescent with a high quantum yield when partitioned into the lipid bilayer of erythrocyte ghosts and its fluorescence is destroyed upon reaction with free radicals. The time-dependent loss of fluorescence was then observed after initiation of free-radical generation by cumene hydroperoxide and cupric ion. As shown in Fig. 5, the disappearance of PnA fluorescence accelerated in the early phases of the reaction, and past the inflection point became purely first order. The predominance of the latter phase indicated that the concentration of free radicals remained in a steady state throughout most of the reaction. With this condition—and assuming that the spontaneous photobleaching of parinaric acid is negligible—one can show that, in the presence of inhibitor at the concentration of L_o with the rate constant for initiation, k_i, the experimental rate constant of the first-order portion of the reaction, k_{exp}, is

$$k_{exp} = \frac{k_p}{2k_t}\left(\sqrt{4k_ik_tC_oM_o + k_i^2L_o^2} - k_iL_o\right) \qquad (7)$$

where v_i of Eq. (6) is replaced by $k_iC_oM_o$, to reflect the fact that the initiation occurs by a metal-ion-catalyzed (M_o) decomposition of cumene hydroperoxide C_o. When the antioxidant is absent, i.e., $L_o = 0$, the experimental rate constant for the control reaction, k^*_{exp}, may be defined as

$$k^*_{exp} = \frac{k_p}{k_t}\sqrt{k_ik_tC_oM_o} \qquad (8)$$

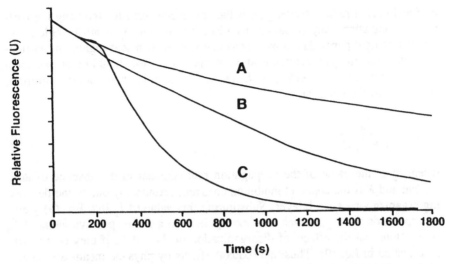

Figure 5 Decay of *cis*-parinaric acid fluorescence. The time-dependent fluorescence decay of *cis*-parinaric acid in the presence or absence of cumene hydroperoxide (1.7 mM) and cupric ion (20 µM). Cuvettes contained erythrocyte ghosts (250 µM phospholipid) ± antioxidant (added in a small amount of ethanol in a final volume of 0.01 M phosphate, 150 mM NaCl, pH 7.4, after 2 min of preincubation). Individual curves are as follows: curve A, blank, no cumene hydroperoxide; curve B, full reaction mixture + U-78,517F (3.8 mole%); curve C, full reaction mixture.

Substituting k^*_{exp} into Eq. (7), we obtain an expression for k_{exp} which may be used for assessing the antioxidant activity of added L_o:

$$k_{exp} = \frac{k_p k_l}{2k_t}\left(\sqrt{L_o^2 + \left[2\frac{k_t}{k_p k_l}k^*_{exp}\right]^2} - L_o \right) \tag{9}$$

This form is suitable for analysis of the dose-response curves by a nonlinear least-squares method. The analysis yields the parameter $k_p k_l/2k_t$, which provides a relative measure of the intrinsic reactivity of the antioxidant in a free-radical-scavenging reaction. Also, measuring k_{exp} as a function of initiator concentration should yield a typical square-root dependency [Eq. (8)]. For the parinaric acid system used, the consistency of the data with Eq. (8) shows that the free-radical-mediated lipid peroxidation is indeed the major process operative in this system.

Equation (9) predicts that scavenging by an antioxidant is a saturable process: The dose-response curve k^*_{exp} versus L_o should extrapolate to 0 at $L_o =$

∞ if and only if radical scavenging is the sole mechanism for reducing the free-radical population. Any deviation from Eq. (9) presumably is attributable to inhibition of lipid peroxidation by mechanisms other than scavenging. In particular, a decrease in k_p by disorganization of the lipid phase should be proportional to L_o. In that case, the effect of L_o on k_p is a simple, hyperbolic inhibition, according to the equation

$$k_p = \frac{k_p^u}{1 + L_o/K} \tag{10}$$

where k_p^u is the value of the propagation rate constant in the absence of antioxidant and K is the apparent inhibition constant, presumably due to the decreasing organization of the bilayer. Substituting this value of k_p into Eq. (9) yields an expression for k_{exp}, which now extrapolates to a finite, positive value at $L_o = \infty$. Note that the effects of disorganization on k_i and k_t, if they occur, will be canceled in Eq. (9). Thus, antioxidant effects by physical means are clearly distinguishable by the shape of the k_{exp} versus L_o curves: In Fig. 6, we show

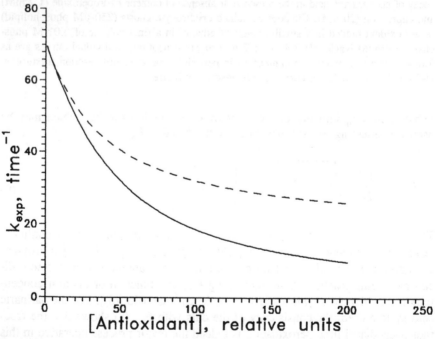

Figure 6 Theoretical curves for inhibition of lipid peroxidation by chemical and physical mechanisms. The theoretical curves for chemical and physical inhibition were generated based on Eq. (9). Solid line—chemical inhibition of lipid peroxidation; dashed line—physical contribution to inhibition.

theoretical curves based on Eq. (9), with and without physical inhibition. The dose-response curve of U-78,517F shows a definite residual k_{exp} value at high L_o (Fig. 7), and this residual rate constant is significantly higher than that of the spontaneous photobleaching reaction of parinaric acid. This then indicates a significant contribution of membrane disorganization to the antioxidant activity of this compound. It could appear paradoxical that a supplemental inhibition results in a higher, instead of a lower, value of the k_{exp}. In fact, k_{exp} measures the steady-state concentration of the free radical, and inhibition of the propagation rate should indeed result in an accumulation of unreacted free radicals.

A number of our synthetic antioxidants were tested and their $k_p k_1/2k_t$ values were compared to that of vitamin E (Table 1). In our system, U-74,006F shows only a moderate free-radical-scavenging activity, 10% that of U-78,517F and 5% that of vitamin E. Of course, moderate scavenging activity also means a longer persistence in the membrane and, thus, a more prolonged membrane-modifying action.

The integrated rate equation for the parinaric acid system describes the full time dependency, including that of the pre–steady state:

$$P = P_o e^{[(k_p(\alpha - k_l L_o)/2k_t)t - k_p/k_t \ln((1 - be^{\alpha t}i)/(1 - b))]} \tag{11}$$

where P is the concentration of parinaric acid:

$$\alpha = \sqrt{4k_i k_t C_o M_o + k_l^2 L_o^2} \tag{12}$$

and

$$b = \frac{k_l L_o + \alpha}{k_l L_o - \alpha} \tag{13}$$

Table 1 Reactivity of Antioxidants in the Erythrocyte Membrane Toward Cumene Hydroperoxide Radicals Relative to That of Vitamin E

Compound	Relative efficiency
Vitamin E	1.00
U-78,517F	0.54
U-78.518E	0.14
U-74,500A	0.05
U-74,006F	0.05
U-75,412E	0.04

Source: From Ref. 24.

The time course of the reaction, with and without U-74,006F, was shown to be consistent with this equation (Fig. 8). Nonlinear least-squares analysis then allowed us to determine the value of the individual rate constants, k_I in particular. In summary, these experiments demonstrate that, in erythrocyte membranes, lipid peroxidation is a chain reaction and its inhibition by U-74,006F occurs by free-radical scavenging with contribution from physical effects.

2. Multilamellar Vesicles

The parinaric acid/erythrocyte ghost system is intrinsically heterogeneous and, thus, does not allow one to assess the importance of membrane localization and orientation within the lipid bilayer to the scavenging activity. Moreover, the system is not conducive to determining the intrinsic radical-scavenging reactivity of large numbers of putative antioxidants. Therefore, a simpler system was developed (26) to compare the relative free-radical-scavenging activity of compounds toward radicals generated from DLPC (dilinoleylphosphatidylcholine)— a chemically well-defined peroxidizable lipid substrate—both in solution and in a lipid bilayer. In the homogeneous system, the rate of lipid peroxidation with and without inhibitors was measured by quantitating by HPLC the linoleate

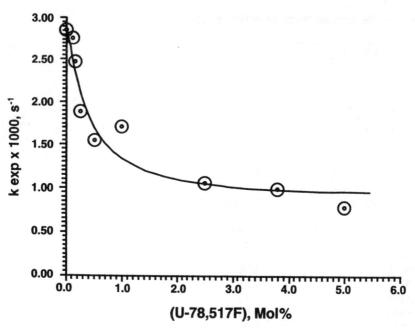

Figure 7 Dose response of the inhibition of *cis*-parinaric acid decay by U-78,517F. The k_{exp}'s of the inhibition by U-78,517F of the time-dependent loss of *cis*-parinaric acid fluorescence in the presence of free radicals are plotted according to Eqs. (9) and (10).

hydroperoxides produced after initiation with AMVN (2',2'-azobis[dimethylval-eronitrile]). In the lipid bilayer system, AAPH (2', 2'-azobis[2-amidino-propane])-initiated peroxidized lipids were quantitated by the color change they produced in a reaction with xylenol orange (27).

As shown, for a steady-state system with a constant initiation rate, constant O_2 levels, with an antioxidant concentration L_o, the apparent first-order rate constant of LOOH formation is given by Eq. (9), after redefining k^*_{exp} as

$$k^*_{exp} = \frac{k_p}{k_t} \sqrt{k_i k_t I_o} \tag{14}$$

where I_o is the initiator concentration (AMVN or AAHP). In the solution system, we chose experimental conditions to measure the linear, initial rate of the reaction, v_o, which is proportional to k_{exp}, since $v_o = k_{exp}$ [Linoleate]$_o$. Equation (9) was then used for analyzing, by nonlinear least-squares fit, v_o as a function of I_o and L_o to yield the parameter $k_i k_p/k_t$, the "relative inhibitory rate constant." After having ascertained that the AMVN concentration dependence of k_{exp} obeyed Eq. (14) (Fig. 9), we showed that U-74,500A inhibits LOOH

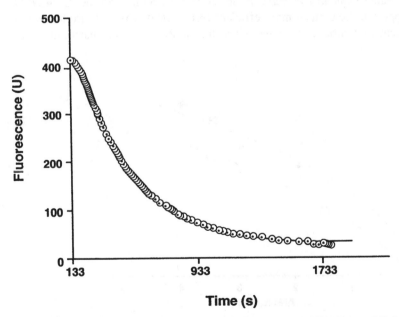

Time (s)

Figure 8 Analysis of the full time course of the time-dependent loss of *cis*-parinaric acid fluorescence induced by free radicals. The digitized data from Fig. 5, curve C, were analyzed according to Eq. (11) by a nonlinear least-squares program. The continuous line represents the theoretical fit to the data points.

formation according to Eq. (9), with the relative inhibitory rate constant having a value of 0.66 ± 0.10 μM^{-1} min^{-1}. We also showed that as the rate of formation of LOOH was inhibited, the inhibitor was also progressively lost from the reaction mixtures in a first-order manner. Thus, we may conclude that, in the solution system, U-74,500A is a relatively effective free-radical-scavenging inhibitor of lipid peroxidation.

After having characterized inhibition in solution, we could then compare it to that in DLPC (dilinoleylphosphatidylcholine) MLVs (multilamellar vesicles). The k_{exp}'s calculated from the initial rates of LOOH formation by the AAPH-initiated peroxidation of DLPC MLVs in the presence of increasing amounts of U-74,500A were also found to be consistent with Eq. (9) (Fig. 10). The agreement between the experimental data and the theoretical curves shows that the proposed kinetic model is sufficient to account for the observations. A comparison of the relative inhibitory constants of three inhibitors is given in Table 2: U-78,517F was, by far, the best free-radical scavenger, followed by U-74,500A and U-74,006F. This rank order of reactivity parallels that observed in the parinaric acid assay (Table 1), indicating that the relative scavenging reactivity is independent of the microenvironment within the lipid bilayer.

At the same *nominal concentration* in the two reaction mixtures, U-74,500A would appear to be a much more efficient free-radical scavenger in the vesicle system than in solution. However, since the inhibitor is fully partitioned into

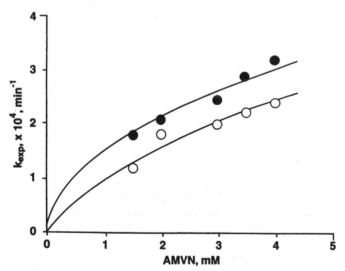

Figure 9 Effect of AMVN concentration on the kinetics of linoleate hydroperoxide formation in methanolic solution. Free-radical reactions were initiated using various concentrations of AMVN, and hydroperoxides were quantitated by HPLC. Control (solid circles); +50 μM U-74,500A (open circles).

Table 2 Inhibitory Constants for Antioxidants

Compound	$k_p k_I / 2k_t \cdot 10^4$ (mol%$^{-1}$ min^{-1})
U-78,517F	771 ± 107
U-74,500A	66 ± 13
U-74,006F	2.6 ± 0.5

Source: From Ref. 25.

the lipid bilayer, its reactivity must be calculated on the basis of its local concentration. On that basis, the scavenging efficiency is much less in the bilayer than in solution. From Eqs. (7) and (8) one can show that, when the rate of initiation is kept constant, the IC$_{50}$ of the inhibitor, i.e., the value of L_o at which $k_{exp}/k_{exp}^* = 0.5$, is proportional to $\sqrt{k_t/k_I}$ but independent of k_p. Thus, compared to the solution, the bilayer promotes chain termination over scavenging. The reason for this apparent loss of reactivity of the antioxidant should be sought in the strict alignment of the bisallytic groups in the bilayer, which favors chain termination by recombination of lipid peroxy free radicals. Thus, only a scavenger concentrated at the precise location of the bisallylic groups could maintain its efficacy displayed in solution.

Perhaps more importantly, the dose-response curves displayed in Fig. 10 do not show a finite k_{exp} asymptote. Thus, unlike in the erythrocyte membrane system, in the model MLV system inhibition by physical means is not an important contributor to the repression of lipid peroxidation. This difference in kinetic behavior must have its root in the high fluidity of the DLPC MLVs as

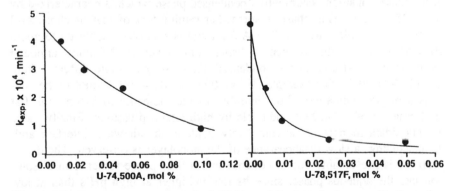

Figure 10 Dose-response curves for inhibition of linoleate hydroperoxide formation in DLPC MLVs. The rates of LOOH formation in DLPC MLVs were determined as a function of U-74,500A and U-78,517F. The data were analyzed using Eq. (9).

compared to that in erythrocyte ghosts and the composition of pure lipids in the model system as opposed to the presence of a large portion of proteins and cholesterol in the erythrocyte membrane. Thus, in the pure lipid system, diffusion of the lipid free radicals is not a critical contributor to the value of k_p.

IV. PHYSICOCHEMICAL MECHANISMS

The preceding sections showed that, in biological membranes, the antioxidant efficacy of U-74,006F is attributable only partially to its free-radical-scavenging ability and that inhibition by physical means must be an important part of its mode of action. In the next sections we will discuss what kind of physical effects enter into play when U-74,006F is inserted into the membrane.

A. BEHAVIOR OF U-74,006F IN MONOMOLECULAR FILMS (28)

The simplest experimental system for investigating the physical properties of amphiphilic agents is the insoluble monomolecular layer at the air/water interface. This type of study yields information concerning the orientation, the cross-sectional area, the rigidity, and the compressibility of the amphiphile, and the flow properties of the lipid layer. Being amphiphilic, U-74,006F forms at the air/water interface a relatively stable monolayer at neutral pH's where the molecule is electrically neutral. Of course, at acidic pH's, the protonated form is too water-soluble to form a stable insoluble monolayer. Figure 11 shows the surface isotherm of U-74,006F at pH 7.4; at low compressions, the curve is typical for a two-dimensional gas, and analysis of the isotherm yields a molecular exclusion area of $A_o = 73.5 \pm 1.4$ Å^2/mol. At $\pi = 10.4$ dynes/cm a first-order phase transition occurs into a condensed phase which is characterized by $A_o = 35$ Å^2/mol. This phase is somewhat reminiscent of that of cholesterol which has a limiting area of 40.5 Å^2/mol and is relatively stable up to about 50 dynes/cm. All other congeners display similar surface behavior, although the cross-sectional area of the condensed phases may be quite different from that of U-74,006F. For example, U-74,500A has $A_o = 60$ Å^2/mol in the condensed phase, indicative of the inability of the molecule to assume the flat conformation of U-74,006F inducible by high lateral pressures. Finally, none of antioxidant monolayers is completely stable at the air/water interface and, upon standing, a slow disappearance of the monolayer is observed. The disappearance is presumably due to collapse into a lipid phase rather than dissolution into the aqueous phase, since its rate is higher at high pH's than at low pH's.

The x-ray crystal structure of U-78,517F sheds some light on the possible orientation of this molecule in monolayers. Figure 2 shows that the most stable

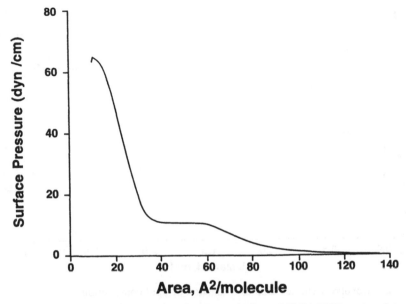

Figure 11   Force area isotherm for the spreading of U-74,006F at the air/water interface.

conformation of the U-78,517F is a V-shaped structure where the two hydrophobic wings are linked by the piperazine ring. The polarity of the latter is somewhat comparable to that of the bis-pyrollidinopyrimidine wing of U-74,006F. Because of this ambiguity, locating this molecule at the interface at low surface pressure should result in an orientation where the piperazine ring is fully hydrated and the two wings are in the hydrophobic layer. This is the high cross-sectional orientation. At high compression, however, the molecule could be reoriented into a low-cross-section position where the pyrimidine ring is hydrated and the rest of the molecule is located in the lipid layer. The x-ray structure of U-74,006F (not shown) indicates a more open V which should normally lead to a preferential orientation perpendicular to the surface.

U-74,006F also forms a mixed monolayer with the liquid phases of long-chain lecithin monolayers such as EYPC (egg yolk phosphatidylcholine). As shown in Fig. 12, the molecular areas are nonadditive; that is, at a given surface pressure, the area of the mixed layer is greater than the sum of the areas of its constituents at the same surface pressure. In other words, U-74,006F has both a thinning and an expanding effect on lipid membranes, and this in striking contrast to the condensing effect of cholesterol. In a liquid lipid bilayer, thinning of the membrane entails disorganization, both within each leaflet and between leaflets. This disorganization should, then, decrease the efficiency of free-radical propagations.

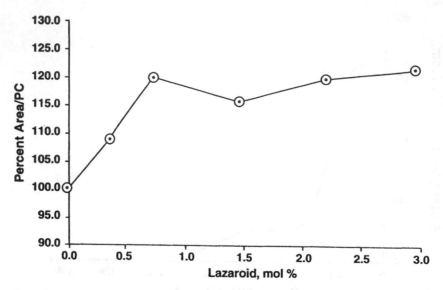

Figure 12   Monolayer surface area as a function of lipid composition.

Since disorganization of the lipid layer leads to decreased interaction between the individual lipid molecules, the effect of U-74,00F on phospholipid mono-layers is also evidenced by a decrease in the surface sheer (flow) viscosity of the mixed monolayer (Fig. 13). This contrasts with the action of cholesterol which is to increase shear viscosity of the monolayer. Again, the effect of the other 21-aminosteroid antioxidants is qualitatively comparable to that of U-74,006F.

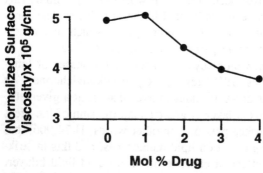

Figure 13   Effect of U-74,006F on the surface viscosity of EYPC monolayers.

B. Differential Scanning Calorimetry (DSC) of U-74,006F and Congeners in DPPC MLVs

The study of monolayers, although informative as to the orientation and fluidity of the antioxidants at the oil/water interface, is unable to show possible effects of the antioxidant on the interdigitation of the two leaflets of the bilayer. The physical study of the bilayer is best accomplished using DSC. In this way, it has been shown that cholesterol, which contains an isoprenyl side chain, forms interlamellar tail-to-tail dimers in DPPC (dipalmitoylphosphatidylcholine) bilayers (29). This type of dimer then should counteract the lateral displacement of one leaflet with respect to the other. Before DSC studies of our antioxidants could be undertaken, we had to ascertain that the molecules did, in fact, partition into the DPPC bilayer. This proved to be the case (Table 3). Addition of U-74,006F to DPPC MLVs lowers the T_m of the phospholipid bilayer, as shown in Fig. 14. Increases in the drug concentration results in progressive broadening of the transition peak and the lowering of T_m, but, in marked contrast with cholesterol, the antioxidants decrease the cooperative unit of chain melting which is proportional to the number of noncovalent bonds made cooperatively during the freezing process. U-74,006F, U-7,532, and vitamin E display relatively sharp initial drops in the sizes of the cooperative unit followed by a concentration region where the size of the cooperative unit is relatively independent of drug concentration. The antioxidant-induced decrease in the size of the cooperative unit persists even in the presence of 10 mole% cholesterol (29). Thus, it appears that all the drugs tested counteract the destabilizing effect of cholesterol on the l_d phase. The cause of the opposing effects must be sought in the different molecular shapes and the different locations of cholesterol and the drugs within the bilayer. Because cholesterol is shorter than the extended phospholipid molecule, its insertion into the hydrophobic lamella of

Table 3 Partition of Antioxidants into DPPC MLVs and Freezing-Point Depressions

Compound	Partition coefficient c_{pl}/c_w^a	M_r apparent	Aggregation state	M_r
U-74,006F	∞	1202 ± 101	Dimer	720.98
U-7532	1.68 × 10^6	610 ± 72	Dimer	374.48
Cholesterol	∞	715 ± 119	Dimer	386.64
Vitamin E	n.d.	280 ± 25	Monomer	430.69

$^a c_{pl}$ and c_w are the mol% drug in the MLVs and in water, respectively.
Source: From Ref. 28.

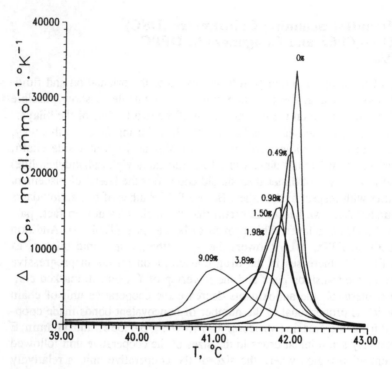

Figure 14 DSC thermograms for DPPC/U-74,006F mixed MLVs. DSC thermograms were recorded in the heating mode for MLVs composed of DPPC and indicated mole percent of U-74,006F.

the liquid-phase bilayer results in a weakening of the phospholipid headgroup/headgroup interactions. This effect is analogous to the decrease in the sheer or flow viscosity produced by U-74,006F in EYPC lipid monolayers. In contrast, the drugs are genuinely amphiphilic, with a large portion of the molecule be-

Table 4 Freezing-Point Depression by the Antioxidants in the Presence of Cholesterol

Compound	M_r	M_r apparent at 4 mol% chol.	M_r apparent at 10 mol% chol.	Aggregation state
U-74,006F	720.98	777 ± 506	388 ± 24	Monomer
U-7532	374.48	746 ± 249	791 ± 52	Dimer
			6	
Vitamin E	430.69	—	382 ± 118	Monomer

[a]Unpublished.

ing located near the phospholipid headgroups and their insertion may actually stabilize the packing of the headgroups.

The results of cryoscopic measurements in DPPC MLVs without cholesterol are shown in Table 3. U-74,006F and U-7,532 both yielded an apparent molecular weight twice that of the true one; i.e., the compounds are fully dimerized. This dimerization does not happen in the presence of 10 mole% cholesterol (Table 4). We propose that the dimerization of the antioxidants is an interleaflet one and it can only happen within bilayers of decreased thickness; cholesterol increases the thickness of the bilayer and thereby negates antioxidant dimerization.

C. Localization of Fluorescent Antioxidants in Lipid Bilayers

1. Acrylamide and Iodide Quenching of Fluorescent Groups

The action of antioxidants on the monolayer and bilayer properties shows the orientation and gross conformation of the antioxidant in the membrane, but does not show its exact location, i.e., its depth of penetration. For determining that property, fluorescence quenching is an ideal tool. Fortunately, vitamin E and three of our antioxidants, U-78,517F, U-78,518E, and U-75,412E, which are structurally similar to U-74,006F (see structures, Fig. 1), contain fluorescent groups. Fluorescence quenching with the penetrating quencher acrylamide, the ionic quencher iodide, and the depth-dependent doxyl fatty acids were then used to locate the antioxidants in egg lecithin/cholesterol bilayer vesicles (30). Energy transfer from the antioxidants to fatty acids with anthroyloxy groups attached to various carbons in the acyl chain was also used to localize the fluorescent moiety of the antioxidants.

The neutral quencher, acrylamide, penetrates to some extent into the lipophilic interior of membranes, whereas the ionic quencher, iodide, is confined to the aqueous phase. Acrylamide quenches all membrane-incorporated antioxidants (Fig. 15). The Stern-Volmer plots for U-78,517F and U-75,412E are linear and hence consistent with the full quenching of a single fluorophore, whereas that for U-78,518E curves upward, thereby indicating the presence of a static component in quenching. The quenching constants given in Table 5 show that U-78,517F is the compound most readily quenched by acrylamide and that the other two compounds are about equally quenched. Thus, the fluorescent moiety of U-78,517F is closer to the aqueous surface of the bilayer than those of the latter two compounds. But, since the quenching constants are relatively low, we conclude that the fluorescent moiety of each component is still imbedded in the bilayer and less accessible to the acrylamide. The conclusions reached from acrylamide quenching experiments were further tested using the ionic quencher iodide which senses fluorophores at least partially in the aque-

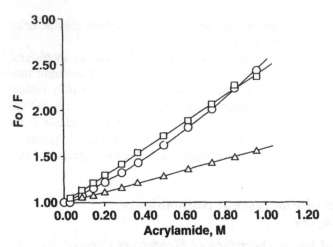

Figure 15 Acrylamide quenching of membrane-bound antioxidants. Data from experiments where membrane-bound antioxidants were quenched by acrylamide were analyzed with the Stern-Volmer equation. The symbols represent the experimental data points, and the solid curves the theoretical fits to the data. △---△, U-75,412E; ○---○, U-78,518E; ◇---◇, U-78,517F.

ous environment or in direct contact with the aqueous phase. Quenching by iodide is also sensitive to local electrostatic forces in the microenvironment of the fluorophore. As shown in Fig. 16, only U-78,517F showed simple, linear dynamic quenching which was consistent with the standard linear form of the Stern-Volmer equation, while the plots for the other two compounds had a downward curvature, indicative of multiple emitting fluorophores. The data were consistent with a model of one quenchable and one unquenchable fluorophore (Fig. 16). The parameters obtained from the fits to the iodide quenching models are given in Table 5.

Both U-78,518E and U-75,412E contain the ionizable piperazinylpyridinyl fluorophore, and acrylamide quenching showed that the two compounds are

Table 5 Quenching of Antioxidants in Phospholipid Bilayers

Quencher	Parameter	Antioxidant		
		U-75,412E	U-78,518E	U-78,517F
Acrylamide	K_{sv} (M^{-1})	0.58 ± 0.002	0.56 ± 0.12	1.44 ± 0.01
KI	K_{sv2} (M^{-1})	0.46 ± 0.06	0.70 ± 0.06	0.25 ± 0.01
	f_1	0.43 ± 0.05	0.33 ± 0.03	<0.05

Source: From Ref. 29.

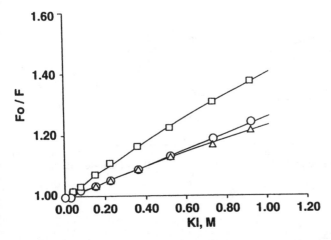

Figure 16 Iodide quenching of membrane-bound antioxidants. Iodide quenching of the intrinsic fluorescence of the membrane-bound antioxidants was performed and the data calculated by the Stern-Volmer equation. The symbols represent the experimental data points, and the solid curves the theoretical fits to the data. Δ---Δ, U-75,412E; O--O, U-78,518E; ◇---◇, U-78,517F.

located deep within the bilayer. Since acrylamide quenches about equally these two antioxidants, they must be located at about the same depth in the bilayer. While acrylamide sees them as single species, iodide distinguishes two forms, one of them inaccessible. It is then likely that the two species are the protonated and deprotonated forms of the piperazinylpyridinyl group and that the charged species is either located close to the interface or it forms an ion pair with the iodide ion in the hydrophobic region of the bilayer. Also, the fluorescent moiety of U-78,517F should be located so close to the aqueous surface that both its protonated and deprotonated forms are equally quenchable by acrylamide.

2. Quenching by Depth-Dependent Spin-Labeled Fatty Acids

The depth at which the fluorophores are located in the membrane can be estimated by using depth-dependent doxyl fatty acid quenchers. As shown in Fig. 17, this quenching is somewhat more complex than that observed with aqueous quenchers. The upward curvature of these Stern-Volmer plots is inconsistent with either dynamic or static quenching. The fact that the fluorescence emission is first increased by the addition of the quencher suggests formation of a strongly fluorescent "bright complex" between the doxyl fatty acid and the fluorophore. The experimental data are consistent (Fig. 17 and Table 6) with the equation describing this model:

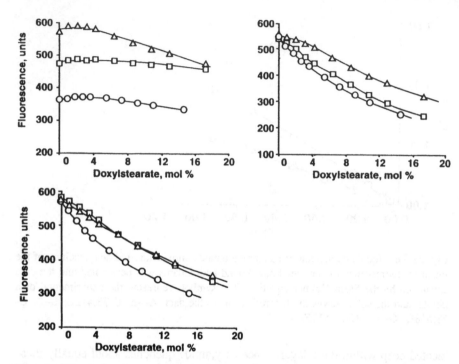

Figure 17 Quenching of membrane-bound antioxidants by depth-dependent doxyl-stearates. Quenching of the antioxidants was studied, and the data were fitted to Eq. (15). The symbols represent the experimental data points, and the solid curves the theoretical fits to the data. O---O, 5-doxylstearate; □---□, 7-doxylstearate; △---△, 10-doxylstearate. Top left: U-75,412E; top right: U-78,518E; bottom: U-78,517E.

$$F = \frac{L_o f_c}{(1 + K_d/[Q])(1 + [Q]K_c} + \frac{L_o f_L}{(1 + [Q]/K_d)(1 + [Q]K_L} \tag{15}$$

where F is the observed fluorescence, L and Q are the antioxidant and quencher concentrations, respectively, K_d is the dissociation constant of their complex, C is the concentration of that complex, K_c and K_L are the dynamic quenching constants for L and C, and f_C and f_L are the molar emissivities of the antioxidant and antioxidant/quencher complexes, respectively. The complex is most easily formed with U-78,518E, followed by U-78,517F and U-75,412E. Since all three compounds carry partially protonated amine functions, the complex is an ion pair. The ion pair is probably further stabilized by hydrogen bonding and π-complex formation. The increase in quantum yield of the bright complex could result from increased structural rigidity. In this regard, both 7-doxylstearic

Table 6 Quenching of Antioxidants by Doxyl Fatty Acids

Quencher	Parameter	U-75,412E	U-78,518E	U-78,517F
5-DS	K_L (l/mol%)	≈0.05	0.28 ± 0.09	0.18 ± 0.06
	K_C (l/mol%)	≈0.07	0.17 ± 0.03	0.17 ± 0.02
	K_d (mol%)	≈40	4.8 ± 1.9	16.6 ± 6.3
	f_c/f_1	3.8	1.9	2.9
7-DS	K_L (l/mol%)	≈0.7	0.29 ± 0.15	0.23 ± 0.02
	K_C (l/mol%)	≈0.07	0.17 ± 0.04	0.18 ± 0.006
	K_d (mol%)	≈25	4.3 ± 1.8	11 ± 0.6
	f_c/f_1	2.8	2.1	3.3
10-DS	K_L (l/mol%)	≈0.02	0.32 ± 0.09	20.07 ± 0.05
	K_C (l/mol%)	≈0.04	0.13 ± 0.01	0.09 ± 0.003
	K_d (mol%)	≈100	4.2 ± 1.1	≈30
	f_c/f_1	3.8	2.1	2.2

Source: From Ref. 29.

acid and stearic acid were found to increase the intrinsic fluorescence of the salt and free base forms of U-78,518E in cyclohexane solution.

The dynamic quenching of the free fluorophore by the membrane-bound quencher is measured by K_L. The values of K_L for U-75,412E and U-78,518E are consistent with the depth of the fluorophore indicated by the aqueous quenchers, but the K_L's are independent of the depth of the doxyl group. We interpret this to mean that, in the fluid bilayer, the polar doxyl groups within the immediate vicinity of the antioxidants are relocated in the plane of the bilayer independently of their position on the fatty acyl chain. The quenching of the complex by the free quencher follows the pattern observed for that of the free fluorophore, but K_C is smaller than the corresponding K_L. Thus, complex formation does not relocate the fluorophore of the antioxidant, but the acyl chain protects it from interaction with the quencher group.

3. Fluorescence Energy Transfer to Depth-Dependent Fluorescent Fatty Acids

Energy transfer between an emitting, donor fluorophore, and an acceptor depends on the distance, r, between the two transition dipoles. For energy transfer to occur this distance must be less than a critical value R_o, which may range from 10 to 70 Å, depending on the chemical nature of the compounds. In fluid bilayers, the efficiency of energy transfer will reach 100% at infinite donor concentration in the membrane since then at least one donor molecule will always be at $r < R_o$ from the acceptor. Thus, in bilayers, the energy transfer efficiency is dependent not only on the distance between the donor and the plane defining the average position of the acceptors but also on the concentration of donor molecules in this plane (31). Anthroyloxy fatty acids with their acceptor group at varying depths in the bilayer can therefore assess the depth of the antioxidant fluorophore.

The fluorescent chromanol moiety of vitamin E inserted into phospholipid bilayers has been probed by this method using n-AS (anthroylstearate) probes (32). The quenching depended on the depth of the probe, and it was proposed that the fluorophore was at a depth of about somewhere between carbon 2 and carbon 7 of the acceptor. However, at very low acceptor concentrations, energy transfer far exceeded the theoretical limit. These experimental data are consistent with a model where a reversible complex is formed between the vitamin E and the fatty acid acceptor. Thus, the depth of vitamin E measured by the probes is not that of the free molecule but that of the complex.

We observed similar behavior with our three antioxidants partitioned into mixed phospholipid/cholesterol bilayers (Fig. 18). These data are, again, fully consistent with the model where radiative transfer occurs via formation of a reversible donor/acceptor complex. The calculated best-fit parameters are given in Table 6. In all cases, f_c is $>> f_a$; that is, the radiative energy transfer within the complex is fully efficient. Complex formation was also observed with anthroyloxy fatty acid acceptors; the dissociation constants determined in this system are reported in Table 7.

Since the three antioxidants form complexed with fatty acids, they also should complex with acidic phospholipids, the ubiquitous constituents of cell membranes. This then may represent an important mechanism by which to enhance the ability of the antioxidants to capture free radicals: If the complex contains a polyunsaturated fatty acid, it is still able to participate in the free-radical chain reaction, but will readily pass the unpaired electron to the antioxidant. Moreover, once the fatty acid in the complex is oxidized, it will sterically hinder the reaction of the antioxidant free radical with other lipids.

Once in the bilayer, the three antioxidants can still remain in intimate contact with the aqueous phase, since this lipid moiety can be anchored by a protonated amino group. [The tertiary amino group of the piperazine ring has an estimated pK_a of 9 (33).] These amino groups link two hydrophobic moieties of about equal size which are located on the same side of the piperazine ring. The antioxidant thus assumes in the bilayer a V-shaped conformation, also observed in the x-ray crystal structure. Since the length of the hydrophobic portions is less than that of a fully extended stearate chain, the antioxidants must have a considerable effect on the packing of the phospholipid headgroups, but no direct effect on the hydrophobic tails. On the other hand, cholesterol is inserted among the hydrophobic tails and it has little if any interaction with the

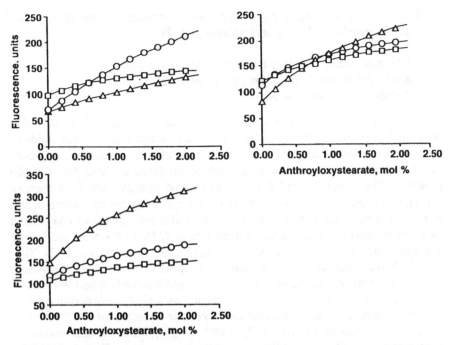

Figure 18 Energy transfer from fluorescent antioxidants to anthroyloxystearates. A fixed amount of fluorescent antioxidant, 0.33 mole% in 300 μM vesicles was titrated with increasing amounts of anthroyloxystearate. The curves represent the theoretical fits to the experimental data points. In all figures, O---O, 6-anthroyloxystearate; Δ---Δ, 9-anthroyloxystearate; □---□, 12-anthroyloxystearate. Top left: U-75,412E; top right: U-78,518E; bottom: U-78,517E.

Table 7 Energy Transfer from Antioxidants to Anthroyloxy Fatty Acids

Antioxidant		U-75,412E	U-78,518E	U-78,517F
Acceptor	Method	K_d (mole%)		
6-AS	a	5.1 ± 0.6	1.0 ± 0.4	1.2 ± 0.4
	b	2.0 ± 0.1	3.7 ± 0.3	2.0 ± 0.1
9-AS	a	6.0 ± 0.5	1.4 ± 0.7	2.1 ± 0.6
	b	2.0 ± 0.1	2.8 ± 0.3	2.0 ± 0.1
12-AS	a	1.5 ± 0.1	1.5 ± 0.4	3.5 ± 0.9
	b	2.1 ± 0.3	2.5 ± 0.2	2.1 ± 0.3

[a]Calculated using Eq. (17) in Ref. 29.
[b]Calculated using Eq. (18) in Ref. 29.
Source: From Ref. 29.

headgroups. It is this difference in shape which accounts for the opposite effects on the membrane by the antioxidants and cholesterol.

D. Effects and Localization of Antioxidants and Oxidized Lipids in Natural and Artificial Membranes

Fluorescence polarization, anisotropy, and/or lifetime measurements have also been used to probe the ability of lipophilic antioxidants and oxidized lipids to reorganize membrane structure and the physical effects of antioxidants which could counteract the membrane modifications produced by lipid peroxidation products. The structural and dynamic effects of hydroxy and hydroperoxy phospholipids on artificial lipid membranes were investigated by a combination of time- and angle-resolved fluorescence depolarization techniques (AFD) and by electron paramagnetic resonance experiments (34). In the same study, the orientation and structural effects of α-tocopherol, the aminosteroids, U-74,006F and U-75,412E, and the trolox derivative, U-78,518E, were also determined.

AFD experiments in planar PLPC (palmitoyllinoleylphosphatidylcholine) membranes with DPH (1,6-diphenylhexatriene) as a probe revealed that phospholipid oxidative products decrease the molecular orientational order, as measured by the order parameters $<P_2>$ and $<P_4>$, while the reorientational dynamics measured by the diffusion coefficient were not affected. The EPR (electron paramagnetic resonance) spectra of CSL (cholestane) in planar PLPC membranes are fully consistent with this conclusion: The presence of 10 mole% oxidized phospholipids in the PLPC bilayers again decreased molecular orientational order without affecting the reorientational dynamics measured by the rotational correlation time. Thus, all measurements agree that phospholipid peroxidation products induce disorder by permitting novel conformers and not by increasing mobility through a decrease in microviscosity.

As measured by AFD or DPH inserted in the membranes, α-tocopherol had no effect on the organization of DOPC (diolzoylphosphatidylcholine) bilayer membranes, although it decreased chain mobility. Time-resolved fluorescence anisotropy measurements showed that α-tocopherol, at low concentrations, is immobilized in DOPC bilayers. Analysis of AFD measurements revealed that, at a DOPC/α-tocopherol ratio of 50:1, the trolox ring of α-tocopherol is preferentially oriented perpendicular to the membrane plane, whereas at a ratio of 100:1 it is oriented parallel to the bilayer plane. Orientation within the bilayer should control the reaction efficiency of α-tocopherol with lipid peroxy radicals, even though α-tocopherol, at low concentrations, does not influence the organization of biological membranes.

As measured by the $<P_2>$ and $<P_4>$ order parameters of membrane-localized TMA (trimethylammonium)-DPH or DPH, U-74,006F decreased the orientational order in all bilayers formed from a variety of phospholipids. Similarly, U-75,412E decreased the orientational order in all phospholipid membranes except those formed with DOPC. In contrast, and in agreement with the vitamin E results, the trolox-based antioxidant, U-78,518E, the most potent free-radical scavenger tested, had little or no effect on the orientational order of TMA-DPH and decreased the motional relaxation of the probe in most lipid systems. All three antioxidants decreased the average fluorescence lifetimes of DPH and increased them for TMA-DPH, indicating that they caused polarity changes in the both the upper (TMA-DPH) and lower (DPH) parts of the bilayer. The observation of disorder induced by U-74,006F and U-75,412E is in agreement with the DSC results discussed earlier, but seem to disagree with the fluorescence measurements of Audus et al. (35) in BMECs (bovine microvessel endothelial cells). However, in the later study, only $<P_2>$ was measured and higher-order parameters were not considered.

The fluorescence measurements reveal major differences between the steroidal antioxidants and vitamin E with respect to their physical effects on membranes. Vitamin E has no effect on the orientational order of bilayers in contrast to the actions of the steroidal compounds. The membrane effects of the latter may derive from their molecular shape, an inverted cone in the membrane which is opposite to that of cholesterol. The cone shape should selectively relieve geometrical constraints in the lipid portion of the bilayer and thus allow for many more conformers. This disordering effect of U-74,006F and U-75,412E on the bilayer may make a significant a contribution to their ability to inhibit peroxidation in vitro and in vivo.

E. Delivery to Cells

U-74,006F is lipophilic; its delivery to tissues from the hydrophilic environment of the plasma is not a trivial problem and its biological activity will be conditioned as well by delivery as by intrinsic reactivity. The method by which the drug is presented to a target tissue also determines whether it is partitioned into the cell membrane or simply remains associated with the external cellular surface. Although an injectable citrate-based formulation is currently used to administer U-74,006F to patients, a lipid-based delivery system may be more efficient. Triglyceride emulsions stabilized by phospholipids have been used previously to deliver lipophilic compounds to whole cells and animals (35–37) and such systems are now under investigation for transfer of U-74,006F to target tissues. The challenges are to formulate an emulsion that is both chemically and physically stable and to characterize the kinetics and thermodynamic nature of drug delivery by such a system.

The kinetics and thermodynamics of delivery of two antioxidants, U-745,00 and U-74,006, were investigated (40). Biocompatible emulsions composed of a triolein/antioxidant core stabilized with egg yolk phosphatidylcholine emulsifier were found to conform to a normal distribution of particle sizes with a single mean value of 280–300 nm. The kinetics of delivery of U-74,006 from these emulsions to N18 cells in monolayer culture are shown in Fig. 19. The delivery was found to be very efficient and the kinetics conformed to a first-order reaction model. The rate constants for the delivery were the same for all experiments and independent of the amount of U-74,006 in the lipid phase of the emulsion. The amount of drug delivered was directly proportional to the concentration of the compound in the lipid phase (Fig. 20), thereby showing that a partition equilibrium had been reached. The partition coefficient for U-74,006 from the emulsion into the cell membranes were calculated to be 1.1. When delivery of U-74,006 from emulsions to cells suspended in culture medium was measured, the rate constant for delivery was still independent of the concentration of compound in the emulsion, but, because of the increased surface area of the cells available for delivery, the rate constants for delivery were approximately three times higher than those measured with cells in monolayer culture.

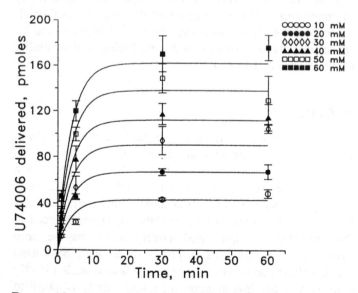

Figure 19 Time course of the transfer of U-74,006 from emulsions to N18 cells for a single concentration of emulsion. The values shown are the means ± s.d. for triplicate determinations of the cell-associated U-74,006. The lines shown are the best fits for a first-order process and were determined by nonlinear least-squares analysis.

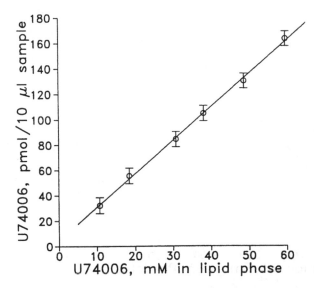

Figure 20 Partitioning of U-74,006 from emulsions to cells. The concentration of U-74,006 in N18 cells at equilibrium was determined from the data in Fig. 19. The values shown are the means ± s.d. for the amount of U-74,006 associated with the cells at equilibrium and were calculated by nonlinear least-squares analysis. The partition coefficient was calculated using the mole% U-74,006 in the lipid phase of the cells and in the lipid phase of the emulsions. The line shown was determined by linear regression, and the value of the Y-intercept is 4 ± 2, which is not significantly different from zero.

The effect of the concentration of emulsion particles on the kinetics of association was also investigated. While at equilibrium, the amount of U-74,006 delivered to the cells depended only upon the concentration of compound in the emulsion, the kinetics of delivery did depend on the concentration of the emulsion particles themselves. As shown in Fig. 21, the relationship between the experimental rate constants and the concentration of carrier particles is linear. Thus, the delivery is a collisional rather than a diffusional process; the kinetics of U-74,500 delivery from emulsions to cells were also consistent with the collisional mechanism. We have observed increased efficacy and delivery of U-74,006, using emulsion formulations in the in vitro cell system and in animal models.

Because the amount of compound delivered depends directly on the amount in the lipid phase of the emulsion and the rate constant for delivery depends on the number of emulsion particles present, delivery of an antioxidant to cells or tissues can be maximized through optimization of these two parameters. These principles are important for the efficient transfer of any lipophilic drug

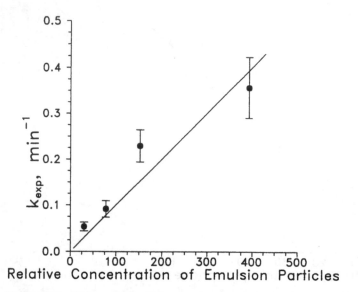

Figure 21   Effect of the relative concentration of emulsion particles on the rate of delivery of U-74,006 from emulsions to cells. The rate constants for the delivery of U-74,006 to cells as a function of particle concentration are plotted vs. the relative concentration of emulsion particles. Linear regression of the relative concentration of particles results in a correlation coefficient of 0.96 and a Y-intercept of 0.048 $\pm$ 0.035 (not significantly different from zero). The line shown is the best fit of a straight line with the Y-intercept = 0; all of the values are within 2 s.d. of this line.

to tissues, not just antioxidants. The increase in the number of lipophilic compounds being studied as potential therapeutic agents both here and at other pharmaceutical companies means that lipid delivery systems will become an even more important area of investigation in the future.

F.   Partitioning of U-74,006F across Cellular Barriers

The efficacy of these antioxidants as potential therapeutic agents for neuronal protection from oxidative damage may depend upon their ability to leave the blood and diffuse into various compartments, especially across the blood brain barrier. Both U-74,006F and U-78,517F are protective in brain and spinal cord injuries (4), but they do not accumulate at appreciable levels within the central nervous system, despite their high lipid solubility. Several possible causes for this have been identified. It was reported that transbarrier diffusion rates of U-74,006F and U-78,517F are limited by the association of these compounds with serum proteins and their high cellular partition coefficients (34). In a model system of continuous monolayers of cultured epithelial cells representing blood-to-endothelium (i.e., uptake) and endothelium-to-tissue (i.e., efflux) partitioning in vivo, the permeabilities are consistent with a model of two kinetic steps involving separate membrane domains of the cellular barrier. Anatomical considerations lead one to conclude that the limiting step for transbarrier permeation is not translocation of the uncharged drug between the aqueous headgroup interfaces of the membrane outer leaflets in the two membrane domains. Rather, uptake kinetics are limited by the equilibrium between serum protein-bound and free drug and diffusion across the unstirred water layer. The rate of uptake was 1000-fold faster than the rate of efflux due to large, intrinsic membrane log partition coefficients of 3.0–3.3 and low aqueous solubility (39). Because transbarrier permeability is dependent upon the concentration of drug in the barrier, these results predict that low free-drug concentrations in the blood and slow diffusion of barrier-associated drug away from the barrier do contribute to low brain levels. Sawada et al. (40) show that subsequent diffusion of a lipophilic compound from the initial blood-tissue barrier into the underlying tissue is dependent upon the mechanism of transport and whether or not it involves aqueous partitioning or a high-affinity carrier molecule.

The data from these in vitro experiments are consistent with first-pass extraction of [^{3}H]-labeled drugs by rat brain measured using the brain uptake index method (42). Extraction efficiencies of $5.4 \pm 0.7\%$ for U-74,006F and $7.5 \pm 1.5\%$ for U-78,517F are greater than the $2.5 \pm 0.6\%$ for the membrane-impermeant sucrose, but are relatively low for lipophilic compounds (43), indicating low blood-to-barrier partitioning and poor brain penetration (Raub et al., 1993). These values were obtained using 3% BSA and are identical to extraction efficiencies obtained in the presence of dilute (20%) rat serum (44), indicating that brain uptake from blood would be even lower. These results may be underestimates, however, since they involve a single bolus dose. Even after daily dosing of rats at 50 mg/kg for 4 weeks, brain levels were very low (45). This suggests that repeated exposure through multiple passes provides little improvement in brain delivery and implies that rapid clearance of U-74,006F by first-pass metabolism in the liver contributes to even lower total blood levels.

The in vitro uptake kinetics are also consistent with permeability coefficients (P_e) measured in vivo using a rat brain perfusion method (48). In situ P_e values (60-s perfusion) for [^{3}H]U-78,517F in 0.1–3% BSA were $8-13 \times 10^{-3}$ cm/min, and the in vitro values were $4-12 \times 10^{-3}$ cm/min (39,46). This indicates that the parameters govern partitioning in vitro and in vivo are the same and are independent of the cell type used.

It is important to realize that neither the BUI nor the perfusion method determines whether the drug actually partitions beyond the blood brain barrier penetrating deep into the parenchyma. It is postulated that U-74,006F acts directly on the blood brain barrier in trauma situations (4), but its distribution there is only inferred. Since U-74,006F lacks suitable fluorescence properties for this purpose, localization efforts have relied on autoradiography methods, but have been difficult primarily because of postsampling diffusion artifacts (49). One study was undertaken (50) to determine if U-74,006F accumulates in endothelium comprising the blood brain barrier. Mice were injected intravenously (tail vein) with various radiolabeled tracers and a crude brain vasculature fraction was recovered at 5 min using the capillary depletion method (51). The capillary fraction contained 56% of the total brain radioactivity of [^{125}I]-acetylated low-density lipoprotein, which is a macromolecule accumulated by brain capillary endothelial cells by receptor-mediated endocytosis, but not transported across the blood brain barrier. This value represents the efficiency of recovery. [^{3}H]BSA which serves as a vascular volume marker was 0% associated with the capillary fraction, because blood contents are lost upon mechanical disruption of the brain, indicating that blood-associated drug is not contaminating the recovered capillary fraction. Association of 21% of [^{3}H]U-74006F with the capillary fraction indicates that drug is accumulated within endothelium. Moreover, this accumulation appears to continue since a slightly greater percentage of radioactivity is recovered in the capillary fraction 60 min postinjection. This result is more profound when compared to the neuroleptic-anesthetic chlorpromazine (CPZ), which has a first-pass brain extraction efficiency of 87% (40). Only between 1.5% and 3% of [^{3}H]CPZ is recovered in the capillary fraction at 5 and 60 min postinjection, suggesting that it is not accumulated within blood brain barrier endothelium, but must diffuse readily into the parenchyma. Taken in toto, the results of Raub et al. show that U-74,006F does not efficiently cross the blood brain barrier and, thus, its site of action is probably in the endothelial cell layer.

V. CONCLUSIONS

The study of U-74,006F and congeners as antioxidant drugs mitigating lipid peroxidation yields several conclusions which can be generalized to the design and mode of action of antioxidant drugs. Perhaps the most revealing observation is that antioxidant action is not restricted to lipid free-radical scavenging. Rather, subtle modifications in the organization of membrane lipids can and do have profound effects on the rates of lipid peroxidation.

The physical antioxidant effects highlighted by our studies are disorganization of the lipid chains, compression of the superficial regions of the bilayer, and alteration of membrane thickness. We emphasize that in order to have an efficacious antioxidant, it is not enough to produce these effects indiscriminately across the lipid portion of the membrane; the effects should be localized at the precise depth in the membrane where the bisallyic functions reside. Of course, localization in the lipid layer requires that the antioxidant be intrinsically hydrophobic and—in order to fix it within the membrane—have a hydrophilic anchor.

Free-radical scavenging is a double-edged sword: Depending on the precise environment, a free-radical scavenger may, in fact, promote lipid peroxidation, since the scavenger molecule usually becomes a new radical, albeit with increased stability (23). Furthermore, too high a reactivity toward free radicals is translated into such a short lifetime that the compound may be exhausted even before it reaches its target. The natural antioxidant vitamin E is efficacious in spite of being highly reactive toward free radicals because a specialized recycling apparatus is in place to back up its action. Thus, we feel that an optimal antioxidant drug should have only a limited reactivity toward lipid free radicals.

The intrinsically hydrophobic nature of the drugs designed to mitigate lipid peroxidation creates two challenges concerning the pharmacokinetics, namely formulation and maintenance of effective plasma concentration: For IV use, the drug may be formulated in an aqueous vehicle or lipid carrier system such as microemulsions or liposomes. Alternatively, the drug may be formulated as a hydrophilic prodrug which would release in situ the hydrophobic active form. The oral delivery of hydrophobic drugs, on the other hand, is still under intense scrutiny, and no ready-made answers have been forthcoming. The second challenge derives from the fact that like other hydrophobic drugs such as warfarin, ibuprofen, diazepam, etc., U-74,006F binds extensively to circulating lipoproteins and plasma albumin, thereby limiting the concentration of its active, free form. Such binding, if moderate, is actually beneficial since it acts as a natural carrier and serves as a slow-releasing reservoir of the drug, thereby facilitating its delivery and prolonging its action. Too tight a binding, on the other hand, would prevent the efficient delivery of the drug to the target membranes.

Thus, the example of U-74,006F teaches us that an optimal balance between chemical and physical modes of action, together with optimization of delivery, is the essential requirement for an efficacious drug acting as a moderator of lipid peroxidation.

ACKNOWLEDGMENTS

We wish to thank Ms. Tillie Peterson and Dr. Greg Amidon from the Pharmacia and Upjohn Co. for providing us with their unpublished data. We also thank Dr. T.J. Raub of the Upjohn Co. for contributing the section concerning cellular permeability, and Drs. Constance Chidester and Fusan Han for providing the crystal structures. Finally, we owe an immense debt of gratitude to Dr. Ferenc J. Kézdy for his help in putting the chapter together and for his insight into the complex processes discussed.

REFERENCES

1. Deby C. In: Curtis-Prior PB, ed. Prostaglandins: Biology and Chemistry of Prostaglandins and Related Eicosanoids. New York: Churchill/Livingstone, 1988: 11–36.
2. Werns SW, Shea MJ, Luchesi BR. Free Radic Biol Med 1985; 6: 103–110.
3. Recknagel, RO, Glende, EA, Hruszkewicz, AM. In: Pryor WA, (ed.) Free Radicals in Biology. Vol. 3. New York: Academic Press, 1977.
4. Hall, ED, Braughler, JM. Free Radic Biol Med 1989; 6: 303–313.
5. Floyd, RA. FASEB J (1990); 4: 2587–2597.
6. O'Brien, PJ. In: (Chan HWS, ed.) Autoxidation of Unsaturated Lipids. London: Academic Press, 1987: 233–280.
7. Hall ED, McCall JM, Means ED. Free Radic Biol Med 1989; 6: 303–313.
8. McCall JM, TenBrink RE. Synthesis 1975; 443–444.
9. Church DE, Althaus JS, Koeplinger KA, Dugas TD, VonVoigtlander PF, McCall JM. J Free Radic Biol Med. Abstracts of the 7th Biennial Meeting of the International Society for Free Radical Research. Abstract D6, 1994.
10. Jacobsen EJ, McCall JM, Ayer DE, VanDoornik FJ, Palmer JR, Belonga KL, Braughler JM, Hall ED, Houser DJ, Krook MA, Runge TA. J Med Chem 1990; 33: 1145–1151.
11. Linseman KL, McCall JM. Unpublished observations.
12. Scherch HM. Unpublished observations.
13. Ryan TP, Steenwyk RC, Pearson PG, Petry TW. Biochem Pharm 1993; 46: 877–884.
14. Fernandes AC. Eur J Pharm 1992; 220: 211–216.
15. Taylor BM, Fleming WE, Sun FF. Abstract at the regional ASPET meeting, Chicago, IL, 1992.
16. Althaus JS, Fici GJ, VonVoigtlander PF, Zhang JR, Hall ED. ISFRR Satellite Meeting, Christchurch, New Zealand, 1994.
17. Salahudeen A, Badar K, Morrow J, Roberts J. ASN Abstract, 27th annual meeting, 1994.
18. Aeba R, Killinger WA, Keenan RJ, Yousem SA, Hamamoto I, Hardesty RL, Grifith BP. J Thorac-Cardiovasc Surg 1992; 104: 1333–1339.
19. Killinger WA, Dorofi DB, Keagy BA, Johnson G. Jr. Transplantation 1992; 53: 983–986.
20. Kay MMB, Gieljan JC, Bosman GM, Shapiro SS, Bendich A, Bassel P. Proc Nat Acad Sci USA 1986; 83: 2463–2467.
21. Chiu D, Lubin B. Semin Hematol 1989; 26: 128–135.
22. Miura T, Ogiso T. Chem Pharm Bull 1991; 39: 1507–1509.
23. Epps DE, Knechtel TJ, Baczynskyj O, Decker D, Guido DM, Buxser SE,

Mathews WR, Buffenbarger SL, Lutzke BS, McCall JM, Oliver LK, Kézdy FJ. Chem Phys Lipids 1994; 74: 163–174.

24. Bowry VW, Mohr D, Cleary J, Stocker R. J Biol Chem 1995; 270: 5756–5763 and refs within.
25. McKenna R, Kézdy FJ, Epps DE. Anal Biochem 1991; 196: 443–450.
26. Horan KL, Lutzke BS, Cazers AR, McCall JM, Epps DE. Free Radic Biol Med 1994; 17: 587–596.
27. Jiang Z.-Y, Woollard ACS, Wolff SP. Lipids 1991; 26: 853–856.
28. Peterson T. Unpublished observations.
29. Harris JS, Epps DE, Davio SR, Kézdy FJ. Biochemistry (1995); 34: 3851–3857.
30. Hinzmann JS, McKenna RL, Pierson TS, Han F, Kézdy FJ, Epps DE. Chem Phys Lipids 1992; 62: 123–138.
31. Wolber PR, Hudson BS. Biophys J 1979; 28: 197–210.
32. Gomez-Fernandez JC, Villalin J, Aranda FJ, Ortiz A, Micol V, Countinho A, Berberan-Santos MN, Prieto MJE. Ann NY Acad Sci 1989; 570: 109–120.
33. Amidon G. Unpublished observations.
34. van Ginkel G, Muller JM, Siemsen F, van't Veld AA, Korstanje LJ, van Zandvoort MAM, Wratten ML, Sevanian A. J Chem Soc Faraday Trans 1992; 88: 1901–1912.
35. Audus KI, Guillot FI, Braughler JM. Free Radic Biol Med 1991; 11: 361–371.
36. Collins Gold LC, Lyons RT, Bartholow LC. Adv Drug Del Rev 1990; 5: 189–208.
37. Singh M, Ravin LJ. J Parenter Sci Tech 1986; 40: 34–41.
38. Tarr BD, Sambandan TG, Yalkowsky SH. Pharm Res 1987; 4: 162–165.
39. Decker DE, Vroegop SM, Goodman TG, Peterson T, Buxser SE. Chem Phys Lipids 1995.
40. Raub TJ, Barsuhn CL, Williams LR, Decker DE, Sawada GA, Ho NFH. Drug Target 1993; 1: 269–286.
41. Sawada GA, Ho NFH, Williams LR, Barsuhn CL, Raub TJ. Pharm Res 1994; 11: 665–673.
42. Ohldendorf WH. Res Methods Neurochem 1981; 5: 91–112.
43. Ohldendorf WH. Proc Soc Exp Biol Med 1974; 147: 813–816.
44. Raub TJ, Williams LR. Unpublished observations.
45. Williams LR, et al. Unpublished observations.
46. Takasato Y, Rapoport SI, Smith QR. Am J Physiol 1984; 247: H484–H493.
47. Chikale E, Raub TJ, Borchardt RT. Unpublished observations.
48. Raub TJ, Douglas SL, Melchior GW, Charman WN, Morozowich W. In: Charman WN, Stella VJ, eds. Lymphatic Transport of Drugs. Boca Raton: CRC Press, 1992: 63–111.
49. Triguero D, Buciak J, Pardridge WM. J Neurochem 1990; 54: 1882–1888.
50. Raub TJ. Unpublished observations.

4

Anionic Tocopherol Esters as Antioxidants and Cytoprotectants

Marc W. Fariss
College of Pharmacy, Washington State University,
Pullman, Washington

I. INTRODUCTION

In light of the enormous cost of health care in the United States and the considerable role of oxidative stress in causing and exacerbating human disease, it is clear that therapeutic strategies to combat oxidative injury are required. To develop pharmaceuticals that protect us from the adverse effects of oxidative stress, the critical cellular and molecular processes that (1) are responsible for acute and chronic oxidative damage and disease and (2) are responsible for protection against such damage, must be defined. By augmenting the intracellular content of an "essential" protective component at the appropriate concentration, time, and subcellular site, it may be possible to diminish or eliminate oxidative stress-mediated injury and disease. This chapter reviews the scientific evidence suggesting that anionic tocopherol esters provide mammalian cells with a form of d-α-tocopherol that enhances the effectiveness of this important endogenous antioxidant. Postulated mechanisms to explain anionic tocopherol-mediated cytoprotection will also be examined. The anionic tocopherol esters considered in this chapter include d-α-tocopheryl hemisuccinate (TS), and dl-α-tocopheryl phosphate (TP). The structures of these compounds are illustrated in Fig. 1.

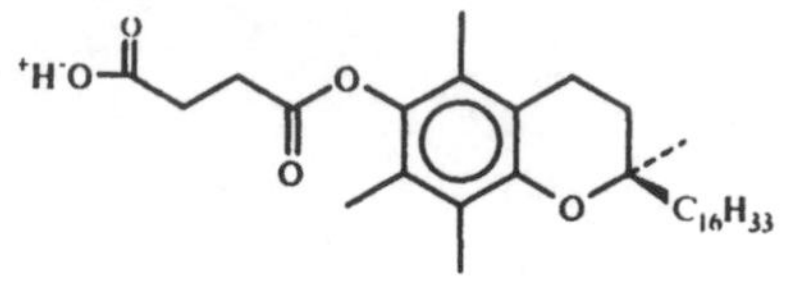

d-α-Tocopheryl Hemisuccinate

(TS)

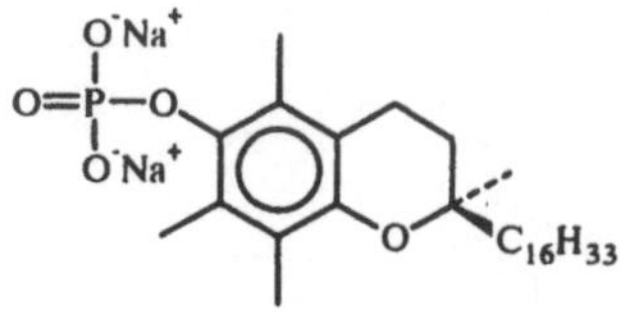

dl-α-Tocopheryl Phosphate, Disodium Salt

(TP)

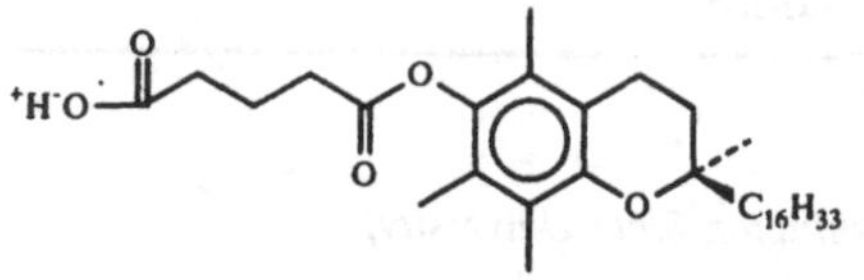

d-α-Tocopheryl Hemiglutarate

(TG)

Figure 1 Structures of anionic tocopherol esters.

A. Antioxidant Role of *d*-α-Tocopherol

Though the critical molecular events that lead to toxic cell death remain unclear, it is well established that reactive oxygen intermediates generated from both endogenous and exogenous insults (e.g., drug, chemical, hyperoxia, hypoxia, radiation, ischemia/reperfusion, aging, inflammation) play an important part in the toxic injury process (1,2). To prevent the oxidation of important cellular lipids, proteins, and nucleic acids, cells normally contain a battery of endogenous protective systems (antioxidants) to ensure the maintenance of viability as well as metabolic and functional performance (2,3). An example of such a protective system is unesterified *d*-α-tocopherol (T). Though T appears to function as the predominant chain-breaking antioxidant in cellular membranes (4–6), this lipophilic compound is not synthesized by mammalian cells but rather is derived solely from exogenous sources. The antioxidant properties of T result from its ability to trap reactive peroxyl radicals by donating a hydrogen atom, becoming a tocopheroxyl radical in the process. In order to preserve cellular T and its membrane antioxidant activity, other cellular hydrophilic reductants such as ascorbate, glutathione, and possibly NADPH can regenerate active T by donating an hydrogen atom to the tocopheroxyl radical (7–9). However, if the tocopheroxyl radical is attacked by another peroxyl radical resulting in a two-electron oxidation of T (e.g., tocopherylquinone formation), the cell's ability to regenerate active T is apparently lost (7,8). Thus the continual need in cellular membranes for the replacement of consumed oxidized T with dietary active T suggest that the cellular uptake and subcellular distribution of this important antioxidant is crucial to its ability to protect membrane constituents and cellular integrity (especially during an oxidative challenge).

B. Cytoprotective Role of *d*-α-Tocopherol

Numerous in vitro and in vivo studies clearly demonstrate that the depletion of cellular T, by eliminating exogenous sources of T, does indeed increase tissue susceptibility to toxic injury from the reactive oxygen-mediated insults mentioned previously (3). Based on these findings and our knowledge of the antioxidant function and consumption of T, it is often suggested and assumed that increasing the cellular content of T (above normal levels) will protect the cell against toxic injury. The scientific evidence to support this claim, however, is contradicting and controversial.

Though there are a considerable number of reports demonstrating the cytoprotective abilities of T administration (10–12), studies from the author's laboratory have demonstrated (as have many other investigators) that the acute in vitro or in vivo administration of T results in a dramatic increase in cellular T levels [4- to 100-fold increase (13–18)], but does not protect liver microsomes, hepatocytes, or rats from the toxic effects of a variety of different insults, including the alkylating agent ethyl methanesulfonate (13,19), calcium ionophore A-23187 (13), cadmium (14), hyperoxia (15), carbon tetrachloride (16,18) adriamycin in combination with 1,3-bis(2-chloroethyl)-1-nitrosourea (BCNU) (17), ADP-iron (18), cumene hydroperoxide (18). It is not the author's intent to provide a comprehensive review of the evidence (or lack of) for T-induced cytoprotection; however, it seems important to emphasize that the ability of exogenously added vitamin E to protect cells, tissue, and organisms against a toxic oxidative insult remains uncertain. It is the author's opinion that the conflicting data on tocopherol-mediated cytoprotection may be related to the ability or inability of administered vitamin E derivatives to deliver active T to a critical tissue, cell type, or subcellular compartment, in any given study. This explanation will be examined in greater detail in Sec. IV.B.

In contrast to T administration, the acute in vitro or ion vivo administration of an anionic ester of T, TS, has been shown to protect liver microsomes, hepatocytes and rats from the toxic insults listed above (13–19). These findings suggest that TS is a more effective cytoprotective agent than T. Furthermore, reports from other laboratories (in addition to those described above) clearly demonstrate that TS-mediated cytoprotection is not selective for a particular drug, chemical, or toxic insult, cell type, or species. In fact, the diversity of these insults suggests that the observed cytoprotection may result from the ability of TS to intervene in the critical cellular event(s) that leads to toxic cell death. These reports on TS cytoprotection will be discussed in greater detail in Sec. II.

II. ANIONIC TOCOPHEROL ESTERS AS ANTIOXIDANTS AND CYTOPROTECTANTS

Traditionally, anionic T esters (TS and TP) have been used as convenient T forms (more water-soluble) for treating cells or subcellular fractions with this lipid-soluble vitamin. Consequently, in the past, investigators have assumed that the biological activity of these anionic T esters are identical to T and thus often report using T or vitamin E instead of the specific anionic T derivative. This practice has probably limited the number of reports (and certainly our ability to identify these reports) describing the biological actions of these anionic T esters. More recently, there has been an increase in the number of studies demonstrating unique biological properties for these T derivatives. It is hoped that this review will further promote this awareness.

A. Tocopheryl Hemisuccinate

1. *Introduction*

The synthetic anionic ester analog of T, TS is commercially available for experimental use and as a dietary supplement in humans. The free acid of the hemisuccinate ester of tocopherol is available as a dry powder with a solubility in complete medium (without serum) of approximately 50 μM (20). The water solubility of TS can be increased by synthesizing the K^+ or tris salt forms (16) or the polyethylene glycol 1000 ester (Eastman Kodak) of TS. Having isolated TS from extracts of the green barley leaf, Badamchian et al. (21) recently reported that TS is also a natural product. This finding suggests the possibility that TS as well as other anionic T derivatives may also be endogenous components of mammalian cells. Such a role for TS in mammalian systems, however, has not yet been demonstrated.

The majority of toxicity/lipid peroxidation studies that report using anionic tocopherol esters have used TS, free acid for in vitro experimentation. As mentioned, the cytoprotective and antioxidant properties of TS observed in these studies do not appear selective for a particular toxic insult, cell type, or species. Investigators have demonstrated TS-mediated protection against 20 different toxic insults including toxic chemicals (13,18,19,22–25), therapeutic drugs (17,22,26–30), heavy metals (14,31–33), hyperoxia (15), UV irradiation (34), and endogenous compounds (18,35-40). These TS cytoprotection studies were conducted with seven different cell or tissue types [liver (13-15,17-19,22-24,26,27,36-38,40), lymphocytes (30), bone marrow (28,29), keratinocytes (39), adrenal (35), ovary cells (33) and serum (25)] from six different species [mouse (28,29), rat (13-15,17-19,22-24,26,27,36-38,40), hamster (31-34), guinea pig (35), canine (13), and human (25,30,39)]. Collectively these experimental findings suggest that TS treatment provides cells with (or maintains in cells) a critical component that is required for the maintenance of viability during a toxic oxidative insult. A brief summary of the scientific evidence demonstrating the antioxidant and cytoprotective properties of TS treatment is described below.

2. In Vitro Studies

Hepatocytes The majority of studies demonstrating TS cytoprotection have been conducted using freshly isolated rat hepatocytes in suspension, with several additional studies using primary cultures of rat hepatocytes and rat liver microsomes.

Using freshly isolated rat hepatocytes in suspension, Fariss et al. (22) and Pascoe et al. (26) reported that cells incubated with 25 μM TS, in the absence of extracellular calcium, were protected against the toxic effects (cell death and lipid peroxidation) induced by the alkylating agent ethyl methanesulfonate (EMS), the calcium ionophore A-23187, the antitumor agents adriamycin/BCNU, and the diuretic agent ethacrynic acid. Thomas and Reed (23) also reported that isolated rat hepatocytes treated with 50 μM TS are protected from the oxidative damage (loss of viability and lipid peroxidation) induced by the omission of extracellular calcium. Interestingly, when the concentration of free extracellular calcium was increased above physiological levels (>1.5 mM), the cytoprotective and antioxidant abilities of TS treatment were diminished or eliminated (13,17,22,26).

In 1989, the author's laboratory reported that the addition of 25 μM TS to rat and canine hepatocyte suspensions, incubated in medium containing physiological free extracellular calcium levels (approximately 1 mM), protects these cells from the toxic effects (cell death, loss of intracellular K^+ and lipid peroxidation) of EMS (13), A-23187 (13), hyperoxia (15), and cadmium chloride (14). More recently these studies have been extended to include TS cytoprotection against carbon tetrachloride- (CCl_4) and hypoxia-induced toxicity (Table 1). To examine the cytoprotective effects of TS we routinely use a rapid centrifugation technique (41) to separate viable hepatocytes from nonviable cells and medium prior to analysis for cell death (LDH leakage), cell injury (intracellular $[K^+]$), lipid peroxidation (TBA reactants), and cellular tocopherol and tocopherol ester content (Fig. 2). This separation technique enables the exami-

Table 1 Protective Effect of Tocopheryl Succinate in Hepatocytes Exposed to a Variety of Toxic Insults

Toxic insults[a]	% Cell death[b]	
	Medium without alpha-TS	Medium with alpha-TS (25μM)
Control (vehicle only)	21%	21%
Doxorubicin/BCNU (100/50 μM)	62%	28%
Ethyl methanesulfonate (50 mM)	68%	25%
Ionophore A23187 (5 μM)	73%	24%
Cadmium (0.1 mM)	70%	24%
95% oxygen (hyperoxia)	57%	23%
Carbon tetrachloride (2.5 mM)	63%	38%
Hypoxia (accomplished by 4 h under N_2 and 2 h under O_2)	70%	49%

[a]Freshly isolated rat hepatocyte suspensions were pretreated with 25 μM TS or vehicle (ethanol, 10 μl) for 15 min prior to treatment with a toxic insult.
[b]Cell death (% LDH leakage) was measured hourly in each suspension for up to 6 h.

nation of cellular events that occur prior to cell death, thus eliminating the artifactual events that occur after cell death (an important consideration in defining critical molecular events that are responsible for toxic injury). In our experiments (including those reported in this chapter) freshly isolated hepatocytes in suspension are prepared by a collagenase perfusion method, suspended in a modified Waymouths medium (1 mM Ca^{2+}) at a concentration of 2×10^6 cells/ml with 10 to 20 ml per 125 ml boiling flask (13–15). Each flask is slowly rotated at 37°C under an air atmosphere for up to 10 h. Each hour samples of the hepatocyte suspension are analyzed as shown in Fig. 2 (13–15).

Additional studies that demonstrate TS cytoprotective and antioxidant activity in hepatocyte suspensions include reports by Dogterom et al. (27) and Sokol et al. (40). These investigators demonstrated TS (100 and 200 μM) protection against cell death and lipid peroxidation induced by the anti-Alzheimer's drug Tacrine and bile acids respectively. Albano et al. (36) found that TS treatment (100 μM) of hepatocyte suspensions inhibited ferrous chloride–induced lipid peroxidation. Carini et al. (18) also reported that TS treatment (100 μM) inhibited ADP-iron- or ascorbate-iron-induced lipid peroxidation in hepatocyte suspensions and rat liver microsomes respectively. These investigators also observed that TS inhibits lipid peroxidation induced by CCl_4 and cumene hydroperoxide treatment of hepatocytes. Interestingly, Carini et al. (18) found that the polyethylene glycol 1000 ester of TS (TS-PEG) was as effective, and in some cases more effective, as an antioxidant than TS, free acid.

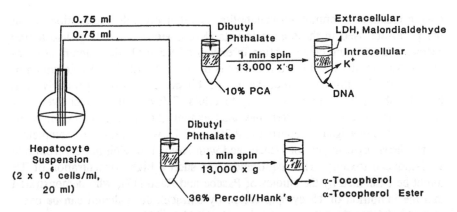

Figure 2 Cell sampling and analysis of rat hepatocyte suspensions for viability, lipid peroxidation and cellular T and T ester content using the dibutyl phthalate separation method.

The cytoprotective and antioxidant properties of TS treatment have also been demonstrated using primary cultures of adult rat hepatocytes. Koch et al. (24) reported that 100 μM TS treatment protected these cultured hepatocytes (isolated from phenobarbital treated male rats) from the toxic effects (as measured by MTT reduction) of CCl_4, $BrCCl_3$ and cadmium chloride. Mikkelsen et al. (38) observed that the addition of 50 μM of TS to primary cultures of rat hepatocytes (isolated from 48-h starved female rats) inhibited cell death (LDH leakage) and lipid peroxidation induced by exogenously added polyunsaturated fatty acids. Finally, Glascott et al. (37) reported that cultured rat hepatocytes supplemented with 1 μM TS overnight were protected from the cytotoxic effects of t-butyl hydroperoxide (tBHP) as compared to cells exposed to vehicle (DMSO).

There are also several reports demonstrating that TS pretreatment does not protect hepatocytes from toxic insults. Goldlin and Boelsterli (42) found that primary cultures of adult rat hepatocytes, pretreated with 100 μM TS-PEG for 1 h, were not protected from the cytotoxic (as measured by LDH leakage) effects of a 24-h cocaine treatment. To explain the inability of TS-PEG to protect hepatocytes, the authors suggested that cocaine-induced toxicity is not dependent on lipid peroxidation. In studies examining the hepatotoxic effects of nitrofurantoin in isolated perfused rat liver, Hoener and Hjalmarson (43) found that the perfusion of the liver with a perfusate containing 25 μM TS for up to 300 min prior to exposure to the toxic drug did not protect the liver (as measured by LDH leakage). Possible explanations offered by the authors (43) for the lack of TS protection include differences between hepatocytes and isolated liver preparations in terms of TS uptake (lower T levels in the isolated liver preparations were observed), and that T does not quench the toxic species, a reactive nitrofurantoin metabolite. Another possible explanation is that the perfusate used in these experiments (Krebs-Henseleit buffer) had a free extracellular calcium concentration in excess of 1.5 mM (the calcium concentration was not disclosed but this buffer is often used with 3 mM calcium chloride). Assuming this to be true, the use of this perfusate would probably limit both the ability of TS to enhance hepatocellular T levels and protect liver cells (22,44). The use of this buffer may also explain the findings of Sokol et al. (40). These investigators reported that 200 μM TS was required to completely protect hepatocyte suspensions (incubated with Krebs-Henseleit buffer) from bile acid–induced toxicity. The requirement for such a high concentration of TS might be explained by the studies of Pascoe and Reed (17), who demonstrated that the inhibition of TS cytoprotection by extracellular calcium can be overcome by raising the TS concentration above 150 μM.

Other Cell Types   Using Chinese hamster V-79 cells, Sugiyama and co-workers showed that a 24-h TS pretreatment (25 μM) inhibited sodium chromate–induced cell death (32) and DNA single-strand breaks (31). This laboratory also observed that this same Chinese hamster cell line, when pretreated with TS as described above, was resistant to the lethal effects of ultraviolet B irradiation but not to the mutational effects of the irradiation (34). Staats et al. (35), using mitochondria isolated from guinea pig adrenal glands, showed that the addition of 20 and 200 μM TS to this preparation significantly inhibited ferrous sulfate–induced lipid peroxidation (maximal protection with 200 μM). In studies examining the cytotoxicity of Zidovudine (AZT) in murine bone marrow cells (CFU-E), Gogu and co-workers (28,29) reported that the addition of approximately 8 μM of TS to these cultures protected these cells from the toxic effect of AZT (as measured by the loss of colony formation). Interestingly, exposure to 8 μM TS (without AZT) was also shown to dramatically stimulate the growth of murine CFU-E cells in culture (29). In experiments examining the genotoxic and cytotoxic effects of nickel sulfate on Chinese hamster ovary cells, Lin et al. (33) demonstrated that ovary cells pretreated with 25 μM TS for 24 h were protected from the lethality and chromosomal damage induced by this carcinogenic heavy metal. Trizna et al. (30) demonstrated that bleomycin-induced chromosomal breakage in primary cultures of human lymphocytes could be inhibited by pretreating the cultures for 24 h with 10 μM TS (optimal concentration tested). Using a recently developed method to measure total oxygen-radical absorbing capacity in human serum samples, Cao et al. (25) found excellent antioxidant activity with 1 μM TS. These investigators reported that the oxygen-radical (peroxyl) absorbing capacity of TS was comparable to the more water-soluble chromanol derivative, Trolox. It is assumed (data not shown) that hydrolysis of TS to T occurs in human serum (in this study) since intact TS is not an antioxidant. Finally, Wey et al. (39) reported that cultured human keratinocytes incubated with 2 μM TS for 4 days were protected against *t*BHP-induced cell death (LDH leakage) and lipid peroxidation.

3. *In Vivo Studies*

There are only a handful of published reports documenting the protective abilities of TS administered in vivo. Possible explanations for the absence of studies on TS cytoprotection include the following: (1) Water-soluble derivatives of T offer few advantages in terms of supplementing the diet of experimental animals with vitamin E. Nonionic forms of tocopherol [unesterified T, and tocopheryl acetate (TA)] are readily available, can be easily formulated for oral consumption or administration and thus are more commonly used. (2) Most in vivo studies administer vitamin E by the oral route. Our experimental findings indicate that the cytoprotective properties of TS (administered in a single dose of 100 mg/kg) are lost when this anionic T ester is given orally (16). (3) The unique and enhanced cytoprotective activity of TS have not been recognized (and thus not used or reported) or do not exist in vivo (and thus are not reported).

Of the studies that have been reported, Svingen and co-workers (45) demonstrated in an experimental rat model that the topical application of TS dissolved in DMSO was more effective in reducing adriamycin-induced skin necrosis than DMSO alone. These investigators reported that ip injections of TS (100 mg/kg for 2 to 7 days following adriamycin treatment, corn oil vehicle) were ineffective in reducing skin necrosis. Fariss et al. (16) reported that the lethal and hepatotoxic effects of a single oral dose of CCl_4 (2.9 g/kg) could be eliminated or diminished by a single ip dose of TS, free acid (100 mg/kg) given to rats 24 h prior to the insult. The protective effect of TS administration on CCl_4-induced hepatotoxicity appears to depend on the hepatocellular accumulation of TS. This conclusion was supported by the findings that the oral administration of TS (single 100 mg/kg dose) eliminates both protection and liver and tissue accumulation of TS but not T (16). Using a lower hepatotoxic dose of CCl_4 (1 g/kg), the pretreatment of rats with TS, free acid provided partial protection (45%, though not statistically significant) against this hepatotoxic dose [Fig. 3A (16)]. However, rats pretreated with an identical dose (100 mg/kg, ip) of the tris salt of TS demonstrated statistically significant protection [75% reduction in serum alanine aminotransferase (ALT) levels] against 1 g/kg CCl_4-induced hepatotoxicity [Fig. 3a (16)]. The enhanced hepatoprotection observed with TS, tris salt was confirmed by measuring the extent of liver necrosis by histological examination [Fig. 3b (16)]. This enhanced cytoprotection appears to be related to an increased accumulation of TS (and release of T) in the liver. As shown in Table 2, the administration of TS, K^+ salt (a protective form of TS) resulted in a fivefold increase in liver T and TS levels, 24 h following

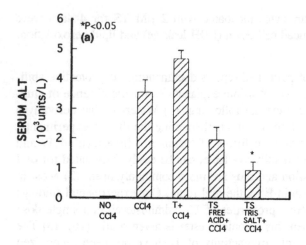

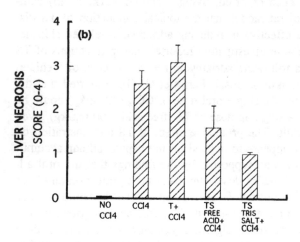

Figure 3   Effect of nonionic and anionic tocopherol derivatives on CCl_4-induced hepatotoxicity (a) and liver necrosis (b). Tocopherol derivatives or vehicle were administered to rats (100 mg/kg, ip) 24 h before a single oral dose of CCl_4 (1 g/kg). Rats were fasted 24 h before CCl_4 administration. Serum ALT levels and liver necrosis were determined in specimens obtained 48 h after CCl_4 treatment. Values are the means ± SEM (N = 4–5). *p < 0.05 as compared to the CCl_4 alone treatment group.

Table 2   Tissue Concentrations of Tocopherol, Tocopheryl Acetate, and Tocopheryl Succinate, 24 h After Tocopherol Analog Administration

Tocopherol analog administered[a]		Tissue concentration (nmol/g)[b]			Protection
		Liver	Blood	Plasma	
Vehicle	T	28 ± 3	10 ± 1	13 ± 1	No
	T ester	ND	ND	ND	
Tocopherol, (100 mg/kg, ip)	T	128 ± 10	16 ± 1	20 ± 2	No
	T ester	ND	ND	ND	
Tocopheryl acetate, (100 mg/kg, ip)	T	56 ± 10	22 ± 1	18 ± 1	No
	T ester	92 ± 6	7 ± 2	6 ± 2	
Tocopheryl succinate, free acid (100 mg/kg, ip)	T	51 ± 5	13 ± 0	14 ± 2	Partial
	T ester	119 ± 16	7 ± 1	8 ± 1	
Tocopheryl succinate, K$^+$ salt (100 mg/kg, ip)	T	260 ± 32	21 ± 2	22 ± 2	Yes
	T ester	518 ± 40	19 ± 2	24 ± 2	

[a]Male Sprague-Dawley rats received a single ip injection of the tocopherol analog or vehicle, fasted for 24 h, and tissue samples were analyzed for T and T ester content (16).

[b]Values are the mean ± SEM (n = 3–4). ND, not detected.

treatment, as compared to TS, free acid administration. Investigating the effect of antioxidants on the pesticide endrin-induced lipid peroxidation, Bagchi (46) observed that the oral pretreatment of rats with 100 mg/kg TS for 3 days followed by 40 mg/kg on day 4 significantly reduced the endrin-induced increase in mitochondrial and microsomal lipid peroxidation by approximately 60% and 40%, respectively. Unfortunately, it is unknown whether an oral dose of TS, given daily, can overcome TS hydrolysis in the gut, resulting in the accumulation of hepatic TS. Thus it is unclear whether the observed protective effect of TS pretreatment in this study (46) resulted from the actions of TS.

B. Tocopheryl Phosphate

1. Introduction

Another commercially available synthetic anionic tocopherol ester reported to be cytoprotective is dl-α-tocopheryl phosphate, disodium salt (TP). This vitamin E ester is a constituent of Williams E medium [18 nM TP (37)] but otherwise has received little reported use. Though studies from the author's laboratory have demonstrated TP-mediated cytoprotection in rat hepatocyte suspensions and cultures, additional investigations have been hampered by the lack of a sensitive analytical method to measure tissue, cellular, and subcellular TP levels. Thus the author's laboratory has concentrated efforts on defining the mechanism and experimental conditions required for TS cytoprotection.

2. In Vitro Studies

Using an isolated rat heart preparation to examine the effect of TP (0.1 μM in the cardioplegic perfusate) on ischemia/reperfusion injury, Conorev et al. (47) found that TP protects the heart from ischemic cell death (as measured by creatine kinase release) and from mechanical (contractile) injury. Koch et al. (24) reported that 100 μM TP treatment protected primary cultures of adult rat hepatocytes from the toxic effects (as measured by MTT reduction) of CCl_4, $BrCCl_3$, and cadmium chloride. Using the same cultured hepatocyte preparation, Lamb et al. (48) demonstrated that cells treated with 25 μM TP (incorporated into liposomes) are protected from ethanol-induced cell death (as measured by MTT reduction and ALT release) and lipid peroxidation. Glascott and co-workers (37) reported that cultured rat hepatocytes supplemented with 1 μM TP overnight were protected from the cytotoxic effects of tBHP as compared to cells exposed to vehicle (DMSO). Studies from the author's laboratory have also demonstrated that the addition of TP (25 to 250 μM) to the incubation medium protects both primary cultures of adult rat hepatocytes (Fig. 4) and freshly isolated rat hepatocyte suspensions (Fig. 5) from the toxic effects (cell death and lipid peroxidation) of the alkylating agent EMS. Interestingly, as observed with TS, TP-mediated cytoprotection is eliminated by increasing the extracel-

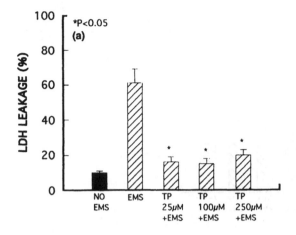

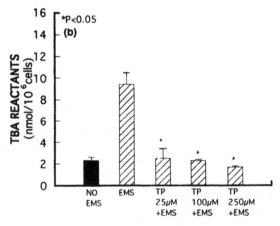

Figure 4 Protective effect of TP on EMS-induced cell death (a) and lipid peroxidation (b) in cultured rat hepatocytes. Four-hour-old primary cultures of adult rat hepatocytes were incubated in serum-free Waymouth MB 751/1 medium containing several concentrations of TP or vehicle (saline). Each culture (except the vehicle only) was then exposed to 50 mM EMS and the %LDH leakage and the concentration of TBA reactants were measured after 6 h of incubation. Values are the means ± SEM ($n = 3$–5). *$p < 0.05$ as compared to EMS alone.

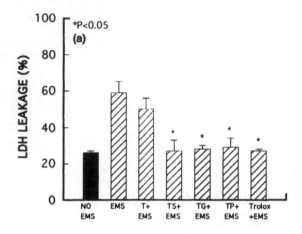

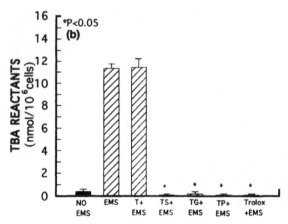

Figure 5 Effect of anionic T esters on EMS-induced cell death (a) and lipid per-oxidation (b) in hepatocyte suspensions. Freshly isolated rat hepatocytes suspensions were incubated with 25 μM tocopherol analog (except TP concentration was 50 μM) or vehicle for 15 min. Each suspension (except the vehicle only) was then exposed to 50 mM EMS and the %LDH leakage and the concentration of TBA reactants were measured after 4 h of incubation. Values are the means ± SEM ($n = 3$). *$p < 0.05$ as compared to EMS alone.

lular calcium concentration from physiological conditions (1 mM) to 3.5 mM (20).

In several published studies investigators have compared the cytoprotective abilities of TP and TS. Koch and co-workers (24) report that TP (as compared with TS) is better able to protect cultured rat hepatocytes from chemical-dependent cell injury. In contrast, Mikkelsen et al. (38) reported that, although the addition of 50 μM of TS to primary cultures of rat hepatocytes inhibited cell death (LDH leakage) and lipid peroxidation induced by exogenously added polyunsaturated fatty acids, the addition of 50 μM TP was not protective. Studies from the author's laboratory using rat hepatocyte suspensions (20) suggest that TS is the superior protective agent (in terms of potency and effectiveness).

3. In Vivo Studies

Only one published study was found that investigated the cytoprotective abilities of TP when administered in vivo. Paranich and co-workers (49) reported that the administration of TP to rats protects these experimental animals from a lethal dose of gamma irradiation. Paranich et al. (50) also demonstrated that TP, as a water-soluble form of T, is more effective than TA (oil-soluble form) in supplementing tissue T levels. Unfortunately, detailed information on these studies is limited since only English abstracts are available from these Russian publications.

C. Anionic Versus Nonionic Tocopherol Esters

1. Introduction

The experimental evidence support the conclusion that the anionic tocopherol esters, TS and TP, protect cells and organisms against a wide variety of toxic insults, most of which also induce lipid peroxidation (oxidative stress). The critical question now becomes: Are the cytoprotective properties observed with TS and TP treatment unique to these anionic T esters or do other nonanionic T derivatives possess comparable activity? In this chapter the nonanionic T derivatives are collectively referred to as nonionic T esters or derivatives and include unesterified T and the T ester TA (see structures in Fig. 6). As mentioned, nonionic forms of tocopherol (T and TA) are the preferred formulations for most experimental investigations with vitamin E and for human dietary supplementation. These nonionic forms of T, as the name implies, have a slight, or no, ionic charge and thus are less water-soluble than the anionic T esters. Several investigators have compared the cytoprotective and antioxidant activity of nonionic and anionic T esters using the same toxic insult. Their experimental findings are described below.

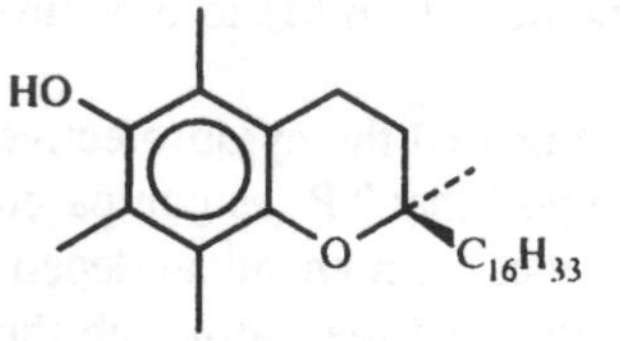

d-α-Tocopherol

(T)

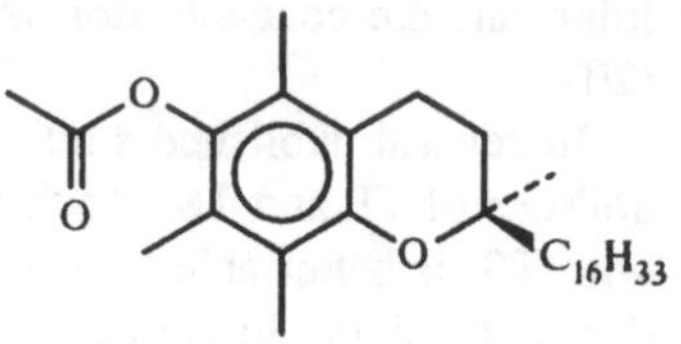

d-α-Tocopheryl Acetate

(TA)

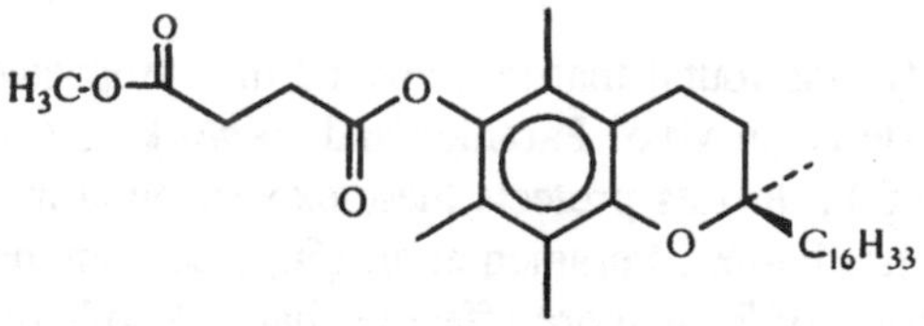

d-α-Tocopheryl Hemisuccinate Methyl Ester

(TS-ME)

Figure 6   Structures of nonionic T derivatives.

2.   In Vitro Studies

Investigations comparing anionic T esters to nonionic T derivatives as cytoprotectants and antioxidants have been reported using both rat hepatocytes in suspension and in culture. The data from freshly isolated hepatocyte suspension preparations clearly demonstrate that TS treatment is more effective than T or TA treatment in terms of protecting cells and microsomes from cell death, and lipid peroxidation caused a variety of different oxidative insults. For example, Pascoe and Reed (17) found that, unlike TS, the addition of T (25 μM) to calcium-free medium did not protect hepatocytes from the lethal and lipid-peroxidizing effects of adriamycin in combination with BCNU treatment. Likewise, Carini et al. (18) demonstrated that TS (either as the free acid or PEG ester), added exogenously to hepatocyte suspensions and rat liver microsomes, provided greater protection against ADP-iron-, CCl_4-, cumene hydroperoxide-, and ascorbate iron–induced lipid peroxidation than a comparable amount of T (100 μM). Finally, Fariss et al. (13–15) demonstrated that 25 μM TS provides hepatocyte suspensions with greater protection against EMS-, hyperoxia-, or cadmium-induced cell death and lipid peroxidation than that observed with T and TA supplementation (25 to 250 μM).

Recently we examined the cytoprotective activity of several new synthetic anionic and nonionic T derivatives (Figs. 1, 6). As shown in Fig. 5, the addition of the anionic T esters, TP (50 µM) and d-α-tocopheryl hemiglutarate (TG, 25 µM), and the water-soluble anionic chromanol derivative, Trolox (25 µM), to the incubation medium protected hepatocyte suspensions against EMS-induced cell death and lipid peroxidation. As expected, these hepatoctyes were not protected against EMS-induced toxicity when pretreated with the nonionic T derivatives (25 µM), T, and TA (Figs. 5, 7). In agreement with these findings, we also discovered that the antioxidant and cytoprotective activity of TS are eliminated by forming the methyl ester of TS (TS-ME), thus eliminating the anionic charge on the molecule and increasing the lipophilicity of the compound (Fig. 7). Interestingly, when the anionic charge of TS was eliminated by forming the PEG ester of TS (a more water-soluble derivative of TS), the antioxidant and cytoprotective properties of TS were retained (Fig. 7). These data suggest that the critical component in determining T-mediated cytoprotection and antioxidant activity in vitro is the water solubility of the administered T derivative—hence, the more effective cytoprotective activity of anionic T esters observed in the studies mentioned.

Using primary cultures of adult rat hepatocytes, Koch et al. (24) observed that cells treated with 100 µM vitamin E esters (in the absence of serum) differentially (TP > TS > TA) reduce cadmium chloride-, CCl_4- and $BrCCl_3$-induced toxicity. In contrast, Glascott et al. (37) reported that cultured hepatocytes pretreated for 24 h with 1 µM T, TA, TP, or TS in the presence of 9% fetal calf serum demonstrated similar cytoprotective activities against tBHP-induced cell death. A possible explanation (as noted by the authors) for the similar cytoprotective abilities of nonionic and anionic T derivatives is the presence of serum (and lipoproteins) in the culture medium. It was proposed that serum might serve as a transport mechanism for the nonionic derivatives (similar to the in vivo situation), thus promoting cellular T uptake during the 24-h pretreatment. Another explanation could be the low concentration of vitamin E derivatives (1 µM) tested in this study. Our studies with hepatocyte suspensions demonstrate that 10 µM is the lowest concentration of TS that resulted in optimal cytoprotection against EMS-induced cell death (Fig. 8).

3. *In Vivo Studies*

Very few investigators have compared the cytoprotective activity of nonionic and anionic T esters administered in vivo (studies using oral administration are not considered since TS is not protective by this route). Svingen et al. (45) observed that the topical daily application of TS was superior to TA in terms of reducing adriamycin-induced skin necrosis. Paranich and co-workers (49) reported that the administration of TP as a water-soluble form of T is more effective than TA (oil-soluble form) in protecting these experimental animals

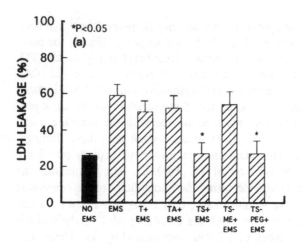

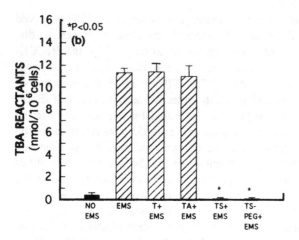

Figure 7 Effect of nonionic T esters on EMS-induced cell death (a) and lipid peroxidation (b) in hepatocyte suspensions. Freshly isolated rat hepatocyte suspensions were incubated with 25 μM tocopherol analog or vehicle for 15 min. Each suspension (except the vehicle only) was then exposed to 50 mM EMS and the %LDH leakage and the concentration of TBA reactants were measured after 4 h of incubation. Values are the means ± SEM ($n = 3$). *$p < 0.05$ as compared to EMS alone.

from a lethal dose of gamma irradiation. In contrast to these studies, Yao et al. (51) demonstrated that a single intravenous dose of cationic liposomes containing TS (2 h prior to CCl_4) protected mice from CCl_4-induced hepatotoxicity (as measured by serum ALT levels). However, liposomes containing vita-

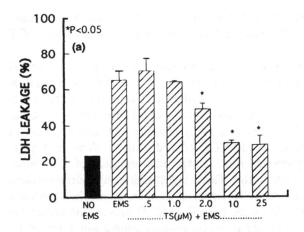

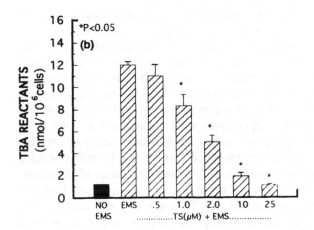

Figure 8 Effect of TS concentration on EMS-induced cell death (a) and lipid peroxidation (b) in hepatocyte suspensions. Freshly isolated rat hepatocytes suspensions were incubated with various concentrations of TS or vehicle for 15 min. Each suspension (except the vehicle only) was then exposed to 35 mM EMS and the %LDH leakage and the concentration of TBA reactants were measured after 6 h of incubation. Values are the means ± SEM ($n = 3$). *$p < 0.05$ as compared to EMS alone.

min E (the specific compound was not disclosed) were more effective. Interestingly, these investigators found that a single iv dose of Rocavit E (Hoffmann LaRoche), a vitamin E–containing emulsion, was also very effective in protecting the liver from CCl_4-induced damage (51). These studies are difficult to interpret since the hepatocellular accumulation of T and TS were not reported. However, they do suggest that water-soluble derivatives or formulations of T are cytoprotective.

Fariss et al. (16) demonstrated that supplementation of endogenous T or TA levels (with a single ip administration of 100 mg/kg T or TA, 24 h prior to CCl$_4$) did not protect rats from the toxic (lethal or liver damage) effects of oral CCl$_4$ administration. In contrast, the administration of TS (using the same treatment protocol described above) resulted in the hepatocellular accumulation of TS and afforded rats protection against CCl$_4$-induced toxicity. When the hepatic accumulation of TS was prevented (oral gavage administration of TS), hepatoprotection was also eliminated (16). These investigations also demonstrated only modest hepatoprotection from TS free acid administration when sublethal concentrations (1 g/kg) of CCl$_4$ were administered. However, the protective activity of TS was dramatically enhanced by administering a salt form of TS (data for TS, tris salt is shown in Fig. 3). Interestingly, the ip administration of a salt form of TS (data for TS, K$^+$ salt is shown in Table 2) also dramatically enhanced the hepatocellular TS and T levels resulting in a fivefold increase as compared to TS, free acid, T, or TA administration.

Our in vitro studies with nonionic and anionic T derivatives suggest that the water solubility of the T analog is an important factor in determining cytoprotective activity. It has been suggested that the in vitro administration of the water-insoluble nonionic T forms does not permit an active "physiological" intercalation with hepatocyte membranes and thus cytoprotection and antioxidant activities are not observed. To overcome this potential artifact, we administered T or TS-PEG to rats (single ip administration of 100 mg/kg T or 200 mg/kg TS-PEG) and 24 h later hepatocytes were isolated and examined for cytoprotective and antioxidant activity against EMS-induced toxicity (19). The results from these studies clearly demonstrated that the in vivo administration of T did not afford hepatocytes protection while protection against EMS-induced cell death and lipid peroxidation was observed in cells obtained from TS-PEG-treated rats (19). These studies confirm that TS (especially the water-soluble forms) possesses unique cytoprotective properties that are not observed with nonionic T derivatives, whether administered in vitro or in vivo.

III. MECHANISM OF ANIONIC TOCOPHEROL CYTOPROTECTION

A. Introduction

The available experimental evidence does indeed support the conclusion that anionic T esters possess more effective or enhanced cytoprotective and antioxidant activities as compared with their nonionic T counterparts. The observed cytoprotection using both in vitro and in vivo experimental models appear to depend on the cellular accumulation of the intact anionic T molecule (data are only available for TS). Thus, the molecular mechanism by which TS treatment results in enhanced cytoprotection must be related to one of the following: (1) unique biological properties of the intact anionic tocopherol molecule, (2) the release and intracellular accumulation of the anionic component, and (3) the release and intracellular accumulation of T. Our findings from the investigation of each of these possible mechanisms for TS cytoprotection are reviewed below.

B. Intact Anionic Tocopherol Molecule

One possible explanation for TS-mediated cytoprotection is that the anionic lipophilic TS molecule itself is the protective agent. In support of this explanation, the cellular accumulation of TS, but not T, appears to be required for protection. Furthermore, recent studies demonstrate that TS exhibits a wide range of biological activities that do not appear to be related to the release of T. For example, once in the cell the amphipathic TS molecule might alter membrane stability or function by interacting with phospholipid and protein components of the membrane. Numerous reports indicate that TS administration does indeed stabilize membranes (52,53), alters membrane enzymatic activity (54,55), inhibits NF-κB DNA binding activity (56), induces differentiation in HL-60 human promyelocytic leukemia cells (57), inhibits tumor cell growth (57–61), induces apoptosis in RL human B lymphoma cells (59), enhances the secretion of TGFβ from human B lymphoma and breast cancer cells (59,60), reduces the expression of oncogenes and adenylate cyclase to external ligands (61), inhibits phospholipase A2 activity (62,63), and stimulates growth hormone and prolactin release from rat anterior pituitary cells (21).

To determine the role of the intact anionic T molecule in cytoprotection, we examined the effect of the esterase inhibitors, diethyl *p*-nitrophenyl phosphate (DENP) and bis(*p*-nitrophenyl) phosphate (BNPP), on TS hydrolysis and TS-mediated cytoprotection (19). Hepatocyte suspensions were exposed to 100 μM DENP or BNPP, 15 min prior to the addition of TS. The addition of these esterase inhibitors did not significantly influence cell viability as compared to vehicle control cells. However, these inhibitors did prevent the release of T and succinate from the TS molecule and eliminated TS cytoprotection against EMS-induced toxicity (19). Similar results were also reported by Carini and co-workers (18) when they examined the effect of BNPP on ADP-iron-dependent lipid peroxidation in rat hepatocytes exposed to TS and TS-PEG. These investigators demonstrated that the antioxidant activity of TS and TS-PEG were eliminated by preventing TS hydrolysis. These data suggest that the intact TS molecule is not responsible for TS cytoprotection.

Another possible explanation for the ability of DENP and BNPP pretreatments to prevent TS cytoprotection is that these lipophilic esterase inhibitors may block the ability of TS to intercalate in the cell membrane in an active conformation. To confirm the importance of TS hydrolysis in cytoprotection, we synthesized and examined the ability of several nonhydrolyzable TS derivatives, d-α-tocopheryloxybutyrate [TSE (58)], and d-α-tocopheryl 3-methyl hemisuccinate (TS-3M) (Fig. 9) to protect hepatocytes from EMS-induced cell death and lipid peroxidation. These nonhydrolyzable TS derivatives are structurally similar to TS and thus would be expected to interact with cellular membranes in a similar fashion. The results from these studies are shown in Fig. 10 and clearly demonstrate that TSE and TS-3M treatment (25 µM) are not as effective as TS in protecting cells from EMS toxicity. In fact, TSE, a TS ether derivative that is not susceptible to hydrolysis (58), demonstrated no cytoprotective or antioxidant activity (Fig. 10). These data support the hypothesis that the hydrolysis of TS, releasing succinate and tocopherol, is required for the TS-mediated cytoprotection and antioxidant activity observed in vitro.

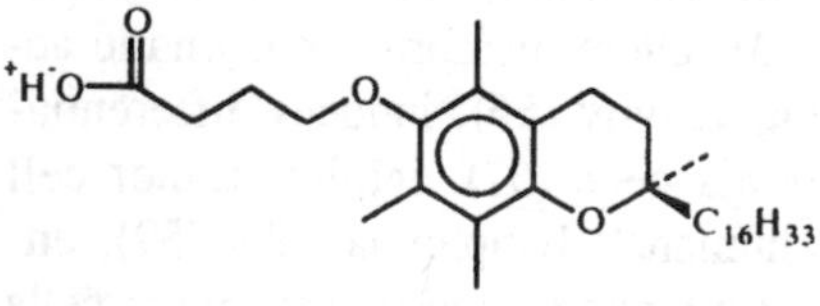

d-α-Tocopheryloxybutyrate

(TSE)

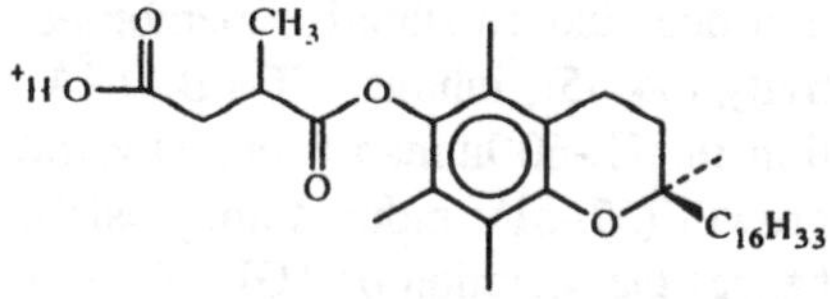

d-α-Tocopheryl 3-Methyl Hemisuccinate

(TS-3M)

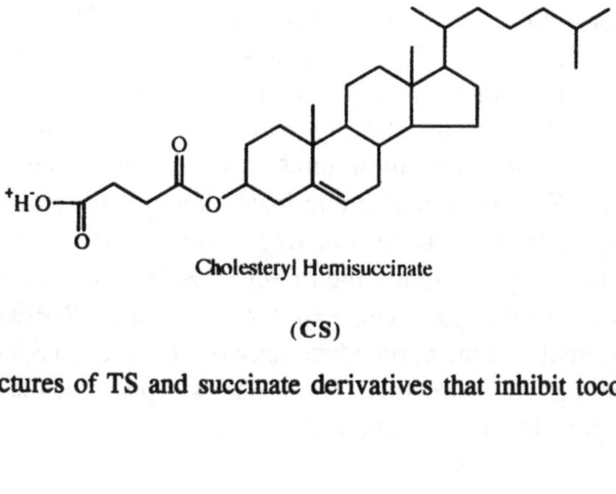

Cholesteryl Hemisuccinate

(CS)

Figure 9 Structures of TS and succinate derivatives that inhibit tocopherol release.

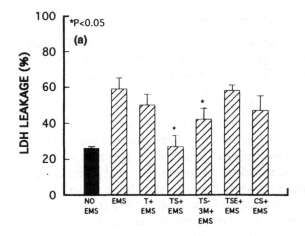

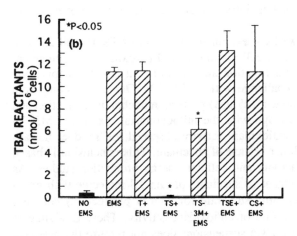

Figure 10 Effect of nonhydrolyzable TS derivatives on EMS-induced cell death (a) and lipid peroxidation (b) in hepatocyte suspensions. Freshly isolated rat hepatocytes suspensions were incubated with 25 μM T or succinate analog or vehicle for 15 min. Each suspension (except the vehicle only) was then exposed to 50 mM EMS and the %LDH leakage and the concentration of TBA reactants were measured after 4 h of incubation. Values are the means ± SEM (n = 3). *p < 0.05 as compared to EMS alone.

C. Release of Anionic Component

Since the in vitro and in vivo supplementation of hepatocytes with T (using T or TA administration) does not protect these cells from toxic oxidative challenges, another explanation for TS-mediated cytoprotection is that the release of the anionic component (e.g. succinate) from cellular TS is the protectant. Previous studies indicate that the administration of succinate can protect mitochondria, cells, and organisms from a variety of toxic insults (64–67). In fact, succinate is a preferred mitochondrial substrate with respect to energy production (oxidative phosphorylation) and the reduction of pyridine nucleotides (68,69) and succinate treatment stimulates gluconeogenesis and glycolysis in rat hepatocytes (70). It has been estimated (15) that the addition of TS (25 μM) to hepatocyte suspensions results in a 20% increase in cellular succinate concentration/hour (released from TS). Thus, assuming that cellular energy production is important to the maintenance of viability during a toxic insult, hepatocytes or mitochondria supplemented with succinate might have an advantage in terms of survival. Because the cellular uptake and accumulation of succinate is severely limited by its hydrophilicity (71), TS may serve as a lipophilic carrier for succinate, thus promoting intracellular accumulation and energy production.

To test this possibility, we have examined the effect of TS as well as tocopherol and succinate analogs on cell viability, lipid peroxidation, cellular energy status and mitochondrial function and structure in isolated hepatocyte suspensions during a toxic challenge with EMS (19). Our conclusions from these studies are that TS treatment (25 μM) protects mitochondrial function and structure during a toxic challenge with the mitochondrial toxicant, EMS. In addition, the protective and antioxidant activity observed for TS does not appear to be related to the cellular release of succinate. This conclusion is based on the following structure/activity and cellular energy production studies. As shown in Fig. 5, the replacement of succinate on the TS molecule with phosphate or glutarate (not a potential substrate for energy production) does not result in the loss of cytoprotection or antioxidant activity. These data suggest that TS protection of rat hepatocyte suspensions does not require the presence or release of succinate. To confirm these findings, the cytoprotective and antioxidant activity of several lipophilic succinate esters (that are not tocopherol derivatives) were also examined. Results from these studies indicate that the elimination of tocopherol from the succinate ester molecule also eliminates hepatocyte protection. As shown in Fig. 10, the treatment of rat hepatocyte suspensions with cholesteryl hemisuccinate (Fig. 9) did not afford these cells protection from EMS-induced toxicity. Likewise, the lipophilic monomethyl and dimethyl esters of succinate (25 μM to 5 mM) were not cytoprotective even though these derivatives did stimulate energy production (enhanced glucose and lactate formation) in rat hepatocyte suspensions (19). In contrast, the exogenous addition of 25 μM TS (cytoprotective concentration) to these hepatocytes did not enhance glucose or lactate formation in these cells (19).

D. Release of Tocopherol

The experimental evidence presented above supports the explanation that TS-mediated cytoprotection results from the cellular accumulation of TS followed by esterase hydrolysis of TS and the release of T. These data suggest that the antioxidant activity of the released T is responsible for the observed protection with TS (and other anionic T esters). In agreement with this explanation, data shown in Table 3 demonstrate that the cytoprotective activity of anionic T esters in hepatocyte suspensions correlates with cellular accumulation of the ester and subsequent release of T. If the cellular release of tocopherol is inhibited or prevented [as demonstrated with the following treatments: TS-3M (Table 3 and Fig. 10), TSE (Fig. 10), CS (Fig. 10), 3.5 mM extracellular calcium (44), and less than 2 μM TS (Fig. 8 and Table 3)], the cytoprotective and antioxidant activity of these treatments is also severely diminished or eliminated

Table 3 Effect of Tocopherol Analog Dose on the Concentration of T and T Ester in Rat Hepatocytes Exposed to EMS for 3 h

Treatment	Hepatocyte concentration[a] (nmol/10^6 cells)		Protection from toxic injury
	Toc	Toc ester	
Control	0.2 ± 0.1	N.D.	No
Toc			
25 μM	2.1 ± 0.7	N.D.	No
50 μM	6.1 ± 0.7	N.D.	No
250 μM	23.7 ± 2.4	N.D.	No
TS			
0.5 μM	0.2 ± 0.1	N.D.	No
2.0 μM	0.4 ± 0.1	0.2 ± 0.1	Partial
10 μM	0.8 ± 0.1	0.7 ± 0.1	Yes
25 μM	0.9 ± 0.1	2.5 ± 0.1	Yes
250 μM	0.9 ± 0.1	7.6 ± 1.7	Yes
TS-3M			
25 μM	0.2 ± 0.1	1.0 ± 0.1	Partial
TG			
25 μM	0.7 ± 0.1	2.1 ± 0.4	Yes
TP			
25 μM	1.1 ± 0.4	Not determined	Yes

[a]Expressed as mean ± SEM, $n = 3$, N.D., not detected.

[as compared to 25 μM TS, TP, or TG (Table 3 and Fig. 5)]. The importance of the antioxidant action of T released from anionic T esters in cytoprotection is also supported by our finding that the antioxidant Trolox also completely protects hepatocytes from EMS toxicity (Fig. 4). Our in vivo studies demonstrating that the more effective protective activity of the salt (tris or K$^+$) form of TS (as compared with TS, free acid, T, or TA administration) correlates with an enhanced cellular accumulation of TS and released T also supports the role of T in TS cytoprotection (Table 2). The inability of T or nonionic T esters, administered acutely in vitro or in vivo, to protect hepatocytes from toxic oxidative insults [even though cellular T levels are enhanced (Table 2 and 3, Figs. 3 and 7)] suggests that the intracellular accumulation and/or membrane incorporation of these T derivatives differs from TS and anionic tocopherol esters. The experimental evidence to support this conclusion is examined below.

IV. MECHANISM OF ENHANCED CYTOPROTECTION WITH ANIONIC TOCOPHEROLS

A. Biomembrane Incorporation of Tocopherol

1. *Nonionic Tocopherol Derivatives*

To explain the ineffective antioxidant and cytoprotective properties of directly added T and nonionic tocopherol esters, investigators (17,18,72) have suggested that the extreme lipophilicity of T and its nonionic esters prevents incorporation of these compounds in biomembranes in an active configuration (i.e., possibly bound to the plasma membrane in aggregates). In support of this explanation, Kagan et al. (73) demonstrated, using fluorescent measurements of T incorporated in biomembranes and in phospholipid bilayers (fluorescence of T is quenched if present in clusters or aggregates), that almost half of the added T is localized in membranes in the form of clusters. Results from this study suggest that when T is introduced in membrane suspensions that its binding, incorporation, and distribution occur mainly in the outer layer of the membrane (73,74). This apparent failure of exogenously added T to intercalate between phospholipid molecules of the bilayer in an even and monomeric fashion may indeed be responsible for its limited effectiveness as an antioxidant and cytoprotective agent.

2. *Anionic Tocopherol Esters*

Unfortunately, the incorporation of TS and other anionic T esters in biomembranes cannot be measured as described above for T because the intrinsic fluorescence of the T molecule is lost upon esterification. However, there is evidence that anionic tocopherol esters do interact with biomembranes differently than unesterified T and nonionic T esters. Using techniques such as differential scanning calorimetry [measuring phase transition (52)], freeze-fracture electron microscopy (53), and x-ray diffraction (53), investigators have reported that TS can dissolve in phospholipid bilayers far beyond the limits of the parent compound, T. For example, Lai et al. (52) demonstrated that lecithin liposomes can be prepared with 90 mol% TS, as compared with no more than 30 to 40 mol% for T (52,53) or TA (75). In fact, the amphiphile TS is capable of forming multilamellar vesicles by itself, 100 mol% (52,53). Boni et al. (53) demonstrated that the incorporation of TS into the bilayer phase can be prevented by neutralizing the ionic charge of TS with the addition of greater than 1.25 mM Ca^{2+}. It is interesting to note that each of the conditions that limit the incorporation of T (as compared with TS) into lipid bilayers (e.g., nonionic forms such as T or TA and the exposure of TS to high concentrations of Ca^{2+}) also prevents cytoprotection against toxic oxidative injury (13). Though the studies mentioned suggest that anionic T esters possess chemical/physical properties that facilitate the incorporation of these tocopherol precursors into biomembranes (in comparison to nonionic T analogs), additional studies are required to confirm this possibility.

B. Intracellular Accumulation of d-α-Tocopherol

1. Unesterified Tocopherol

Unfortunately very little is known about the intracellular handling of T. Previous studies have demonstrated that T is an extremely lipophilic molecule and as such is absorbed from the intestine in chylomicrons through the lymphatic system and is transported in plasma by a tocopherol binding protein incorporated in lipoproteins (76). The cellular uptake of T has been reported to be mediated by both lipoprotein receptor-dependent and -independent pathways (77). It is generally accepted that in lipid bilayers and biomembranes, T intercalates between phospholipids with the chroman head group (phenolic hydrogen) toward the surface (in close proximity to water-soluble reducing agents for regeneration) and with the hydrophobic phytyl chain buried in the hydrocarbon region (78). Interestingly, in biological membranes only one tocopherol molecule is present for every 500 to 1000 polyunsaturated fatty acids (PUFA)(79). This concentration of membrane-bound T is thought to be close to the threshold of T required to effectively protect phospholipid bilayers [0.2 mol% (80)] and biomembranes [0.4 mol% (81)] against oxidative damage, thus again emphasizing the importance of replenishing oxidized T by regenerating T with hydrophilic reductants or by incorporating new active T into the membrane.

Biomembranes or lipid bilayers are not limited to this ratio of T to PUFA. In fact, Lai et al. (52) have shown that lecithin liposomes can be prepared with up to 40 mol% T, while numerous reports indicate that increasing the T content of biomembranes decreases the susceptibility of these membranes to lipid peroxidation (79,82). At present it is not understood why the concentration of T in biomembranes is kept at such a low mol% (close to the threshold). The amount of T embedded in intracellular membranes appears to result from the concentration of T available from the diet and its intracellular transport as well as from the rate of consumption by oxidation and by transport out of the cell.

During an oxidative challenge when membrane-bound T is being rapidly consumed, a rate-limiting factor in providing intracellular membranes with additional active T may be the requirement for a tocopherol transporting protein. Niki et al. (83) have demonstrated using artificial phospholipid membranes that the extreme water insolubility conferred on T by the phytyl tail greatly inhibits its ability to exchange between membranes in the absence of any transporting factors. Other investigators (84–86) have also suggested that the intracellular transport of T requires a tocopherol transport protein that can carry T to subcellular locations. Such a protein has been identified in rat and rabbit liver (84,85) and heart (86) for the transfer of T to the nucleus (87), mitochondria (88), and microsomes (89), and the binding of T to this protein is saturable. However, the ability of this tocopherol transfer protein to rapidly supplement intracellular membranes with active T remains unclear and seems doubtful based on our limited knowledge of T transport. In fact the observed inability of T to freely exchange between intracellular membranes may limit the ability of acute T administration, given in vitro or in vivo, to protect cells, organs, and organisms from the toxic effects of oxidative stress. This apparently limited ability of dietary T to be incorporated intracellularly may also be responsible for the low toxicity observed with this fat-soluble vitamin. Unfortunately, the subcellular distribution of active T following acute and chronic T administrations and its relationship to cytoprotection have not been adequately examined and represent an important area for future investigations.

2. Anionic Tocopherol Esters

We suggest that increasing the anionic nature or hydrophilicity of T increases the access and retention of T (hydrophilic anionic T esters and releasable T) in intracellular membranes. To explain the ineffective antioxidant and cytoprotective properties of directly added T and nonionic T esters, investigators (17,18,72) have also suggested that the extreme lipophilicity of T and its nonionic esters result in T aggregates bound to the plasma membrane which prevents the distribution of T to intracellular organelles and membranes in need of additional T during a toxic oxidative challenge. The inadequate subcellular distribution of T (due to aggregate formation or to the limitations inherent in intracellular T transport, as previously discussed) following the acute administration of this hydrophobic antioxidant might also explain our observation that a single 100 mg/kg dose of a nonionic T derivative (T, TA, or d-α-tocopheryl nicotinate) does not protect rats (16) or hepatocytes (19) from the hepatotoxic effects of CCl_4 or EMS (unlike TS).

In support of the idea that the enhanced intracellular distribution and retention of TS (with the gradual release of T) in cellular membranes is responsible for TS-induced cytoprotection, we found a 33% increase in the T equivalents (T and/or TS) present in the liver of rats 24 h following the administration of TS-free acid (protective conditions) as compared to an equivalent dose of T [nonprotective conditions (16)]. Similar results were also observed for the accumulation of TS and T in isolated hepatocytes (13). Of particular significance is our finding that when a more cytoprotective and water-soluble form of TS (K^+ salt) was administered to rats, the amount of T equivalents found in the liver 24 h later was over 500% greater than that observed with an equivalent dose of T (Table 2). Thus, we speculate that the ability of water-soluble anionic tocopherol esters to promote the cellular uptake of T derivatives by a mechanism other than the tocopherol binding protein may be responsible for the observed enhanced cytoprotection.

Unfortunately, very little information is available concerning the subcellular distribution of T and TS following TS administration or the influence of TS hydrophilicity (free acid and salts) on the tissue and subcellular distribution of T and TS. Investigators, including ourselves, have postulated that the unique cytoprotective properties of TS (as compared with T) may be related to its ability to accumulate at a "critical" subcellular organelle or site (19,90). In fact, we suggest that TS cytoprotection results from this compound's ability to prevent mitochondrial dysfunction during a toxic insult (19, and see next section). In the only reported study to examine the subcellular distribution of TS and T, Slack and Proulx (91) demonstrated that neuroblastoma cells incubated with TS (25 μM) for 48 h accumulated significantly greater amounts of TS in mitochondria (245%) and microsomal (345%) fractions, as compared to the subcellular accumulation of T (100%) in cells incubated with T (25 μM) for the same period of time.

One possible explanation for this enhanced intracellular distribution of a water-soluble derivative of T is that the intracellular transport of TS is not dependent on a tocopherol binding protein (as described for T). In support of this claim, the studies of Catignani and Bieri (84) demonstrated that the tocopherol binding protein binding is very specific for d-α-tocopherol and that tocopherol esters, tocopherol quinone or Trolox were ineffective competitors of this binding. In addition, Traber et al. (92) have demonstrated that Caco-2 human intestinal cells exposed to the water-soluble TS ester, TS-PEG, accumulate this compound intracellularly as the intact TS-PEG molecule. Because the water-soluble PEG 1000 molecule does not readily enter cells, these investigators suggest that the TS-PEG molecule, since it readily forms micelles, "dissolves" in the plasma membrane and thus gains intracellular access (92). Unfortunately, information on the intracellular partitioning of hydrophilic esters of T have not been reported. However, it is well known that intracellular esterases, which are capable of releasing T from hydrophilic T esters such as TS and TP, are rather ubiquitous with activity found in most species (13,93), organ systems (16,93), and subcellular fractions examined (93) especially in microsomes (73,74) and mitochondria (73,74). Consequently, regardless of the intracellular location of the anionic T ester, active T will most likely be continuously generated at that site from this hydrophilic T precursor (16).

C. Anionic Tocopherol Derivatives as Mitochondrial Protectants

The vast number and diversity of toxic insults that cause mitochondrial dysfunction are striking (1,9,19,40,90,94,95). The importance of maintaining mitochondrial function for cellular energy production, metabolic performance, and cell viability is well known. Accordingly, studies on the role and mechanism of mitochondrial dysfunction in oxidative injury have received considerable attention in recent years. It is interesting to note that mitochondria appear to be more susceptible to oxidative damage than other cellular organelles (1,96). For example, mitochondrial DNA from rat liver has more than 10 times the level of oxidative DNA damage than does nuclear DNA from the same tissue (1). The high rate of oxidants generated in mitochondria during oxidative phosphorylation has been proposed as a possible mechanism for this age-related enhancement of DNA damage. Another explanation or consideration is that the rapid depletion of mitochondrial T, the major membrane antioxidant, during an endogenous or exogenous oxidative challenge could enhance mitochondrial oxidative damage and dysfunction. In support of this later hypothesis, Tappel (4) has demonstrated that mitochondria isolated from livers of vitamin E–deficient rabbits oxidize at twice the rate of those from animals given vitamin E. This enhanced oxidation resulted in an increase in mitochondrial lipid peroxidation, inactivation of mitochondrial enzymes and cytochromes of the electron transport chain and was prevented by the addition of T (97). Vitamin E deficiency in man or experimental animals also results in myopathy in which myocyte mitochondria appear damaged and respiratory chain activities at complexes I and IV are significantly diminished (98,99). Bjorneboe et al. (100) studied the effects of long-term administration of ethanol on the distribution of T and found that hepatic mitochondrial T content was reduced by 55% in these treated animals as compared to the controls, whereas no significant difference was observed in microsomes, light mitochondria, or cytosol (100). These studies emphasize both the enhanced susceptibility of this organelle to oxidative damage and the importance of mitochondrial T levels in protecting mitochondria and cells from oxidative injury.

Our suggestion that mitochondria may be the critical site for TS-mediated protection against toxic chemical injury was derived not only from our experimental findings but from the reports of others as well. Studies from Don Reed's laboratory have demonstrated that TS treatment protects cells from a variety of mitochondrial toxicants such as calcium ionophore A-23187 (101), extracellular Ca^{2+} omission (23), and adriamycin (17). Thomas and Reed (94) reported that the loss of mitochondrial membrane potential, induced by extracellular Ca^{2+} omission, could be prevented by the addition of TS to the medium. In addition the ability of TS to prevent O_2-induced permeability transition in rat liver mitochondria was recently demonstrated by Guidoux and coworkers (102). Pascoe and Reed (90) published an excellent review on TS cytoprotection and speculated that the observed TS cytoprotection may be related to the ability of TS to rapidly enter the cell (unlike T) and preferentially localize T to the mitochondrial membrane. As mentioned, our data and the reports of others agree with this hypothesis; however, additional studies are required to define the role of TS as a mitochondrial protectant.

V. CONCLUSIONS

Numerous studies have clearly demonstrated that the administration of anionic T esters (especially TS) protect subcellular fractions, cells, tissue, and experimental animals from oxidative damage induced by a wide variety of toxic insults. There appear to be sufficient scientific evidence to conclude that anionic T esters (especially TS) are more effective antioxidants and cytoprotective agents than nonionic T derivatives (T and TA) when administered in vitro. Though the in vivo administration of TS (especially the more water-soluble forms) also appears to be more effective than T or TA, additional studies are required to support this claim.

Our investigations on the mechanism of TS-mediated cytoprotection indicate that the cellular accumulation of TS with the concomitant release of T are required for protection. The mechanism by which TS treatment provides more effective protection against oxidative damage than nonionic T derivatives is unknown. We speculate that anionic T esters like TS, being more water-soluble forms of T, possess the unique ability (not observed with nonionic T forms) to rapidly supplement intracellular membranes with a reservoir of active T (in the releasable form of TS), thus ensuring protection of these membranes during a toxic oxidative insult. Future investigations will test these hypotheses, with the aim of developing therapeutic benefits from anionic T derivatives that can rapidly target active T to critical intracellular sites.

ACKNOWLEDGMENTS

The author is grateful to Kay F. Bryson, Sharon J. Fariss, Sidhartha D. Ray, Jeyananthan Chelliah, Xiao-Guang Zhao, Xiao-Yan Gu, H. Robert Lippman, Donna D. Li, Philip S. Guzelian, and J. Doyle Smith for their assistance with the experiments presented in this chapter. The author also thanks Sharon J. Fariss for preparing the graphic illustrations. This project and the experiments presented in this review were supported by Grant 5R01 ES05452 from NIEHS/ NIH.

REFERENCES

1. Ames BN, Shigenaga MK, Hagen TM. Oxidants, antioxidants, and the degenerative diseases of aging. Proc Natl Acad Sci USA 1993; 90:7915–7922.
2. Kehrer JP. Free radicals as mediators of tissue injury and disease. Crit Rev Toxicol 1993; 23:21–48.
3. Reed DJ. Cellular defense mechanisms against reactive metabolites. In: Anders MW, ed. Bioactivation of Foreign Compounds. New York: Academic Press, 1985:71–108.
4. Tappel AL. Vitamin E as the biological lipid antioxidant. Vitam Horm 1962; 20:493–510.
5. Burton GW, Joyce A, Ingold KU. Is vitamin E the only lipid-soluble, chain breaking antioxidant in human blood plasma and erythrocyte membranes? Arch Biochem Biophys 1983; 221: 281–290.
6. Cheeseman KH, Emery S, Maddix SP, Slater TF, Burton GW, Ingold KU. Studies on lipid peroxidation in normal and tumor tissues. Biochem J 1988; 250:247–252.
7. Burton GW, Ingold KU. Vitamin E as an in vitro and in vivo antioxidant. Ann NY Acad Sci 1989; 570:7–22.
8. Liebler DC. The role of metabolism in the antioxidant function of vitamin E. Crit Rev Toxicol 1993; 23:147–169.
9. Kehrer JP, Lund LG. Cellular reducing equivalents and oxidative stress. Free Rad Biol Med 1994; 17:65–75.
10. Gallagher CH. Protection by antioxidants against lethal doses of carbon tetrachloride. Nature 1961; 192:881–882.
11. Meyers CE, McGuire W, Young R. Adriamycin: amelioration of toxicity by α-tocopherol. Cancer Treat Rep 1976; 60:961–962.
12. Kramer JH, Misik V, Weglicki WB. Magnesium deficiency potentiates free radical production associated with postischemic injury to rat hearts: vitamin E affords protection. Free Rad Biol Med 1994; 16:713–723.
13. Fariss MW, Merson MH, O'Hara TM. α-Tocopherol succinate protects hepatocytes from chemical-induced toxicity under physiological calcium conditions. Toxicol Lett 1989; 47:61–75.
14. Fariss MW. Cadmium toxicity: unique cytoprotective properties of alpha tocopheryl succinate in hepatocytes. Toxicology 1991; 69:63–77.

15. Fariss MW. Oxygen toxicity: unique cytoprotective properties of vitamin E succinate in hepatocytes. Free Rad Biol Med 1990; 9:333–343.

16. Fariss MW, Bryson KF, Hylton EE, Lippman HR, Stubin CH, Zhao X. Protection against carbon tetrachloride-induced hepatotoxicity by pretreating rats with the hemisuccinate esters of tocopherol and cholesterol. Environ Health Perspect 1993; 101:528–536.

17. Pascoe GA, Reed DJ. Vitamin E protection against chemical-induced cell injury. II. Evidence for a threshold effect of cellular α-tocopherol in prevention of adriamycin toxicity. Arch Biochem Biophys 1987; 256:159–166.

18. Carini R, Giuseppe P, Dianzani MU, Maddix SP, Slater TF, Cheeseman KH. Comparative evaluation activity of α-tocopherol, α-tocopherol polyethylene glycol 1000 succinate and α-tocopherol succinate in isolated hepatocytes and liver microsomal suspensions. Biochem Pharmacol 1990; 39:1597–1601.

19. Ray SD, Fariss MW. Role of cellular energy status in tocopherol hemisuccinate cytoprotection against ethyl methanesulfonate-induced toxicity. Arch Biochem Biophys 1994; 311:180–190.

20. Fariss M. Unpublished work.

21. Badamchian M, Spangelo BL, Bao Y, Hagiwara Y, Hagiwara H, Ueyama H, Goldstein AL. Isolation of a vitamin E analog from a green barley leaf extract that stimulates release of prolactin and growth hormone from rat anterior pituitary cells in vitro. J Nutr Biochem 1994; 5:145–150.

22. Fariss MW, Pascoe GA, Reed DJ. Vitamin E reversal of the effect of extracellular calcium on chemically induced toxicity in hepatocytes. Science 1985; 227:751–754.

23. Thomas CE, Reed DJ. Effect of extracellular Ca^{++} omission on isolated hepatocytes. I. Induction of oxidative stress and cell injury. J Pharmacol Exp Therap 1988; 245:493–500.

24. Koch J, Mathews S, McKenzie R, Williams M, McElderry J, Bush R, Lamb R. Vitamin E esters differentially alter hepatotoxin-induced changes in cultured hepatocyte phosphatidylcholine metabolism [abstract]. FASEB J 1991; 5:A1578.

25. Cao G, Alessio HM, Cutler RG. Oxygen-radical absorbance capacity assay for antioxidants. Free Rad Biol Med 1993; 14:303–311.

26. Pascoe GA, Olafsdottir K, Reed DJ. Vitamin E protection against chemical-induced cell injury. I. Maintenance of cellular protein thiols as a cytoprotective mechanism. Arch Biochem Biophys 1987; 256:150–158.

27. Dogterom P, Nagelkerke JF, Mulder GJ. Hepatoxicity of tetrahydroamminoacridine in isolated rat hepatocytes: Effect of glutathione and vitamin E. Biochem Pharmacol 1988; 37:2311–2313.

28. Gogu SR, Beckman BS, Rangan SRS, Agrawal KC. Increased therapeutic efficacy of Zidovudine in combination with vitamin E. Biochem Biophys Res Comm 1989; 165:401–407.

29. Gogu SR, Lertora JJL, George WJ, Hyslop NE, Agrawal KC. Protection of Zidovudine-induced toxicity against murine erythroid progenitor cells by vitamin E. Exp Hematol 1991; 19:649–652.

30. Trizna Z, Hsu TC, Schantz SP. Protective effects of vitamin E against bleomycin-

induced genotoxicity in head and neck cancer patients in vitro. Anticancer Res 1992; 12:325–328.

31. Sugiyama M, Ando A, Furuno A, Furlong NB, Hidaka T, Ogura R. Effects of vitamin E, vitamin B2 and selenite on DNA single strand breaks induced by sodium chromate (VI). Cancer Lett 1987; 38:1–7.

32. Sugiyama M, Ando A, Ogura R. Effect of vitamin E on survival, glutathione reductase and formation of chromium (V) in chinese hamster V-79 cells treated with sodium chromate (VI). Carcinogenesis 1989; 10:737–741.

33. Lin X, Sugiyama M, Costa M. Differences in the effect of vitamin E on nickel sulfide or nickel chloride-induced chromosomal aberrations in mammalian cells. Mutation Res 1991; 260:159–164.

34. Sugiyama M, Tsuzuki K, Matsumoto K, Ogura R. Effect of vitamin E on cyto-toxicity, DNA single strand breaks, chromosomal aberrations, and mutation in chinese hamster V-79 cells exposed to ultraviolet-B light. Photochem Photobiol 1992; 56:31–34.

35. Staats DA, Lohr D, Colby HD. Relationship between mitochondrial lipid pero-xidation and α-tocpoherol levels in the guinea-pig adrenal cortex. Biochem Biophys Acta 1988; 961:279–284.

36. Albano E, Bellomo G, Parola M, Carini R, Dianzani MU. Stimulation of lipid peroxidation increases the intracellular calcium content of isolated hepatocytes. Biochem Biophys Acta 1991; 1091:310–316.

37. Glascott PA, Gilfor E, Farber JL. Effects of vitamin E on the killing of cultured hepatocytes by *tert*-butyl hydroperoxide. Mol Pharmacol 1992; 41:1155–1162.

38. Mikkelsen L, Hansen HS, Grunnet N, Dich J. Inhibition of fatty acid synthesis in rat hepatocytes by exogenous polyunsaturated fatty acids is caused by lipid peroxidation. Biochem Biophys Acta 1993; 1166:99–104.

39. Wey HE, Pyron L, Woolery M. Essential fatty acid deficiency in cultured hu-man keratinocytes attenuates toxicity due to lipid peroxidation. Toxicol Appl Pharmacol 1993; 120:72–79.

40. Sokol RJ, Devereaux M, Khandwala R, O'Brien K. Evidence for involvement of oxygen free radicals in bile acid toxicity to isolated rat hepatocytes. Hepatology 1993; 17:869–881.

41. Fariss MW, Brown MK, Schmitz JA, Reed DJ. Mechanism of chemical-induced toxicity. 1. Use of a rapid centrifugation technique for the separation of viable and nonviable hepatocytes. Toxicol Appl Pharmacol 1985; 79:283–295.

42. Goldlin CR, Boelsterli UA. Reactive oxygen species and non-peroxidative mecha-nisms of cocaine-induced cytotoxicity in rat hepatocyte cultures. Toxicology 1991; 69:79–91.

43. Hoener B, Hjalmarson AC. α-Tocopherol succinate does not mitigate nitrofuran-toin-induced changes in the glutathione and protein thiol status of the isolated perfused rat liver. Toxicology 1991; 67:165–170.

44. Pascoe GA, Reed DJ. Relationship between cellular calcium and vitamin E me-tabolism during protection against cell injury. Arch Biochem Biophys 1987; 253:287–296.

45. Svingen BA, Powis G, Appel PL, Scott M. Protection against adriamycin-induced

skin necrosis in the rat by dimethyl sulfoxide and α-tocopherol. Cancer Res 1981; 41:3395-3399.

46. Bagchi D, Hassoun EA, Bagchi M, Stohs SJ. Protective effects of antioxidants against endrin-induced hepatic lipid peroxidation, DNA damage, and excretion of urinary lipid metabolites. Free Rad Biol Med 1993; 15:217-222.

47. Conorev EA, Sharov VG, Saks VA. Improvement in contractile recovery of isolated rat heart after cardioplegic ischaemic arrest with endogenous phosphocreatine: involvement of antiperoxidative effect? Cardiovasc Res 1991; 25:164-171.

48. Lamb RG, Koch JC, Snyder JW, Huband SM, Bush SR. An in vitro model of ethanol-dependent liver cell injury. Hepatology 1994; 19:174-182.

49. Paranich AV, Oleksieiev SM, Tarabrin MB, Godonu Zh, Savchenko GV, Traore M. Comparison of the effectiveness of oil and water solutions of alpha-tocopherol as radiation protection agent. Fiziol Zh 1993; 39:89-91.

50. Paranich AV, Oleksieiev SM, Tarabrin MB, Godonu Zh, Savchenko GV, Ortis L. Effectiveness of various pharmaceutical forms of alpha-tocopherol. Fiziol Zh 1993; 39:91-93.

51. Yao T, Esposti SD, Huang L, Arnon R, Spangenberger A, Zern MA. Inhibition of carbon tetrachloride-induced liver injury by liposomes containing vitamin E. Am J Physiol 1994; 267:G476-G484.

52. Lai M, Duzgunes N, Szoka FC. Effects of replacement of the hydroxyl group of cholesterol and tocopherol on the thermotropic behavior of phospholipid membranes. Biochemistry 1985; 24:1646-1653.

53. Boni LT, Perkins WR, Minchey SR, Bolcsak LE, Gruner SM, Cullis PR, Hope MJ, Janoff AS. Polymorphic phase behavior of alpha-tocopherol hemisuccinate. Chem Phys Lipids 1990; 54:193-203.

54. Brase DA, Westfall TC. Stimulation of rat liver phenylalanine hydroxylase activity by derivatives of vitamin E. Biochem Biophys Res Comm 1972; 48:1185-1191.

55. Chelliah J, Smith JD, Fariss MW. Inhibition of cholinesterase activity by tetrahydroaminoacridine and the hemisuccinate esters of tocopherol and cholesterol. Biochem Biophys Acta 1994; 1206:17-26.

56. Suzuki YJ, Packer L. Inhibition of NF-κB DNA binding activity by α-tocopheryl succinate. Biochem Mol Biol Int 1993; 31:693-700.

57. Turley JM, Sanders BG, Kline K. RRR-α-tocopheryl succinate modulation of human promyelocytic leukemia (HL-60) cell proliferation and differentiation. Nutr Cancer 1992; 18:201-213.

58. Fariss MW, Fortuna MB, Everett CK, Smith JD, Trent DF, Djuric Z. The selective antiproliferative effects of α-tocopheryl hemisuccinate and cholesteryl hemisuccinate on murine leukemia cells result from the action of the intact compounds. Cancer Res 1994; 54:3346-3351.

59. Turley JM, Funakoshi S, Ruscetti FW, Kasper J, Murphy WJ, Longo DL, Birchenall-Roberts MC. Growth inhibition and apoptosis of RL Human B Lymphoma cells by vitamin E succinate and retinoic acid: Role for transforming growth factor β. Cell Growth Differentiation 1995; 6:655-663.

60. Charpentier A, Groves S, Simmons-Menchaca M, Turley J, Zhao B, Sanders BG,

Kline K. RRR-α-tocopheryl succinate inhibits proliferation and enhances secretion of transforming growth factor-β (TGF-β) by human breast cancer cells. Nutr Cancer 1993; 19:225–239.

61. Prasad KN, Edwards-Prasad J. Vitamin E and cancer prevention: Recent advances and future potentials. J Am Coll Nutr 1992; 11:487–500.

62. Toivanen JL. Effects of selenium, vitamin E and vitamin C on human prostacyclin and thromboxane synthesis in vitro. Prostaglandins Leukotrienes Med 1987; 26:265–280.

63. El Attar TMA, Lin HS. Effect of vitamin C and vitamin E on prostaglandin synthesis by fibroblasts and squamous carcinoma cells. Prostaglandins Leukotrienes Essential Fatty Acids 1992; 47:253–257.

64. Sanders AP, Hall JH, Woodhall B. Succinate: protective agent against hyperbaric oxygen toxicity. Science 1965; 150:1830–1833.

65. Maliuk VI, Molotkov VN, Korzhov VI. Metabolic disorders in acute poisoning and their correction by sodium succinate. Vrach Delo 1981; 8:66–72.

66. Bindoli A, Covallini L, Jocelyn P. Mitochondrial lipid peroxidation by cumene hydroperoxides and its prevention by succinate. Biochem Biophys Acta 1982; 681:496–503.

67. Meszaros L, Tihanyi K, Horvath I. Mitochondrial substrate oxidation-dependent protection against lipid peroxidation. Biochem Biophys Acts 1982; 713:675–677.

68. Chance B, Hollunger G. The interaction of energy and electron transfer reactions in mitochondria. I. General properties and nature of the products of succinate linked reduction of pyridine nucleotides. J Biol Chem 1961; 236:1534–1538.

69. Kondrashova MN, Gogvadze VG, Medvedev BI, Babsky AM. Succinic acid oxidation as the only energy support of intensive Ca^{2+} uptake by mitochondria. Biochem Biophys Res Commun 1982; 109:376–383.

70. Rognstad R. Gluconeogenesis in rat hepatocytes from monomethyl succinate and other esters. Arch Biochem Biophys 1984; 230:605–609.

71. Mapes JP, Harris RA. On the oxidation of succinate by parenchymal cells isolated from rat liver FEBS Lett 1975; 51:80–83.

72. Cadenas E, Ginsberg M, Rabe U, Sies H. Evaluation of α-tocopherol antioxidant activity in microsomal lipid peroxidation as detected by low-level chemiluminescence. Biochem J 1984; 223:755–759.

73. Kagan VE, Bakalova RA, Serbinova EE, Stoytchev TS. Fluorescence measurements of incorporation and hydrolysis of tocopherol and tocopherol esters in biomembranes. Meth Enzymol 1990; 186:355–367.

74. Kagan VE, Serbinova EA, Bakalova RA, Stoytchev Ts S, Erin AN, Prilipko LL, Evstigneeva RP. The role of the hydrocarbon chain in the inhibition of lipid peroxidation. Biochem Pharmacol 1990; 40:2403–2413.

75. Schmidt D, Steffen H, Von Planta C. Lateral diffusion, order parameter and phase transition in phospholipid bilayer membranes containing tocopherol acetate. Biochim Biophys Acta 1976; 443:1–9.

76. Traber MG. Determinants of plasma vitamin E concentrations. Free Rad Biol Med 1994; 16:229–239.

77. Cohn W, Gross P, Grun H, Loechleiter F, Muller DPR, Zulauf M. Tocopherol transport and absorption. Proc Nutr Soc 1992; 51:179–188.

78. Perly BI, Smith CP, Hughes L, Burton GW, Ingold KU. Estimation of the location of natural alpha tocopherol in lipid bilayers by C-NMR spectroscopy. Biochim Biophys Acta 1985; 819:131–135.

79. Kornburst DJ, Mavis RD. Relative susceptibility of microsomes from lung heart, liver, kidney, brain and testes to lipid peroxidation: Correlation with vitamin E content. Lipids 1980; 15:315–322.

80. Liebler DC, Kling DS, Reed DJ. Antioxidant protection of phospholipid bilayers by α-tocopherol. J Biol Chem 1986; 261:12114–12119.

81. Leedle RA, Aust SD. Importance of the polyunsaturated fatty acid to vitamin E ratio in the resistance of rat lung microsomes to lipid peroxidation. J Free Rad Biol Med 1986; 2:397–403.

82. Murphy DJ, Mavis RD. A comparison of the in vitro binding of α-tocopherol to microsome of lung, liver, heart, and brain of the rat. Biochim Biophys Acta 1981; 663:390–400.

83. Niki E, Kawakami A, Saito M, Yamamoto Y, Tsuchiya J, Kamiya Y. Effect of phytl side chain of vitamin E on its antioxidant activity. J Biochem 1985; 260:2191–2196.

84. Catignani GL, Bieri JG. Rat liver α-tocopherol binding protein. Biochim Biophys Acta 1977; 497:349–357.

85. Murphy DJ, Mavis RD. Membrane transfer of α-tocopherol: Influence of soluble α-tocopherol-binding factors from the liver, lung, heart, and brain of the rat. J Biol Chem 1981; 256:10464–10468.

86. Dutta-Roy AK, Gordon MJ, Leishman DJ, Paterson BJ, Duthie GG, James WPT. Purification and partial characterization of an α-tocopherol-binding protein from rabbit heart cytosol. Mol Cell Biochem 1993; 123:139–144.

87. Guarnieri C, Flamigni F, Caldarera CM. A possible role of rabbit heart cytosol tocopherol binding in the transfer of tocopherol into nuclei. Biochem J 1980; 190:469–471.

88. Mowri H, Nakagawa Y, Inoue K, Nojima S. Enhancement of the transfer of α-tocopherol between liposomes and mitochondria by rat liver protein(s). Eur J Biochem 1981; 117:537–542.

89. Behrens WA, Madere LT. Transfer of α-tocopherol to microsomes mediated by a partially purified liver α-tocopherol binding protein. Nutr Res 1982; 2:611–618.

90. Pascoe GA, Reed DJ. Cell calcium, vitamin E, and the thiol redox system in cytotoxicity. Free Rad Biol Med 1989; 6:209–224.

91. Slack R, Proulx P. Studies on the effects of vitamin E on neuroblastoma N1E 115. Nutr Cancer 1989; 12:75–83.

92. Traber MG, Thellman CA, Rindler MJ, Kayden HJ. Uptake of intact TPGS (d-α-tocopherol polyethylene glycol 1000 succinate) a water-miscible form of vitamin E by human cells in vitro. Am J Clin Nutr 1988; 48:605–611.

93. Heymann E. Carboxylesterases and amidases. In: Jaakoby WB, ed. Enzymatic Basis of Detoxication. Vol. 2. New York: Academic Press, 1980:291–323.

94. Thomas CE, Reed DJ. Effect of extracellular Ca^{++} omission on isolated hepatocytes. II. Loss of mitochondrial membrane potential and protection by inhibitors of uniport Ca^{++} transduction. J Pharmacol Exp Therap 1988; 245:501–507.

95. Jones DP, Lash LH. Introduction: Criteria for assessing normal and abnormal mitochondrial function. In: Lash LH, Jones DP, eds. Methods in Toxicology. Vol. 2. Mitochondrial Dysfunction. San Diego: Academic Press, 1993:1–7.

96. Vatassery GT, Smith WE, Quach HT. Increased susceptibility to oxidation of vitamin E in mitochondrial fractions compared with synaptosomal fractions from rat brains. Neurochem Int 1994; 24:29–35.

97. Tappel AL, Zalkin H. Inhibition of lipid peroxidation in mitochondria by vitamin E. Arch Biochem Biophys 1959; 80:333–336.

98. Thomas PK, Cooper JM, King RHM, Workman JM, Schapiria AHV, Goss-Sampson MA, Muller DPR. Myopathy in vitamin E deficient rats: Muscle fibre necrosis associated with disturbances of mitochondrial function. J Anat 1993; 183:451–461.

99. Wiesner R, Ludwig P, Schewe T, Rapoport SM. Reversibility of the inhibition of cytochrome c oxidase by reticulocite lipoxygenase. FEBS Lett 1981; 123:123–126.

100. Bjorneboe GE, Bjorneboe A, Hagen BF, Morland J, Drevon CA. Reduce hepatic α-tocopherol content after long-term administration of ethanol to rats. Biochim Biophys Acta 1987; 918:236–241.

101. Olafsdottir K, Pascoe GA, Reed DJ. Mitochondrial glutathione status during Ca^{2+} ionophore-induced injury to isolated hepatocytes. Arch Biochem Biophys 1988; 263:226–235.

102. Guidoux R, Lambelet P, Phoenix J. Effects of oxygen and antioxidants on the mitochondrial Ca-retention capacity. Arch Biochem Biophys 1993; 306:139–147.

5

Phenolic Antioxidants: Physiological and Toxicological Aspects

Regine Kahl
University of Düsseldorf, Düsseldorf, Germany

In the last decade, evidence from abundant animal and in vitro experimentation has suggested a role of oxidative stress in the pathophysiology of disease. This has given rise to the anticipation that antioxidants could be powerful agents to be used in the prevention and/or therapy of oxidant-related disease. In the past, the term "antioxidant" was used to characterize a compound with chain-breaking properties in the process of lipid peroxidation, and when biological actions of synthetic phenolic antioxidants were first examined it may frequently have been presumed that this type of action was involved. In recent years, the meaning of the term antioxidant has expanded to comprise virtually all activities directed toward the formation and action of oxidants. This adds considerably to the explanations of biological effects of synthetic phenolic antioxidants but still does not exhaust the possibilities by which these compounds can interact with biological material and which include "paradoxical" prooxidative actions. This chapter intends to give insight into the complicated and divergent mechanisms by which synthetic phenolic antioxidants may exert their effects in living organisms.

I. PREVENTION OF ATHEROSCLEROSIS

Atherosclerosis and coronary heart disease are among the most widespread diseases for which a causal link to oxidative stress is assumed to exist and for which chemoprevention by antioxidants is currently investigated. The focus of this discussion is on micronutrients, especially vitamin E and β-carotene. Two recent retrospective studies on 87,245 women (1) and 39,910 men (2) indicating a reduction of coronary events by supplementation with vitamin E have stimulated the interest in this topic. Synthetic antioxidants have also been proposed to be of potential value.

More than one mechanism may be involved in the adverse effect of oxidants on arterial vessels. The most advanced hypothesis concerns the oxidation of LDL and the enhanced uptake of oxidized LDL by macrophages leading to enhanced cholesterol deposition in and damage to the arterial wall (3). The mechanisms of LDL oxidation have been reviewed, and data on in vitro protection of LDL against oxidation by the food additives BHT, BHA, and PG and the hypocholesterolemic BHT analog probucol have been compiled (4). Parthasarathy et al. (5) demonstrated the prevention of LDL oxidation by probucol. The role of probucol as an antioxidant in the atherosclerotic process has recently been reviewed (6). The drug is an inhibitor of lipid peroxidation (7,8); whether it is a scavenger of O_2^- has not been settled (9,10). The probucol phenoxyl radical can be reduced by ascorbic acid providing the basis for a synergistic action of the two antioxidants (11). Besides the LDL lipids, the protein component of LDL, apolipoprotein B (apo B) which is critical for receptor interaction of LDL is also modified by oxidants. The fragmentation of apo B was reported to be inhibited by BHT and PG in an early paper by Schuh et al. (12). More recently, vitamin E and probucol were compared in their ability to prevent Cu(II)-induced apo B fragmentation and vitamin E was found to be superior; in this study, only vitamin E was able to inhibit uptake of oxidized LDL into macrophages (13). Others did not detect inhibition of apo B fragmentation by probucol, though the drug was very efficient in inhibiting lipid peroxidation in the LDL particle (14).

A recent review concentrates on pharmacological studies performed in animals, predominantly with probucol (15). The reason why most studies have been performed with probucol is probably its established role as a hypocholesterolemic agent in man allowing for use in therapeutic trials. In addition, it has favorable pharmacokinetic properties in that it partitions into the lipid phase where oxidation takes place and is transported in the plasma primarily with LDL itself (16). The review (15) cites 12 studies with probucol in either cholesterol-fed rabbits or the WHHL (Watanabe heritable hyperlipidemic) rabbit 10 of which came up with decreased progression of atherosclerotic lesions. A regression of preexisting atherosclerotic lesions in WHHL rabbits was demonstrated in two high-dose probucol studies, while a third one performed with a low dose in an attempt to mimic the situation in humans was negative. Studies in primates have also been performed. In *Macaca nemestrina* a marked reduction of aortic atherosclerosis was found while the number of foam cells was increased (17).

The protective effect of probucol in the animal studies could have been due to the reduction of cholesterol levels; however, in some of the studies, plasma cholesterol was not significantly decreased in spite of the antiatherogenic effect. BHT is a potentially suitable tool to discriminate between both mechanisms because it does not share the hypocholesterolemic action of probucol. Administration of BHT to cholesterol-fed rabbits resulted in a marked antiatherogenic effect along with an increase in plasma cholesterol and triglycerides (18) and a reduction of the microcirculatory changes in the conjunctivae (19). These findings argue for a genuine role of the antioxidant action in the prevention of atherosclerosis. On the other hand, protection by the antioxidative probucol analogs MDL 29.311 and BM 15.0639 which have no effect on plasma cholesterol did not reach statistical significance in WHHL rabbits (20,21). In a recent study in cholesterol-fed rabbits, vitamin E and probucol were both efficient in inhibiting LDL oxidation, but only probucol was capable to lower the extent of atherosclerotic lesions in the aorta, indicating that antioxidant activity was not sufficient to reduce atherosclerosis (22).

A number of probucol effects on endothelial function and integrity have been identified in animal and cell studies. Probucol ameliorates oxidative injury to endothelial cells (23) and may suppress the release of increased amounts of ROS which in turn will oxidize LDL (6). In a rabbit aortic ring model, probucol prevented the inhibition of acetylcholine-induced EDRF-dependent relaxation after cholesterol feeding (24,25,26). A decrease of the expression of interleukin-1β in vitro (27) and in vivo (28) has been described but was not confirmed in another in vitro study (29). The increased adhesion of mononuclear cells to endothelium in cholesterol-fed animals in vivo and to endothelial cell monolayers ex vivo was inhibited by administration of probucol for 5 weeks (30). Pretreatment of human umbilical vein endothelium with vitamin E or probucol-inhibited agonist-stimulated adhesion of monocytes, and evidence was provided with vitamin E that an increase in mRNA levels and cell surface expression of E-selectin was responsible for this effect (31). These actions may contribute to the antiatherosclerotic activity of probucol.

In humans, probucol has been tested in a clinical trial (PQRST, Probucol Quantitative Regression Swedish Trial) involving 300 patients treated for 3 years with either diet + colestyramine + placebo or diet + colestyramine + probucol (500 mg). No statistically significant difference in lumen volume, arterial edge roughness, and aortofemoral atherosclerosis between treatment groups was found (32). LDL from patients from this (33) and other studies (34–37) do, however, demonstrate increased resistance of LDL to oxidation in vivo and/or ex vivo. Interestingly, ex vivo experiments also demonstrated prevention of oxidation of lipoprotein (a), an important determinant of coronary heart disease, after probucol treatment (38). Probucol has successfully been used for the prevention of restenosis after percutaneous transluminal coronary angioplasty in humans (39), and this effect has also been demonstrated in animal models of balloon injury (40–43). The doses applied in human studies were considerably lower than those used in animal experiments and probably cannot be increased much because probucol has adverse effects on the ECG in some patients (44), although in general the rate of adverse effects in a postmarketing surveillance study has been low (45).

II. MODULATION OF TUMORIGENESIS

The extensive research on mechanisms of biological action of synthetic phenolic antioxidants carried out in the 1970s and 1980s was triggered by a paper in 1972 by Wattenberg demonstrating inhibition of tumor formation from benzo-(a)pyrene and 7,12-dimethylbenz(a)anthracene in a variety of mouse tissues by BHA, BHT and EQ (46). Table 1 gives a compilation of positive findings elaborated since then. The early enthusiasm was, however, considerably dampened when it became apparent that the outcome of such studies was not always beneficial. Enhancement of chemical carcinogenesis could be obtained with the very carcinogens and often in the same tissues for which protection had been observed in other studies (Table 1). As a rule, pretreatment or concomitant treatment resulted in protection, while posttreatment could, but need not necessarily, result in enhancement. This can be mimicked in a cell transformation model of two-stage carcinogenesis using benzo(a)pyrene as the initiating carcinogen and BHA or BHT as the modifier (Fig. 1). Administration of the antioxidants together with the carcinogen decreased transformation frequency, while administration of the antioxidants subsequent to the carcinogen increased it (47). In animal studies, posttreatment protection may, however, also be obtained (48,49,50), and in the bladder enhancement of tumor formation occurs regardless of the sequence of administration (51). Witschi has recently provided a metaanalysis of 33 studies performed with dietary BHT (51). Statistically significant effects were obtained in only 41% of the experiments in rats and 22% of the experiments in mice. Overall tumor incidence with BHT was 41.0% and without BHT, 43.5%. This overall figure was derived from 26% (8%) protective experiments in rats (mice) and 15% (14%) enhancing experiments in rats (mice). Simultaneous exposure was invariably protective in rat liver, rat mamma, rat stomach, rat pancreas, rat adrenals, mouse stomach and mouse lung. Postfeeding enhanced tumor formation in rat liver, rat colon, rat pancreas, rat thyroid, rat esophagus and mouse lung. Protection occurred in rat mammal and rat ear duct regardless of the feeding protocol, and enhancement only was found in rat bladder regardless of the feeding protocol. Hirose et al. (52) in a multiorgan carcinogenesis model in rats administered a combination of six carcinogens (N-methyl-N'-nitro-N-nitrosoguanidine, N-ethyl-N-hydroxyethyl-nitrosamine, N-methylbenzylnitrosamine, 1,2-dimethylhydrazine, N-dibutyl-nitrosamine and 2,2'-dihydroxy-di-n-propylnitrosamine) followed by posttreat-

Table 1 Modulation of Chemically Induced Tumorigenesis by Synthetic Phenolic Antioxidants

Tissue	Protection	Antioxidant	Ref.	Enhancement	Antioxidant	Ref.
Rat liver	Nitrosamines	BHA, BHT, EQ	265, 266, 268, 269	Nitrosamines	BHA, BHT	267
	Aflatoxin B_1	BHA, BHT	270			
	2-Acetylaminofluorene	BHT, PG	271, 272, 275, 276	2-Acetylaminofluorene	BHT	273, 274
	N-hydroxy-2-acetylaminofluorene					
	Ciprofibrate	BHA	277			
Rat esophagus	Diethylnitrosamine	BHT	269	Dibutylnitrosamine	BHT, EQ	278
Rat forestomach				N-Methyl-N'-nitro-N-nitrosoguanidine	BHA, BHT, EQ	279, 280, 281, 282
				Methylnitrosourea	BHA	283, 284
				Dibutylnitrosamine	BHA	278
Mouse forestomach	Benzo(a)pyrene	BHA, BHT	46, 285			
	Benzo(a)pyrene-7,8-dihydrodiol	BHA	46			
	7,12-Dimethylbenz(a)anthracene	BHT	286			
	Diethylnitrosa ine	BHT	287			
Rat stomach	N-Methyl-N'-nitro-N-nitrosoguanidine	BHT	288, 289	1,2-Dimethylhydrazine	BHT, EQ	289, 290
Rat intestine	1,2-Dimethylhydrazine	BHT		1,2-Dimethylhydrazine	BHT	294
Mouse intestine	Azoxymethane	BHT, BHA	291, 292, 293			
	Methylazoxymethanol acetate					
	1,2-Dimethylhydrazine					
Rat pancreas	Azaserine	BHA	295	Azaserine	BHT	296
Hamster pancreas	Azaserine	BHA	296			
	Dihydroxy-n-propylnitrosamine	BHA	297, 50	Dihydroxy-n-propyl-nitrosamine	BHA	298

(continued)

Table 1 Continued

Tissue	Protection	Antioxidant	Ref.	Enhancement	Antioxidant	Ref.
Mouse lung	Benzo(a)pyrene	BHA	46, 285, 299	Benzo(a)pyrene	BHT	300
	Benzo(a)pyrene-7,8-dihydrodiol					
	Dibenz(a)anthracene	BHA	301	3-Methylcholanthrene	BHT	300, 302
	7,12-Dimethylbenz(a)anthracene	BHA	46, 301			
	N-hydroxy-7,12-dimethyl-benz(a)anthracene					
	Nitrosamines	BHA	303, 304	Nitrosamines	BHT	300
	Isoniazid	BHA, BHT	305			
	4-Nitroquinoline-N-oxide	BHA	303	N-Nitrosopyrrolizidine	BHA	304
	Urethan	BHT	299, 301, 307, 308,	Urethan	BHT	300, 306, 308
Rat lung	Dihydroxy-n-propylnitrosamine	BHA, BHT, EQ	309			
Rat mamma	7,12-Dimethylbenz(a)anthracene	BHA, BHT, EQ, TBHQ, PG	272, 310, 311, 312, 48, 313			
Rat bladder				Nitrosamines	BHA, BHT, TBHQ	265, 268, 267, 283, 314, 315, 316
				Methylnitrosourea	BHA, BHT	283, 284
Rat thyroid				2-Acetylaminofluorene	BHT	275, 276
				Nitrosamines	BHT	268, 317
				Methylnitrosourea	BHT	283
Rat ear duct	Azoxymethane	BHT	292			
	7,12-Dimethylbenz(a)anthracene	BHT	318			
Mouse skin	7,12-Dimethylbenz(a)anthracene	BHA, BHT	319			

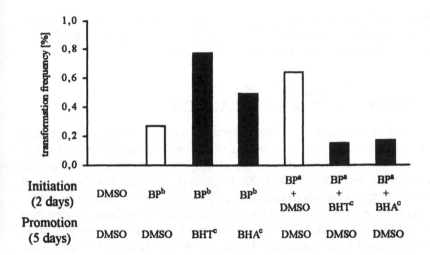

Figure 1 Potentiation and inhibition of morphological transformation of cultured Syrian hamster embryo cells by BHT and BHA. [a]BP 0.1 μg × ml⁻¹; [b]BP 1 μg × ml⁻¹; [c]BHT, BHA 100 μM. (From Ref. 47).

ment with TBHQ, PG, BHT or α-tocopherol starting 3 days after cessation of carcinogen administration. In these experiments, TBHQ enhanced hyperplasia and papillomas in esophagus and forestomach but decreased the multiplicity of colon adenocarcinomas. PG was inhibitory only, reducing kidney atypical tubules. BHT reduced the incident of colon adenocarcinomas but increased thyroid hyperplasia. α-Tocopherol reduced the incidence and multiplicity of kidney atypical tubules but increased the incidence of preneoplastic foci in the stomach.

The procarcinogenic effects of BHA and BHT are most probably due to a tumor-promoting activity residing in one or several active metabolites and/or reactive oxygen species formed during metabolism (see Sec. IV). BHA and BHT fulfill a number of criteria elaborated in vitro for tumor promoters:

1. Enhancement of transformation frequency in cell transformation tests by BHT and BHA (53,47).
2. Induction of ornithine decarboxylase in rat liver by BHT (54).
3. Inhibition of metabolic cell-cell cooperation by BHT and BHA (55–58).

Recently, tumor promotion by BHT was shown to increase the frequency of Ki-ras mutations in 3-methylcholanthrene- or urethan-induced lung tumors in mice (59). Tumor-promoting activity is not ascribed to vitamin E or β-carotene, and these micronutrients are currently tested for chemoprevention of cancer. Published intervention trials with vitamin E and β-carotene came up with inconsistent results (60,61), and further studies are reportedly under way. For the synthetic phenolic antioxidants BHA and BHT, the dual nature of their influence on carcinogenesis precludes their use for chemoprevention.

III. POSSIBLE MECHANISMS OF ANTICARCINOGENIC ACTION

Induction of xenobiotic-metabolizing enzymes is the most thoroughly elaborated explanation for the protective properties of antioxidants against mutagens and carcinogens. A wide variety of enzymes directly or indirectly involved in drug metabolism can be induced (Table 2). In general, the inducer effect is much more marked with nonmonooxygenation enzymes than with cytochrome P-450-dependent enzyme activities. Therefore, the antioxidants have been grouped with the "monofunctional inducers," i.e., inducers of phase II enzymes and other enzymes engaged in detoxication only (62). A shift to detoxication may, however, also occur during phase I metabolism by the induction of specific cytochromes P-450 forming innoxious metabolites of a given compound. Early reports suggest that in rat liver antioxidants act on a phenobarbital-inducible cytochrome P-450 species (63,64). In mouse lung and intestine but not in mouse liver BHA increased the content of cytochrome P_1-450 mRNA, while cytochrome P_3-450 mRNA was found to respond to BHA in all three tissues (65). Benzo(a)pyrene metabolism has been used as a tool to study shifts in the cytochrome P-450 pattern by antioxidant treatment (66–69). However, these findings are complicated by the fact that epoxide hydrolase, an enzyme which was among the first postmonooxygenation enzymes reported to be induced by syn-

Table 2 Enzymes Inducible by Synthetic Phenolic Antioxidants

Cytochrome P-450, NADPH-cytochrome P-450 reductase
Cytochrome b_5 content, NADH-cytochrome b_5-reductase
Epoxide hydrolase
Glutathione transferase
Glutathione content
γ-Glutamyl transpeptidase
γ-Glutamylcysteine synthetase
Glutathione reductase
NAD(P)H-quinone oxidoreductase
Glucuronyl transferase
UDP-glucuronic acid content
UDP-glucose dehydrogenase
Aldehyde dehydrogenase

---

thetic antioxidants (63,70,64) plays a dual role in the toxication/detoxication balance at least for polycyclic aromatic hydrocarbons with a bay region because the bay region dihydrodiols are precursors of the carcinogenic diol epoxides as first shown for benzo(a)pyrene (71).

Among phase II enzymes, the glutathione transferases (GSTs) have received most attention with respect to induction by antioxidants. The cytosolic GSTs can be divided into four families: alpha, mu, pi (72) and theta (73). They are homodimers or heterodimers of family-specific subunits; 13 subunits have been characterized so far. Benson et al. (74) were the first to report induction of GST activity in mice by pretreatment with BHA in 1978. More recently, tissue-specific inducibility of GSTs by BHA was studied in the mouse (Table 3). While mu- and pi-class GSTs are readily inducible by BHA in all tissues examined, most investigators have described selective induction of one or two class alpha subunits, while another form was nonresponsive to BHA treatment. In rat liver induction of the rat alpha class GST subunit Yc2 (active in the conjugation of aflatoxin B1-8,9-epoxide) by EQ, BHA and BHT was detected; EQ was the most active inducer for Yc2 and for some other subunits from different GST families, Yc1, Yb1, and Yf (75).

In addition to GSTs, a number of other enzymes related to glutathione formation and function, γ-glutamylcysteine synthetase (76,77), glutathione reductase (78,79) and γ-glutamyl transpeptidase (80,81) are induced by synthetic phenolic antioxidants. Induction of γ-glutamylcysteine synthetase is the most likely reason for the well-known elevation of the glutathione content (early findings reviewed in (82).

Inducibility of NAD(P)H-quinone-oxidoreductase (DT diaphorase, quinone reductase) by BHA in liver was first described in 1980 by Benson et al. (83). Quinones react with cellular nucleophiles leading to functional impairment by arylation. A major target is glutathione; depletion of glutathione will render protein thiols more sensitive to oxidation and alkylation. The two-electron transfer type of the reaction circumvents the formation of semiquinone radicals and thus prevents the formation of $O_2^{\cdot-}$. It has, however, been pointed out that the protective outcome depends on the redox stability of the hydroquinones formed (84). In cases where the hydroquinones formed undergo superoxide dismutase-supported autoxidation or are capable of rearranging into arylating agents, quinone reductase can mediate toxicity. Protective actions of the enzyme have been reviewed (85). Protection can also be achieved by indirect mechanisms: The inhibition of ethanol-induced triglyceride accumulation in the liver by BHA administration was attributed to increased supply of the limiting cofactor in ethanol metabolism, NAD^+, by increased activity of the enzyme (86).

Increase of glucuronyl transferase activity has been shown in the early period of induction research with synthetic antioxidants (87,88) and is now known to depend on increased gene expression (89). In addition, the hepatic content of UDP-glucuronic acid (90) and the activity of UDP-glucose dehydrogenase

Table 3　Tissue-Specific Expression of Glutathione Transferases by BHA in the Mouse

GST form	Inducible in	Noninducible in	Ref.
Class alpha			
mRNA detected with pGT41 (Ya)			
GT-10.3	Liver, intestine, kidney	Lung	65
GT-10.6	Liver		320
Ya₁-Ya₁	Liver[a]	Liver	321
Ya₃Ya₃			
Ya₂ subunit	Liver	Liver	322
Ya subunit	Intestine, kidney, (female > male)	Lung[a]	323
Yk subunit	Intestine, kidney (female > male)	Lung	
Ya mRNA	Intestine, kidney, liver[a]		324
Yc mRNA		Liver, lung, intestine[a], kidney[a]	
Class mu			
mRNA detected with pGT875 (GT-8.7)			
GT-8.7	Liver, lung, intestine, kidney		65
GF-9.3	Liver		320
Yb subunit	Liver[a]	323	321
	Liver, lung, intestine, kidney		
Class pi			
pi mRNA	Liver, lung, intestine, spleen	65	
GT-9.0	Liver (females only)		320
Yf subunit	Liver		321
	Intestine, Kidney	323	

[a]Constitutive expression low or absent.

(87) are also elevated. In spite of the prominent function of glucuronyl transferase in the detoxication of xenobiotics, its inducibility by antioxidants has not attracted as much attention as that of glutathione transferase and quinone reductase.

Aldehyde dehydrogenase (AIDH) is the most recent member of the antioxidant-inducible enzyme group. TBHQ caused an increase in AIDH mRNA in mouse Hepa-1 cells (91), and induction by EQ, BHA and BHT of the form which metabolizes the cytotoxic dialdehydic form of aflatoxin B1 was observed in rat liver (75). In human breast adenocarcinoma cells, induction of AIDH-3 mRNA, protein and activity by catechol paralleled the development of resistance against the alkylating cytostatic agent mafosfamide. BHT and hydroquinone were also active as inducers in this tumor cell line (92). This example illustrates that cancer prevention properties of an antioxidant may well coincidence with impairment of cancer chemotherapeutic potential.

From a number of findings it appears likely that enzyme induction is only one of several mechanisms contributing to protection against chemical carcinogens:

1. Antimutagenic activity and inhibition of covalent binding to DNA have been achieved under in vitro conditions excluding induction as a protective mechanism (82).
2. In some animal studies showing protection, the antioxidant has been administered subsequent to the carcinogen, again precluding an induction effect (48,49,50,93).
3. The antioxidant vitamins C and E also protect against chemical carcinogenesis but lack the inducer properties of the synthetic phenolic antioxidants (94) indicating that the radical-scavenging character per se may be important in the inhibition of carcinogenesis.

A review on the various processes by which radicals may be involved in the initiation phase of chemical carcinogenesis has been provided by Trush and Kensler (95). Radical intermediates may be formed which rearrange or are further metabolized by DNA-binding or toxic species. These radicals might be removed by the scavenger action of the antioxidant; however, experimental data on this possibility are lacking. For the 6-oxybenzo(a)pyrene radical, no reaction with BHA, BHT, EQ and PG was detected (96). Facilitation of the metabolic activation of the precarcinogen by radicals or ROS may take place, e.g., in the cooxygenation of various substrates during the prostaglandin H synthase reaction. BHA and BHT can inhibit the latter reaction, but since they are themselves subject to cooxygenation (97) the interpretation is difficult. N-Methyl-N'-nitro-N-nitrosoguanidine (MNNG) forms a radical upon reaction with hydrogen peroxide which can be scavenged by BHT. This action has been suggested to be responsible for the protective action of BHT against MNNG-induced gastric cancer (98).

Inhibition of cytochrome P-450 mediated metabolic activation of precarcinogens may also confer protection in the initiation phase of chemical carcinogenesis. Interference of antioxidants with monooxygenation was first described in 1974 by Yang et al. (99). BHA metabolites inhibit electron transfer from NADPH cytochrome P-450 reductase to cytochrome P-450 (100). In vivo administration of BHA or EQ leads to decreased concentrations of oxyferrous cytochrome P-450 (101), and in a reconstituted system of the bacterial camphor-metabolizing cytochrome P-450 CIA1 BHA causes decomposition of substrate-bound oxyferrous cytochrome P-450, indicating that the antioxidant may perform its inhibitory action via enhanced autooxidation of this species (102). In rat liver microsomes inhibition was more efficient with phenobarbital-inducible cytochrome P-450 than with polycyclic hydrocarbon-inducible cytochrome P-450 (63,64). However, inhibition of the secondary epoxidation of benzo(a)-pyrene-7,8-diol by BHA could also be obtained (103).

In the studies using concomitant treatment with the carcinogen and the antioxidant for prolonged time as well as in the few posttreatment experiments available, postinitiation targets of the antioxidant may additionally or exclusively be operative. Antipromotional activity was ascribed to antioxidants (104). An obvious explanation would be the removal of ROS; a role of ROS in tumor promotion has been suggested by their capability to mimic various biochemical actions of tumor promoters, by the ability of tumor promoters to produce ROS in a variety of cells, by the upregulation of antioxidant enzymes by tumor promoters and by the suppressive effect of antioxidants (105,106). A number of indicator effects for tumor-promoting activity can be inhibited by synthetic phenolic antioxidants, including induction of ornithine decarboxylase (107,108,109), cell transformation by 12-O-tetradecanoylphorbol-13-acetate (TPA) (110) and suppression of PKC activity (111,112).

IV. FORMATION OF REACTIVE METABOLITES AND ROS

The oxidative metabolism of BHA and BHT has been studied extensively. These studies provided evidence for the formation of reactive metabolites and of ROS both of which are assumed to be linked to the biological actions of these compounds. The main routes of oxidative metabolism of BHA are depicted in Fig. 2. BHA undergoes O-demethylation to the hydroquinone metabolite, TBHQ, followed by glucuronidation and sulfation, in the dog (113), in the rat (114), and in humans (115). TBHQ is further oxidized to *tert*-butylquinone (TBQ) in rat liver microsomes (100) and by a horseradish peroxidase/H_2O_2 system (116).

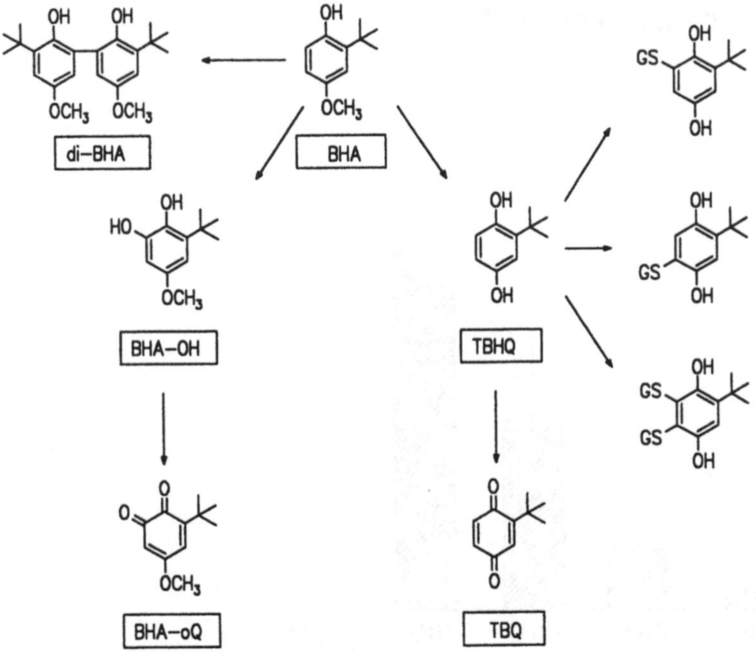

Figure 2 Oxidative metabolism of 3-BHA. TBHQ: *tert*-butylhydroquinone; TBQ: *tert*-butyl-1,4-benzoquinone; 3-BHA-OH: 3-*tert*-butyl-4,5-dihydroxyanisole; 3-BHA-oQ: 3-*tert*-butyl-5-methoxy-1,2-benzoquinone; di-3-BHA: 2,2'-dihydroxy-3,3'-di-*tert*-butyl-5,5'-dimethoxydiphenyl.

TBQ and/or other BHA metabolites appear extensively bound to proteins in microsomes (117). TBQ can be converted to the TBQ epoxide by H_2O_2 (116). BHA causes formation of O_2^-, H_2O_2 and $HO^.$ and increases NADPH consumption to liver microsomes (100,101,118–121). This requires metabolic conversion to TBHQ and further oxidation to the semiquinone radical TBQ, which upon autoxidation yields TBQ and O_2^- (121,122). Upon addition of TBQ to microsomes, $TBQ^.$ is detected (122), indicating that redox cycling occurs if reducing enzymes are present. TBQ is much more potent in eliciting the O_2^- burst than TBHQ both in rat liver microsomes and rat forestomach (Fig. 3), a target organ of BHA toxicity (see Sec. VI) (121). Three glutathione conjugates have recently been described to be formed from TBHQ. This cannot be considered a detoxication in terms of oxygen activation since all of them possess a higher redox cycling activity than the parent compound itself, the 6-GS-conjugate being the most efficient one with an increase of oxygen consumption by a factor of 11 over TBHQ (123).

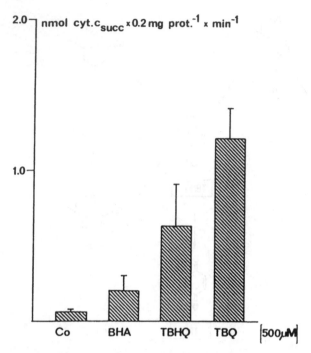

Figure 3 Formation of O_2^- in rat forestomach homogenate by BHA, TBHQ, and TBQ as measured by the stimulation of superoxide dismutase-inhibitable reduction of succinoylated cytochrome c. (From Ref. 121.)

Besides TBHQ, the ring-hydroxylated catechol metabolites (BHA-OH) and the corresponding quinones (BHA-oQ) are also found in liver microsomes and may contribute to oxygen activation (124). 3-*tert*-BHA-OH was more cytotoxic to rat hepatocytes than BHA itself and TBHQ, and BHA-oQ and TBQ were more efficient in inhibiting state 3 respiration in isolated hepatic mitochondria than BHA (125). Peroxidatic attack by an enzyme located in rat intestinal mucosa (126,127,128) or by horseradish peroxidase or prostaglandin H synthase (126,117,129) leads to the formation of a dimer, 2,2'-dihydroxy-3,3'-di-*tert*-butyl-5,5'-dimethoxydiphenyl (Di-BHA), from 3-BHA. The BHA phenoxyl radical is formed upon the reaction of BHA with organic radicals; it was recently shown to arise also from reaction of BHA with oxyhemoglobin, a process resulting in methemoglobin formation (130).

Figure 4 shows the oxidative metabolism of BHT (for a review 131). Three routes are apparent, starting with (1) oxidation of the methyl group, (2) pi-electron oxidation of the benzyl ring, and (3) hydroxylation of one of the butyl groups. Route 1, the main metabolic pathway in rat and man, is considered a detoxication pathway. Route 2 leads to at least two reactive metabolites linked

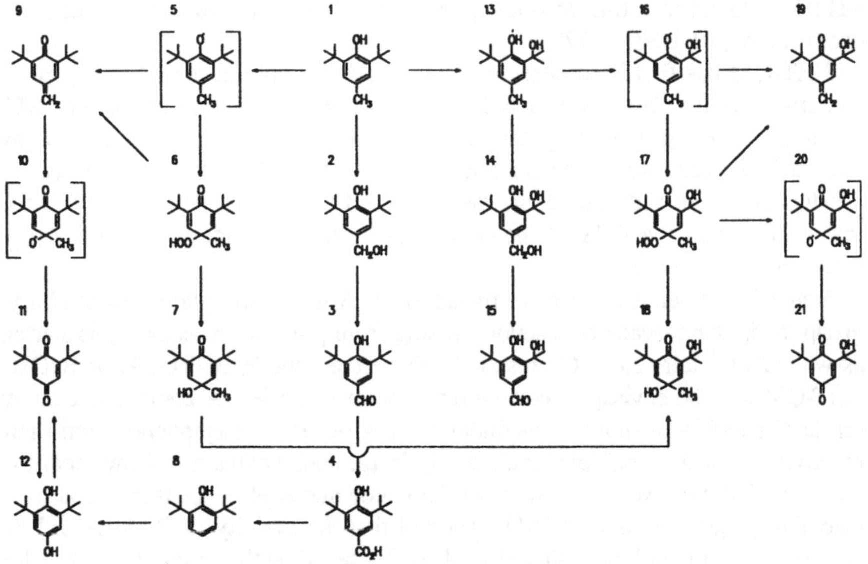

Figure 4 Oxidative metabolism of BHT. 1: BHT; 2: BHT-MeOH; 3: BHT-CHO; 4: BHT-COOH; 5: BHT phenoxyl radical; 6: BHT-OOH; 7: BHT-OH; 8: 2,6-di-*tert*-butylphenol (DBP); 9: BHT-QM; 10: BHT quinoxyl radical; 11: 2,6-di-*tert*-butyl-1,4-benzoquinone (DBQ); 12: 2,6-di-*tert*-butylhydroquinone (DBHQ); 13: BuOH-BHT; 14: BuOH-BHT-MeOH; 15: BuOH-BHT-CHO; 16: BuOH-BHT phenoxyl radical; 17: BuOH-BHT-OOH; 18: BuOH-BHT-OH; 19: BuOH-BHT-QM; 20: BuOH-BHT quinoxyl radical; 21: BuOH-1,4-benzoquinone.

to BHT toxicity, the hydroperoxide (BHT-OOH) and the quinone methide (BHT-QM). The formation of BHT-OOH by a monooxygenase reaction has been described in an early paper (132). BHT-OOH is metabolically reduced to the corresponding quinol, BHT-OH, which can undergo butyl group hydroxylation (133,134). BHT-OH will also be formed by attack of BHT-OOH on lipids; however, BHT-OOH causes little lipid peroxidation relative to other peroxides, e.g., *tert*-butyl hydroperoxide, because of the ease of intramolecular rearrangement as a competing reaction (131). Homolytic scission of BHT-OOH results in the expulsion of a methyl radical and formation of the quinoxyl radical (134) which could be one of the ultimate toxic species derived from BHT-OOH. The quinoxyl radical can be converted into the quinone, DBQ, or may rearrange to ring-expanded or ring-contracted products (not shown in Fig. 4) (134). BHT leads to increased formation of ROS in rat liver and more distinctly in rat lung microsomes (101,135,119). This may be due to redox cycling of DBQ in a process similar to that described above for TBQ (136). Recently, it has been proposed that DBQ can also arise in rats from route 1 by decarboxylation of

BHT-MeCOOH to 2,6-di-*tert*-butylphenol (DBP) and subsequent aromatic hydroxylation to DBHQ (137).

BHT-QM has first been observed as a BHT metabolite in the rat (138). BHA can stimulate the formation of BHT-QM from BHT and thus aggravate BHT pneumotoxicity, probably by the reduction of the BHA phenoxyl radical by BHT and the concomitant formation of BHT-QM (120,139). H_2O_2 formed by TBHQ-dependent redox cycling was suggested to provide the cosubstrate for the formation of the BHT phenoxyl radical by the peroxidase function of cytochrome P-450 (140).

Route 3 also leads to reactive metabolites. After hydroxylation of the butyl group, a similar cascade of reactions arising from pi-electron oxidation is started as with BHT itself (131). Correspondingly, either BuOH-BHTOOH or BuOH-BHT-QM may be envisaged as the ultimately toxic species. A comparative study on the formation of quinone methides from a series of alkylphenols and their reactivity toward glutathione and toxicity in rat hepatocytes has shown that the reaction of BuOH-BHT-QM with the indicator nucleophile glutathione is considerably faster than that of BHT-QM and that the toxicity of Bu-OH-BHT to hepatocytes is much higher than that of BHT itself though the quinone methides are formed by the same rate (141). For the other compounds in this series which had less than two bulky alkyl substituents adjacent to the oxo group and thus could be stabilized in a charge-separated state, the reaction with the nucleophile was very fast and not rate-limiting so that an excellent correlation was obtained between the rate of quinone methide formation and cytotoxicity.

There is evidence that the tumor-promoting activity of BHA and BHT is related to reactive products or ROS produced during their metabolism. H_2O_2 formed from the BHA metabolite TBHQ may well be involved in its tumor-promoting action. However, direct evidence for this is at present lacking. BHT-OOH is active as a tumor promoter in DMBA-initiated mouse skin (142) and as an inducer of ornithine decarboxylase in mouse epidermis (143), while BHT itself is inactive. BHT-QM has been proposed to be the ultimate tumor-promoting (144) and ornithine decarboxylase-inducing (145) species derived from BHT-OOH; a role of PKC in the induction of ornithine decarboxylase by the active metabolite was demonstrated. BuOH-BHT is four times more potent as a lung tumor promoter than BHT itself (97); the active species may be BuOH-BHTOOH or BuOH-BHT-QM.

Reactive intermediates of synthetic antioxidants other than BHA and BHT have not been described. In hepatocytes PG was more active than its metabolites in causing cell death and inhibition of state 3 mitochondrial respiration (146). EQ, like PG (146) and BHA (125), is also an inhibitor of mitochondrial respiration (147). H_2O_2 formation due to PG and its congeners and to EQ was observed in rat liver microsomes (101), but suppression rather than enhancement of HO˙ formation was found with gallic acid ester antioxidants (119). On the other hand, a recent paper reports that PG stimulated the formation of a deoxyribose degrading species, probably HO˙, from a Fenton system and accelerated DNA damage by bleomycin (148).

V. INFLUENCE ON THE CONTROL OF PHYSIOLOGICAL PROCESSES

In the last few years, there has been growing awareness that the low levels of oxidants formed by normal tissues may play an important role in the control of physiological processes such as signal transduction, calcium homeostasis, gene expression, cell proliferation and apoptosis and that elevated oxidant levels cause functional damage to cells before oxidative damage of tissue constituents is apparent. Antioxidants may protect cellular control mechanisms, but it is also conceivable that they interfere with essential oxidant functions. These possibilities have preferentially been examined with physiological antioxidants such as α-tocopherol or thiol compounds, but a few topics have also been addressed with synthetic phenolic antioxidants.

A. Calcium Homeostasis, Contractility, and Platelet Function

ROS increase free cystolic calcium concentration. A number of mechanisms have been identified: Ca^{2+}-ATPases are inhibited by oxidation of a critical SH group (149), possibly via the formation of HO˙ (150), and translocation of calcium from the cytosol into the extracellular space and into internal stores is thus impaired (151,152). In addition to reuptake inhibition, the release of calcium from the endoplasmic reticulum can be stimulated (153,154,155). Calcium from mitochondrial stores is also released (156). Calcium influx from the extracellular space both by L-type channel activity (157) and by agonist-sensitive pathways (154) has been reported to be increased, but inhibition of agonist-dependent influx can also be caused by oxidants (152).

While these effects have mostly been detected at subtoxic ROS concentrations, it is conceivable that a physiological oxidant tone contributes to the maintenance of calcium homeostasis. In this case, antioxidants should be expected to lower the calcium concentration in the cytosol and to impair calcium-dependent functions. The inhibition of voltage-operated calcium channels by BHA, BHT, or PG in brain synaptosomes (158), by BHA in the rat ileum (159), and with 2,5-di-*tert*-butylhydroquinone (2,5-TBHQ) in pituitary cells (160) could be an example, but the relation of this action to endogenous oxidants is obscure. The interference of 2,5-TBHQ and BHT with the mobilization of calcium from the endoplasmic reticulum is an example for an effect which may not be related to antioxidant properties. Similar to oxidants, they act as inhibitors of the

Ca^{2+}-ATPase, as shown with the purified enzyme from skeletal and cardiac muscle (161,162). IC_{50} values of 5 μM for BHT (163) and 1.5 μM for 2,5-TBHQ (161) were calculated. The inhibition of the Ca^{2+}-ATPase may result in a sustained elevation of cytosolic calcium concentrations as shown for 2,5-TBHQ in hepatocytes (164). BHA only transiently enhanced cytosolic calcium in endothelial cells, BHK cells, and granulocytes, but in two cell types possessing voltage-operated calcium channels, cardiomyocytes, and pituitary cells, a sustained increase was observed in the presence of extracellular calcium (165). Eventually, the suppression of ATPase function may, however, lead to depletion of intracellular calcium stores and to the failure of agonist-induced calcium release as shown in pituitary cells with TRH stimulation after 2,5-TBHQ pretreatment (160) and in platelets with thrombin stimulation after BHT pretreatment (166).

The concentration of free cystolic calcium regulates many functions in secretory and contractile cells, and antioxidant-induced depletion of calcium stores or inhibition of calcium channels may be responsible for a variety of antioxidant effects in these cells, e.g., inhibition of GABA and glutamate release in brain synaptosomes by BHT (158), inhibition of thrombin-induced platelet aggregation by BHT (166), and possibly the decreased contractility of smooth muscle by BHA (167,159), BHT (168), and gallates (169), and the premature termination of spontaneous contraction of isolated rabbit atria by BHT (168) though calcium concentration was not measured in the latter studies. On the other hand, the acute elevation of calcium concentration by Ca^{2+}-ATPase inhibition and the subsequent increase of thromboxane synthesis was assumed to be the cause of bronchoconstriction and vasoconstriction in the perfused rat lung by 2,5-TBHQ (170).

A direct antioxidant action also contributes to the effects on eicosanoid synthesis. Inhibition of lipoxygenase activity has been found by PG (171), BHA (108,172), and the *tert*-butylphenol derivatives E-5110 (173) and R-830 (174); however, a stimulation by BHT and probucol of membrane oxidation by lipoxygenase has also been described recently (175). Prostaglandin H synthase requires a basal peroxide tone (176); low levels of H_2O_2 are also stimulatory in the subsequent step of thromboxane formation (177). PG inhibited thromboxane synthesis in leukocytes (178) and the aminomethylphenol derivative ONO-3144 in platelets (179). In contrast, PGI_2 synthesis is sensitive to peroxides (180) and can be protected by antioxidants as shown with PG (181). The resulting shift in the ratio between TXA_2 and PGI_2 has been suggested to be responsible for the platelet inhibitory activity of antioxidants. Mechanisms independent of arachidonic acid metabolism such as inhibition of protein kinase C (PKC) are also involved (166). The activity and translocation from the cytosol to the membrane of PKC is stimulated by ROS (182). Inhibition of PKC activation by diacylglycerol, TPA, or thrombin has been demonstrated with a variety of antioxidants including BHT (183,184) and probucol (112). In the study of Ruzzene et al. (183,184) the inhibitory effect toward agonist-induced PKC activation required preincubation with BHT. Upon concomitant administration, potentiation of PKC activation was observed instead, and BHT caused translocation and transient activation of PKC on its own. This was substantiated by the finding that purified brain PKC was also activated. The latter observation argues against a role of ROS generated by BHT metabolism in the activation process. The authors conclude that BHT causes desensitization toward subsequent phospholipase C activation rather than primary inhibition of PKC. Another mechanism of BHT action on PKC was reported for mouse lung: In a strain responsive to the tumor-promoting action of BHT, administration of BHT in vivo led to a sustained downregulation of PKC content and activity. In a nonpromotable strain, downregulation was reversible upon repeated BHT dosing (185).

BHA and the BHT metabolite BHT-MeOH inhibited the release of arachidonic acid and the induction of cytoxicity by TNF-α, possibly by preventing phospholipase A_2 activation following the TNF-α induced late increase in cytosolic calcium. α-Tocopherol and BHT were inactive, indicating that the effect could be independent of the antioxidant properties of the inhibitors (186). Notably, in a previous study, BHT was also active in preventing TNF-α-induced cytotoxicity (187).

B. Cell Proliferation and Expression of Growth-Related Genes

Current assumptions on the role of ROS in cell proliferation have recently been reviewed by Burdon (188). Briefly, O_2^- and H_2O_2 at low (nanomolar to submicromolar) concentrations will elicit growth responses in many cells. ROS act on a variety of redox-sensitive signal transduction pathways and transcription factors involved in cell growth, either directly or by a decrease in the cellular antioxidant (especially glutathione) pool. ROS can be considered a "life signal" because antioxidant enzymes can drive the cell into apoptosis. Subtoxic ROS concentrations will induce apoptosis by themselves. On the other hand, a role of ROS in tumor promotion has been established (see Sec. II).

Targets for the growth controlling action of ROS may be enzymes switching on proliferation genes, e.g., the oxidative activation of PKC described above. PKC activity is required for expression of the growth-related proto-oncogene c-fos (189). Transcription factors such as NF-κB and AP-1 are also modulated by the redox state. NF-κB is activated by a variety of stimuli such as endotoxin, cytokines, TPA, radiation, and H_2O_2. These stimuli can be abrogated by antioxidants including BHA (190), indicating that ROS are the second messenger common to all these activation pathways. NF-κB exists in an inactive form as a complex with I-κ B and must be released for DNA binding. This is achieved by the action of a protein kinase on I-κ B (190). Accordingly, phosphatase inhibitors can also activate NF-κ B, and this activation pathway has been shown to be insensitive to antioxidants (191) pointing to an antioxidant-inhibitable step upstream I-κ B phosphorylation.

AP-1, a dimeric transcription factor consisting of proteins from the Fos and the Jun family, also responds to changes in redox conditions. ROS induces c-fos and c-jun expression (189), but reducing conditions must be present for efficient DNA-binding of AP-1 (192). Synthetic phenolic antioxidants may contribute to such conditions; however, they are also able to increase the for-mation of AP-1 by themselves via activation of the serum-responsive element in the promotor region of the c-fos gene (193,194). BHA, BHT, catechol, hydroquinone and 1,2,3-trihydroxybenzene were capable of inducing the expres-sion of c-fos, the binding of AP-1 to the AP-1 site of the c-jun gene and the expression of a reporter gene coupled to the c-fos promoter, while resorcinol and 1,3,5-trihydroxybenzene were not (193). This structure-activity relationship is highly reminiscent of the structural requirements for the activation of the ARE (Sec. V.C) and raises the question whether ROS produced by redox cycling of the antioxidants are involved. However, point mutants and deletions in the promoter revealed that distinct sequences are required for activation by H_2O_2 and antioxidants.

In endothelial cells growth was induced not only by BHT but also by probucol and α-tocopherol, substances which do not produce ROS by them-selves (195). No data are available on a role of reactive antioxidant metabo-lites, e.g., BHT hydroperoxides and quinone methides, on c-fos expression.

The action of ROS and antioxidants on enzymes and transcription factors essential for proliferation control may well be linked to the dual role of BHA and BHT in the promotion phase of chemical carcinogenesis: TBHQ has been shown to induce DNA binding activity of AP-1 by itself but to inhibit the tran-scriptional activity of AP-1 induced by the tumor promoter TPA because it is active in inducing the Fra protein which combines with Jun proteins to form an AP-1-like heterodimer with lower transactivation potential than the Jun-Fos heterodimers (196). An indirect mechanism by which BHT might elevate the activity of AP-1 has been demonstrated in mouse lung: in a strain responsive to the tumor-promoting action of BHT, downregulation by BHT of calpain II, an enzyme capable of hydrolyzing the Jun and Fos protein constituents of AP-1, was observed (185).

C. The Antioxidant-Responsive Element

The antioxidant-specific pattern of induction of enzymes involved in xenobiotic metabolism (Sec. III) has forwarded the idea that these enzymes may form a group of coordinately regulated proteins. A regulatory motif in the 5'-flanking region of a number of antioxidant-inducible genes was identified, which has been named "antioxidant responsive element" (ARE) (197), "electrophile responsive element" (EpRE) (198), or GPE1 (199). To simplify matters, the term "ARE" is used throughout this chapter. The enzymes up to now shown to possess this regulatory element are the rat glutathione S-transferase Ya subunit (rGST-Ya) (200), the mouse glutathione S-transferase Ya subunit (mGST-Ya) (198), the rat glutathione transferase P (rGST-P) (199), the rat NADPH-quinone oxidoreductase (rNQO1) (201), the human NADPH-quinone oxidoreductase (hNQO1) (202), the human γ-glutamylcysteine synthetase heavy subunit (hGCSh) (203), the mouse metallothioneine-I (MT-1) (204), and the mouse manganous superoxide dismutase (MnSOD) (205). No ARE or ARE-like motif has been found in the human GST Ya subunit gene so far (206).

Transfection experiments with plasmids containing mutated ARE sequences and a reporter gene into suitable cells such as the HepG2 cell line have revealed that two motifs identical or closely related to the AP-1 element contribute to the transcriptional activation. The elements may be oriented forward or reverse. The AP-1 element is a TPA-responsive element (TRE) with the sequence TGACTCA which binds the Jun family proteins or a heterodimer of Jun and Fos proteins. A perfect AP-1 sequence with a 3'-GC box is present in the hNQO1 gene (207). Consequently, TPA also induces the reporter gene activity regulated by the hNQO1 ARE. In the rNQO1 and rGST Ya gene, the motif TGACnnnGC is considered the core ARE (208). A second AP-1-like sequence located three or eight nucleotides upstream the core ARE sequence in five of the genes characterized so far is required for full inducibility. The hGCSh gene has an extreme 3'-AP-1-like element instead. An extreme 3'-AP-1-like element in addition to the 5'-AP-1-like element is present in the rat and human NQO1 gene (207). The murine MnSOD and the MT-1 genes do not fit this pattern in that only one ARE-like sequence is present.

The *trans* acting factors binding to the ARE appear to differ among genes, and more than one nuclear protein can be bound. The Jun- and Fos-type proteins are obvious candidates, and gel mobility shift assays have demonstrated c-Jun and c-Fos binding to the mGST Ya. In addition, the reporter gene was expressed upon cotransfection of c-jun and c-fos into the AP-1-deficient F9 cell (209). Evidence for Jun-D and c-Fos binding to the hNQO1 ARE has also been provided in supershift analysis (207). However, the ARE of the rGST Ya gene which does not possess a perfect AP-1 element binds to a HepG2 cell nuclear extract which is not completed with a TRE sequence, although binding of purified c-Jun to this ARE has also been shown (210). The *trans* acting factor in this case is a heterodimer consisting of a 28 kd protein and a 45 kd protein. The ARE of the hNQO1 gene forms two complexes with nuclear extracts from HeLa cells, one competed with c-Jun and c-Fos antibodies and a second one with a 160 kd protein which does not contain c-Jun or c-Fos. The binding of this novel complex requires the complete TGACnnnGCA sequence of the core motif of the ARE (211).

The signal transduction pathway leading to activation of the ARE has not yet been elucidated. The mere fact that antioxidants can start it would point to the involvement of a reductive process. However, there is ample evidence that the antioxidant property of the inducers is not operating in activating the ARE. Rather, the induction of the "ARE gene battery" appears to be an adaptation to oxidant stress. In mechanistic studies on the induction of glutathione transferase and quinone reductase, TBHQ and its precursor BHA have been used extensively; the potential of these compounds to produce $O_2^{.-}$ and H_2O_2 has been discussed in Sec. IV. A study on polyhydroxylated benzenes related to TBHQ has defined the minimal structural properties required for activity (208). Only 1,2- or 1,4-hydroxylated compounds such as hydroquinone, catechol, 1,2,3-trihydroxybenzene and 1,2,4-trihydroxybenzene were active, while 1,3-hydroxylation of the benzene ring as in resorcinol or 1,3,5-trihydroxybenzene was not sufficient. This correlates with the potential of the compounds to drive redox cycling and concomitant $O_2^{.-}$ release. BHT and EQ which have not served as tools in studies on ARE-mediated gene expression fit the concept in that they act as inducers of phase II enzymes, on the one hand, and produce reactive oxygen, on the other hand.

If reactive oxygen species are important for activating the ARE H_2O_2 may by itself be capable of doing so. Indeed, it was shown that reporter gene expression via the rGST Ya ARE (208) and the rNQO1 ARE (212) can be achieved by H_2O_2. The MT-1 gene was also responsive to H_2O_2 (204). The mechanism by which H_2O_2 contributes to activation of the ARE has not been clarified. Lowering of the cellular glutathione content or treatment with diamide stimulates AP-1 binding activity and expression of the mGST Ya ARE sequence (213) and dithiotreitol inhibits the binding of the 160 kd nuclear protein described by Wang and Williamson (211) to the hNQO1 ARE, suggesting that thiol oxidation may be involved in the activation process. Notably, this appears to be opposite to AP-1-directed TRE activation where the reduction of a cysteine residue in the DNA binding domain of the Jun and Fos proteins by chemicals or by the Ref-1 protein increases rather than decreases the binding of the transcription factor to the DNA (192). Interpretation is further complicated by the fact that antioxidants increase the GSH content of the cell, possibly by ARE-mediated induction of γ-glutamylcysteine synthetase as suggested for the phenolic antioxidant BHA (214) and the indolic antioxidant 5,10-dihydroindeno(1,2-b)indole (77). If, indeed, in the case of the ARE thiol, oxidation activates binding of nuclear proteins, the increase of thiol-reducing power by the very action of the ARE could act as a negative feedback mechanism.

In addition to alteration of the thiol oxidation state, thiol alkylation may also play a role in the binding of *trans* acting factors (215). This would correspond to the concept that the induction of the anticarcinogenic enzymes glutathione transferase and quinone reductase is regulated by a common chemical signal, the electrophilic nature of the inducer (216). β-Naphthoflavone and 3-methylcholanthrene have frequently been used in experiments on ARE function. They are of special interest because they belong to the bifunctional inducers, which on the one hand act through the Ah-receptor-regulated xenobiotic responsive element (XRE) and thus induce CYP1A1 and CYP1A2 and, on the other hand, via electrophilic metabolites, activate the ARE. The requirement for metabolism of these polycyclics in order to gain the ARE-activating property has been demonstrated in experiments with mutant cells lacking either a functional CYP1A1 or the Ah receptor translocase (217). In similar experiments, it has been shown that the indolic antioxidant 5,10-dihydroindeno(1,2-b)indole requires metabolic activation for ARE-activating activity (77).

VI. TOXICITY OF BHA AND BHT

When Ito and his coworkers first reported the occurrence of papillomas and squamous cell carcinomas in rats and hamsters by high dietary BHA exposure (218,219), a long-lasting debate on the safety of BHA as a food additive was initiated. A recent series of papers has reviewed the main issues in this discussion (220–223). The following aspects proved to be the most important in the risk assessment process.

1. *Genotoxicity*

Generally, no mutagenic activity could be ascribed to BHA (58,220,223) and no DNA adducts were found (224,225). However, BHA exhibited initiating activity in a cell transformation assay (226). TBQ, but not BHA-oQ, was also active. TBHQ was clastogenic in a number of experiments (227,228), although sometimes the effect was only sporadic and weak (229). ROS formation from TBHQ was assumed to be involved (228). TBHQ and particularly TBQ cause $O_2^{\cdot-}$ formation in rat forestomach homogenates (121) (Fig. 3). In the forestomach BHA and TBHQ did not cause DNA single-strand breaks, but TBQ did (230). In one study, TBQ was not genotoxic in V79 cells but caused gene conversion and reverse mutation in *Saccharomyces cerevisiae* (231). For the induction of chromosomal aberrations BHA, TBHQ and TBQ but not BHA-oQ and TBQ oxide required metabolic activation (232). TBQ was detected in rat forestomach after feeding of 3-BHA (230). Moreover, BHA can interact with nitrite in an acidic environment to form TBQ (233). Summarizing, BHA is not mutagenic but is clastogenic via products formed during its metabolism, and most authors assume that its carcinogenic action is due to a promoting effect on "spontaneously" initiated cells by these products. However, an initiating potential via formation of oxidized DNA bases as shown in lymphocytes and glandular stomach (325) has also been suggested (234).

2. Development and Reversibility of the Lesion

The tumors develop from a transient inflammatory and a sustained hyperplastic response of the forestomach (235). BHA-induced cellular proliferation has extensively been studied (236,221). The superficial hyperplasia and many of the papillomas regressed within a few months after BHA withdrawal but basal cell hyperplasia and dysplasia persisted for much longer (237,238,239). Notably, the BHA metabolite TBHQ was highly active in enhancing the thickness of the forestomach mucosa (240).

3. Occurrence in Other Tissues and in Nonrodent Species

Two findings suggest a tumorigenic action of BHA in tissues other than the rodent forestomach: formation of lung adenomas in the Japanese musk shrew (241) and formation of hepatocellular carcinomas in the fish Rivulus ocellatus marmoratus (242). Both studies have been subject to criticism (221). In rodents, no tumor localization other than the forestomach was found; however, increased cell proliferation was observed in the rat esophagus, glandular stomach, small intestine, colon/rectum (236), and urinary bladder (243). The esophagus is of particular interest because squamous epithelium similar to that of the rodent forestomach is present in human esophagus. In pigs, parakeratosis and proliferative changes have been claimed to occur in the esophagus and the esophageal part of the stomach in a few animals (244). Hyperplastic changes were not observed in dogs (245); in cynomolgus monkeys a slight elevation of the mitotic index in the esophagus was observed (246).

Taking the above into consideration, it has been argued by most authors that BHA is a nongenotoxic carcinogen to which threshold considerations can be applied. An ADI value of 0.5 mg/kg body weight has been established on the basis of the hyperplastic changes in the rat forestomach (247). A NOEL of 0.1% in the diet corresponding to a daily dose of 50 mg/kg body weight and an uncertainty factor of 100 were applied. NOELs for other endpoints are thymidine incorporation 0.25% (125 mg/kg), tumor promotion 0.3% (150 mg/kg), papilloma formation 0.5% (250 mg/kg), and carcinoma formation 1% (500 mg/kg) (223,222). The daily intake in the U.S. population has been estimated to be less than 0.1 mg/kg (248).

With BHT, forestomach carcinogenesis has not been observed. BHT caused increased tumor incidence in the liver of very old rats (249) and mice (250), which was only manifest when the control animals had already died, while the BHT animals still survived due to a life-prolonging effect of the antioxidant.

Similar as with BHA, the carcinogenic effect may be explained with tumor-promoting activity on latent tumor cells. BHT was positive in an assay for replicative DNA synthesis in which animals were treated in vivo and hepatocytes were then isolated and tested in vitro (251). This assay has been proposed as a suitable method for detection of nongenotoxic carcinogens. BHT has in most tests proven to be nonmutagenic (82) but it can enhance the mutagenicity of other compounds (252). The liver is also a target organ of acute high-dose BHT toxicity in rats (253), exhibiting mainly centrilobular, but to a lesser degree periportal, necrosis (254). A reactive metabolite is assumed to be responsive for liver necrosis. BHT-OOH is much more toxic to rat hepatocytes than BHT (131). BHT-QM leads to glutathione depletion in lung (255) and liver (256), and the covalent binding of BHT-derived material to hepatic proteins is assumed to be due to BHT-QM (257).

Inhibition of coagulation is the toxic effect which constitutes the basis for the ADI value for BHT of 0.125 mg/kg (WHO) or 0.05 mg/kg (EEC), respectively. The main mechanism is inhibition of vitamin K epoxide reductase, resulting in decreased synthesis of the clotting factors II, VII, IX, and X (258). The rat is especially susceptible to this effect, the LOEL being 0.017% in the food (259). The hemorrhagic action of BHT is ascribed to the formation of BHT-QM (260). In mice, but not in rats, BHT leads to lung injury. Though species-specific, this effect has become a model for pneumotoxicity (261). Lung injury is ascribed to reactive metabolites. An early report states that only BHT analogs capable of forming quinone methides will elicit a toxic response (262). More recently, BHT-BuOH has been linked with pneumotoxicity. This metabolite is more potent in vivo (263) and in isolated mouse Clara cells (131) than BHT itself. There is one additional intriguing argument in favor of BuOH-BHT as the active metabolite responsible for lung toxicity (and lung tumor promotion) in the mouse: Rats are refractory to these effects and do not form BuOH-BHT from BHT in lung microsomes (264).

When related to the amounts taken up via food consumption, the concentrations of BHA and BHT eliciting toxic responses in animals are high. Most authors assume that a threshold exists for these toxic responses, making it unlikely that the general population is endangered by the consumption of BHA and BHT as food additives. A use of pharmacological doses of these compounds in the therapy or prevention of oxidant-related diseases is, however, not recommended.

ACKNOWLEDGMENT

The author gratefully acknowledges the financial support by the Deutsche Forschungsgemeinschaft, Bonn, Germany.

ABBREVIATIONS

ARE	antioxidant-responsive element
BHA	butylated hydroxyanisole
BHT	butylated hydroxytoluene
EQ	ethoxyquin
GST	glutathione-S-transferase
H_2O_2	hydrogen peroxide
HO·	hydroxyl radical
$O_2^{\cdot-}$	superoxide anion radical
PG	propyl gallate
PGI_2	prostaglandin I_2 (prostacyclin)
PKC	protein kinase C
ROS	reactive oxygen species
TBHQ	*tert*-butylhydroquinone
TBQ	*tert*-butyl-1,4-quinone
TNF-α	tumor necrosis factor α
TPA	12-*O*-tetradecanoylphorbol-13-acetate

REFERENCES

1. Stampfer MJ, Hennekens CH, Manson JE, Colditz GA, Rosner B, Willett WC. Vitamin E consumption and the risk of coronary disease in women. New Engl J Med 1993; 328:1444–1449.
2. Rimm EB, Stampfer MJ, Ascherio A, Giovannucci E, Colditz GA, Willett WC. Vitamin E consumption and the risk of coronary heart disease in men. New Engl J Med 1993; 328:1450–1456.
3. Steinberg D, Parthasarathy S, Carew TE, Khoo JC, Witztum JL. Beyond cholesterol. Modifications of low-density lipoprotein that increase its atherogenicity. New Engl J Med 1989; 320:915–924.
4. Gebicki JM, Jürgens G, Esterbauer H. Oxidation of low-density lipoprotein in vitro. In: Sies H, ed. Oxidative Stress: Oxidants and Antioxidants. London: Academic Press, 1991:371–397.
5. Parthasarathy S, Young SG, Witztum JL, Pittman RC, Steinberg D. Probucol inhibits oxidative modification of low density lipoprotein. J Clin Invest 1986; 77:641–644.
6. Kuzuya M, Kuzuya F. Probucol as an antioxidant and antiatherogenic drug. Free Rad Biol Med 1993; 14:67–77.
7. Pryor WA, Strickland T, Church DF. Comparison of the efficiencies of several natural and synthetic antioxidants in aqueous sodium dodecyl sulfate micelle solutions. J Am Chem Soc 1988; 110:2224–2229.
8. Minakami H, Sotomatsu A, Ohma C, Nakano M. Antioxidant activity of probucol. Drug Res 1989; 39:1090–1091.

9. Bridges AB, Scott NA, Belch JJ. Probucol, a superoxide free radical scavenger in vitro. Atherosclerosis 1991; 89:263-265.

10. Hiramatsu M, Liu J, Edamatsu R, Ohba S, Kadowaki D, Mori A. Probucol scavenged, 1,1-diphenyl-2-picrylhydrazyl radicals and inhibited formation of thiobarbituric acid reactive substances. Free Rad Biol Med 1994; 16:201-206.

11. Kalyanaraman B, Darley-Usmar VM, Wood J, Joseph J, Parthasarathy S. Synergistic interaction between the probucol phenoxyl radical and ascorbic acid in inhibiting the oxidation of low density lipoprotein. J Biol Chem 1992; 267:6789-6795.

12. Schuh J, Fairclough GF, Haschemeyer RH. Oxygen-mediated heterogeneity of apo-low-density lipoprotein. Proc Natl Acad Sci USA 1978; 75:3173-3177.

13. Hunt JV, Bottoms ME, Taylor SE, Lyell V, Mitchinson MJ. Differing effects of probucol and vitamin E on the oxidation of lipoproteins, ceroid accumulation and protein uptake by macrophages. Free Rad Res 1994; 20:189-201.

14. Noguchi N, Gotoh N, Niki E. Effects of ebselen and probucol on oxidative modifications of lipid and protein of low density lipoprotein induced by free radicals. Biochim Biophys Acta 1994; 1213:176-182.

15. Daugherty A, Roselaar SE. Lipoprotein oxidation as a mediator of atherogenesis: insights from pharmacological studies. Cardiovasc Res 1995; 29:297-311.

16. Marshall FN. Pharmacology and toxicology of probucol. Artery 1982; 10:7-21.

17. Sasahara M, Raines EW, Chait A, Carew TE, Steinberg D, Wahl PW, Ross R. Inhibition of hypercholesterolemia-induced atherosclerosis in the nonhuman primate by probucol. I. Is the extent of atherosclerosis related to resistance of LDL to oxidation? J Clin Invest 1994; 94:155-164.

18. Björkhem I, Henriksson-Freyschuss A, Breuer O, Diczfalusy U, Berglund L, Henriksson P. (1991) The antioxidant butylated hydroxytoluene protects against atherosclerosis. Arterioscler Thromb 1991; 11:15-22.

19. Xiu RJ, Freyschuss A, Ying X, Berglund L, Henriksson P, Björkhem I. The antioxidant butylated hydroxytoluene prevents early cholesterol-induced microcirculatory changes in rabbits. J Clin Invest 1994; 93:2732-2737.

20. Mao SJT, Yates MT, Parker RA, Chi EM, Jackson RL. Attenuation of atherosclerosis in a modified strain of hypercholesterolemic Watanabe rabbits with use of a probucol analogue (MDL 29,311) that does not lower serum cholesterol. Arterioscler Thromb 1991; 11:1266-1275.

21. Fruebis J, Steinberg D, Dresel HA, Carew TE. A comparison of the antiatherogenic effects of probucol and of a structural analogue of probucol in low density lipoprotein receptor-deficient rabbits. J Clin Invest 1994; 94:392-398.

22. Morel DW, de la Llera-Moya M, Friday KE. Treatment of cholesterol-fed rabbits with dietary vitamin E and C inhibits lipoprotein oxidation but not development of atherosclerosis. J Nutr 1994; 124:2123-2130.

23. Kuzuya M, Naito M, Funaki C, Hayashi T, Asai K, Kuzuya F. Probucol prevents oxidative injury to endothelial cells. J Lipid Res 1991; 32:197-204.

24. Simon BC, Haudenschild CC, Cohen RA. Preservation of endothelium-dependent relaxation in atherosclerotic rabbit aorta by probucol. J Cardiovasc Pharmacol 1993; 21:893-901.

25. Plane F, Jacobs M, McManus D, Bruckdorfer KR. Probucol and other antioxidants prevent the inhibition of endothelium-dependent relaxation by low density lipoproteins. Atherosclerosis 1993; 103:73–79.
26. Keanay JF, Xu AM, Cunningham D, Jackson T, Frei B, Vita JA. Dietary probucol preserves endothelial function in cholesterol-fed rabbits by limiting vascular oxidative stress and superoxide generation. J Clin Invest 1995; 95:2520–2529.
27. Akeson AL, Woods CW, Mosher LB, Thomas CE, Jackson RL. Inhibition of IL-1 beta expression in THP-1 cells by probucol and tocopherol. Atherosclerosis 1991; 86:261–270.
28. Ku G, Doherty NS, Schmidt LF, Jackson RL, Dinerstein RJ. Ex vivo lipopolysaccharide-induced interleukin-1 secretion from murine peritoneal macrophages inhibited by probucol, a hypocholesterolemic agent with antioxidant properties. FASEB J 1990; 4:1645–1653.
29. Li SR, Forster L, Änggård E, Ferns G. The effects of LPS and probucol on interleukin 1 (IL-1) and platelet-derived growth factor (PDGF): gene expression in the human monocytic cell line U-937. Biochim Biophys Acta 1994; 1225:271–274.
30. Ferns GAA, Forster L, Stewart-Lee A, Nourooz-Zadeh J, Änggård EE. Probucol inhibits mononuclear cell adhesion to vascular endothelium in the cholesterol-fed rabbit. Atherosclerosis 1993; 100:171–181.
31. Faruqi R, Motte de la C, DiCorleto PE. alpha-Tocopherol inhibits agonist-induced monocytic cell adhesion to cultured human endothelial cells. J Clin Invest 1994; 94:592–600.
32. Walldius G, Erikson U, Olsson AG, Bergstrand L, Hådell K, Johansson J, Kaijser L, Lassvik C, Mölgaard J, Nilsson S, Schäfer-Elinder L, Stenport G, Holme I. The effect of probucol on femoral atherosclerosis: The Probucol Quantitative Regression Swedish Trial (PQRST). Am J Cardiol 1994; 74:875–883.
33. Regnstrom J, Walldius G, Carlson LA, Nilsson J. Effect of probucol treatment on the susceptibility of low density lipoprotein isolated from hypercholesterolemic patients to become oxidatively modified in vitro. Atherosclerosis 1990; 82:43–51.
34. Cristol LS, Jialal I, Grundy SM. Effect of low-dose probucol therapy on LDL oxidation and the plasma lipoprotein profile in male volunteers. Atherosclerosis 1992; 97:11–20.
35. Bittolo-Bon G, Cazzolato G, Avogaro P. Probucol protects low-density lipoproteins from in vitro and in vivo oxidation. Pharmacol Res 1994; 29:337–344.
36. Dujovne CA, Harris WS, Gerrond LLC, Fan J, Muzio F. Comparison of effects of probucol versus vitamin E on ex vivo oxidation succeptibility of lipoproteins in hyperlipoproteinemia. Am J Cardiol 1994; 74:38–42.
37. Reaven PD, Parthasarathy S, Beltz WF, Witztum JL. Effect of probucol dosage on plasma lipid and lipoprotein levels and on protection of low density lipoprotein against in vitro oxidation in humans. Arterioscler Thromb 1992; 12:318–324.
38. Naruszewicz M, Selinger E, Dufour R, Davignon J. Probucol protects lipoprotein (a) against oxidative modification. Metabolism 1992; 41:1225–1228.

39. Setsuda M, Inden M, Hiraoka N, Okamoto S, Tanaka H, Okinaka T, Nishimura Y, Okano H, Kouji T, Konishi T. Probucol therapy in the prevention of restenosis after successful percutaneous transluminal coronary angioplasty. Clin Ther 1993; 15:374–382.

40. Kisanuki A, Asada Y, Hatakeyama K, Hayashi T, Sumiyoshi A. Contribution of the endothelium to intimal thickening in normocholesterolemic and hypercholesterolemic rabbits. Arterioscler Thromb 1992; 12:1198–1205.

41. Shinomiya M, Shirai K, Saito Y, Yoshida S. Inhibition of intimal thickening of the carotid artery of rabbits and of outgrowth of explants of aorta by probucol. Atherosclerosis 1992; 97:143–148.

42. Ferns GAA, Forster L, Stewart-Lee A, Konneh M. Nourooz-Zadeh J, Änggård EE. Probucol inhibits neointimal thickening and macrophage accumulation after balloon injury in the cholesterol-fed rabbit. Proc Natl Acad Sci USA 1992; 89:11312–11316.

43. Schneider JE, Berk BC, Gravanis MB, Santoian EC, Cipolla GD, Tarazona N, Lassegue B, King III SB. Probucol decreases neointimal formation in a swine model of coronary artery balloon injury. Circulation 1993; 88:628–637.

44. Ohya Y, Kumamoto K, Abe I, Tsubota Y, Fujishima M. Factors related to QT interval prolongation during probucol treatment. Eur J Clin Pharmcol 1993; 45:47–52.

45. Kosasayama A, Yoshida M, Okada S. Post-marketing surveillance of probucol (Sinlestal) in Japan. Artery 1992; 19:147–161.

46. Wattenberg LW. Inhibition of carcinogenic and toxic effects of polycyclic hydrocarbons by phenolic antioxidants and ethoxyquin. J Natl Cancer Inst 1972; 48:1425–1429.

47. Potenberg J, Schiffmann D, Kahl R, Hildebrandt AG, Henschler D. Modulation of benzo[a]pyrene-induced morphological transformation of Syrian hamster embryo cells by butylated hydroxytoluene and butylated hydroxyanisole. Cancer Lett 1986; 33:189–198.

48. McCormick DL, Major N, Moon RC. Inhibition of 7,12-dimethylbenz[a]-anthracene-induced rat mammary carcinogenesis by concomitant or postcarcinogen antioxidant exposure. Cancer Res 1984; 44:2858–2863.

49. Hirose M, Masuda A, Fukushima S, Ito N. Effects of subsequent antioxidant treatment on 7,12-dimethylbenz[a]anthracene-initiated carcinogenesis of the mammary gland, ear duct and forestomach in Sprague-Dawley rats. Carcinogenesis 1988; 9:101–104.

50. Mizumoto K, Ito S, Kitazawa S, Tsutsumi M, Denda A, Konishi Y. Inhibitory effect of butylated hydroxyanisole administration on pancreatic carcinogenesis in Syrian hamsters initiated with N-nitrosobis(2-oxopropyl)amine. Carcinogenesis 1989; 10:1491–1494.

51. Witschi HP. Modulation of tumor development by butylated hydroxytoluene in experimental animals. Toxicol Industr Health 1993; 9:259–281.

52. Hirose M, Yada H, Hakoi K, Takahashi S, Ito N. Modification of carcinogenesis by alpha-tocopherol, t-butylhydroquinone, propyl gallate and butylated hy-

droxytoluene in a rat multi-organ carcinogenesis model. Carcinogenesis 1993; 14:2359–2364.

53. Djurhuus R, Lillehaug JR. Butylated hydroxytoluene: tumor-promoting activity in an in vitro two-stage carcinogenesis assay. Bull Environ Contam Toxicol 1982; 29:115–120.

54. Kitchin KT, Brown JL. Biochemical effects of two promotors of hepatocarcinogenesis in rats. Fd Chem Toxicol 1987; 25:603–607.

55. Trosko JE, Yotti LP, Warren ST, Tsushimoto G, Chang C. Inhibition of cell-cell communication by tumor promoters. Carcinogenesis 1982; 7:565–585.

56. Williams GM. Epigenetic promoting effects of butylated hydroxyanisole. Fd. Chem. Toxicol 1986; 24:1163–1166.

57. Masui T, Fukushima S, Katoh F, Yamasaki H, Ito N. Effects of sodium L-ascorbate, uracil, butylated hydroxyanisole and extracellular pH on junctional intercellular communication of BALB/c 3T3 cells. Carcinogenesis 1988; 9:1143–1146.

58. Williams GM, McQueen CA, Tong C. Toxicity studies of butylated hydroxyanisole and butylated hydroxytoluene. I. Genetic and cellular effects. Fd. Chem. Toxicol 1990; 28:793–798.

59. Wang XR, Witschi H. Mutations of the Ki-ras protooncogene in 3-methylcholanthrene and urethan-induced and butylated hydroxytoluene promoted lung tumors of strain A/J and SWR mice. Cancer Letters 1995; 91:33–39.

60. Blot WJ, Li J.-Y, Taylor PR, Guo W, Dawsey S, Wang G.-Q, Yang CS, Zheng S.-F, Gail M, Li G.-Y, Yu Y, Liu B.-Q, Tangrea J, Sun Y.-H, Liu F, Fraumeni JF Jr, Zhan Y-H, Li B. Nutrition intervention trials in Linxian, China: Supplementation with specific vitamin/mineral combinations, cancer incidence, and disease-specific mortality in the general population. J Natl Cancer Inst 1993; 85:1483–1491.

61. Heinonen OP. The alpha-tocopherol, beta-carotene lung cancer prevention study: design, methods, participant characteristics, and compliance. Ann Epidemiol 1994; 4:1–10.

62. Prochaska HJ, Talalay P. Regulatory mechanisms of monofunctional and bifunctional anticarcinogenic enzyme inducers in murine liver. Cancer Res 1988; 48:4776–4782.

63. Kahl R, Netter KJ. Ethoxyquin as an inducer and inhibitor of phenobarbital-type cytochrome P-450 in rat liver microsomes. Toxicol Appl Pharmacol 1977; 40:473–483.

64. Kahl R, Wulff U. Induction of rat hepatic epoxide hydratase by dietary antioxidants. Toxicol Appl Pharmacol 1979; 47:217–227.

65. Pearson WR, Reinhart J, Sisk SC, Anderson KS, Adler PN. Tissue-specific induction of murine glutathione transferase mRNAs by butylated hydroxyanisole. J Biol Chem 1988; 263:13324–13332.

66. Lam LKT, Fladmoe AV, Hochalter JB, Wattenberg LW. Short time interval effects of butylated hydroxyanisole on the metabolism of benzo[a]pyrene. Cancer Res 1980; 40:2824–2828.

67. Dock L, Cha Y.-N, Jernström B, Moldéus P. Effect of 2(3)-tert-butyl-4-hy-

droxyanisole on benzo[a]pyrene metabolism and DNA-binding of benzo[a]pyrene metabolites in isolated mouse hepatocytes. Chem Biol Interact 1982; 41:25–37.

68. Kahl R, Kahl GF. Effect of dietary antioxidants on benzo[a]pyrene metabolism in rat liver microsomes. Toxicology 1983; 28:229–233.

69. Depner M, Kahl GF, Kahl R. Influence of gallic acid esters on drug metabolizing enzymes of rat liver. Fd Chem Toxicol 1982; 20:507–511.

70. Cha Y.-N, Martz F, Bueding E. Enhancement of liver microsome epoxide hydratase activity in rodents by treatment with 2(3)-*tert*-butyl-4-hydroxyanisole. Cancer Res 1978; 38:4496–4498.

71. Kapitulnik J, Wislocki PG, Levin W, Yagi H, Jerina DM, Conney AH. Tumorigenicity studies with diol epoxides of benzo[a]pyrene which indicate that (±)-*trans*-7β, 8α-dihydroxy-9α,10α-epoxy-7,8,9,10-tetrahydrobenzo[a]pyrene is an ultimate carcinogen in newborn mice. Cancer Res 1978; 38:354.

72. Mannervik B, Ålin P, Guthenberg C, Jensson H, Tahir MK, Warholm M, Jörnvall H. Identification of three classes of cytosolic glutathione transferase common to several mammalian species: Correlation between structural data and enzymatic properties. Proc Natl Acad Sci USA 1985; 82:7202–7206.

73. Meyer DJ, Coles B, Pemble SE, Gilmore KS, Fraser GM, Ketterer B. Theta, a new class of glutathione transferases purified from rat and man. Biochem J 1991; 274:409–414.

74. Benson AM, Batzinger RP, Ou S.-YL, Bueding E, Cha Y.-N, Talalay P. Elevation of hepatic glutathione S-transferase activities and protection against mutagenic metabolites of benzo[a]pyrene by dietary antioxidants. Cancer Res 1978; 38:4486–4495.

75. McLellan LI, Judah DJ, Neal GE, Hayes JD. Regulation of aflatoxin B-1-metabolizing aldehyde reductase and glutathione S-transferase by chemoprotectors. Biochem J 1994; 300:117–124.

76. Eaton DL, Hamel DM. Increase in gamma-glutamylcysteine synthetase activity as a mechanism for butylated hydroxyanisole-mediated elevation of hepatic glutathione. Toxicol Appl Pharmacol 1994; 126:145–149.

77. Liu R.-M, Vasiliou V, Zhu H, Duh J.-L, Tabor MW, Puga A, Nebert DW, Sainsbury M, Shertzer HG. Regulation of [Ah] gene battery enzymes and glutathione levels by 5,10-dihydroindeno[1,2-b]indole in mouse hepatoma cell lines. Carcinogenesis 1994; 15:2347–2352.

78. Cha Y-N, Heine HS, Moldéus P. Differential effects of dietary and intraperitoneal administration of antioxidants on the activities of several hepatic enzymes of mice. Drug Metab Disp 1982; 10:434–435.

79. Nakagawa Y, Hiraga K, Suga T. Effects of butylated hydroxytoluene (BHT) on the level of glutathione and the activity of glutathione-S-transferase in rat liver. J Pharm Dyn 1981; 4:823–826.

80. Furukawa K, Maeura Y, Furukawa NT, Williams GM. Induction by butylated hydroxytoluene of rat liver gamma-glutamyl transpeptidase activity in comparison to expression in carcinogen-induced altered lesions. Chem Biol Interact 1984; 48:43–58.

81. Davies R, Edwards RE, Green JA, Legg RF, Snowden RT, Manson MM. An-

tioxidants can delay liver cell maturation which in turn affects gamma-glutamyl-transpeptidase expression. Carcinogenesis 1993; 14:47–52.

82. Kahl R. Synthetic antioxidants: Biochemical actions and interference with radiation, toxic compounds, chemical mutagens and chemical carcinogens. Toxicology 1984; 33:185–228.

83. Benson AM, Hunkeler MJ, Talalay P. Increase of NAD(P)H: quinone reductase by dietary antioxidants: Possible role in protection against carcinogenesis and toxicity. Proc Natl Acad Sci USA 1980; 77:5216–5220.

84. Cadenas E. Antioxidant and prooxidant functions of DT-diaphorase in quinone metabolism. Biochem Pharmacol 1995; 49:127–140.

85. Prochaska HJ, Talalay P. The role of NAD(P)H: quinone reductase in protection against the toxicity of quinones. In: Sies H, ed. Oxidative Stress: Oxidants and Antioxidants. London: Academic Press, 1991:195–211.

86. Chung JH, Cha YN, Rubin RJ. Role of quinoine reductase in in vivo ethanol metabolism and toxicity. Toxicol Appl Pharmacol 1994; 124:123–130.

87. Cha Y.-N, Bueding E. Effect of 2(3)-*tert*-butyl-4-hydroxyanisole administration on the activities of several hepatic microsomal and cytoplasmic enzymes in mice. Biochem Pharmacol 1979; 28:1917–1921.

88. Bock KW, Kahl R, Lilienblum W. Induction of rat hepatic UDP-glucuronosyltransferases by dietary ethoxyquin. Naunyn-Schmiedebergs Arch Pharmacol 1980; 310:249–252.

89. Kashfi K, Yang EK, Chowdhury JR, Chowdhury NR, Dannenberg AJ. Regulation of uridine diphosphate glucuronosyltransferase expression by phenolic antioxidants. Cancer Res 1994; 15; 5856–5859.

90. Gregus Z, Klaassen CD. Effect of butylated hydroxyanisole on hepatic glucuronidation and biliary excretion of drugs in mice. J Pharm Pharmacol 1988; 40:237–242.

91. Vasiliou V, Puga A, Nebert DW. Mouse class 3 aldehyde dehydrogenases: Positive and negative regulation of gene expression. In: Weiner H, ed. Enzymol. Mol. Biol. Carbonyl Metabolism 4. New York: Plenum Press, 1993:131–139.

92. Sreerama L, Rekha GK, Sladek NE. Phenolic antioxidant-induced overexpression of class-3 aldehyde dehydrogenase and oxazaphosphorine-specific resistance. Biochem Pharmacol 1995; 49:669–675.

93. Hirose M, Yada H, Hakoi K, Takahashi S, Ito N. Modification of carcinogenesis by alpha-tocopherol, *t*-butylhydroquinone, propyl gallate and butylated hydroxytoluene in a rat multi-organ carcinogenesis model. Carcinogenesis 1993; 14:2359–2364.

94. Kahl R. The dual role of antioxidants in the modification of chemical carcinogenesis. J Environ Sci Health 1986; C4:47–92.

95. Trush MA, Kensler TW. Role of free radicals in carcinogen activation. In: Sies H, ed. Oxidative Stress: Oxidants and Antioxidants. Academic Press, London. 1991:277–318.

96. Sullivan PD, Calle LM, Shafer K, Nettleman M. Effect of antioxidants on benzo[a]pyrene free radicals. In: Jones PW, Freudenthal RI, eds. Carcinogenesis, 3: Polynuclear Aromatic Hydrocarbons. New York: Raven Press, 1978:1–8.

97. Thompson, JA, Schullek KM, Fernandez CA, Malkinson AM. A metabolite of butylated hydroxytoluene with potent tumor-promoting activity in mouse lung. Carcinogenesis 1989; 10:773-775.

98. Mikuni T, Tatsuta M, Kamachi M. Scavenging effect of butylated hydroxytoluene on the production free radicals by the reaction of hydrogen peroxide with *N*-methyl-*N'*-nitro-*N*-nitrosoguanidine. J Natl Cancer Inst 1987; 79:281-283.

99. Yang CS, Strickhart FS, Woo GK. Inhibition of the monooxygenase system by butylated hydroxyanisole and butylated hydroxytoluene. Life Sci 1974; 15:1497-1505.

100. Cummings SW, Prough RA. Butylated hydroxyanisole-stimulated NADPH oxidase activity in rat liver microsomal fractions. J Biol Chem 1983; 258:12315-12319.

101. Rössing D, Kahl R, Hildebrandt AG. Effect of synthetic antioxidants on hydrogen peroxide formation, oxyferro cytochrome P-450 concentration and oxygen consumption in liver microsomes. Toxicology 1985; 34:67-77.

102. Gettings SD, Brewer CB, Pierce WM Jr, Peterson JA, Rodrigues AD, Prough RA. Enhanced decomposition of oxyferrous cytochrome P450CIA1 (P450cam) by the chemopreventive agent 3-*t*-butyl-4-hydroxyanisole. Arch Biochem Biophys 1990; 276:500-509.

103. Sydor W, Chou MW Jr, Yang SK, Yang CS. Regioselective inhibition of benzo-[a]pyrene metabolism by butylated hydroxyanisole. Carcinogenesis 1983; 4:131-136.

104. Perchellet J-P, Perchellet EM. Antioxidants and multistage carcinogenesis in mouse skin. Free Rad Biol Med 1989; 7:377-408.

105. Troll W, Wiesner R. The role of oxygen radicals as a possible mechanism of tumor promotion. Ann Rev Pharmacol Toxicol 1985; 25:509-528.

106. Kensler TW, Taffe BG. Free radicals in tumor promotion. Adv Free Rad Biol Med 1986; 2:347-387.

107. Kozumbo WJ, Seed JL, Kensler TW. Inhibition by 2(3)-*tert*-butyl-4-hydroxyanisole and other antioxidants of epidermal ornithine decarboxylase activity induced by 12-*O*-tetradecanoylphorbol-13-acetate. Cancer Res 1983; 43:2555-2559.

108. Nakadate T, Yamamoto S, Aizu E, Kato R. Effects of flavonoids and antioxidants on 12-*O*-tetradecanoyl-phorbol-13-acetate-caused epidermal ornithine decarboxylase induction and tumor promotion in relation to lipoxygenase inhibition by these compounds. Gann 1984; 75:214-222.

109. Taniguchi S, Kono T, Mizuno N, Ishii M, Matsui-Yuasa I, Otani S, Hamada T. Effects of butylated hydroxyanisole on ornithine decarboxylase activity and its gene expression induced by phorbol ester tumor promoter. J Invest Dermatol 1991; 96:289-291.

110. Nakamura Y, Colburn NH, Gindhart TD. Role of reactive oxygen in tumor promotion: Implication of superoxide anion in promotion of neoplastic transformation in JB-6 cells by TPA. Carcinogenesis 1985; 6:229-235.

111. Malkinson AM, Beer DS, Sadler AJ, Coffman DS. Decrease in the protein kinase C-catalyzed phosphorylation of an endogenous lung protein (M, 36,000)

following treatment of mice with the tumor modulatory agent butylated hydroxytoluene. Cancer Res 1985; 45:5751–5756.

112. Kunisaki M, Bursell SE, Umeda F, Nawata H, King GL. Normalization of diacylglycerol-protein kinase C activation by vitamin E in aorta of diabetic rats and cultured rat smooth muscle cells exposed to elevated glucose levels. Diabetes 1994; 43:1372–1377.

113. Astill BD, Mills J, Fassett DW, Roudabush RL, Terhaar CJ. Fate of butylated hydroxyanisole in man and dog. Agric Fd Chem 1962; 10:315–319.

114. Verhagen H, Thijssen HHW, ten Hoor F, Kleinjans JCS. Disposition of single oral doses of butylated hydroxyanisole in man and rat. Fd Chem Toxicol 1989; 27:151–158.

115. El-Rashidy R, Niazi S. A new metabolite of butylated hydroxyanisole in man. Biopharm Drug Disp 1983; 4:389–396.

116. Tajima K, Hashizaki M, Yamamoto K, Mizutani T. Metabolism of 3-*tert*-butyl-4-hydroxyanisole by horseradish peroxidase and hydrogen peroxide. Drug Metab Disp 1992; 20:816–820.

117. Rahimtula A. In vitro metabolism of 3-*t*-butyl-4-hydroxyanisole and its irreversible binding to proteins. Chem Biol Interact 1983; 45:125–135.

118. Cummings SW, Ansari GAS, Guengerich FP, Crouch LS, Prough RA. Metabolism of 3-*tert*-butyl-4-hydroxyanisole by microsomal fractions and isolated rat hepatocytes. Cancer Res 1985; 45:5617–5624.

119. Weinke S, Kahl R, Kappus H. Effect of four synthetic antioxidants on the formation of ethylene from methional in rat liver microsomes Toxicol Lett 1987; 35:247–251.

120. Thompson DC, Trush MA. Enhancement of butylated hydroxytoluene-induced mouse lung damage by butylated hydroxyanisole. Toxicol Appl Pharmacol 1988; 96:115–121.

121. Kahl R, Weinke S, Kappus H. Production of reactive oxygen species due to metabolic activation of butylated hydroxyanisole. Toxicology 1989; 59:179–194.

122. Bergmann B, Dohrmann JK, Kahl R. Formation of the semiquinone anion radical from *tert*-butylquinone and from *tert*-butylhydroquinone in rat liver microsomes. Toxicology 1992; 74:127–133.

123. Van Ommen B, Koster A, Verhagen H, Bladeren van PJ. The glutathione conjugates of *tert*-butyl hydroquinone as potent redox cycling agents and possible reactive agents underlying the toxicity of butylated hydroxyanisole. Biochem Biophys Res Comm 1992; 189:309–314.

124. Armstrong KE, Wattenberg LW. Metabolism of 3-*tert*-butyl-4-hydroxyanisole to 3-*tert*-butyl-4,5-dihydroxyanisole by rat liver microsomes. Cancer Res 1985; 45:1507–1510.

125. Nakagawa Y, Nakajima K, Moore G, Moldéus P. On the mechanisms of 3-*tert*-butyl-4-hydroxyanisole- and its metabolites-induced cytotoxicities in isolated rat hepatocytes. Eur J Pharmacol Environ Toxicol Pharmacol 1994; Section 270:341–348.

126. Sgaragli GP, Della Corte L, Puliti R, De Sarlo F, Francalanci R, Guarna A. Oxidation of 2-*t*-butyl-4-methoxyphenol (BHA) by horseradish and mammalian peroxidase systems. Biochem Pharmacol 1980; 29:763–769.

127. Guarna A, Della Corte L, Giovannini MG, De Sarlo F, Sgaragli GP. 2,2'-dihydroxy-3,3'-di-*t*-butyl-5,5'-dimethoxydiphenyl, a new metabolite of 2-*t*-butyl-4-methoxyphenol in the rat. Drug Metab Disp 1983; 11:581–584.

128. Valoti M, Della Corte L, Tipton KF, Sgaragli GP. Purification and characterization of rat intestinal peroxidase. Biochem J 1988; 250:501–507.

129. Thompson DC, Cha Y-N, Trush MA. The peroxidase-dependent activation of butylated hydroxyanisole and butylated hydroxytoluene (BHT) to reactive intermediates. J Biol Chem 1989; 264:3957–3965.

130. Stolze K, Nohl H. Methemoglobin formation from butylated hydroxyanisole and oxyhemoglobin. Comparison with butylated hydroxytoluene and *p*-hydroxyanisole. Free Rad Res Comms 1992; 16:159–166.

131. Thompson JA, Bolton JL, Malkinson AM. Relationship between the metabolism of butylated hydroxytoluene (BHT) and lung tumor promotion in mice. Exp Lung Res 1991; 17:439–453.

132. Shaw Y-S, Chen C. Ring hydroxylation of di-*tert*-butylhydroxytoluene by rat liver microsomal preparation. Biochem J 1972; 128:1285–1291.

133. Thompson JA, Wand MD. Interaction of cytochrome P-450 with a hydroperoxide derived from butylated hydroxytoluene. Mechanism of isomerization. J Biol Chem 1985; 260:10637–10644.

134. Wand MD, Thompson JA. Cytochrome P-450-catalyzed rearrangement of a peroxyquinol derived from butylated hydroxytoluene. J Biol Chem 1986; 261:14049–14056.

135. Kahl R, Weimann A, Weinke S, Hildebrandt AG. Detection of oxygen activation and determination of the activity of antioxidants towards reactive oxygen species by use of the chemiluminigenic probes luminol and lucigenin. Arch Toxicol 1987; 60:158–162.

136. Kahl, unpublished results.

137. Yamamoto K, Tajima K, Takemura M, Mizutani T. Further metabolism of 3,5-di-*tert*-butyl-4-hydroxybenzoic acid, a major metabolite of butylated hydroxytoluene, in rats. Chem Pharm Bull 1991; 39:512–514.

138. Takahashi O, Hiraga K. 2,6-Di-*tert*-butyl-4-methylene-2,5-cyclohexadienone: a hepatic metabolite of butylated hydroxytoluene in rats. Fd Cosmet Toxicol 1979; 17:451–454.

139. Thompson DC, Trush MA. Enhancement of the peroxidase-mediated oxidation of butylated hydroxytoluene to a quinone methide by phenolic and amine compounds. Chem Biol Interactions 1989; 72:157–173.

140. Thompson DC, Trush MA. Studies on the mechanism of enhancement of butylated hydroxytoluene-induced mouse lung toxicity by butylated hydroxyanisole. Toxicol Appl Pharmacol 1988; 96:122–131.

141. Bolton JL, Valerio LG Jr, Thompson JA. The enzymatic formation and chemical reactivity of quinone methides correlate with alkylphenol-induced toxicity in rat hepatocytes. Chem Res Toxicol 1992; 5:816–822.

142. Taffe BG, Kensler TW. Tumor promotion by a hydroperoxide metabolite of butylated hydroxytoluene, 2,6-di-*tert*-butyl-4-hydroperoxy-4-methyl-2,5-cyclohexadienone, in mouse skin. Res Commun Chem Pathol Pharmacol 1988; 61:291–303.

143. Taffe BG, Zweier JL, Pannell LK, Kensler TW. Generation of reactive intermediates from the tumor promoter butylated hydroxytoluene hydroperoxide in isolated murine keratinocytes or by hematin. Carcinogenesis 1989; 10:1261–1268.

144. Guyton KZ, Bhan P, Kuppusamy P, Zweier JL, Trush MA, Kensler TW. Free
radical-derived quinone methide mediates skin tumor promotion by butylated
hydroxytoluene hydroperoxide: expanded role for electrophiles in multistage
carcinogenesis. Proc Natl Acad Sci USA 1991; 88:946–950.

145. Guyton KZ, Dolan PM, Kensler TW. Quinone methide mediates in vitro induction of ornithine decarboxylase by the tumor promotor butylated hydroxytoluene
hydroperoxide. Carcinogenesis 1994; 15:817–821.

146. Nakagawa Y, Nakajima K, Tayama S, Moldéus P. Metabolism and cytotoxicity
of propyl gallate in isolated rat hepatocytes: Effects of a thiol reductant and an
esterase inhibitor. Mol Pharmacol 1995; 47:1021–1027.

147. Reyes JL, Hernandez ME, Melendez E, Gomez-Lojero C. Inhibitory effect of the
antioxidant ethoxyquin on electron transport in the mitochondrial respiratory
chain. Biochem Pharmacol 1995; 49:283–289.

148. Aruoma OI, Evans PJ, Kaur H, Sutcliffe L, Halliwell B. An evaluation of the
antioxidant and potential pro-oxidant properties of food additives and of trolox
C, vitamin E and probucol. Free Rad Res Commun 1990; 10:143–157.

149. Scherer NM, Deamer DW. Oxidative stress impairs the function of sarcoplasmic
reticulum by oxidation of sulfhydryl groups in the Ca^{2+}-ATPase. Arch Biochem
Biophys 1986; 246:589–601.

150. Lee CI, Okabe E. Hydroxyl radical-mediated reduction of Ca^{2+}-ATPase activity of masseter muscle sarcoplasmic reticulum. Jpn J Pharmacol 1995; 67:21–28.

151. Grover AK, Samson SE. Effect of superoxide radical on Ca^{2+} pumps of coronary artery. Am J Physiol 1988; 255:C297–C303.

152. Schilling WP, Elliott SJ. Ca^{2+} signaling mechanisms of vascular endothelial cells
and their role in oxidant-induced endothelial cell dysfunction. Am J Physiol 1992;
262:H1617–H1630.

153. Suzuki YJ, Ford GD. Superoxide stimulates IP_3-induced Ca^{2+} release from vascular smooth muscle sarcoplasmic reticulum. Am J Physiol 1992; 262:H114–
H116.

154. Doan TN, Gentry DL, Taylor AA, Elliott SJ. Hydrogen peroxide activates agonist-sensitive Ca^{2+}-flux pathways in canine venous endothelial cells. Biochem
1994; 297:209–215.

155. Boraso A, Williams AJ. Modification of the gating of the cardiac sarcoplasmic
reticulum Ca^{2+}-release channel by H_2O_2 and dithiothreitol. Am J Physiol 1994;
267:H1010–H1016.

156. Weis M, Kass GE, Orrenius S. Further characterization of the events involved
in mitochondrial Ca^{2+} release and pore formation by prooxidants. Biochem
Pharmacol 1994; 15:2147–2156.

157. Josephson RA, Silverman HS, Lakatta EG, Stern MD, Zweier JL. Study of the
mechanisms of hydrogen peroxide and hydroxyl free radical-induced cellular
injury and calcium overload in cardiac myocytes. J Biol Chem 1991; 266:2354–
2361.

158. Zoccarato F, Pandolfo M, Deana R, Alexandre A. Inhibition by some phenolic antioxidants of Ca^{2+} uptake and neurotransmitter release from brain synaptosomes. Biochem Biophys Res Commun 1987; 146:603–610.

159. Sgaragli GP, Valoti M, Palmi M, Mantovani P. Nifedipine-like activity of 2-t-butyl-4-methoxphenol (BHA) on rat ileum longitudinal muscle preparation. Pharmacol Res 1989; 21:649–650.

160. Nelson EJ, Li CCR, Bangalore R, Benson T, Kass RS, Hinkle PM. Inhibition of L-type calcium-channel activity by thapsigargin and 2,5-t-butylhydroquinone, but not by cyclopiazonic acid. Biochem J 1994; 302:147–154.

161. Nakamura H, Nakasaki Y, Matsuda N, Shigekawa M. Inhibition of sarcoplasmic reticulum Ca^{2+}-ATPase by 2,5-di-(tert-butyl)-1,4-benzohydroquinone. J Biochem 1993; 112:750–755.

162. Wictome M, Holub M, East JM, Lee AG. The importance of the hydroxyl moieties for inhibition of the Ca^{2+}-ATPase by trilobolide and 2,5-di(tert-butyl)-1,4-benzohydroquinone. Biochem Biophys Res Commun 1994; 199:916–921.

163. Sokolove PM, Albuquerque EX, Kauffman FC, Spande TF, Daly JW. Phenolic antioxidants: potent inhibitors of the $(Ca^{2+} + Mg^{2+})$-ATPase of sarcoplasmic reticulum. FEBS Lett 1986; 203:121–126.

164. Kass GEN, Duddy SK, Moore GA, Orrenius S. 2,5-Di-(tert-butyl)-1,4-benzohydroquinone rapidly elevates cytosolic Ca^{2+} concentration by mobilizing the inositol 1,4,5-trisphosphate-sensitive Ca^{2+} pool. J Biol Chem 1989; 264:15192–15198.

165. David M, Horvarth G, Schimke I, Mueller MM, Nagy I. Effects of the antioxidant butylated hydroxyanisole on cytosolic free calcium concentration. Toxicology 1993; 77:115–121.

166. Alexandre A, Doni MG, Padoin E, Deana R. Inhibition by antioxidants of agonist evoked cytosolic Ca^{++} increase, ATP secretion and aggregation of aspirinated human platelets. Biochem Biophys Res Commun 1986; 139:509–514.

167. Franchi-Micheli S, Della Corte L, Puliti R, Giovannini MG, Zilletti L, Marconi M, Sgaragli GP. Studio sui meccanismi del'azione tossica del butil-idrossi-anisolo (BHA): effetti "in vitro" sul muscolo liscio di mammifero. Bull Soc It Biol Sper 1980; LVI; 2521–2524.

168. Gad SC, Leslie SW, Acosta D. Inhibitory actions of butylated hydroxytoluene on isolated ileal, atrial, and perfused heart preparations. Toxicol Appl Pharmacol 1979; 49:45–52.

169. Posati LP, Fox KK, Pallansch MJ. Inhibition of bradykinin by gallates. J Agr Fd Chem 1970; 18:632–635.

170. Atzori L, Bannenberg G, Corriga AM, Ryrfeidt Å, Moldéus P. Vasoconstriction and bronchoconstriction induced by 2,5-di-(tert-butyl)-1,4-benzohydroquinone, an endoplasmic reticular Ca^{2+}-ATPase inhibitor, in isolated and perfused rat lung. Agents Actions 1992; 36:33–38.

171. van Wauwe J, Goossens J. Effects of antioxidants on cyclooxygenase and lipoxygenase activities in intact human platelets: comparison with indomethacin and ETYA. Prostaglandins 1983; 26:725–730.

172. Schilderman PAEL, Engels W, Wenders JJM, Schutte B, Tenhoor F, Kleinjans

JCS. Effects of butylated hydroxyanisole on arachidonic acid and linoleic acid metabolism in relation to gastrointestinal cell proliferation in the rat. Carcinogenesis 1992; 13:585–591.

173. Katayama K, Shirota H, Kobayashi S, Terato K, Ikuta H, Yamatsu I. In vitro effect of N-methoxy-3-(3,5-di-tert-butyl-4-hydroxybenzylidene)-2-pyrrolidone (E-5110), a novel nonsteroidal antiinflammatory agent, on generation of some inflammatory mediators. Agents Actions 1987; 21:269–271.

174. Moore GGI, Swingle KF. 2,6-Di-tert-butyl-4-(2'-thenoyl)phenol(R-830): a novel nonsteroid anti-inflammatory agent with antioxidant properties. Agents Actions 1982; 12:674–683.

175. Schnurr K, Kuhn H, Rapoport SM, Schewe T. 3,5-Di-t-butyl-4-hydroxytoluene (BHT) and probucol stimulate selectively the reaction of mammalian 15-lipoxygenase with biomembranes. Biochim Biophys Acta 1995; 1254:66–72.

176. Hemler ME, Lands WEM. Evidence for a peroxide-initiated free radical mechanism of prostaglandin biosynthesis. J Biol Chem 1980; 255:6253–6261.

177. Ambrosio G, Golino P, Pascucci I, Rosolowsky M, Campbell WB, Declerck F, Tritto I, Chiariello M. Modulation of platelet function by reactive oxygen metabolites. Am J Physiol 1994; 267:H308–H318.

178. Laughton MJ, Evans PJ, Moroney MA, Hoult JRS, Halliwell B. Inhibition of mammalian 5-lipoxygenase and cyclo-oxygenase by flavonoids and phenolic dietary additives. Biochem Pharmacol 1991; 42:1673–1681.

179. Aishita H, Morimura T, Obata T, Miura Y, Miyamoto T, Tsuboshima T, Mizushima Y. ONO-3144, a new anti-inflammatory drug and its possible mechanism of action. Arch Int Pharmacodyn 1983; 261:316–327.

180. Moncada S, Gryglewski RJ, Bunting S, Vane JR. A lipid peroxide inhibits the enzyme in blood vessel microsomes that generate from prostaglandin endoperoxides the substance (prostaglandin x) which prevents platelet aggregation. Prostaglandins 1976; 12:715–737.

181. Beetens JR, Claeys M, Herman AG. Antioxidants increase the formation of 6-oxo-PGF1a by ram seminal vesicle microsomes. Biochem Pharmacol 1981; 30:2811–2815.

182. Larsson R, Cerutti P. Translocation and enhancement of phosphotransferase activity of protein kinase C following exposure in mouse epidermal cells to oxidants. Cancer Res 1989; 49:5627–5632.

183. Ruzzene M, Donella-Deana A, Alexandre A, Francesconi MA, Deana R. The antioxidant butylated hydroxytoluene stimulates platelet protein kinase C and inhibits subsequent protein phosphorylation induced by thrombin. Biochim Biophys Acta 1991; 1094:121–129.

184. Ruzzene M, Francesconi M, Donella-Deana A, Alexandre A, Deana R. The antioxidant butylated hydroxytoluene (BHT) inhibits the dioctanoylglycerol-evoked platelet response but potentiates that elicited by ionomycin. Arch Biochem Biophys 1992; 294:724–730.

185. Miller ACK, Dwyer LD, Auerbach CE, Miley FB, Dinsdale D, Malkinson AM. Strain-related differences in the pneumotoxic effects of chronically administered butylated hydroxytoluene on protein kinase C and calpain. Toxicology 1994; 90:141–159.

186. Brekke O-L, Espevik T, Bjerve KS. Butylated hydroxyanisole inhibits tumor necrosis factor-induced cytotoxicity and arachidonic acid release. Lipids 1994; 29:91–102.

187. Schulze-Osthoff K, Bakke AC, Vanhesebroeck B, Beyaert R, Jacob WA, Fiers W. Cytotoxic activity of tumor necrosis factor is mediated by early damage of mitchondrial functions. J Biol Chem 1992; 267:5317–5323.

188. Burdon RH. Superoxide and hydrogen peroxide in relation to mammalian cell proliferation. Free Rad Biol Med 1995; 18:775-794.

189. Amstad PA, Krupitza G, Cerutti PA. Mechanism of c-fos induction by active oxygen. Cancer Res 1992; 52:3952–3960.

190. Schreck R, Albermann K, Baeuerle PA. Nuclear factor kappa B: An oxidative stress-responsive transcription factor of eukaryotic cells (a review). Free Rad Res Comms 1992; 17:221-237.

191. Suzuki YJ, Mizuno M, Packer L. Signal transduction for nuclear factor-kappa B activation—proposed location of antioxidant-inhibitable step. J Immunol 1994; 153:5008–5015.

192. Xanthoudakis S, Miao G, Wang F, Pan Y-CE, Curran T. Redox activation of Fos-Jun DNA binding activity is mediated by a DNA repair enzyme. EMBO J 1992; 11:3323–3335.

193. Choi H-S, Moore DD. Induction of c-fos and c-jun gene expression by phenolic antioxidants. Mol Endocrinol 1993; 7:1596–1602.

194. Meyer M, Schreck R, Baeuerle PA. H_2O_2 and antioxidants have opposite effects on activation of NF-kappaB and AP-1 in intact cells: AP-1 as secondary antioxidant-responsive factor. EMBO J 1993; 12:2005–2015.

195. Kuzuya M, Naito M, Funaki C, Hayashi T, Yamada K, Asai K, Kuzuya F. Antioxidants stimulate endothelial cell proliferation in culture. Artery 1991; 18:115–124.

196. Yoshioka K, Deng TL, Cavigelli M, Karin M. Antitumor promotion by phenolic antioxidants: inhibition of AP-1 activity through induction of fra expression. Proc Natl Sci USA 1995; 23:4972–4976.

197. Rushmore TH, Pickett CB. Transcriptional regulation of the rat glutathione S-transferase Ya subunit gene. J Biol Chem 1990; 265:14648–14653.

198. Friling RS, Bensimon A, Tichauer Y, Daniel V. Xenobiotic-inducible expression of murine glutathione S-transferase Ya subunit gene is controlled by an electrophile-responsive element. Proc Natl Acad Sci USA 1990; 87:6258–6262.

199. Okuda A, Imagawa M, Maeda Y, Sakai M, Muramatsu M. Structural and functional analysis of an enhancer GPEI having a phorbol 12-O-tetradecanoate 13-acetate responsive element-like sequence found in the rat glutathione transferase P gene. J Biol Chem 1989; 264:16919–16926.

200. Rushmore TH, King RG, Paulson KE, Pickett CB. Regulation of glutathione S-transferase Ya subunit gene expression: Identification of a unique xenobiotic-responsive element controlling inducible expression by planar aromatic compounds. Proc Natl Acad Sci USA 1990; 87:3826–3830.

201. Favreau LV, Pickett CB. Transcriptional regulation of the rat NAD(P)H:Quinone reductase gene. J Biol Chem 1991; 266:4556–4561.

202. Jaiswal AK. Human NAD(P)H: quinone oxidoreductase (NQO_1) gene structure and induction by dioxin. Biochemistry 1991; 30:10647–10653.

203. Mulcahy RT, Gipp JJ. Identification of a putative antioxidant response element in the 5'-flanking region of the human gamma-glutamyl-cysteine synthetase heavy subunit gene. Biochem Biophys Res Comm 1995; 209:227–233.

204. Dalton T, Palmiter RD, Andrews GK. Transcriptional induction of the mouse metallothionein-I gene in hydrogen peroxide-treated Hepa cells involves a composite major late transcription factor/antioxidant response element and metal response promoter elements. Nucl Acid Res 1994; 22:5016–5023.

205. Jones PL, Kucera G, Gordon H, Boss JM. Cloning and characterization of the murine manganous superoxide dismutase-encoding gene. Gene 1995; 153:155–161.

206. Suzuki T, Smith S, Board PG. Structure and function of the 5'flanking sequences of the human alpha class glutathione S-transferase genes. Biochem Biophys Res Commun 1994; 200:1665–1671.

207. Xie T, Belinsky M, Xu Y, Jaiswal AK. ARE- and TRE-mediated regulation of gene expression. J Biol Chem 1995; 270:6894–6900.

208. Rushmore TH, Morton MR, Pickett CB. The antioxidant responsive element. J Biol Chem 1991; 266:11632–11639.

209. Friling BS, Bergelson S, Daniel V. Two adjacent AP-1-like binding sites form the electrophile-responsive element of the murine glutathione S-transferase Ya subunit gene. Proc Natl Acad Sci USA 1992; 89:668–672.

210. Nguyen T, Pickett CB. Regulation of rat glutathione S-transferase Ya subunit gene expression. J Biol Chem 1992; 267:13535–13539.

211. Wang B, Williamson G. Detection of a nuclear protein which binds specifically to the antioxidant responsive element (ARE) of the human NAD(P)H:quinone oxidoreductase gene. Biochim Biophys Acta 1994; 1219:645–652.

212. Favreau LV, Pickett CB. Transcriptional regulation of the rat NAD(P)H:quinone reductase gene: Characterization of a DNA-protein interaction at the antioxidant responsive element and induction by 12-O-tetradecanoylphorbol 13-acetate. J Biol Chem 1993; 268:19875–19881.

213. Bergelson S, Pinkus R, Daniel V. Intracellular glutathione levels regulate fos/jun induction and activation of glutathione S-transferase gene expression. Cancer Res 1994; 54:36–40.

214. Borroz KI, Buetler TM, Eaton DL. Modulation of gamma-glutamylcysteine synthetase large subunit mRNA expression by butylated hydroxyanisole. Toxicol Appl Pharmacol 1994; 126:150–155.

215. Jaiswal AK. Antioxidant response element. Biochem Pharmacol 1994; 48:439–444.

216. Talalay P, DeLong MJ, Prochaska HJ. Identification of a common chemical signal regulating the induction of enzymes that protect against chemical carcinogenesis. Proc Natl Acad Sci USA 1988; 85:8261–8265.

217. Rushmore TH, Pickett CB. Xenobiotic responsive elements controlling inducible expression by planar aromatic compounds and phenolic antioxidants. Meth Enzymol 1991; 206:409–420.

218. Ito N, Hagiwara A, Shibata M, Ogiso T, Fukushima S. Induction of squamous cell carcinoma in the forestomach of F344 rats treated with butylated hydroxyanisole. Gann 1982; 73:332–334.

219. Ito N, Fukushima S, Imaida K, Sakata T, Masui T. Induction of papilloma in the forestomach of hamsters by butylated hydroxyanisole. Gann 1983; 74:459–461.

220. Brusick D. Genotoxicity of phenolic antioxidants. Toxicol Industr Health 1993; 9:223–230.

221. Clayson DB, Iverson F, Nera EA, Lok E. The importance of cellular proliferation induced by BHA and BHT. Toxicol Industr Health 1993; 9:231–242.

222. Hayashi Y, Morimoto K, Miyata N, Sato H. Quantitative cancer risk analysis of BHA based on integration of pathological and biological/biochemical information. Toxicol Industr Health 1993; 9:243–258.

223. Whysner J. Mechanism-based cancer risk assessment of butylated hydroxyanisole. Toxicol Industr Health 1993; 9:283–293.

224. Hirose M, Asamoto M, Hagiwara A, Ito N, Kaneko H, Saito K, Takamatsu Y, Yoshitake A, Miyamoto J. Metabolism of 2- and 3-*tert*-butyl-4-hydroxyanisole (2- and 3-BHA) in the rat (II): metabolism in forestomach and covalent binding to tissue macromolecules. Toxicology 1987; 45:13–24.

225. Saito K, Nakagawa S, Yoshitake A, Miyamoto J, Hirose M, Ito N. DNA-adduct formation in the forestomach of rats treated with 3-*tert*-butyl-4-hydroxyanisole and its metabolites as assessed by an enzymatic 32P-postlabelling method. Cancer Lett 1989; 48:189–195.

226. Sakai A, Miyata N, Takahashi A. Initiating activity of 3-*tert*-butyl-4-hydroxyanisole (3-BHA) and its metabolites in two-stage transformation of BALB/3T3 cells. Carcinogenesis 1990; 11:1985–1988.

227. Giri AK, Sen S, Talukder G, Sharma A, Banerjee TS. Mutachromosomal effects of *tert*-butylhydroquinone in bone-marrow cells of mice. Fd Chem Toxicol 1984; 22:459–460.

228. Phillips BJ, Carroll PA, Tee AC, Anderson D. Microsome-mediated clastogenicity of butylated hydroxyanisole (BHA) in cultured Chinese hamster ovary cells: the possible role of reactive oxygen species. Mutat Res 1989; 214:105–114.

229. Rogers CG, Boyes BG, Matula TI, Stapley R. Evaluation of genotoxicity of *tert*-butylhydroquinone in an hepatocyte-mediated assay with V79 chinese hamster lung cells and in strain D7 of *Saccharomyces cerevisiae*. Mutat Res 1992; 280:17–27.

230. Morimoto K, Tsuji K, Iio T, Miyata N, Uchida A, Osawa R, Kitsutaka H, Takahashi A. DNA damage in forestomach epithelium from male F344 rats following oral administration of *tert*-butylquinone, one of the forestomach metabolites of 3-BHA. Carcinogenesis 1991; 12:703–708.

231. Rogers CG, Boyes BG, Matula TI, Neville G, Stapley R. Cytotoxic and genotoxic properties of *tert*-butyl-*p*-quinone (TBQ) in an in vitro assay system with Chinese hamster V79 cells and in strain D7 of *Saccharomyces cerevisiae*. Mutat Res 1993; 299:9–18.

232. Matsuoka A, Matsui M, Miyata N, Sofuni T, Ishidate M Jr. Mutagenicity of 3-

tert-butyl-4-hydroxyanisole (BHA) and its metabolites in short-term test in vitro. Mutat Res 1990; 241:125–132.

233. Phillips BJ, Tee AC, Carroll PA, Purchase R, Walters DG. Toxicity to Chinese hamster ovary (CHO) cells of the products of reaction of butylated hydroxyanisole with nitrite at low pH. Toxicology in vitro 1994; 8:117–123.

234. Verhagen H, Schilderman PAEL, Kleinjans JCS. Butylated hydroxyanisole in perspective. Chem Biol Interactions 1991; 80:109–134.

235. Altmann H-J, Wester PW, Matthiaschk G, Grunow W, van der Heijden CA. Induction of early lesions in the forestomach of rats by 3-*tert*-butyl-4-hydroxyanisole (BHA). Fd Chem Toxicol 1985; 23:723–731.

236. Verhagen H, Furnée C, Schutte B, Bosman FT, Blijham GH, Henderson PT, ten Hoor F, Kleinjans JCS. Dose-dependent effects of short-term dietary administration of the food additive butylated hydroxyanisole on cell kinetic parameters in rat gastro-intestinal tract. Carcinogenesis 1990; 11:1461–1468.

237. Masui T, Asamoto M, Hirose M, Fukushima S, Ito N. Regression of simple hyperplasia and papillomas and persistence of basal cell hyperplasia in the forestomach of F344 rats treated with butylated hydroxyanisole. Cancer Res 1987; 47:5171–5174.

238. Hirose M, Masuda A, Hasagawa R, Wada S, Ito N. Regression of butylated hydroxyanisole (BHA)-induced hyperplasia but not dysplasia in the forestomach of hamsters. Carcinogenesis 1990; 11:239–244.

239. Tatematsu M, Ogawa K, Mutai M, Aoki T, Hoshiya T, Ito N. Rapid regression of squamous cell hyperplasia and slow regression of basal cell hyperplasia in the forestomach of F344-rats treated with *N*-methyl-*N'*-nitro-*N*-nitrosoguanidine and/or butylated hydroxyanisole. Cancer Res 1991; 51:318–323.

240. Kawabe M, Takaba K, Yoshida Y, Hirose M. Effects of combined treatment with phenolic compounds and sodium nitrite on two-stage carcinogenesis and cell proliferation in the rat stomach. Jpn J Cancer Res 1994; 85:17–25.

241. Amo H, Kubota H, Lu J, Matsuyama M. Adenomatous hyperplasia and adenomas in the lung induced by chronic feeding of butylated hydroxyanisole of Japanese house musk shrew (*Suncus murinus*). Carcinogenesis 1990; 11:151–154.

242. Park E-H, Chang H-H, Cha Y-N. Induction of hepatic tumors with butylated hydroxyanisole in the self-fertilizing hermaphroditic fish *Rivulus ocellatus marmoratus*. Jpn J Cancer Res 1990; 81:738–741.

243. Nera EA, Iverson F, Lok E, Armstrong CL, Karpinski K, Clayson DB. A carcinogenesis reversibility study of the effects of butylated hydroxyanisole on the forestomach and urinary bladder in male Fischer 344 rats. Toxicology 1988; 53:251–268.

244. Würtzen G, Olsen P. BHA study in pigs. Fd Chem Toxicol 1986; 24:1229–1233.

245. Tobe M, Furuya T, Kawasaki Y, Naito K, Sekita K, Matsumoto K, Ochiai T, Usui A. Six-month toxicity study of butylated hydroxyanisole in beagle dogs. Fd Chem Toxicol 1986; 24:1223–1228.

246. Iverson F, Trulove J, Nera E, Lok E, Clayson DB, Wong J. A 12-week study of BHA in the cynomolgus monkey. Fd Chem Toxicol 1986; 24:1197–1200.

247. JECFA. Thirty-third Report of the Joint FAO/WHO Expert Committee on Food Additives. Technical Report Series 776, WHO, Geneva.

248. IARC. Some naturally occurring and synthetic food components, furocoumarins and ultraviolet radiation. In: IARC Monographs on the Evaluation of the Carcinogenic Risk of Chemicals to Humans. Vol 40. IARC, 1985:126. Lyon.

249. Olsen P, Meyer O, Bille N, Würtzen G. Carcinogenicity study on butylated hydroxytoluene (BHT) in Wistar rats exposed in utero. Fd Chem Toxicol 1986; 24:1-12.

250. Inai K, Kobuke T, Nambu S, Takemoto T, Kou E, Nishina H, Fujihara M, Yonehara S, Suehiro S-I, Tsuya T, Horiuchi K, Tokuoka S. Hepatocellular tumorigenicity of butylated hydroxytoluene administered orally to B6C3F1 mice. Gann 1988; 79:49-58.

251. Uno Y, Takasawa H, Miyagawa M, Inoue Y, Murata T, Yoshikawa K. An in vivo-in vitro replicative DNA synthesis (RDS) test using rat hepatocytes as an early prediction assay for nongenotoxic hepatocarcinogens: Screening of 22 known positives and 25 noncarcinogens. Mutat Res 1994; 320:189-205.

252. von der Hude W, Bauszus M, Basler A, Kahl R. Enhancement and inhibition of benzo[a]pyrene-induced SOS function in E. coli by synthetic antioxidants. Mutat Res 1988; 207:7-11.

253. Nakagawa Y, Tayama K, Nakao T, Hiraga K. On the mechanism of butylated hydroxytoluene-induced hepatic toxicity in rats. Biochem Pharmacol 1984; 33:2669-2674.

254. Powell CJ, Connolly AK. The site specificity and sensitivity of the rat liver to butylated hydroxytoluene-induced damage. Toxicol Appl Pharmacol 1991; 108:67-77.

255. Mizutani T, Nomura H, Yamamoto K, Tajima K. Modification of butylated hydroxytoluene-induced pulmonary toxicity in mice by diethyl maleate, buthionine sulfoximine, and cysteine. Toxicol Lett 1984; 23:327-331.

256. Nakagawa Y. Effects of buthionine sulfoximine and cysteine on the hepatotoxicity of butylated hydroxytoluene in rats. Toxicol Lett 1987; 37:251-256.

257. Nakagawa Y, Suga T, Hiraga K. Preventive effect of cysteine on butylated hydroxytoluene-induced pulmonary toxicity in mice. Biochem Pharmacol 1984; 33:502-505.

258. Takahashi O. Feeding of butylated hydroxytoluene to rats caused a rapid decrease in blood coagulation factors II (prothrombin). VII, IX and X. Arch Toxicol 1986; 58:177-181.

259. Takahashi O, Hiraga K. Effects of low levels of butylated hydroxytoluene on the prothrombin index of male rats. Fd Cosmet Toxicol 1978; 16:475-477.

260. Takahashi O. 2,6-Di-tert-butyl-4-methylene-2 5-cyclohexadienone (BHT quinone methide): an active metabolite of BHT causing haemorrhages in rats. Arch Toxicol 1988; 62:325-327.

261. Witschi HP, Malkinson AM, Thompson JA. Metabolism and pulmonary toxicity of butylated hydroxytoluene (BHT). Pharmac Ther 1989; 42:89-113.

262. Mizutani T, Ishida I, Yamamoto K, Tajima K. Pulmonary toxicity of butylated hydroxytoluene and related alkylphenols: structural requirements for toxic potency in mice. Toxicol Appl Pharmacol 1982; 62:273-281.

263. Malkinson AM, Thaete LG, Blumenthal EJ, Thompson JA. Evidence for a role

of *tert*-butyl-hydroxylation in the induction of pneumotoxicity in mice by butylated hydroxytoluene. Toxicol Appl Pharmacol 1989; 101:196–204.

264. Thompson JA, Malkinson AM, Wand MD, Mastovich SL, Mead EW, Schullek KM, Laudenschlager WG. Oxidative metabolism of butylated hydroxytoluene by hepatic and pulmonary microsomes from rats and mice. Drug Metab Dispos 1987; 15:833–840.

265. Imaida K, Fukushima S, Shirai T, Ohtani M, Nakanishi K, Ito N. Promoting activities of butylated hydroxyanisole and butylated hydroxytoluene on 2-stage urinary bladder carcinogenesis and inhibition of gamma-glutamyl transpeptidase-positive foci development in the liver of rats. Carcinogenesis 1983; 4:895–899.

266. Tsuda H, Fukushima S, Imaida K, Sakata T, Ito N. Modification of carcinogenesis by antioxidants and other compounds. Acta Pharmacol Toxicol 1984; 55:125–143.

267. Imaida K, Fukushima S, Inoue K, Masui T, Hirose M, Ito N. Modifying effects of concomitant treatment with butylated hydroxyanisole or butylated hydroxytoluene on *N,N*-dibutylnitrosamine-induced liver, forestomach and urinary bladder carcinogenesis in F344 male rats. Cancer Lett. 1988; 43:167–172.

268. Thamavit W, Fukushima S, Kurata Y, Asamoto M, Ito N. Modification by sodium L-ascorbate, butylated hydroxytoluene, phenobarbital and pepleomycin of lesion development in a wide-spectrum initiation rat model. Cancer Lett 1989; 45:93–101.

269. Balansky RM, Blagoeva PM, Mircheva ZI, Deflora S. Modulation of diethylnitrosamine carcinogenesis in rat liver and oesophagus. J Cell Biochem 1994; 56:449–454.

270. Williams GM, Tanaka T, Maeura Y. Dose-related inhibition of aflatoxin B$_1$ induced hepatocarcinogenesis by the phenolic antioxidants, butylated hydroxyanisole and butylated hydroxytoluene. Carcinogenesis 1986; 7:1043–1050.

271. Ulland BM, Weisburger JH, Yamamoto RS, Weisburger EK. Antioxidants and carcinogenesis: butylated hydroxytoluene, but not diphenyl-*p*-phenylenediamine, inhibits cancer induction by *N*-2-fluorenyl acetamide and by *N*-hydroxy-*N*-2-fluorenylacetamide in rats. Fd Cosmet Toxicol 1973; 11:199–207.

272. McCay PB, King M, Rikans LE, Pitha JV. Interactions between dietary fats and antioxidants on DMBA-induced mammary carcinomas and on AAF-induced hyperplastic nodules and hepatomas. J Environ Path Toxicol 1979; 3:451–465.

273. Peraino C, Fry RJM, Staffeldt E, Christopher JP. Enhancing effects of phenobarbitone and butylated hydroxytoluene on 2-acetylaminofluorene-induced hepatic tumorigenesis in the rat. Fd Chem Toxicol 1977; 15:93–96.

274. Maeura Y, Williams GM. Enhancing effect of butylated hydroxytoluene on the development of liver altered foci and neoplasms induced by *N*-2-fluorenylacetamide in rats. Fd Chem Toxicol 1984; 22:191–198.

275. Maeura Y, Weisburger JH, Williams GM. Dose-dependent reduction of *N*-2-fluorenylacetamide-induced liver cancer and enhancement of bladder cancer in rats by butylated hydroxytoluene. Cancer Res 1984; 44:1604–1610.

276. Williams GM, Tanaka T, Maruyama H, Maeura Y, Weisburger JH, Zang E. Modulation by butylated hydroxytoluene of liver and bladder carcinogenesis in-

duced by chronic low-level exposure to 2-acetylaminofluorene. Cancer Res 1991; 51:6224–6230.

277. Rao MS, Lalwani ND, Watanabe TK, Reddy JK. Inhibitory effect of antioxidants ethoxyquin and 2(3)-*tert*-butyl-4-hydroxyanisole on hepatic tumorigenesis in rats fed ciprofibrate, a peroxisome proliferator. Cancer Res 1984; 44:1072–1076.

278. Fukushima S, Sakata T, Tagawa Y, Shibata M-A, Hirose M, Ito N. Different modifying response of butylated hydroxyanisole, butylated hydroxytoluene, and other antioxidants in *N,N*-dibutylnitrosamine esophagus and forestomach carcinogenesis in rats. Cancer Res 1987; 47:2113–2116.

279. Shirai T, Fukushima S, Ohshima M, Masuda A, Ito N. Effects of butylated hydroxyanisole, butylated hydroxytoluene, and NaCl on gastric carcinogenesis initiated with *N*-methyl-*N'*-nitro-*N*-nitrosoguanidine in F344 rats. J Natl Cancer Inst 1984; 72:1189–1198.

280. Newberne PM, Charnley G, Adams K, Cantor CM, Roth D, Supharkarn V. Gastric and oesophageal carcinogenesis: Models for the identification of risk and protective factors. Fd Chem Toxicol 1986; 24:1111–1119.

281. Takahashi M, Furukawa F, Toyoda K, Sato H, Hasegawa R, Hayashi Y. Effects of four antioxidants on *N*-methyl-*N'*-nitro-*N*-nitrosoguanidine initiated gastric tumor development in rats. Cancer Lett 1986; 30:161–168.

282. Whysner J, Wang CX, Zang E, Iatropoulos MJ, Williams GM. Dose response of promotion by butylated hydroxyanisole in chemically initiated tumours of the rat forestomach. Fd Chem Toxicol 1994; 32:215–222.

283. Imaida K, Fukushima S, Shirai T, Masui T, Ogiso T, Ito N. Promoting activities of butylated hydroxyanisole, butylated hydroxytoluene and sodium L-ascorbate on forestomach and urinary bladder carcinogenesis initiated with methylnitrosourea in F344 male rats. Gann 1984; 75:769–775.

284. Tsuda H, Sakata T, Shirai T, Kurata Y, Tamano S, Ito N. Modification of *N*-methyl-*N*-nitrosourea initiated carcinogenesis in the rat by subsequent treatment with antioxidants, phenobarbital and ethinyl estradiol. Cancer Lett 1984; 24:19–27.

285. Wattenberg LW, Jerina DM, Lam LKT, Yagi H. Neoplastic effects of oral administration of ($\pm$)-*trans*-7,8-dihydroxy-7,8-dihydrobenzo[a]pyrene and their inhibition by butylated hydroxyanisole. J Natl Cancer Inst 1979; 62:1103–1106.

286. Clapp NK, Tyndall RL, Satterfield LC, Klima WC, Bowles ND. Selective sex related modification of diethylnitrosamine-induced carcinogenesis in BALB/c mice by concomitant administration of butylated hydroxytoluene. J Natl Cancer Inst 1978; 61:177.

287. Tatsuta M, Mikuni T, Taniguchi H. Protective effect of butylated hydroxytoluene against induction of gastric cancer by *N*-methyl-*N'*-nitro-*N*-nitrosoguanidine in Wistar rats. Int J Cancer 1983; 32:253–254.

288. Jones FE, Komorowski RA, Condon RE. The effects of ascorbic acid and butylated hydroxyanisole in the chemoprevention of 1,2-dimethylhydrazine-induced large bowel neoplasms. J Surg Oncol 1984; 25:54–60.

289. Shirai T, Ikawa E, Hirose M, Thamavit W, Ito N. Modification by five antioxi-

dants of 1,2-dimethylhydrazine-initiated colon carcinogenesis in F344 rats. Carcinogenesis 1985; 6:637–639.

290. Lindenschmidt RC, Tryka AF, Witschi HP. Modification of gastroinestinal tumor development in rats by dietary butylated hydroxytoluene. Fund Appl Toxicol 1987; 8:474–481.

291. Reddy BS, Maeura Y, Weisburger JH. Effect of various levels of dietary butylated hydroxyanisole on methylazoxymethanol acetate-induced colon carcinogenesis in CF1 mice. J Natl Cancer Inst 1983; 71:1299–1305.

292. Weisburger EK, Evarts RP, Wenk ML. Inhibitory effect of butylated hydroxytoluene (BHT) on intestinal carcinogenesis in rats by azoxymethane. Fd Chem Toxicol 1977; 15:139–141.

293. Clapp NK, Bowles ND, Satterfield LC, Klima WC. Selective protective effect of butylated hydroxytoluene against 1,2-dimethylhydrazine carcinogenesis in BALB/c mice. J Natl Cancer Inst 1979; 63:1081–1085.

294. Lindenschmidt RC, Tryka AF, Goad ME, Witschi HP. The effects of dietary butylated hydroxytoluene on liver and colon tumor development in mice. Toxicology 1986; 38:151–160.

295. Roebuck BD, MacMillan DL, Bush DM, Kensler TW. Modulation of azaserine-induced pancreatic foci by phenolic antioxidants in rats. J Natl Cancer Inst 1984; 72:1405–1410.

296. Thornton M, Moore MA, Ito N. Modifying influence of dehydroepiandrosterone or butylated hydroxytoluene treatment on initiation and development stages of azaserine-induced acinar pancreatic preneoplastic lesions in the rat. Carcinogenesis 1989; 10:407–410.

297. Moore MA, Tsuda H, Thamavit W, Masui T, Ito N. Differential modification of development of preneoplastic lesions in the Syrian golden hamster initiated with a single dose of 2,2'-dioxo-N-nitrosodipropylamine: Influence of subsequent butylated hydroxyanisole, a-tocopherol, or carbazole. J Natl Cancer Inst 1987; 78:289–293.

298. Moore MA, Thamavit W, Hyasa Y, Ito N. Early lesions induced by DHPN in Syrian golden hamsters: Influence of concomitant Opisthorchis infestation, dehydroepiandrosterone or butylated hydroxyanisole administration. Carcinogenesis 1988; 9:1185–1189.

299. Witschi HP, Doherty DG. Butylated hydroxyanisole and lung tumor development in A/J mice. Fund Appl Toxicol 1984; 4:795–801.

300. Witschi HP, Morse CC. Enhancement of lung tumor formation in mice by dietary butylated hydroxytoluene: Dose-time relationships and cell kinetics. J Natl Cancer Inst 1983; 71:859–865.

301. Wattenberg LW. Inhibition of chemical carcinogen-induced pulmonary neoplasia by butylated hydroxyanisole. J Natl Cancer Inst 1973; 50:1541–1544.

302. Witschi HP. Enhancement of lung tumor formation in mice. In: Mass MJ, Kaufmann DG, Siegfried JM, Steele VE, Nesnow S, eds. Carcinogenesis 8: Cancer of the Respiratory Tract. New York: Raven Press, 1985:147–158.

303. Wattenberg LW. Inhibition of carcinogenic effects of diethylnitrosamine (DEN) and 4-nitroquinoline-N-oxide (NQO) by antioxidants. Fed Proc 1972; 31:633.

304. Chung F-L, Wang M, Carmella SG, Hecht SS. Effects of butylated hydroxyanisole on the tumorigenicity and metabolism of *N*-nitrosodimethylamine and *N*-nitrosopyrrolidine in A/J mice. Cancer Res 1986; 46:165–168.

305. Maru GB, Bhide SV. Effect of antioxidants and antitoxicants of isoniazid on the formation of lung tumours in mice by isoniazid and hydrazine sulphate. Cancer Lett. 1982; 17:75–80.

306. Witschi HP, Williamson D, Lock S. Enhancement of urethan tumorigenesis in mouse lung by butylated hydroxytoluene. J Natl Cancer Inst 1977; 58:301–305.

307. Witschi HP. Enhancement of tumor formation in mouse lung by dietary butylated hydroxytoluene. Toxicology 1981; 21:95–104.

308. Malkinson AM, Beer DS. Pharmacologic and genetic studies on the modulatory effects of butylated hydroxytoluene on mouse lung adenoma formation. J Natl Cancer Inst 1984; 73:925–933.

309. Hasegawa R, Furukawa F, Toyoda K, Takahashi M, Hayashi Y, Hirose M, Ito N. Inhibitory effects of antioxidants on *N*-bis(2-hydroxypropyl)nitrosamine-induced lung carcinogenesis in rats. Jpn J Cancer Res 1990; 81:871–877.

310. King MM, Bailey DM, Gibson DD, Pitha JV, McCay PB. Incidence and growth of mammary tumors induced by 7,12-dimethylbenz[a]anthracene as related to the dietary content of fat and antioxidant. J Natl Cancer Inst 1979; 63:657–663.

311. Hirose M, Masuda A, Inoue T, Fukushima S, Ito N. Modification by antioxidants and *p,p'*-diaminodiphenylmethane of 7,12-dimethylbenz[a]anthracene-induced carcinogenesis of the mammary gland and ear duct in CD rats. Carcinogenesis 1986; 7:1155–1159.

312. Hirose M, Fukushima S, Kurata Y, Tsuda H, Tatematsu M, Ito N. Modification of *N*-methyl-*N'*-nitro-*N*-nitrosoguanidine-induced forestomach and glandular stomach carcinogenesis by phenolic antioxidants in rats. Cancer Res 1988; 48:5310–5315.

313. Cohen LA, Polansky M, Furuya K, Reddy M, Berke B, Weisburger JH. Inhibition of chemically induced mammary carcinogenesis in rats by short-term exposure to butylated hydroxytoluene (BHT). Interrelationships among BHT concentration, carcinogen dose, and diet. J Natl Cancer Inst 1984; 72:165–173.

314. Sakata T, Shirai T, Fukushima S, Hasegawa R, Ito N. Summation and synergism in the promotion of urinary bladder carcinogenesis initiated by *N*-butyl-*N*-(4-hydroxybutyl)-nitrosamine in F344 rats. Gann 1984; 75:950–956.

315. Fukushima S, Ogiso T, Kurata Y, Hirose M, Ito N. Dose-dependent effects of butylated hydroxyanisole, butylated hydroxytoluene and ethoxyquin for promotion of bladder carcinogenesis in *N*-butyl-*N*-(4-hydroxybutyl)-nitrosamine-initiated, unilaterally ureter-ligated rats. Cancer Res 1987; 34:83–90.

316. Tamano S, Fukushima S, Shirai T, Hirose M, Ito N. Modification by α-tocopherol, propyl gallate and tertiary butylhydroquinone of urinary bladder carcinogenesis in Fischer 344 rats pretreated with *N*-butyl-*N*-(4-hydroxybutyl)nitrosamine. Cancer Lett 1987; 35:39–46.

317. Moore MA, Thamavit W, Tsuda H, Ito N. The influence of subsequent dehydroepiandrosterone, diaminopropane, phenobarbital, butylated hydroxyanisole and butylated hydroxytoluene treatment on the development of preneoplastic

and neoplastic lesions in the rat initiated with di-hydroxy-di-*N*-propyl nitrosamine. Cancer Lett 1986; 30:153–160.

318. Ito N, Hirose M, Fukushima S, Tsuda H, Shirai T, Tatematsu M. Studies on antioxidants: Their carcinogenic and modifying effects on chemical carcinogenesis. Fd Chem Toxicol 1986; 24:1071–1082.

319. Slaga TJ, Solanki V, Logani M. Studies on the mechanism of action of antitumor promoting agents: Suggestive evidence for the involvement of free radicals in promotion. In: Nygaard OF, Simic MG, eds. Radioprotectors and Anticarcinogens New York: Academic Press, 1983:471–485.

320. Benson AM, Hunkeler MJ, York JL. Mouse hepatic glutathione transferase isoenzymes and their differential induction by anticarcinogens. Biochem J 1989; 261:1023–1029.

321. McLellan LI, Hayes JD. Differential induction of class alpha glutathione S-transferases in mouse liver by the anticarcinogenic antioxidant butylated hydroxyanisole. Biochem J 1989; 263:393–402.

322. McLellan LI, Kerr LA, Cronshaw AD, Hayes JD. Regulation of mouse glutathione S-transferases by chemoprotectors. Biochem J 1991; 276:461–469.

323. McLellan LI, Harrison DJ, Hayes JD. Modulation of glutathione S-transferases and glutathione peroxidase by the anticarcinogen butylated hydroxyanisole in murine extrahepatic organs. Carcinogenesis 1992; 13:2255–2261.

324. Buetler TM, Eaton DL. Complementary DNA cloning, messenger RNA expression, and induction of a-class glutathione S-transferases in mouse tissues. Cancer Res 1992; 52:314–318.

325. Schilderman PA, Rhijnsburger E, Zwingmann I, Kleinjans JC. Induction of oxidative DNA damage and enhancement of cell proliferation in human lymphocytes in vitro by butylated hydroxyanisole. Carcinogenesis 1995; 16:507–512.

6

Antioxidant Properties of Probucol

Etsuo Niki and Noriko Noguchi
*Research Center for Advanced Science and Technology,
University of Tokyo, Tokyo, Japan*

I. INTRODUCTION

It is now generally accepted that the oxidative modification of low-density lipoprotein (LDL) is a key initial event in the formation of the foam cells that accumulate in the early fatty-streak lesion of atherosclerosis (1). It was hypothesized (2) that increased LDL levels in the plasma increase the rate of LDL infiltration into the intima, where oxidative modification is thought to take place. It is not known at present how the LDL oxidation is initiated, but all of the three major cell types in the artery wall, endothelial cells, smooth muscle cells and macrophages can cause this oxidative modification. The oxidized LDL can contribute to atherogenesis in four ways (2): (1) it attracts monocytes into the subendothelial space; (2) it is rapidly taken up by macrophages generating foam cells and eventually a fatty streak; (3) it inhibits the mobility of macrophages; and (4) it is toxic to endothelial cells and may cause endothelial injury.

If the hypothesis that the oxidative modification of LDL plays an important role in the early event of atherosclerosis is correct, then one could expect that the inhibition of such oxidation should prevent fatty-streak formation and atherosclerosis (3). In support of this, a recent epidemiological study shows that high intake of vitamin E, a potent antioxidant, reduces the risk of coronary heart disease, and low serum levels of vitamin E seem to correlate with an increased incidence of myocardial infarction (4).

Probucol, 4,4'-(isopropylidenedithio)bis(2,6-di-*tert*-butylphenol), is a drug widely used in the treatment of hypercholesterolemia (5). It is effective in lowering plasma cholesterol levels, but it is also capable of acting as an antioxidant. Probucol has a structure similar to butylated hydroxytoluene (BHT), a well-known synthetic antioxidant for foods and plastics. It is really just two of the structures connected by a sulfur-carbon-sulfur bridge, and it is quite conceivable that it acts as a radical-scavenging antioxidant. In this chapter, the action and role of probucol as an antioxidant will be reviewed.

II. CHEMICAL STRUCTURE AND PROPERTIES

The chemical structure and some physical properties are summarized in Table 1. Probucol is a lipophilic phenolic compound, and it is assumed that it is located in the lipophilic compartment of membranes and lipoproteins.

III. INHIBITION OF OXIDATION IN VITRO BY PROBUCOL

A. Reactivity of Probucol Toward Radicals

Potent radical-scavengng antioxidants react rapidly with stable radicals such as galvinoxyl and 2,2-diphenyl-1-picrylhydrazyl, and this can be used as a simple screening test for the evaluation of antioxidant compound. When equal amounts of probucol and galvinoxyl are reacted, little change is observed in the electron spin resonance (ESR) spectrum of galvinoxyl. However, when 100 times as much probucol as galvinoxyl is reacted, the ESR spectrum of galvinoxyl disappears and a new triplet ESR signal is observed (Fig. 1), the ESR parameters being $g = 2.0058$ and $a^H(2H) = 0.14$ mT (6). The same ESR signal with identical ESR parameters is obtained when probucol is reacted with the *tert*-butoxyl radical. The ESR signal of probucol has been also observed by Aruoma et al. (7), Kagan et al. (8), and Kalyanaraman et al. (9). As shown in Figure 1, the ESR signal of probucol radical is eliminated by the addition of α-tocopherol or 6-*O*-palmitoylascorbic acid, suggesting that the probucol radical is reduced by vitamin E and vitamin C to regenerate probucol. The absorption at 342 nm, probably due to galvinoxyl radical, also disappears when α-tocopherol or 6-*O*-palmitoylascorbic acid is added to a solution containing probucol radical.

The rate of interaction between probucol and galvinoxyl can be followed with a stopped-flow spectrophotometer. The absorption at 428 nm due to galvinoxyl decreases, while that at 342 nm increases first and then declines (6). Probucol reacts with galvinoxyl much slower than α-tocopherol (Fig. 2).

Table 1 Chemical Structure and Physical Properties of Probucol

Name	4,4'-isopropylidenedithio-bis-(2,6-di-*tert*-butylphenol)
Structure	(CH$_3$)$_3$C, HO— , (CH$_3$)$_3$C —S—C(CH$_3$)$_2$—S— C(CH$_3$)$_3$, —OH, C(CH$_3$)$_3$ $C_{31}H_{48}O_2S_2$
Molecular weight	516.84
Melting point	125-128 °C
Appearance	white powder
Solubility	Very soluble to acetone, chloroform and dimethylformamide, soluble to ethanol, ether and carbon tetrachloride, but insoluble to water
pKa	13.5
Ultraviolet absorption	240 ~ 244 nm
Infrared absorption	3260, 2950, 1315, 1240, 1150, 890 cm^{-1}

These data suggest that probucol is capable of acting as a radical-scavenging antioxidant, but its reactivity toward radicals is much smaller than that of α-tocopherol.

B. Inhibition of Oxidation of Lipids in Solution

The activity of probucol as an antioxidant against lipid peroxidation has been measured in several systems (6–11). The kinetic study on the inhibition of oxidation of methyl linoleate induced by free radicals in homogeneous solution serves as a research tool to quantitatively evaluate the chemical activity of antioxidant of biological interest. The oxidation of methyl linoleate can be

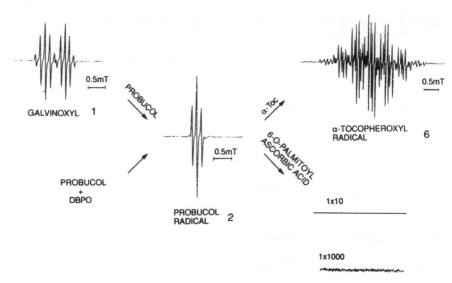

Figure 1 ESR spectrum of probucol radical (7). When probucol was reacted with galvinoxyl (100/1 by mol/mol), the ESR signal of galvinoxyl 1 disappeared and a new triplet signal 2 appeared. The identical ESR spectrum was observed when probucol was mixed with di-*tert*-butyl peroxyoxalate (DBPO), a *tert*-butoxyl radical generator. When α-tocopherol is added to the mixture of probucol and galvinoxyl, the ESR signal of probucol radical disappears and that of α-tocopheroxyl radical is observed. When 6-*O*-palmitoylascorbic acid is added in place of α-tocopherol, the probucol radical disappears.

measured quantitatively by following oxygen uptake, substrate consumption, hydroperoxide formation, or conjugated diene accumulation (12). Figure 3 shows a typical example of the oxidations of methyl linoleate in hexane initiated with a radical initiator in the absence or presence of antioxidant (6). In the absence of antioxidant, oxidation takes place without any induction period or lag phase, and methyl linoleate hydroperoxide is accumulated at a constant rate. When either probucol or α-tocopherol is added, oxidation is suppressed. In particular, α-tocopherol inhibited the oxidation quite markedly and gave a clear induction period. It is consumed at a constant rate, and when it is completely depleted the induction period is over and the oxidation proceeds rapidly at the same rate as that in the absence of antioxidant. Under these circumstances, the absolute rate constant for the reaction of peroxyl radical and antioxidant, a key step for antioxidation, can be obtained (10,13–15). On the other hand, probucol retards the oxidation much less efficiently than α-tocopherol, and a clear induction period is not observed. This makes the quantitative kinetic analysis difficult. However, the ratio of the oxidation rate in the presence of antioxidant to that of uninhibited oxidation gives relative antioxidant efficacy. This

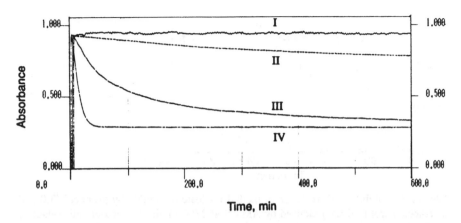

Figure 2 Reduction of galvinoxyl by probucol and α-tocopherol at 37°C in air as studied by stopped-flow spectrophotometer. A methanol solution containing 10 μM galvinoxyl and another methanol solution containing an antioxidant were mixed rapidly, and the consumption of galvinoxyl was measured by following the decay of its absorption at 429 nm. (I) no antioxidant; (II) probucol, 10 μM; (III) α-tocopherol, 10 μM; (IV) probucol, 1 mM.

ratio is obtained from Figure 3 as 0.49 and 0.028 for probucol and α-tocopherol, respectively, suggesting that α-tocopherol is almost 20 times as effective as probucol under these conditions.

Antioxidant activities, however, depend very much on the medium; that is, the antioxidant potency is determined by its chemical reactivity and physical factors such as concentration, location, and mobility of the antioxidant at the microenvironment. It has been found that the antioxidant activity of α-tocopherol is much smaller in the phosphatidylcholine (PC) liposomal membranes than in organic homogeneous solution (16). Barclay et al. (17) found that the rate constant for scavenging of peroxyl radicals by α-tocopherol is much smaller in the PC liposomal membranes than in the homogeneous solution. The ESR spin label study shows that the efficiency of radical scavenging by α-tocopherol decreases as the radical goes deeper into the interior of the membrane (18).

The antioxidant activities of probucol and α-tocopherol against the oxidation of PC liposomal membranes are compared in Figure 4 (6). The ratio of antioxidant to lipid in this membrane system is much higher than that in the homogeneous solution in Figure 3. α-Tocopherol inhibits the oxidation markedly well and produces a clear induction period, while probucol suppresses the oxidation, although not as efficiently as α-tocopherol. However, the difference in the antioxidant activities of the two antioxidants is much smaller in the membrane oxidation than in homogeneous solution. This may well be ascribed to a profound decrease in α-tocopherol mobility in the membranes.

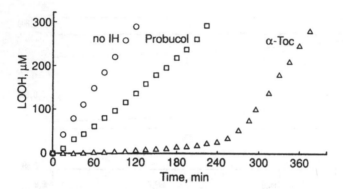

Figure 3 Inhibition of oxidation of methyl linoleate (453 mM) by probucol (3.0 μM) or α-tocopherol (3.0 μM) induced by AMVN at 37°C in air. ○: without antioxidant; □: probucol; Δ: α-tocopherol.

Pryor et al. (10) reported that α-tocopherol was 7.1 times as active as probucol in aqueous sodium dodecyl sulfate micelles.

C. Inhibition of Oxidation of Low-Density Lipoprotein

The oxidation of low-density lipoprotein (LDL) and its inhibition by antioxidants, have been studied extensively, and the action of probucol as an antioxi-

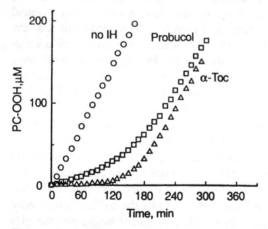

Figure 4 Inhibition of oxidation of soybean phosphatidylcholine (PC; 5.15 mM) liposomal membranes by probucol (3.0 μM) and α-tocopherol (3.0 μM) induced by AMVN (1.02 mM) at 37°C in air. ○ : without antioxidant; □: probucol; Δ: α-tocopherol.

dant has been the subject of numerous investigations. Oxidation of LDL can be induced by various oxidants, such as copper, ion, endothelial cells, peroxynitrite, and azo radical initiator. Since the oxidation proceeds by a free-radical chain mechanism (19), the oxidation of lipids in LDL proceeds independently of the initial active species to give hydroperoxides of cholesterol ester, phosphatidylcholine, and triglyceride as primary products. Probucol taken into LDL endogenously or exogenously suppresses the oxidation of LDL in vitro.

Parthasarathy et al. (20) found that the addition of probucol during the incubation of LDL with endothelial cells or copper ion prevented oxidative modification of LDL. Specifically, it prevented increasing lipid peroxidation, electrophoretic mobility and subsequent susceptibility to macrophage degradation (Table 2). It was also found that the samples of LDL isolated from the plasma of hypercholesterolemic patients under treatment with conventional dosages of probucol were highly resistant to oxidative modification induced by either endothelial cells or cupric ion (20). LDL from control and probucol-treated subjects was incubated with endothelial cells or with copper for 24 h, then tested for modification. LDL isolated from the untreated controls showed an expected increase in the rate of degradation, but LDL isolated from probucol-treated patients showed markedly reduced increase in degradation (20).

Oxidized LDL is toxic to cells (21). The pretreatment of cultured endothelial cells with probucol was found to effectively protect endothelial cells from the toxicity insulted by oxidized LDL or organic hydroperoxide dose-dependently (22). Probucol is especially effective in inhibiting the oxidation of lipids in LDL, but it inhibits the oxidative modification of protein less efficiently (11). Hunt et al. (23) also observed in the oxidation of LDL by copper that probucol was better able to inhibit lipid peroxidation than α-tocopherol, but the reverse was true for protein degradation.

Table 2 Inhibition by Probucol of Oxidation of LDL Induced by Endothelial Cell or Cupric Ion at 37°C, 24 h

Oxidant	Probucol (µM)	TBARS[a] (µM)	Macrophage degradation (µg/5 h/mg)
None	0	0.98	0.64
Endothelial cell	0	4.46	9.62
Endothelial cell	5	1.10	0.73
Cu^{2+}	0	8.12	8.37
Cu^{2+}	5	3.47	1.10

[a]Thiobarbituric acid reactive substances.
Source: Ref. 20.

D. Cooperative Inhibition of Oxidation with Other Antioxidants

LDL contains several kinds of lipophilic antioxidants (Table 3) (24). The concentration of probucol found in LDL isolated from patients taking it regularly is approximately twice as high as α-tocopherol. In the aqueous phase, ascorbic acid, uric acid, bilirubin, and albumin act as hydrophilic antioxidants, although it is not known whether they are present in intima in the same quantity as in plasma. They act not only independently but also cooperatively or even synergistically. It is well known that vitamin E and vitamin C act synergistically (25).

The probucol radical is also reduced by ascorbic acid (6) (Fig. 1), and it has been found that the probucol radical derived from probucol incorporated into LDL was reduced by ascorbate (8,9). Kalyanaraman et al. (9) found that the probucol-dependent inhibition of LDL oxidation was enhanced in the presence of low concentrations of ascorbic acid. They attributed such synergistic enhancement of the antioxidant capacity to the reaction between the phenoxyl radicals of probucol and ascorbate (9). However, as shown in Fig. 5, α-tocopherol is spared by ascorbate almost completely, while probucol is only partially spared in the oxidation of LDL, suggesting that the efficiency of reduction of phenoxyl radical by ascorbate is higher for α-tocopherol than probucol. This is probably because the phenoxyl group of α-tocopherol is located at or near the surface of the LDL particle, whereas that of probucol, which is more lipophilic and lacks a long side chain, is inside the LDL particle. It has been found that oxygen radical reduction by ascorbate becomes less efficient as the radical gets into

Table 3 Lipophilic Antioxidants Contained in Human LDL

	Concentration	
Antioxidant	nmol/mg LDL protein	mol/mol LDL
α-Tocopherol	11.58	6.37
γ-Tocopherol	0.93	0.51
β-Carotene	0.53	0.29
α-Carotene	0.22	0.12
Lycopene	0.29	0.16
Cryptoxanthin	0.25	0.14
Cantaxanthin	0.04	0.02
Lutein + zeaxanthin	0.07	0.04
Phytofluence	0.09	0.05
Ubiquinol-10	0.18	0.10

Source: Ref. 24.

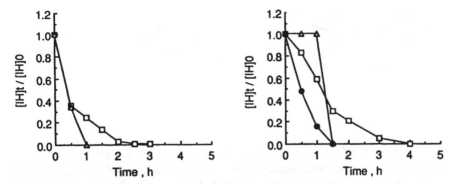

Figure 5 Effect of ascorbic acid on the consumptions of α-tocopherol and probucol during the oxidation of WHHL rabbit LDL at 37°C in air. α-tocopherol (Δ) = 2.8 μM; probucol (□) = 5.9 μM; ascorbic acid (●) = 0 (left), 10 μM (right).

the membrane interior (26) and into LDL (27). Thomas et al. (28) observed no synergy between ascorbate and probucol incorporated into liposomal membrane.

The importance of cooperative interaction between probucol and α-tocopherol in LDL is not clear. When probucol and α-tocopherol are present in the homogeneous solution, α-tocopherol is consumed but only small consumption of probucol is observed, apparently because α-tocopherol is much more reactive toward the radical than probucol and probucol is regenerated from probucol radical by α-tocopherol. On the other hand, both α-tocopherol and probucol are consumed simultaneously during oxidation of human LDL, possibly implying that probucol has as important role in the LDL particle as α-tocopherol does. It also suggests that the reduction of the probucol radical by α-tocopherol in LDL does not proceed as rapidly as in the homogeneous solution.

Jialel and Grundy (29) observed that, although ascorbate preserved α-tocopherol in LDL, probucol in concentrations ranging from 10 to 80 μM failed to protect α-tocopherol in the oxidation of LDL induced by copper or macrophages.

E. Mechanism of Oxidation Inhibition

The mechanism of oxidation inhibition by probucol—that is, the oxidation pathway of probucol—is not fully understood. The initial step is probably the donation of phenolic hydrogen to the peroxyl radical to produce the probucol phenoxyl radical, which has been confirmed by ESR study (6–9). The subsequent reaction of the probucol radical is not known. Barnhart et al. (30) reported the formation of spiroquinone **7** and diphenoquinone **8**.

7   8

The physical effect of probucol and its metabolites on the phase behavior of membranes has also been studied (31–33). Probucol is soluble in cholesteryl esters, and it changes the phase state of cholesteryl ester droplets in cells to a more fluid phase in which the cholesteryl esters are more readily mobilized (34). Its effect should be small under physiological conditions where the molar concentration of probucol to lipids is smaller than 1%.

IV. ANIMAL STUDIES AND CLINICAL STUDIES

Kritchevsky et al. (35) found that adding 1% probucol to a 2% cholesterol–6% corn oil diet of rabbits significantly reduced the severity of aortic atherosclerosis, but to a degree compatible with the degree of plasma cholesterol lowering. More recently, several groups of investigators have found that probucol reduces atherosclerosis in the LDL receptor-deficient Watanabe heritable hyperlipidemic (WHHL) rabbit (36–40), an animal model for familial hypercholesterolemia, and the cholesterol-fed rabbit (41). Kita and his colleagues (36) found that probucol fed for 6 months to WHHL rabbits decreased plaque in the descending thoracic aorta and that LDL isolated from WHHL rabbits treated with probucol was highly resistant to oxidative modification by cupric ion.

Steinberg et al. (37,38) found that the rate of progression of lesions in probucol-treated WHHL rabbits was significantly slower than in a lovastatin-treated group maintained at equal total plasma cholesterol levels, suggesting that probucol functions by acting as an antioxidant not necessarily related to its ability to lower plasma cholesterol levels.

This was more clearly shown by Mao et al. (39,42), who tested probucol and its analogs with the substitution at the disulfide-linked carbon and phenolic rings for their ability to lower total serum cholesterol and prevent aortic atherosclerosis in modified WHHL rabbits and also to inhibit copper-induced oxidation of LDL. Probucol was effective in lowering serum cholesterol, but analogs were not. Probucol and analogs prevented copper-induced oxidation of LDL in vitro to an extent that directly related to their concentrations in LDL.

These results suggest that the antioxidant activity of probucol accounts for the pharmacological activity in reducing atherosclerosis.

Nagano et al. (43) performed long-term studies at both early and late stages. In one study, probucol was administrated to 2-month-old WHHL rabbits for 16 months, and it was found that the percentage area of aorta covered with atherosclerotic plaque was markedly less in probucol-treated rabbits than that in nontreated ones. In another study, probucol was administrated for 6 months with 8-month-old WHHL rabbits in which plaque had already developed and found that probucol was effective in reducing the size of lesions (43). It has also been shown that probucol prevents foam cell formation not by affecting lipoprotein metabolism in macrophages but by inhibiting oxidation of LDL.

Other functions of probucol have been reported. It has been found that probucol suppressed the initial thickening of the carotid artery after injury by balloon catheter (44). It was also observed that probucol pretreatment enhanced macrophage chemotaxis (45). This indicates that probucol may stimulate macrophage egress toward the bloodstream and this effect was proposed as a possible novel mechanism of probucol in preventing atherosclerosis (45). In support of this, O'Brien et al. (46) observed that the atherosclerotic lesions of probucol-treated WHHL rabbits were poor in macrophage-derived foam cells. On the other hand, probucol may retard intimal monocyte recruitment possibly by suppressing the expression of monocyte chemotactic protein 1. Furthermore, probucol may also regulate interleukin-1 expression, which leads to the inhibition of smooth muscle cell proliferation.

High-density lipoprotein (HDL) plays an important role in regulating the cholesterol level. It has been found that the oxidative modification of HDL reduces its ability to stimulate efflux of cholesterol from foam cells (47). Therefore, probucol is able to prevent atherosclerosis by inhibiting the oxidation of HDL as well.

Yamamoto et al. (48) found that regression of xanthomas is enhanced in probucol-treated patients whose cholesterol levels decreased only modestly. This observation also supports the thesis that the major role of probucol is to inhibit oxidation. Although the clinical, preventive, and therapeutic effects of probucol are outside the scope of this chapter, it has been widely used as a drug and its positive effect has been reported (49). Probucol has also received attention in connection with diabetes mellitus (50,51).

REFERENCES

1. Steinberg D. Modified forms of low-density lipoprotein and atherosclerosis. J Int Med 1993; 233:227–232.
2. Steinberg D. Modifications of low-density lipoprotein that increase its atherogenicity. New Engl J Med 1991; 320:815–924.

3. Steinberg D. Antioxidant and atherosclerosis. N Engl J Med 1989; 84:1420–1425.
4. Gey F. Ten-year retrospective on the antioxidant hypothesis of arteriosclerosis: Threshold plasma levels of antioxidant micronutrients related to minimum cardiovascular risk. J Nutr Biochem 1995; 6:206–236.
5. Barkley MM, Goa KL, Price AH, Brogden RN. Probucol: A reappraisal of its pharmacological properties and therapeutic use in hypercholesterolemia. Drugs 1986; 37:761–800.
6. Gotoh N, Shimizu K, Komuro E, Tsuchiya J, Noguchi N, Niki E. Antioxidant activites of probucol against lipid peroxidations. Biochem Biophys Acta 1992; 1128:147–154.
7. Aruoma OI, Evans PJ, Kaur H, Sutcliffe L, Halliwell B. An evaluation of the antioxidant and potential pro-oxidant properties of food additives and of trolox C, vitamin E and probucol. Free Rad Res Comms 1990; 10:143–157.
8. Kagan VE, Freisleben H.-J, Tsuchiya M, Forte T, Packer L. Generation of probucol radicals and their reduction by ascorbate and dihydrolipoic acid in human low density lipoproteins. Free Rad Res Commn 1991; 15:265–276.
9. Kalyanaraman B, Darley-Usmar VM, Woods J, Joseph J, Parthasarathy S. Syndergistic interaction between the probucol phenoxyl radical and ascorbic acid in inhibiting the oxidation of low density lipoprotein. J Biol Chem 1992; 267, 6789–6795.
10. Pryor WA, Cornicelli JA, Devall LJ, Tait B, Trivedi BK, Witiak DT, Wu M. A rapid screening test to determine the antioxidant potencies of natural and synthetic antioxidants. J Org Chem 1993; 58:3521–3532.
11. Noguchi N, Gotoh N, Niki E. Effects of ebselen and probucol on oxidative modifications of lipid and protein of low density lipoprotein induced by free radicals. Biochim Biophys Acta 1994; 1213:176–182.
12. Niki E. Free radical initiators as source of water- or lipid-soluble peroxyl radicals. Methods Enzymol 1990; 186:100–108.
13. Burton GW, Ingold KU. Autoxidation of biological molecules. 1. The antioxidant activity of vitamin E and related chain-breaking phenolic antioxidants in vitro. J Am Chem Soc 1981; 103:6472–6477.
14. Niki E, Saito E, Kawakami A, Kamiya Y. Oxidation of lipids. VI. Inhibition of oxidation of methyl linoleate in solution by vitamin E and vitamin C. J Biol Chem 1984; 259:4177–4182.
15. Barclay LRC. Model biomembranes: quantitative studies of peroxidation, antioxidant action, partitioning, and oxidative stress. Can J Chem 1993; 71:1–16.
16. Niki E, Takahashi M, Komuro E. Antioxidant activity of vitamin E in liposomal membranes. Chem Lett 1986; 6:1573–1576.
17. Barclay LRC, Baskin KA, Dakin KA, Locke SJ, Vinqvist MR. The antioxidant activities of phenolic antioxidants in free radical peroxidation of phospholipid membranes. Can J Chem 1990; 68:2258–2269.
18. Takahashi M, Tsuchiya J, Niki E. Scavenging of radicals by vitamin E in the membranes as studied by spin labeling. J Am Chem Soc 1989; 111:7179–7185.
19. Sato K, Niki E, Shimasaki H. Free radical-mediated chain oxidation of low den-

sity lipoprotein and its synergistic inhibitin by vitamin E and vitamin C. Arch Biochem Biophys 1990; 279:402–405.

20. Parthasarathy S, Young SG, Witztum JL, Pittman RC, Steinberg D. Probucol inhibits oxidative modification of low density lipoprotein. J Clin Invest 1986; 77:641–644.

21. Evensen SA, Galdal KS, Nilsen E. LDL-induced cytotoxicity and its inhibition by antioxidant treatment in cultured human endothelial cells and fibroblasts. Atherosclerosis 1983; 49:23–30.

22. Kuzuya M, Naito M, Funaki C, Hayashi T, Asai K, Kuzuya F. Probucol prevents oxidative injury to endothelial cells. J Lipid Res 1991; 32; 197–204.

23. Hunt JV, Bottoms MA, Taylor SE, Lyell V, Mitchinson MJ. Differing effects of probucol and vitamin E on the oxidative of lipoproteins, ceroid accumulation and protein uptake by macrophages. Free Rad Res 1994; 20:189–201.

24. Esterbauer H, Gebicki J, Puhl H, Jurgens G. The role of lipid peroxidation and antioxidants in oxidative modification of LDL. Free Rad Biol Med 1992; 13:341–390.

25. McCay PB. Vitamin E: Interaction with free radicals and ascorbate. Ann Rev Nutr 1985; 5:323–340.

26. Takahashi M, Tsuchiya J, Niki E, Urano S. Action of vitamin E as antioxidant in phospholipid liposomal membranes as studied by spin label technique. J Nutr Sci Vitaminol 1988; 34:25–34.

27. Gotoh N, Noguchi N, Tsuchiya J, Morita K, Sakai H, Shimasaki H, Niki E. Inhibition of oxidation of low density lipoprotein by vitamin E and related compounds. Free Rad Res. 1996; 24:123–134.

28. Thomas CE, McLean LR, Parker RA, Ohlweiler DF. Ascorbate and phenolic antioxidant interactions in prevention of liposomal oxidation. Lipids 1992; 27:543–550.

29. Jialal I, Grundy SM. Preservation of the endogenous antioxidants in low density lipoprotein by ascorbate but not probucol during oxidative modification. J Clin Invest 1991; 87:597–601.

30. Barnhart RL, Busch SJ, Jackson RL. Concentration-dependent antioxidant activity of probucol in low density lipoproteins in vitro: probucol degradation precedes lipoprotein oxidation. J Lipid Res 1989; 30:1703–1710.

31. McLean LR, Hagaman KA. Antioxidant activity of probucol and its effects on phase transitions in phosphatidylcholine liposomes. Biochim Biophys Acta 1990; 1029:161–166.

32. McLean LR, Hagaman KA. Effect of probucol on the physical properties of low-density lipoproteins oxidized by copper. Biochem 1989; 28:321–327.

33. McLean LR, Hagaman KA. Interactions of MDL 29,311 and probucol metabolites whith cholesteryl esters. Lipids 1994; 29:819–823.

34. McLean LR, Thomas CE, Weintraub B, Hagaman KA. Modulation of the physical state of cellular cholesteryl esters by 4,4′-(isopropylidenedithio)bis(2,6-di-*t*-butylphenol) (Probucol). J Biol Chem 1992; 267:12291–12298.

35. Kietchevsky D, Kirn HK, Tepper SA. Influence of 4′4-(isopropylidene-

dithio)bis(2,6-di-*t*-butylphenol)(DH-581) on experimental atherosclerosis in rabbits. Proc Soc Exp Biol Med 1971; 136:1216–1221.

36. Kita T, Nagano Y, Yokode M, Ishii K, Kume N, Ooshima A, Yoshida H, Kawai C. Probucol prevents the progression of atherosclerosis in Watanabe heritable hyperlipidemic rabbit, an animal model for familial hypercholesterolemia. Proc Natl Acad Sci USA 1987; 84:5928–5831.

37. Carew TE, Schwenke DC, Steinberg D. Antiatherogenic effect of probucol unrelated to its hypocholesterolemic effect: Evidence that antioxidants in vivo can selectively inhibit low density lipoprotein degradation in macrophage-rich fatty streaks and slow the progression of atherosclerosis in the Watanabe heritable hyperlipidemic rabbit. Proc Natl Acad Sci USA 1987; 84:7725–7729.

38. Steinberg D, Parthasarathy S, Carew TE. In vivo inhibition of foam cell development by probucol in Watanabe rabbits. Am J Cardiol 1988; 62:6B–12B.

39. Mao SJT, Yates MT, Rechtin AE, Jackson RL and Van Sicle WA. Antioxidant activity of probucol and its analogues in hypercholesterolemic Watanabe rabbits. J Med Chem 1991; 34:298–302.

40. Nagano Y, Nakamura T, Matsuzawa Y, Cho M, Ueda Y, Kita T. Probucol and atherosclerosis in the Watanabe heritable hyperlipidemic rabbit-long-term antiatherogenic effect and effects on established plaques. Atherosclerosis 1992; 92:131–140.

41. Tawara K. Effect of probucol, pantethine and their combinations on serum lipoprotein metabolism and on the incidence of atheromatous lesions in the rabbit. Jpn J Pharmacol 1986; 41:211.

42. Mao SJT, Yates MT, Parker RA, Chi EM, Jackon RL. Attenuation of atherosclerosis in a modified strain of hypercholesterolemic Watanabe rabbits with use of a probucol analogue (MDL 29,311) that does not lower serum cholesterol. Arterioscl Thromb 1991; 11:1266–1275.

43. Nagano Y, Kita T, Yokode M, Ishii K, Kume N, Otani H, Arai H, Kawai C. Probucol does not affect lipoprotein metabolism in macrophages of Watanabe heritable hyperlipidemic rabbits. Arteriosclerosis 1989; 9:458–461.

44. Shinomiya M, Shirai K, Saito Y, Yoshida S. Inhibition of intimal thickening of the carotid artery of rabbits and of outgrowth of explants of aorta by probucol. Atherosclerosis 1992; 97:143–148.

45. Hara S, Nagano Y, Sasada M, Kita T. Probucol pretreatment enhances the chemotaxis of mouse peritoneal macrophages. Arterioscl Thromb 1992; 12:593–600.

46. O'Brien K, Nagano Y, Gown A, Kita T, Chait A. Probucol treatment affects the cellular composition but not anti-oxidized low density lipoprotein immunoreactivity of plaques from Watanabe heritable hyperlipidemic rabbits. Arterioscl Thromb 1991; 11:751–759.

47. Nagano Y, Arai H, Kita T. High density lipoprotein loses its effect to stimulate efflux of cholesterol from foam cells after oxidative modification. Proc Natl Acad Sci USA 1991; 88:6457–6461.

48. Yamamoto A, Matsuzawa Y, Yokoyama S. Effects of probucol on xanthoma regression in familial hypercholesterolemia. Am J Cardiol 1986; 57:29H–35H.

49. Miettinen TA, Huttunen JK, Nankkarinen V, Strandberg T, Vanhanen H. Long-term use of probucol in the multifactorial primary prevention of vascular diseases. Am J Cardiol 1986; 57:49H–54H.
50. Chisolm GM, Morel DW. Lipoprotein oxidation and cytotoxicity: effect of probucol on streptozotocin-treated rats. Am J Cardiol 1988; 62:20B–26B.
51. Drash AL, Rudert WA, Borquaye S, Wang R, Lieberman I. Effect of probucol on development of diabetes mellitus in β B rats. Am J Carodiol 1988; 62:27B–30B.

7

Antioxidant Properties of the Catechol Derivative Nitecapone

Nobuya Haramaki
Kurume University School of Medicine, Kurume, Japan

Lucia Marcocci
University of California, Berkeley, California, and University of Rome, Rome, Italy

Teruyuki Kawabata
Okayama University Medical School, Okayama, Japan

Lester Packer
University of California, Berkeley, California

I. INTRODUCTION

Nitecapone [3-(3,4-dihydroxy-5-nitrobenzylidene)-2,4-pentanedione] is a catechol derivative with both catechol-*O*-methyltransferase (COMT) inhibitory and gastroprotective properties (1–3). Although the gastroprotective effects of nitecapone could be exerted by the stimulation of duodenal bicarbonate secretion through the inhibition of COMT activity (4), antioxidant properties of nitecapone have been suggested to be related to its gastroprotective action (5). Because reactive oxygen species have been reported to play a possible role in gastrointestinal damage (6,7), the antioxidant properties of nitecapone have been widely analyzed; they are reviewed in this chapter.

Studying how the structure of a compound relates to its antioxidant activity may enable us to improve our understanding of its biological effects. Therefore, information about the antioxidant action of two nitecapone-related catechol derivatives, entacapone [2-cyano-3-(3,4-dihydroxy-5-nitrophenyl)-*N*,*N*,diethyl-2-propenamide] and OR-1246 [3,(3,4-dihydroxy-5-nitrobenzyl)-2,4-pentanedione], whose structures and physicochemical characteristics are shown in Fig. 1 and Table 1, respectively, are also reported and compared to those of nitecapone.

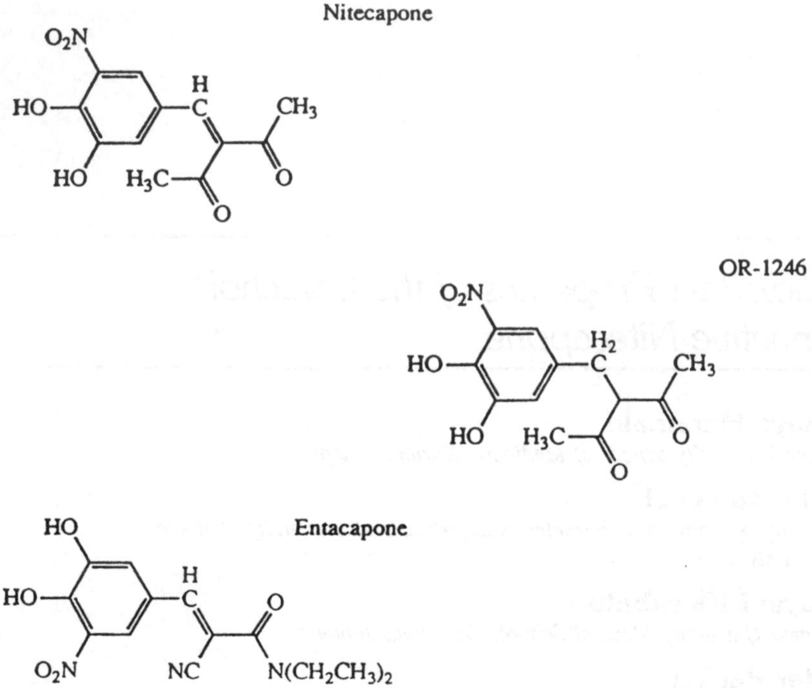

Nitecapone

OR-1246

Entacapone

Figure 1 Molecular structures of nitecapone, entacapone, and OR-1246.

Table 1 Physicochemical Characteristics of Nitecapone, OR-1246, and Entacapone

Property	Nitecapone	OR 1246	Entacapone
Molecular weight	265	267	317
Wavelength of maximum absorbance[a] (nm)	380	444	379
Extinction coefficient (M^{-1} cm^{-1})	24,200	2,900	25,130
pK_a	4.7	6.3	n.d.[d]
Log P[b]	1.32	1.16	n.d
Log D[c]	−1.13	1.68	n.d

[a]Measured in 20 mM phosphase buffer, pH 7.4.
[b]P is the partition coefficient in octanol/100 mM HC1.
[c]D is the partition coefficient in octanol/67 mM phopshate buffer, pH 7.4.
[d]n.d., not determined.

To characterize a compound as an antioxidant, it is necessary to evaluate its efficacy in a wide range of oxidative conditions (8). A compound can exert antioxidant activity in various ways: it may directly scavenge free radicals; it may inhibit formation of free radicals (e.g., by chelating metals); it may modulate the expression or activity of enzymes involved in free-radical generation or scavenging; it may enhance the activities of other antioxidants. In addition, it is now widely accepted that reactive oxygen species may play an important role in the pathogenesis of many diseases (i.e., cancer, myocardial infarction, cataract, neurogeneration, diabetes) and aging (9). Hence, the efficacy of nitecapone in protecting heart tissue against ischemia-reperfusion injury is also reviewed in this chapter to support the hypothesis that nitecapone or its related compounds could be useful therapeutic agents against free-radical dependent pathological conditions.

II. SCAVENGING REACTIVE OXYGEN SPECIES

A. Superoxide Anion

Nitecapone scavenges superoxide anion, as demonstrated by monitoring the reduction of cytochrome c by superoxide anion generated by DMSO plus NaOH (Fig. 2). A rate constant of $1.0 \times 10^4 \ M^{-1} \ s^{-1}$ has been calculated for the reaction of nitecapone with superoxide anion (10). By using the same systems for generating and detecting superoxide anion, the analog OR-1246 has been shown to scavenge superoxide anion as efficiently as nitecapone (11). These data suggest that the double bond in the structure of nitecapone is not crucial for its interaction with superoxide anion.

B. Hydroxyl Radicals

Nitecapone has also been suggested as a scavenger of hydroxyl radicals (10). An electron spin resonance (ESR) spin trapping method with 5,5-dimethyl-1-pyrroline-*N*-oxide (DMPO) as spin trap was used to detect hydroxyl radicals. Hydroxyl radicals were generated by the Fenton reaction (200 µM $FeSO_4$ with 2 mM H_2O_2) in this study. The intensity of the signal of the DMPO-OH adduct, which is formed by reaction between hydroxyl radicals and DMPO, was decreased by addition of nitecapone, in a dose-dependent manner. Fifty percent suppression of the DMPO-OH signal was observed with addition of 300 µM nitecapone, and almost complete inhibition was observed by adding 1 mM nitecapone. However, other possible mechanisms for those results, such as iron chelation or reaction with H_2O_2, remain to be excluded.

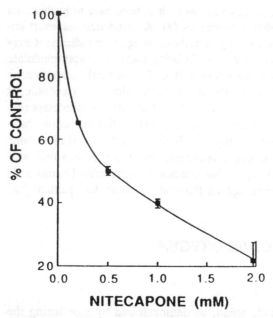

Figure 2 Superoxide radical-scavenging activity of nitecapone. Superoxide radical-scavenging activity of nitecapone was determined by monitoring the competition of nitecapone with 20 μM cytochrome c for superoxide radicals generated by 5 mM DMSO plus 2 μM NaOH in 200 mM phosphate buffer pH 8.6 at 25°C.

C. Peroxyl Radicals Generated in Hydrophilic Environments

By monitoring the effect of nitecapone on the peroxyl radical–dependent increase in luminol chemiluminescence, it has been shown that nitecapone effectively quenches peroxyl radicals generated in hydrophilic solution. A stoichiometric factor of 4.99 has been calculated for the reaction (5). The peroxyl radical–scavenging action of nitecapone was also confirmed by other studies. Suzuki et al. (10) observed the influence of nitecapone on the fluorescence decay of B-phycoerythrin induced by peroxyl radicals generated in hydrophilic solution from the thermal decomposition of the soluble azo initiator, 2,2′-azobis(2-amidinopropane)dihydrochloride (AAPH). The decay of fluorescence was transiently delayed by the addition of 2.5 μM nitecapone (Fig. 3). A stoichiometric factor of 2 for the reaction of nitecapone with peroxyl radicals was obtained in this study (10). In another study, OR-1246 showed the same efficacy as nitecapone in peroxyl radical scavenging (11). Pentikainen et al. tested the effect of nitecapone on peroxyl radicals generated in a hydrophilic

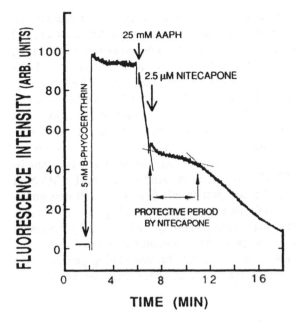

Figure 3 Fluorescence study of the effect of nitecapone on AAPH-generated peroxyl radicals. Fluorescence of 5 nM B-phycoerythrin was monitored at 540 nm excitation and 575 nm emission in 20 mM Tris-HCl pH 7.4 at 40°C. [AAPH] = 25 mM; [nitecapone] = 2.5 µM.

environment, using oxidative modification of human low-density lipoproteins (LDL) as a marker of oxidative damage (12). Nitecapone inhibited the formation of thiobarbituric acid-reactive substance (TBARS) in human LDL exposed to 25 mM AAPH for 60 min at 37°C in a dose-dependent manner. At a concentration of 30 µM, nitecapone completely inhibited LDL oxidation; this inhibition was more efficient than that of Trolox, a soluble form of tocopherol.

D. Peroxyl Radicals Generated in Hydrophobic Environment

The scavenging action of nitecapone against peroxyl radicals generated in hydrophobic environments has also been studied. The radicals were generated in dioleoylphospatidylcholine (DOPC) liposomal membranes by the thermal decomposition of a hydrophobic azo initiator, 2,2'-azobis(2,4)-dimethylvaleronitrile (AMVN), and two different probes, luminol and *cis*-parinaric acid, were used for detection of peroxyl radicals. When peroxyl radicals were generated in DOPC liposomes, the addition of nitecapone prevented luminol-dependent chemiluminescence in a dose-dependent fashion at concentrations of 0.2 μM and higher (Fig. 4) (10). Furthermore, at a concentration of 0.5 μM, nitecapone decreased the rate of *cis*-parinaric acid oxidation induced by AMVN-generated peroxyl radicals in DOPC liposomes by about 50%. The same concentration of OR-1246 inhibited this peroxyl radical–dependent oxidation of *cis*-parinaric with greater efficiency (70%) (11). The efficacies of nitecapone and OR-1246 in interacting with peroxyl radicals generated in the membranes have also been investigated by observing their protective effects against AMVN-induced lipid peroxidation in rat heart microsomes. At concentrations between 10 and 40 μM, they inhibited TBARS formation induced by 2.5 mM AMVN at 40°C in rat heart microsomes (2.5 mg protein/ml) (10,11). At all concentrations used, OR-1246 was more efficient than nitecapone (Fig. 5), confirming that OR-1246 may be a better scavenger of peroxyl radicals in hydrophobic environments. These results also suggest the importance of a small difference in the number of double bonds in their structure for scavenging peroxyl radicals in hydrophobic regions, which was not crucial for their interaction with superoxide anion.

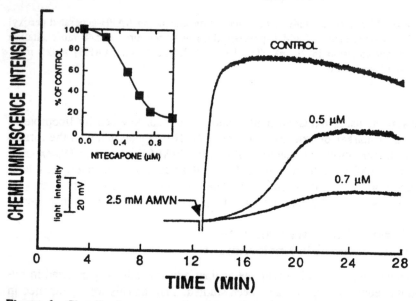

Figure 4 Chemiluminescence study of the effect of nitecapone on AMVN-generated peroxyl radicals. Luminol-enhanced chemiluminescence induced by AMVN-derived peroxyl radicals was monitored in 20 mM Tris-HCl, pH 7.4 at 40°C. [AMVN]=2.5 mM; [luminol]=150 μM; [DOPC]=2.5 mM.

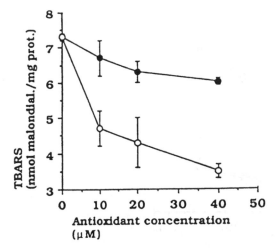

Figure 5   Effect of nitecapone or OR-1246 on AMVN-induced lipid peroxidation in rat liver microsomes. Lipid peroxidation was measured as a product of thiobarbituric acid-reactive substances (TBARS) after incubation of rat microsomes (2.5 mg protein/ml) with 2.5 mM AMVN for 1 h at 40°C in the presence of nitecapone (closed circles) or OR-1246 (open circles). Mean values from four experiments ±SD are reported.

E.   Nitric Oxide

Catechol derivatives may also be good scavengers of nitric oxide generated in vitro from acellular and cellular systems. In concentrations under 50 μM, nitecapone dose-dependently inhibited the oxidation of 1.5 μM oxyhemoglobin by nitric oxide generated from the reaction of Complex I of catalase with hydroxylamine; maximum inhibition (40% the rate of oxidation) was obtained at concentrations of 50 μm and higher (13). In the same study, catechol derivatives also inhibited the rate of nitrite accumulation from the decomposition of 5 mM sodium nitroprusside. Nitecapone (300 μM) inhibited about 80% of the accumulation of nitrite generated from nitroprusside (Fig. 6). An IC_{50} of about 20 μM for nitecapone was calculated. The analog OR-1246 (300 μM) inhibited the accumulation of nitrite less effectively (about 50%) than nitecapone, while entacapone showed the same efficacy as nitecapone. These results again demonstrate the importance of structural differences in scavenging radicals. The ability of nitecapone to scavenge nitric oxide generated from a cellular system was also investigated. Nitecapone suppressed L-arginine-dependent accumulation of nitrite from casein-induced rat peritoneal polymorphonuclear cells, and the inhibition depended on the concentration of nitecapone. At a concentration

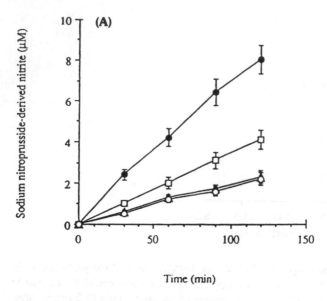

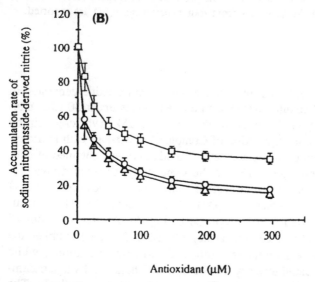

Figure 6 Effect of nitecapone, entacapone, and OR-1246 on nitrite accumulation from sodium nitroprusside. Sodium nitroprusside (5 mM, final concentration) was incubated in PBS at 25°C in the absence (closed circles) or in the presence of (A) 100 μM antioxidant; (B) different concentrations of antioxidants. Nitecapone (open circle); entacapone (open triangle); OR-1246 (open square). Each point represents the mean ±SD of four experiments.

of 200 µM, nitecapone suppressed the accumulation of nitrite about 50%, and the IC_{50} was found to be 120 µM.

III. METAL CHELATION

Transition metals such as iron and copper play an important role in the generation of reactive oxygen species and free radicals (9,14–16).

For biological application of chelators, the chemical properties necessary for sequestering catalytic iron are to have

1. A large stability constant

$$Fe + L \leftrightarrow FeL; \qquad K_f \frac{[FeL]}{[Fe][L]}$$

2. Redox inactivity, which means a resistance to oxidation or reduction

Catechol is a well-known chelator, forming thermodynamically stable (pK_f=40) complexes with iron (17), but the phenolic hydroxyl groups in catechol have a high pK_a ($\approx$ 9.0), and catechol is not an effective chelator at physiological pH. The electrophilic nitro group introduced to the benzene ring (Fig. 1) of the synthesized catechol derivatives might be important in chelation at a pH of around 7.4, because the pK_a value of the phenolic hydroxyl groups is decreased to less than 6 (11).

Ferrous iron is potentially much more dangerous in biological systems than ferric iron. A major source of iron-induced free-radical injury in biological systems is the production of hydroxyl and alkoxyl radicals (18), in which ferrous iron is a reactant:

$$H(L)OOH + Fe^{2+} \rightarrow H(L)O^{\cdot} + OH^- + Fe^{3+}$$

Hence, if iron can be kept ferric by a chelator under physiological conditions, the activity of iron in free-radical reactions will be decreased, suggesting the significance of a low reduction potential for an iron chelator complex. Transferrin and desferal (the methanesulfonate salt of desferrioxamine B) are widely used for protection against iron-induced free-radical injury and they have much lower reduction potentials ($\Delta E = -0.3$ V) than other iron chelators, such as EDTA, citrate, and ADP (19). How can we decrease the reduction potentials of catechol-type iron chelators? Figure 7 shows a scheme of energy levels of an iron-catechol complex. The redox potential is sensitive to the iron $d\pi$ orbitals, which is composed of the iron d orbital and catechol π orbital (20,21). It is possible to decrease the reduction potential by increasing the energy level of the catechol π orbital (22). Substituents with double bonds introduced to the

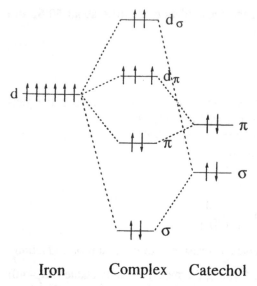

Iron   Complex   Catechol

Figure 7 Energy diagram of iron catechol complex.

catechol moiety in nitecapone and entacapone increase the energy levels of the catechol π orbital.

To test this hypothesis, metal chelation of nitecapone and the metal-induced free-radical generation were compared to OR-1246 and entacapone. Iron binding and the physicochemical properties of the bound iron were studied by UV/vis absorption spectroscopy, IR, ESR, and ^1H NMR to clarify the iron chelation by catechol derivatives at physiological pH. All catechol derivatives formed catechol-type iron chelates [tris(catecholato)ferrate(III)] in aqueous solution at pH 7.4 (Fig. 8) with high-spin state. To check the reactive properties of the iron in the catechol derivative complexes, free-radical reactions such as lipid peroxidation and hydroxyl radical generation were studied. In iron-induced lipid peroxidation using rat liver microsomes, iron in catechol derivative complexes did not promote lipid peroxidation and lipid peroxidation was inhibited as compared with the control. Catechol also inhibited lipid peroxidation in the same manner as the derivatives. Using ESR spin trapping to detect hydroxyl radicals, it was found that catechol enhanced hydroxyl radical generation in Fenton-like reactions compared to citrate, and OR-1246 generated hydroxyl radicals to the same extent as citrate. However, nitecapone and entacapone did not generate a hydroxyl radical. The rates of oxidation and reduction were measured in Hepes-saline (pH 7.4) at 25°C. The ratios of k_{Rd}/k_{Ox}, in order of magnitude, were citrate > catechol > OR-1246 > nitecapone, entacapone. Therefore, the

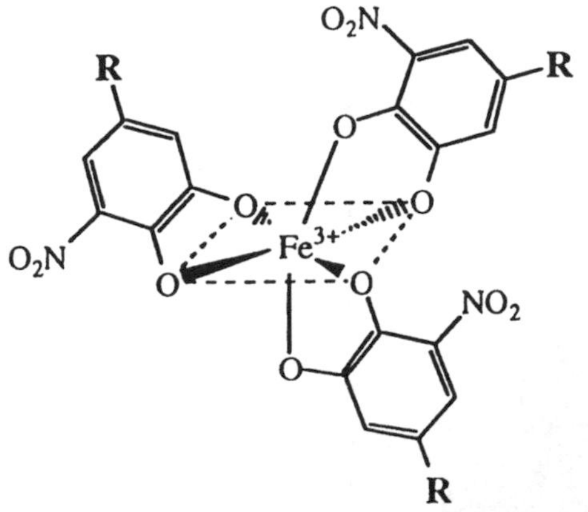

Figure 8 Structure of iron complex of catechol derivatives at pH 7.4.

catechol derivatives with conjugation (nitecapone and entacapone) are also good transition-metal chelators, adding to their antioxidant actions.

The ability of nitecapone to bind transition metals (copper, iron, and zinc) has been tested by using immobilized metal-ion-affinity chromatography (12). It was shown that nitecapone has an affinity for Sepharose-chelated copper ion, a strong affinity for chelated iron, and little or no affinity for chelated zinc ion. Moreover, it has been demonstrated that nitecapone effectively binds copper ions bound to human low-density lipoprotein overloaded with the metal in vitro (12).

The interaction of nitecapone with copper has been also analyzed in our laboratory. Addition of $CuSO_4$ to an aqueous solution of nitecapone (30 μM) caused a decrease in the absorbance at 380 nm and an increase in the absorbance at 320 nm in the UV spectrum of nitecapone. Changes in the UV spectrum of nitecapone upon copper addition were observed until the concentration of the added metal was equimolar with those of nitecapone. Furthermore, nitecapone, in a concentration-dependent manner, inhibited the copper-dependent oxidation of ascorbate as measured by an oxygraphic assay and confirmed by HPLC. The rate of copper-dependent ascorbate oxidation (800 μM ascorbate, 10 μM $CuSO_4$ in phosphate buffer pH = 7.4) was 50% lower in the presence of 60 μM nitecapone compared to the rate in the absence of nitecapone, and no ascorbate oxidation was observed when concentration of nitecapone was raised to 400 μM (Fig. 9).

The efficacy of nitecapone in interacting with copper ions has also been confirmed by observing the protective effects of nitecapone against oxidative damage induced in human LDL upon exposure to copper alone, or copper in

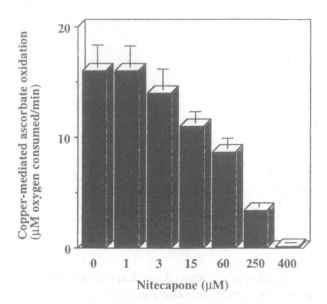

Figure 9 Effect of nitecapone on copper-mediated oxidation of ascorbate. A solution of 0.8 mM ascorbate in 50 mM phosphate buffer, pH = 7.4 was exposed at $T = 37°C$ to 10 μM $CuSO_4$ in the presence of different concentration of nitecapone. Oxygen consumption was measured by a Clark oxygen electrode. Mean values from four experiments ±SD are reported.

the presence of mouse peritoneal macrophages (12). In this system, nitecapone added in 3 to 5 molar eccess of the copper ions (5–25 μM) inhibited the oxidation of LDL promoted by the metal, as measured by various parameters, including TBARS formation, conjugated diene formation, the change in the LDL elecrophoretic mobility, and the uptake of lipoprotein by macrophages. The inhibitory effect of nitecapone was not time-dependent but could be overcome by increasing the concentration of copper. These observations suggest that its inhibitory effect depends on its ability to chelate and to reduce the activity of copper ions without itself being consumed, rather than on its ability to directly scavenge radicals. LDL oxidation induced by exposure to mouse peritoneal macrophages in the presence of 0.3 μM copper was completely inhibited by 5 μM nitecapone.

IV. INHIBITION OF ENZYME ACTIVITY

Nitecapone affects the rate of oxidation of xanthine to uric acid catalyzed by xanthine oxidase. Kinetic analysis of Lineweaver-Burk plots suggests that nitecapone is a competitive inhibitor for xanthine oxidase with an inhibition constant of 8.8 μM (Fig. 10) (10). Although OR-1246 also inhibits the enzyme, the inhibition was weaker than that of nitecapone. Xanthine oxidase activity was completely inhibited in the presence of 150 μM nitecapone, whereas OR-1246 caused only 40% inhibition at the same concentration (11). Nitecapone and its related compounds did not inhibit the enzyme activity of either catalase or glucose oxidase (11).

V. EFFECTS ON GENE EXPRESSION

Recent studies have suggested that reactive oxygen species are involved in the signal transduction pathways for the activation of NF-κB, which is a mammalian transcription factor that regulates genes involved in immune, inflammatory, or acute phase responses (23). The effects of nitecapone and OR-1246 on both

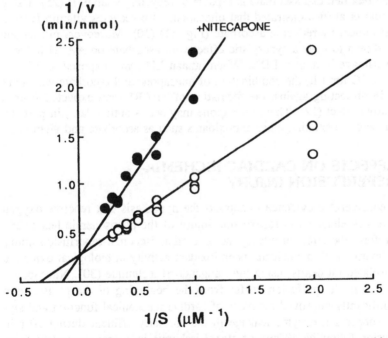

Figure 10 Lineweaver-Burk plots of the effect of nitecapone on xanthine oxidase activity. Xanthine oxidase activity was measured by monitoring the formation of uric acid at 295 nm in 100 mM phopshate buffer, pH 7.4 [nitecapone] (closed circles) = 10 μM; [xanthine oxidase] = 2.5 mU/ml. The reaction was initiated by the addition of xanthine oxidase.

activation and DNA binding of NF-κB were tested in an in vitro system (24). Thirty minutes preincubation of Jurkat T (acute human leukemia) cells with nitecapone or OR-1246 (10–300 μM) produced a concentration-dependent inhibition of NF-κB activation induced by tumor necrosis factor-α (TNF-α). The effects of the two catechol derivatives were comparable at low concentrations; however, nitecapone exhibited better inhibitory effects at higher concentration. Moreover, both compounds also inhibited the binding of NF-κB, activated by TNF-α, to DNA. However, OR-1246 did not inhibit DNA binding activity of NF-κB activated by okadaic acid, whereas nitecapone effectively inhibited it. Thus, nitecapone and OR-1246, which have minimal structural differences, utilize different mechanisms in inhibiting NF-κB.

VI. INTERACTION WITH OTHER ANTIOXIDANTS

It has been reported that ascorbate is capable of recycling vitamin E (25). Using ESR, Suzuki et al. demonstrated that nitecapone also recycles vitamin E from vitamin E radical by reducing ascorbate (Fig. 11) (10). Moreover, nitecapone has been shown to have a synergistic effect with ascorbate on AAPH-induced oxidative damage in human LDL. When human LDL was exposed to 25 mM AAPH at 37°C for 1 h, the combination of nitecapone and ascorbate was more effective in protecting against the formation of TBARS than expected from a mere additive effect (12). Thus, nitecapone may act as antioxidant, in part, by interacting and potentiating other antioxidants such as ascorbate and vitamin E.

VII. EFFECTS ON CARDIAC ISCHEMIA-REPERFUSION INJURY

There is considerable evidence to support the hypothesis that reactive oxygen species are key elements in reperfusion injury of the postischemic heart (26–29). Therefore, the effect of nitecapone on cardiac ischemia-reperfusion injury has been investigated to elucidate its antioxidant activity in biological oxidative stress in isolated rat hearts, using the Langendorff technique (30). Nitecapone, administered in the perfusion buffer from the beginning of the preischemic phase significantly improved recovery of cardiac mechanical function and significantly suppressed enzyme leakage in the coronary effluent during 20 min of reperfusion following 40 min of global ischemia in a dose-dependent fashion, compared to the control group. In addition, nitecapone minimized loss of ascorbate from cardiac tissue, normally observed after ischemia-reperfusion. Since nitecapone may recycle vitamin E, a membrane-bound chain-breaking antioxidant, by way of reduction of ascorbate, it is hypothesized that nitecapone

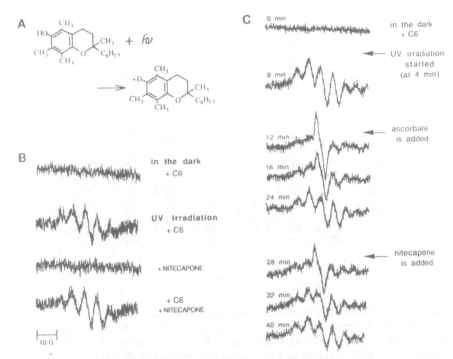

Figure 11 ESR study of the effect of nitecapone on chromanoxyl radicals. (A) Chrom-anoxyl radical is generated from chromanol-α-C6 by UV irradiation. (B) Chromanoxyl radical ESR signal wsa generated by exposing chromanol-α-C6 (20 mM) to UV in 100 mM phosphate buffer, pH 7.4, at 25°C containing DOPC liposomes (20 mg lipid/ml). No effect of nitecapone (2 mM) was observed. (C) The addition of ascorbate (20 μM) in the aforementioned system caused initial appearance of semidehydroascorbyl radical signal followed by chromanoxyl radical signal. The subsequent addition of nitecapone after ascorbate depletion, as observed by the appearance of chromanoxyl radical, resulted in a semidehydroascorbyl radical signal.

may prevent ischemia-reperfusion injury through its effect on maintaining tis-sue ascorbate. To test the hypothesis, rats were fed a diet containing 4% ascor-bate. Although the myocardial content of ascorbate in ascorbate-fed rats after ischemia-reperfusion was higher than that in control rats fed a normal diet without ischemia (Fig. 12), ascorbate did not show any beneficial effects on cardiac mechanical recovery or enzyme leakage (Fig. 13). These results sug-gest that maintenance of tissue ascorbate levels is not the cause, but rather the effect, of the protective effects of nitecapone against cardiac ischemia-reper-fusion injury. The effect of nitecapone on iron chelation was tested in the same

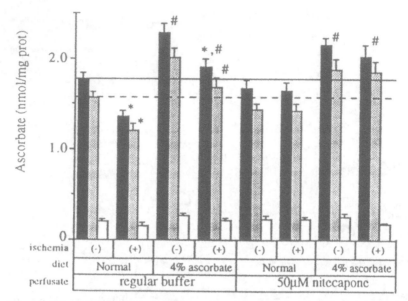

Figure 12 Influence of ischemia, ascorbate supplementation, and nitecapone perfusion on myocardial ascorbate content. Total ascorbate, black bar; ascorbate (reduced form), gray bar; DHA (oxidized form), white bar. Each value is mean ± S.E.M. $n \geq 5$ for all points. *, $p < 0.05$ compared with 60 min control perfusion of each group. #, $p < 0.05$ compared with the same condition in rats fed a normal diet. Total ascorbate and reduced ascorbate content in hearts of rat fed a normal diet and perfused by regular buffer without ischemia are shown as a solid line and a broken line, respectively.

study. It was confirmed by ESR that 50 µM nitecapone easily chelates the same concentration of iron released from the heart into the coronary effluent. Hence, the iron-chelating ability as well as the radical-scavenging properties of nitecapone were suggested to be responsible for its cardioprotective effects in ischemia-reperfusion injury. Since it has been reported that nitecapone is well tolerated and does not affect hemodynamics (3,31,32), this powerful catechol-derived antioxidant may be a good candidate as a therapeutic agent against biological oxidative stress, including cardiac ischemia-reperfusion injury.

VIII. SUMMARY

The catechol derivatives described in this chapter possess a wide range of antioxidant properties:

1. Nitecapone and OR-1246 scavenge superoxide anion.
2. Nitecapone suppresses hydroxyl radicals generated by Fenton reagent.

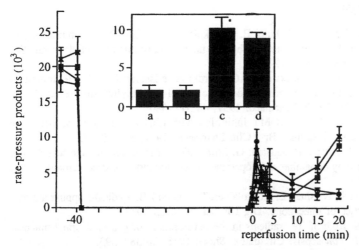

Figure 13 Effect of ascorbate supplementation and nitecapone perfusion on cardiac mechanical recovery. Rate-pressure products which indicate cardiac function were calculated as [HR (bpm)] × [LVDP (mmHg)]. (a) Normal diet/regular buffer, ●; (b) Ascorbate supplemented diet/regular buffer, ○; (c) Normal diet/50 μM nitecapone, x; (d) Ascorbate supplemented diet/50 μM nitecapone, ■. Rate-pressure products at 20 min of reperfusion are shown in the bar graph. Each value is expressed as mean ± S.E.M. $n \geq 5$ for all points. *$p < 0.05$ compared with normal diet/regular buffer.

3. Nitecapone, as well as OR-1246, scavenges peroxyl radicals in hydrophilic environments.

4. Although both nitecapone and OR-1246 inhibit peroxyl radicals, OR-1246 does so more efficiently in hydrophobic environments.

5. All three compounds scavenge nitric oxide. Both nitecapone and entacapone have similar efficiencies in scavenging nitric oxide, which are greater than that of OR-1246.

6. Nitecapone competitively inhibits xanthine oxidase, and the inhibition is stronger than that of OR-1246.

7. Although both nitecapone and OR-1246 inhibit NF-κB transcription, they utilize different mechanisms to do it.

8. Nitecapone may recycle vitamin E by way of reducing ascorbate.

9. All three compounds bind to iron with ratio of 3:1 (chelator:iron); however, nitecapone and entacapone may be better iron chelators than OR-1246.

10. Perfusion with nitecapone protects isolated rat heart against ischemia-reperfusion injury, which is a model of biological oxidative stress.

REFERENCES

1. Nissinen E, Linden IB, Schultz E, Kaakkola S, Mannisto PT. Inhibition of cat-echol-O-methyltransferase activity by two novel disubstituted catechols in rat. Eur J Pharmacol 1988; 153:263-269.
2. Schultz E, Tarpila S, Backstrom AC, Gordin A, Nissinen E, Pohto P. Inhibition of human erythrocyte and gastroduodenal catechol-O-methyltransferase activity by nitecapone. Eur J Clin Pharmacol 1991; 40:577-580.
3. Sundberg S, Gordin A. COMT inhibition with nitecapone does not affect the tyramine pressor response. Br J Clin Pharmacol 1991; 32:130-132.
4. Flemstrom G, Safsten B, Jedstedt G. Stimulation of mucosal alkaline secretion in rat duodenum by dopamine and dopaminergic compounds. Gastroenterology 1993; 104:825-833.
5. Metsa-Ketela T, Nissinen E, Korkolainen T, Linden I-B. Radical-trapping antioxi-dant properties of nitecapone. Dig Dis Sci 1990; 35:1037.
6. Arvidsson S. Role of free radicals in the development of gastrointestinal mucosal damage in $E.$ $coli$ sepsis. Circulation Shock 1985; 16:383-393.
7. Kusterer K, Pihan G, Szabo S. Role of lipid peroxidation in gastric mucosal lisions induced by HC1, NaOh, ischemia. Am J Physiol 1987; 252:G811-G816.
8. Halliwell B. How to characterize a biological antioxidant. Free Rad Res Commun 1990; 9:1-32.
9. Halliwell B, Gatteridge JMC. Free Radicals in Biology and Medicine. 2nd ed. Oxford: Oxford University Press.
10. Suzuki YJ, Tsuchiya M, Safadi A, Kagan VE, Packer L. Antioxidant properties of nitecapone (OR-462). Free Rad Biol Med 1992; 13:517-525.
11. Marcocci L, Suzuki YJ, Tsuchiya M, Packer L. Antioxidant activity of nitecapone and its analog OR-1246: effect of a structural modification on antioxidant action. In: Packer L, ed. Methods in Enzymology. San Diego: Academic Press; 1994:526-541.
12. Pentikainen MO, Lindstedt KA, Kovanen PT. Inhibition of the oxidative modifi-cation of LDl by nitecapone. Arterioscler Thromb Vasc Biol 1995; 15:740-747.
13. Marcocci L, Maguire JJ, Packer L. Nitecapone: a nitric oxide radical scavenger. Biochem Mol Biol Int 1994; 34:531-541.
14. Lauffer RB. Iron and Human Disease. Boca Raton: CRC Press, 1992.
15. Millar DM, Buettner GR, Aust SD. Transition metals as catalysts of "autoxida-tion" reactions. Free Rad Biol Med 1990; 8:95-108.
16. Stadtman ER. Metal ion-catalyzed oxidation of proteins. biochemical mechanism and biological consequences. Free Rad Biol Med 1990; 9:315-325.
17. Avdeef A, Sofen SR, Bregante TL, Raymond KN. Coordination chemistry of microbial iron transport compounds 9. Stability constants for catechol models of enterobactin. J Am Chem Soc 1978; 100:5362-5370.
18. Halliwell B, Gutteridge JMC. Role of free radicals and catalytic metal ions in human disease: an overview. Methods Enzymol 1990; 186: 1-85.
19. Buettner GR. The pecking order of free radicals and antioxidants: Lipid peroxidation, α-tocopherol, and ascorbate. Arch Biochem Biophys 1993; 300:535-543.

20. Gordon DJ, Fenske RF. Theoretical study of o-quinone complexes of iron. Inorg Chem 1982; 21:2916–2923.
21. Pyrz JW, Roe AL, Stern LJ, Que LJ. Model studies of iron-tyrosinate protein. J Am Chem Soc 1985; 107:614–620.
22. Kawabata T, Schepkin V, Haramaki N, Phadke RS, Packer L. Iron coordination by catechol derivative antioxidants. Biochem Pharmacol 1996; 51:1569–1577.
23. Staal FJ, Roederer M, Herzenberg LA, Herzenberg LA. Intracellular thiols regulate activation of nuclear factor kappa B and transcription of human immunodeficiency virus. Proc Natl Acad Sci USA 1990; 87:9943–9947.
24. Suzuki YJ, Packer L. Inhibition of NF-κB transcription factor by catechol derivatives. Biochem Mol Biol Int 1994; 32:299–305.
25. Packer JE, Slater TF, Willson RL. Direct observation of free radical interaction between vitamin E and vitamin C. Nature 1979; 278:737–738.
26. Brown JM, Terada LS, Grosso MA, Whitmann GJ, Velasco SE, Patt A, Harken AH, Repine JE. Xanthine oxidase produces hydrogen peroxide which contributes to reperfusion injury of ischemic, isolared, perfused rat heart. J Clin Invest 1988; 82:1297–1301.
27. Chambers DE, Parks DA, Patterson G, Roy R, McCord JM, Yoshida S, Parmley LF, Downey JM, Xanthine oxidase as a source of free radical damage in myocardial ischemia. J Mol Cell Cardiol 1985; 17:145–152.
28. Terada LS, Rubinstein JD, Lesnefsky EJ, Horwitz LD, Leff JA, Repine JE. Existence and participation of the xanthine oxidase in reperfusion injury of ischemic rabbit myocardium. Am J Physiol 1991; 260:H805–810.
29. Zweier JL, Kuppusamy P, Williams R, Rayburn BK, Smith D, Weisfeldt ML, Flaherty JT. Measurement and characterization of postischemic free radical generation in the isolated perfused heart. J Biol Chem 1989; 264:18890–18895.
30. Haramaki N, Stewart DB, Aggarwal S, Kawabata T, Packer L. Role of ascorbate in protection by nitecapone against cardiac ischemia-reperfusion injury. Biochem Pharmacol 1995; 50:839–843.
31. Sundberg S, Scheinin M, Ojala-Karlsson P, Kaakkola S, Akkila J, Gordin A. Exercise hemodynamics and catecholamine metabolism after catechol-O-methyltransferase inhibition with nitecapone. Clin Pharmacol Ther 1990; 48:356–364.
32. Sundberg S, Scheinin M, Ojala-Karlsson P, Akkila J, Gordin A. The effects of the COMT inhibitor nitecapone for one week on exercise haemodynamics and catecholamine disposition. Eur J Clin Pharmacol 1993; 44:287–290.

8

Neuroprotective Efficacy and Mechanisms of the Lazaroids

Edward D. Hall
Pharmacia & Upjohn, Inc., Kalamazoo, Michigan

I. INTRODUCTION

There is now compelling experimental support for the early occurrence and pathophysiological importance of oxygen radical formation and cell membrane lipid peroxidation (LP) in the pathophysiology of CNS injury and focal and global cerebral ischemia, whether associated with temporary or permanent cerebrovascular occlusion (1–7) or secondary to subarachnoid hemorrhage (SAH) (8,9). The potential sources of oxygen radicals within the ischemic nervous system include the arachidonic acid cascade (i.e., prostaglandin synthase and 5-lipoxygenase activity), catecholamine oxidation, mitochondrial "leak," oxidation of extravasated hemoglobin and, later, infiltrating neutrophils. The radical-initiated peroxidation of neuronal, glial, and vascular cell membranes and myelin is catalyzed by free iron released from hemoglobin, transferrin, and ferritin by either lowered tissue pH or oxygen radicals. If unchecked, LP is a geometrically progressing process that will spread over the surface of the cell membrane, causing impairment to phospholipid-dependent enzymes, disruption of ionic gradients, and, if severe enough, membrane lysis.

The purpose of this chapter is to review the pharmacological properties of a novel series of LP inhibitors, the 21-aminosteroids (commonly referred to as the "lazaroids"), that have been shown to be protective in experimental models of acute spinal cord or head injury, focal and global cerebral ischemia, and

SAH. These compounds were developed subsequent to extensive studies demonstrating the ability of high doses of the glucocorticoid steroid, methylprednisolone, to inhibit posttraumatic CNS LP, an action unrelated to glucocorticoid receptor activation. Antioxidant doses of methylprednisolone have been shown to be significantly neuroprotective in experimental models of brain and spinal cord injury and in some models of focal cerebral ischemia (10). In addition, the early (within first 8 h) administration of a high-dose methylprednisolone dosing regimen to spinal cord–injured patients has been shown to significantly improve motor recovery at 6 weeks, 6 months, and 1 year postinjury (11,12).

The definition of this high-dose, nonglucocorticoid antioxidant action of methylprednisolone initially led to the pursuit of nonglucocorticoid steroid analogs of MP (e.g., U-72099E; Fig. 1) which also weakly inhibited LP in high concentrations, and at high doses were active in models of experimental CNS trauma (13). However, synthetic efforts continued that were aimed at preparation of antioxidant steroids that were even more potent and effective inhibitors of LP, with greater activity in experimental models of CNS trauma, ischemia, and SAH. This resulted in the discovery of the 21-aminosteroids (e.g., U-74006F, U-74389F, U-74500A; Fig. 1) which were designed to be devoid of glucocorticoid receptor interactions while showing an improved propensity for cell membrane localization and LP inhibitory efficacy in comparison to methylprednisolone and the earlier nonglucocorticoid steroid U-72099E (14–16). One of these compounds, U-74006F (generic name is tirilazad mesylate), was selected for clinical development as a parenterally administered acute neuroprotective agent and is currently the focus of Phase III clinical trials in head and spinal cord injury, ischemic stroke, and aneurysmal SAH. The bulk of this chapter will be devoted to a description of the mechanisms by which this compound inhibits LP, its efficacy and physiological mechanisms of action in preclinical models of head and spinal cord injury, cerebral ischemia, and SAH, and results to date from Phase III clinical trials in SAH. In addition, two newer classes of lazaroid-type antioxidants, the 2-methylaminochromans and the pyrrolopyrimidines which have improved antioxidant efficacy and brain penetration compared to the 21-aminosteroids, are briefly reviewed.

A. Antioxidant Mechanisms

As noted in a separate chapter (see D. Epps and J. McCall), tirilazad mesylate is a nonglucocorticoid 21-aminosteroid that is a potent inhibitor of oxygen radical-induced, iron catalyzed LP. It is a very lipophilic compound (log of calculated octanol/water partition coefficient = 8) that distributes preferentially to the lipid bilayer of cell membranes. It appears that the compound exerts its anti-

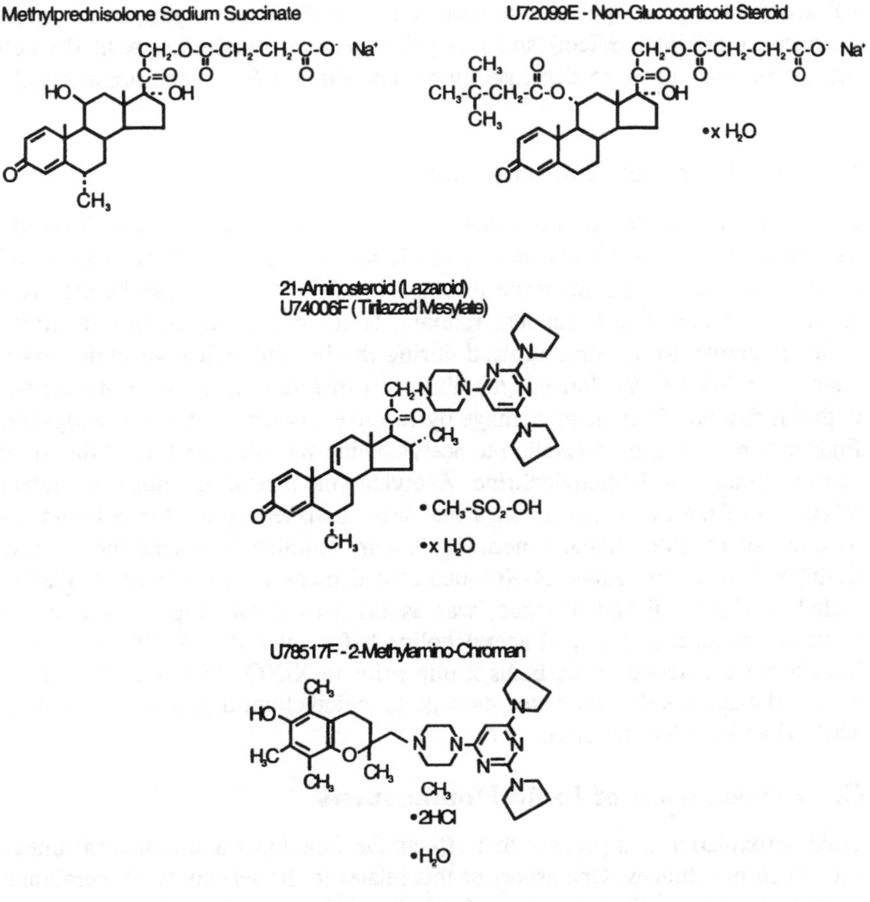

Figure 1 Chemical structures of the glucocorticoid steroid methylprednisolone, the nonglucocorticoid steroid U-72099E, the 21-aminosteroid tirilazad mesylate (U-74006F), the 2-methylaminochroman U-78517F, and the pyrrolopyrimidine U-101033E.

LP action through cooperative mechanisms: a radical scavenging action (i.e., chemical antioxidant effect) and a physicochemical interaction with the cell membrane that serves to decrease membrane fluidity (i.e., membrane stabilization).

B. Protection of Endothelium

Later the effects of tirilazad in models of CNS injury, ischemia, and SAH are described, including effects on post-SAH microvascular hypoperfusion and delayed vasospasm. An important mechanism of these effects may be the preservation of endothelial-dependent relaxing factor production and/or function, which is known to be compromised during the lipid peroxidative injury associated with SAH (8,9). Indeed, the ability of tirilazad mesylate to protect endothelial function from acute damage by reactive oxygen has been investigated. Endothelium-dependent relaxation to acetylcholine was measured in rabbit aortic rings contracted with phenylephrine. Acetylcholine produced a dose-dependent relaxation of the rabbit aortic rings that was abolished by a 30-min treatment with the superoxide radical-generating system xanthine plus xanthine oxidase (X/XO). Protection against X-XO-mediated damage by various antioxidants, including vitamin E and tirilazad, was assessed by comparing the amount of relaxation produced by 1 µM acetylcholine before and after X/XO treatment. Inhibitors were added to the baths 2 min prior to X/XO. Tirilazad (0.05 µM) protected against X/XO-mediated damage to endothelium-dependent relaxation, while vitamin E had no effect (17).

C. Preservation of Ionic Homeostasis

Lipid peroxidation is a process that affects the functional and structural integrity of cell membranes. One aspect of this relates to the sensitivity of membrane Ca^{2+} ATPase (i.e., Ca^{2+} pump) to LP-induced damage which results in intracellular Ca^{2+} accumulation (18). Similarly, oxidative inactivation of the membrane Na^+/K^+-ATPase (18) can lead to intracellular Na^+ accumulation, which will then reverse the direction of the Na^+/Ca^{2+} exchanger (antiporter) and further exacerbate intracellular Ca^{2+} overload. Consequently, LP-damaged cerebral neurons suffer from abnormal membrane-dependent Ca^{2+} and Na^+ homeostasis. As an example of this, gerbils subjected to 3 h of unilateral carotid occlusion manifest a pronounced deficit in postreperfusion recovery of cortical extracellular Ca^{2+} levels (i.e., persistence of intracellular accumulation) which occurs together with postischemic LP (19). Administration of tirilazad was shown to attenuate postreperfusion LP (i.e., decreases LP-associated vitamin E depletion) and to facilitate recovery of extracellular Ca^{2+} together with a reduction in later cortical neuronal damage (20). The likelihood that the improved Ca^{2+} recovery is due to an inhibition of peroxidative inactivation of the membrane Ca^{2+} pumping mechanisms is based on the finding that tirilazad can protect red blood cell membrane Na^+/K^+- and Ca^{2+}-ATPases from iron-induced inhibition simultaneous with inhibition of MDA formation (18).

D. Site of Action

As noted, tirilazad is a very lipophilic compound that localizes in and protects endothelial cell membranes from peroxidative damage. This is manifest in the protection of endothelium-dependent relaxation described in the preceding section (17). Moreover, I later note the ability of tirilazad to protect the blood brain barrier (BBB) against a SAH-induced permeability increase. These actions clearly reflect an action at the level of the endothelial cell.

In addition, tirilazad has been shown to penetrate the BBB in rats very poorly after intracarotid injection (21). In the same study, the compound was examined in regard to permeability across a monolayer of canine kidney epithelial cells which possess diffusional characteristics similar to brain endothelium. Tirilazad displays very low penetration of the epithelial cells in this model and, in fact, becomes highly concentrated in the cell membranes. Thus, this profile, taken together with the demonstrated protection of endothelial function, points to an endothelial site of action. Furthermore, tirilazad has been reported to protect hepatic endothelial cells from structural degeneration in a rat model of hemorrhagic shock (22). Therefore, the endothelial localization and protection is not confined to the central nervous system.

However, despite tirilazad's high affinity for endothelial cell membranes, it has been observed that the penetration of tirilazad into brain parenchyma is enhanced after injury, apparently by virtue of the trauma-induced disruption of the BBB (23). Consequently, it is not possible to rule out a direct neuronal protective effect. Indeed, a direct neuronal action may explain the efficacy of tirilazad in preserving cortical neuronal Ca^{2+} homeostasis in the gerbil 3-h unilateral carotid occlusion ischemia model (19).

II. TIRILAZAD IN MODELS OF SPINAL CORD INJURY

Based on our extensive experience in defining the role of LP in the pathophysiology of acute spinal cord injury and antioxidant neuroprotective actions of high-dose methylprednisolone (10), as well as the demonstrated efficacy of this steroid in human spinal injury (11,12), an initial focus for tirilazad was a definition of its efficacy in acute spinal injury models.

A. Effects on Neurological Recovery

Tirilazad mesylate has been extensively investigated for its ability to promote neurological recovery of cats following a moderately severe compression injury to the lumbar spinal cord. Beginning at 30 min after injury, the animals received a 48-h intravenous regimen of vehicle (sterile water) or tirilazad in a random and blinded protocol. Initial tirilazad doses ranged from 0.01 to 30 mg/kg. At 4 weeks after injury, vehicle-treated animals uniformly remained paraplegic. In contrast, cats that received 48-h doses ranging from 1.6 to 160.0 mg/kg showed significantly better recovery, regaining approximately 75% of normal neurological function (24). Tirilazad has also been reported to improve the subacute (9 day) neurological recovery of rats subjected to a compression spinal injury (25,26).

Recent studies suggest that tirilazad mesylate retains its efficacy in promoting posttraumatic recovery after experimental spinal cord injury even when initiation of treatment is delayed to 4 h (27). Tirilazad has similarly been shown to improve functional recovery in rabbits subjected to a 25-min period of aortic occlusion-induced spinal ischemic injury (28). Thus, tirilazad is effective in models of compression, contusion, and ischemic spinal cord injury.

B. Effects on Posttraumatic Ischemia and Lipid Peroxidation

To pursue a physiological mechanism by which tirilazad may be acting to promote chronic recovery after blunt spinal cord injury, the acute effects of the compound have also been viewed in relation to a possible action to attenuate progressive posttraumatic spinal cord ischemia following spinal cord compression injury (29). In vehicle-treated cats, there was a progressive decline in spinal cord white matter blood flow (SCBF) over the course of the experiment, from normal levels immediately after injury. By 4 h postinjury, SCBF had decreased by 42%. In contrast, the 4-h SCBF in cats that were treated with any of the three highest dose levels of tirilazad mesylate were significantly improved in comparison to the vehicle-treated cats. Tirilazad has also been shown to retard and partially reverse posttraumatic spinal cord ischemia in a severe cat contusion injury model (30). This effect of tirilazad to maintain spinal cord blood flow may be related to preservation of endothelium-dependent relaxation (17) important for microvascular autoregulation.

The cellular mechanism of action of tirilazad in antagonizing posttraumatic ischemia development is believed to involve an inhibition of oxygen radical–mediated microvascular LP. This conclusion is based on the concomitant action of tirilazad to attenuate an injury-induced decline in spinal tissue vitamin

E at the same doses that reduce posttraumatic ischemia (29). Figure 2 illustrates the dose-response correlation between the attenuation of LP (i.e., reduced loss of vitamin E), the improved maintenance of SCBF, and the enhanced neurological recovery.

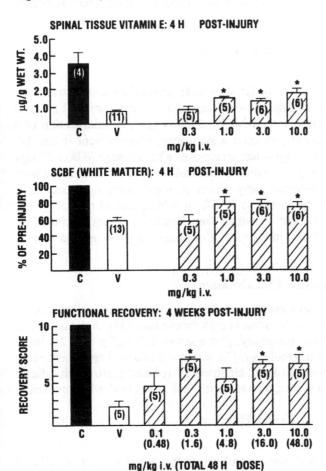

Figure 2 Dose-response correlation in cats of the effects of tirilazad mesylate on posttraumatic (compression injury) spinal cord lipid peroxidation (i.e., loss of reduced form of vitamin E) and on progressive white matter ischemia at 4 h postinjury (data are taken from Ref. 29) versus chronic (4-week) neurological recovery (data are taken from Ref. 24). Doses indicated were administered 30 min after injury. Doses in parentheses under the chronic recovery dose-response curve indicate the total 48-h dosing regimen that these cats received. All values = mean ± standard error. Numbers of animals are given in parentheses in each bar. *$p < 0.05$ versus vehicle-treated animals.

III. TIRILAZAD IN MODELS OF BRAIN INJURY

The activity of tirilazad mesylate has been evaluated in multiple animal models of acute head injury. Across various paradigms, the compound has consistently proven efficacious.

A. Effects on Neurological Recovery and Survival

Initial studies of the efficacy of tirilazad in acute head injury demonstrated the ability of the compound to improve early neurological recovery and survival of mice subjected to severe concussive head injury (31). Administration of a single i.v. dose of tirilazad produced a significant improvement in the 1-h postinjury neurological status (grip test score) over a broad range (0.003–30 mg/ kg). A 1 mg/kg i.v. dose given within 5 min and again at 1.5 h after a severe injury, in addition to improving early recovery, significantly increased the 1 week survival to 78.6% compared to 27.3% in vehicle-treated mice.

Tirilazad has similarly been shown to exert beneficial effects on motor recovery (32) and survival (33) in rat models of moderately severe fluid percussion head injury.

B. Effects on Aerobic Metabolism

Additional experiments have been conducted in severely head-injured cats to assess the effects of tirilazad on brain energy metabolism (34). A 1 mg/kg i.v. dose administered at 30 min postinjury, plus a second 0.5 mg/kg dose 2 h later, resulted in an improved metabolic profile within the injured hemisphere measured at 4 h. Most notably, tirilazad significantly reduced posttraumatic lactic acid accumulation in both the cerebral cortex and subcortical white matter.

C. Effects on Blood Brain Barrier Permeability

Mechanistic data has been obtained which shows that a major effect of tirilazad is to prevent posttraumatic opening of the BBB. In a rat-controlled cortical impact head injury model, a 10 mg/kg i.v. dose at 5 min postinjury was shown to reduce the brain parenchymal extravasation of the protein-bound dye Evan's blue measured at 30 min by 52% (35) (Fig. 3). Using the mouse severe concussive head injury model, tirilazad has been found to blunt the posttraumatic increase in BBB permeability, together with an attenuation of the rise in brain hydroxyl radical levels measured via the salicylate trapping method (23,36). Thus, it is conceivable that the effect of tirilazad to protect the BBB is due to

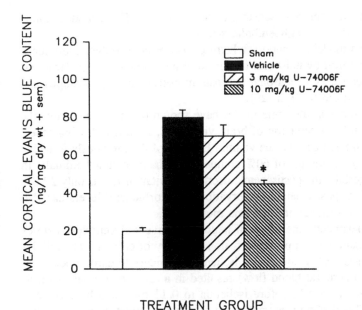

Figure 3 Dose-related (5 min postinjury) attenuation of posttraumatic BBB opening at 30 min after a controlled cortical contusion injury by tirilazad mesylate. Values = means ± standard error for each group. *$p < 0.025$ versus vehicle-treated animals. (Data are taken from Ref. 35.)

either a reduced formation of hydroxyl radicals or perhaps a protection of the microvascular endothelium from hydroxyl radical–induced LP.

IV. TIRILAZAD IN MODELS OF FOCAL CEREBRAL ISCHEMIA

Numerous preclinical evaluations of tirilazad's anti-ischemic efficacy have been carried out in models of focal ischemia. These include studies in models of temporary (ischemia-reperfusion) and permanent focal ischemia.

A. Effects in Temporary Ischemia Models

While increased oxygen radical formation is probably initiated during an is-chemic episode, it is greatly amplified following the reoxygenation of the tis-sue following reperfusion. Thus, the most relevant context for evaluation of antioxidant compounds is the situation of ischemia-reperfusion. Protective ef-fects of the 21-aminosteroid, tirilazad, have been observed in several experimental models of focal cerebral ischemia with reperfusion. These models mimic the clinical situation of thromboembolic stroke.

First of all, in a model of temporary hemispheric cerebral ischemia produced in the Mongolian gerbil by unilateral occlusion of a carotid artery, tirilazad has been shown to reduce 24-h neuronal necrosis in cortex and hippocampus and to promote survival out to 48 h (20).

Mechanistic studies in the same model have shown that tirilazad blunts the postischemic LP-related depletion of brain vitamin E. At 2 h following reperfusion in vehicle-treated gerbils, the vitamin E levels of the previously ischemic hemisphere fell by an average of 60% compared to sham-occluded animals. In tirilazad-treated gerbils, the postischemic decrease in vitamin E was only 27.0% (19). A similar 2-h, postischemic preservation of ascorbic acid levels has also been observed with tirilazad in the same model (37).

In parallel experiments, the effect of neuroprotective and antioxidant doses of tirilazad was examined on postreperfusion recovery of cortical extracellular calcium homeostasis (19). Hemispheric ischemia secondary to carotid occlusion (75% reduction in cerebral blood flow) resulted in a drop in cortical extracellular calcium from 1.05 mM before ischemia to 0.11 mM at 3 h in vehicle-treated gerbils. This decline was unaffected by tirilazad pretreatment. Following 2 h of reperfusion, calcium recovered slightly to 0.22 mM in vehicle-treated animals, while in the tirilazad-treated animals the 2-h recovery reached 0.56 mM ($p < 0.03$ versus vehicle).

These collective results provide evidence that the ability of the 12-aminosteroid antioxidant, tirilazad, to reduce postischemic neuronal degeneration (20) is due to an inhibition of postischemic LP. Secondary to this membrane protective mechanism, the drug acts to preserve cellular processes responsible for the reversal of the ischemia-triggered intracellular calcium accumulation (Fig. 4). Consistent with this mechanism, an in vitro study has demonstrated the ability of tirilazad to protect cell membrane Ca^{2+} and Na^+,K^+-ATPases from iron-induced peroxidative inactivation (18).

In other work with the gerbil 3-h unilateral carotid occlusion model, tirilazad has been shown to attenuate delayed (12–24 h) neutrophil influx into the reperfused hemisphere (38). Considering the production of oxygen radicals and other toxic products by neutrophils, the lessened postischemic infiltration might also serve to contribute to the neuroprotective action. Indeed, these investigators showed a significant correlation between the extent and time course of neutrophil infiltration and neuronal damage over a 24-h period after reperfusion. Perhaps mechanistically related to the decrease in neutrophil accumulation, identical dosing with tirilazad in the same model has been shown to blunt the early postreperfusion-induced increase in tissue levels of the neutrophil chemotaxin leukotriene C4 (39). However, it is possible that the decrease in leuko-

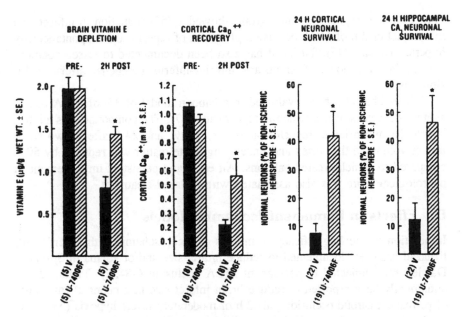

Figure 4 Comparison of the effects of tirilazad (U-74006F) in gerbils on postischemic lipid peroxidation (i.e., depletion of brain vitamin E), recovery of extracellular calcium and postischemic neuronal necrosis in lateral cortex (neurons counted in a 222 μ^2 area and hippocampal CA1 region (neurons counted in a 315-μ length). Numbers in parentheses indicate N. Gerbils received a 10 mg/kg i.p. dose of tirilazad or vehicle at 10 min before unilateral carotid occlusion and again immediately after reperfusion. *$p < 0.05$ versus vehicle. (Data are taken from Refs. 19 and 20.)

triene C4 is simply a manifestation of a decrease in membrane LP since lipid hydroperoxides activate the leukotriene-forming 5-lipoxygenase.

Tirilazad has also been examined in the cat following a 1-h temporary occlusion of the middle cerebral artery (MCA) (40). Postischemic treatment with tirilazad significantly reduced the infarction volume compared to that in vehicle-treated cats. However, in other cat studies that employed longer periods of temporary MDA occlusion, no effect of the compound was observed (41,42). In contrast, a near complete prevention of 3-h postischemic brain edema has been reported in cynomolgus monkeys subjected to 3 h of MCA occlusion but treated with tirilazad administered 10 min prior to reperfusion (43).

The compound has further been shown to reduce infarct size in rat models of temporary MCA occlusion. In Wistar rats pretreated with a 10 mg/kg i.v. dose followed by a 2-h episode of MCA occlusion, there was a 67.3% reduction in the 24-h postischemic infarct volume compared to vehicle-pretreated rats

(44). In a later report, the same group showed a 25% reduction in infarct size in a model of 2 h of MCA occlusion plus 24 h of reperfusion in spontaneously hypertensive rats (45). Tirilazad has also been documented to lessen cochlear damage in a rat model of temporary anterior-inferior cerebellar artery occlusion (46).

Tirilazad has also been evaluated in a baboon model of 3 h of MCA occlusion plus 2 weeks of postischemic reperfusion (47). The compound was administered 15 min prior to reperfusion and 2, 4, 12, and 24 h after. Two-week infarct volume in the cortex and underlying white matter was reduced by 50% compared to vehicle-treated baboons. An early and persistent improvement in neurological score was also associated with tirilazad treatment.

B.   Effects in Permanent Ischemia Models

In addition to showing efficacy in models of focal ischemia with reperfusion, tirilazad has been documented to reduce infarct size and/or edema in Sprague-Dawley rats subjected to permanent MCA occlusion (48–52). Tirilazad has additionally been reported to reduce brain infarct size in a neonatal rat model of permanent carotid occlusion plus 2 h of moderately severe hypoxia (53). This implies that the relevance of free-radical mechanisms, and thus antioxidant protection, is not confined to ischemic situations in which reperfusion takes place. However, the compound has not shown efficacy in permanent MCA occlusion models in spontaneously hypertensive rats (45) and hyperglycemic cats (54). On the other hand, each of these studies only investigated single-dose levels of tirilazad. A dose-response study would be required to completely rule out protective efficacy. An additional study of tirilazad's efficacy in focal ischemia was carried out in a model of photoillumination-induced MCA occlusion in rats (55). Tirilazad treatment lessened peri-infarct edema in parallel with a reduction in lipid peroxide content, which supports once again the association between neuroprotection and the compound's ability to inhibit neural LP.

C.   Effects in Thromboembolic Models

Tirilazad has also been examined in rabbits in two thromboembolic rabbit stroke models, both of which have shown efficacy. In the first, a 3 mg/kg i.v. dose of the compound was administered 30 min before and again 2 h after carotid arterial injection of a 3.5-cm-long embolus. In the treated animals, the infarct measured at 4 h was only 14.8% of the affected hemisphere versus 36.0% in the vehicle-treated rabbits ($p > 0.05$). Interestingly, tirilazad did not affect cerebral blood flow (56).

In the second study, the effect of tirilazad was examined in terms of the number of microspheres that had to be injected into the carotid artery in order to produce a functional deficit in the rabbits. A 3 mg/kg i.v. dose of tirilazad was given 10 min before the beginning of microembolization, with additional 1.5 mg/kg i.v. doses every 5 h beginning 5 h after embolization. In comparison to the vehicle-treated rabbit group, tirilazad doubled the amount of microspheres that were required to produce a 50% functional deficit (57). This latter study suggests that tirilazad may be beneficial in protecting the brain when employed as a pretreatment before coronary artery bypass graft and other open-heart surgeries where there is the risk of cerebral shower of microemboli arising from the heart-lung machine.

V. TIRILAZAD IN MODELS OF GLOBAL CEREBRAL ISCHEMIA

The clinical correlate of models of global cerebral ischemia is the situation of cardiac arrest followed by resuscitation in which the occurrence of secondary brain injury is not uncommon. In view of tirilazad's potential use in this type of ischemic insult, a number of studies have been carried out in animal models of cardiac arrest/resuscitation, global ischemia produced by raised either intracranial pressure or transient, near complete forebrain ischemia produced by vascular occlusion.

A. Effects on Selective Neuronal Vulnerability

In models of global cerebral ischemia with reperfusion, which mimic the clinical situation of cardiac arrest/resuscitation, tirilazad has produced mixed results. A reduction in delayed (7-day) postischemic hippocampal CA_1 damage has been observed in the widely used gerbil brief (10-min) bilateral carotid occlusion model, but only with maintained dosing; acute peri-ischemic dosing is ineffective (58). Using the same model, pretreatment with tirilazad has been shown to reduce the early postreperfusion increase in brain levels of hydroxyl radical measured with the salicylate trapping method (59).

In the rat model of transient (10–12 min) bilateral carotid occlusion plus hypotension, a reduction in cortical neuronal loss has been reported (60,62). In contrast, the highly sensitive hippocampal CA_1 region has not been effectively protected by tirilazad in the same model (60,62–64) or in the rat 15-min, four-vessel occlusion model (44,65). This discrepancy between antioxidant protection of the cortex and the hippocampus has led to the suggestion that the mechanisms of postischemic neuronal damage may differ in those two regions, with LP being more relevant in the cortex. However, the mixed results with tirilazad in regard to its ability to successfully impact the cortical damage associated with brief global ischemia (60), but not the more severe hippocampal damage (particularly the highly vulnerable CA_1 region) (60,62–65), may be due to the observation of accentuated postischemic free-radical production and LP in the hippocampus in comparison to the cortex (66). With this in mind, it is noteworthy that the doses examined thus far in the rat are much lower than the dose level found to salvage the gerbil hippocampus (58).

B. Effects on Neurological, Neurophysiological, and Metabolic Recovery

In a dog model of 10-min normothermic cardiac arrest, tirilazad has been documented to improve 24-h neurological recovery and survival (67,68). In addition, tirilazad has been shown to improve the neurological recovery of dogs subjected to 11-min complete global ischemia produced via raising intracranial pressure above the level of the cerebral perfusion pressure (69). Despite these two reports, tirilazad failed to improve early neurophysiological (i.e., somatosensory-evoked potentials) and metabolic (i.e., magnetic resonance phosphorus spectroscopic measurement of ATP, phosphocreatine, and pH) recovery in a nearly identical model of 10 min of complete global ischemia (70). On the other hand, the latter group of investigators observed a striking improvement in early neurophysiological and metabolic recovery in a more severe dog model of 30 min of incomplete global ischemia in hyperglycemic animals produced via raised intracranial pressure (71). A similar improvement in recovery of brain energy metabolism and acid-base balance has been reported in the rat two-vessel occlusion plus hypotension forebrain ischemia model (72,73).

C. Effects on Cerebral Blood Flow

In regards to providing a physiological mechanism for the beneficial effect of tirilazad on cortical neurophysiological and metabolic recovery and histopathology, the possibility of an effect of the compound on postischemic and postreperfusion blood flow has been studied. Using a cat model of a 7-min near complete global cerebral ischemia produced by neck tourniquet, tirilazad (1 mg/kg i.v. at 15 min postreperfusion) was shown to better maintain blood flow together with an enhanced recovery of somatosensory-evoked potentials (74). However, the improvement in blood flow was paralleled by a prevention of a progressive fall in mean arterial pressure after the ischemic episode in this model, which may explain the maintenance of cerebral perfusion. In contrast, tirilazad does not appear to attenuate postreperfusion cerebral hypoperfusion after a 10-min period of normothermic cardiac arrest in dogs (75) at a dose level associated with improved 24-h postresuscitation neurological recovery and survival (67,68). Thus, the neuroprotective effects of tirilazad in global cerebral ischemia, in the absence of effects on systemic blood pressure, are not associated with direct actions on cerebral blood flow. This is similar to the lack of correlation of blood flow effects with the neuroprotection observed in models of focal cerebral ischemia (45,56).

VI. TIRILAZAD IN MODELS OF SUBARACHNOID HEMORRHAGE

Subarachnoid hemorrhage (SAH) is commonly associated with rupture of cerebrovascular aneurysms or with moderate-to-severe brain injury. The pathophysiology of SAH involves acute disruption of microvascular autoregulatory mechanisms, increased BBB permeability, and delayed spasm of the major cerebral vessels leading to a secondary ischemic insult (8,9). Tirilazad has been examined in a number of animal models of SAH with consistent benefit.

A. Effects on Acute Cerebral Microvascular Hypoperfusion

The ability of tirilazad to antagonize acute progressive cerebral hypoperfusion following experimental SAH has been examined in chloralose-anesthetized cats (76). Subarachnoid hemorrhage was produced by injection of nonheparinized autologous blood into the cisterna magna after prior withdrawal of an equivalent volume of cerebrospinal fluid (CSF). In untreated animals, SAH caused a progressive decline in caudate nuclear blood flow (–51.4% by 3 h) and an increase in intracranial pressure (ICP; +18.5 mmHg by 3 h). In comparison, in cats that received a 1 mg/kg i.v. dose of tirilazad at 30 min after SAH, there was a complete prevention of the fall in blood flow and a significant attenuation of the rise in ICP. While it is believed that much of the preservation of blood flow was the result of a preservation in autoregulatory mechanisms, the support of blood flow was, in part, due to an improved cerebral perfusion pressure (CPP). This was, in turn, the result of a better maintenance of mean arterial pressure (MAP), together with the attenuated rise in ICP (i.e., CPP = MAP – ICP).

B. Effects on Blood Brain Barrier Permeability

In another investigation aimed at studying the effects of tirilazad on post-SAH blood brain barrier disruption and the associated vasogenic edema, SAH was produced in rats by unilateral injection of blood into the cortical subarachnoid space. In animals that received only vehicle injections, the SAH resulted in a substantial amount of barrier disruption as judged by extravasation of the protein-bound dye Evan's blue. The leakage of Evan's blue was significantly attenuated by treatment with tirilazad (1 mg/kg given 15 min before and again 2 h after SAH) (77). An identical action was recently demonstrated for the principal human tirilazad metabolite U-89678 (Fig. 1), which retains the LP-inhibiting efficacy of the parent compound (78).

The suppression of BBB opening may explain the results from the study which showed that tirilazad ameliorated the acute SAH-induced rise in ICP in the cat model (76). Tirilazad has also been shown to attenuate the increased BBB permeability produced by cortical application of either ferrous chloride or arachidonic acid, both of which are known to induce LP (77,79).

C. Effects on Delayed Cerebral Vasospasm

Multiple studies have been carried out to determine tirilazad's ability to prevent the development of delayed cerebral vasospasm after experimental SAH. In the first of these using a rabbit model (80), SAH was produced by percutaneous injection of 4.5 ml of nonheparinized autologous blood into the cisterna magna. Vehicle or tirilazad (1 mg/kg) was injected i.p. every 12 h, starting 12 h before induction of SAH for a total of six doses. After 48 h, the animals were perfused and the basilar artery removed and processed for morphometric analysis of arterial diameter. In vehicle-treated rabbits, the basilar diameter was reduced by nearly half in comparison to non-SAH controls. In contrast, tirilazad treatment preserved normal basilar artery diameter. A second study of the ability of tirilazad to antagonize cerebral vasospasm following SAH in rabbits has confirmed that treatment with i.v. tirilazad, started 30 min after SAH, significantly lessens angiographically demonstrable vasospasm at 72 h and preserves normal cerebral blood flow (81).

The efficacy of tirilazad in the prophylaxis of post-SAH delayed cerebral vasospasm has been further evaluated in a randomized, double-blind, placebo (vehicle)-controlled trial in a cynomolgus monkey SAH paradigm (82). When comparing the effects of tirilazad (1 mg/kg i.v. every 8 h beginning at 20 h post-SAH) to those of placebo at day 7, there was a significant decrease in the degree of angiographic vasospasm in the right extradural internal carotid and right middle cerebral arteries (MCA). Analysis of cortical energy metabolism revealed a 54% decrease in the ATP/ADP ± AMP ratio in vehicle-treated animals, indicative of vasospasm-induced cerebral ischemia. However, the decrease in the energy ratio was only 7% in animals that received tirilazad. A second randomized, placebo-controlled, dose-response analysis verified the ability of tirilazad to antagonize delayed vasospasm in the monkey model (83). Tirilazad has also been shown to reduce vasospasm in a canine SAH model (84).

D. Effects on LP in Subarachnoid Blood

Another study in the cynomolgus monkey SAH model investigated the effect of tirilazad on the increased level of the LP product malondialdehyde (MDA) seen within the subarachnoid clot at 7 days post-SAH (85). Cynomolgus monkeys were divided into three groups. There were two tirilazad-treated groups (0.3 and 1.0 mg/kg) and a placebo-treated group which were dosed every 8 h (beginning at 20 h post-SAH) for 6 days. On day 7, the angiography was repeated, the animals sacrificed, and subarachnoid clot reclaimed for analysis of MDA. In the vehicle-treated group, significant vasospasm occurred in the clot-side MCA. After tirilazad treatment, significantly less vasospasm developed in the clot-side MCA (for both tirilazad-treated groups, $p < 0.01$). In the placebo-treated group, the MDA content in freshly prepared clot was significantly elevated over that seen in fresh blood clots not placed in the subarachnoid space for 7 days. In the 0.3 mg/kg tirilazad group, the MDA content of the clot was significantly less at day 7 compared with clot from the vehicle-treated group. These results support the hypothesis that LP in a subarachnoid clot may play a role in the pathogenesis of vasospasm (8,9), and that the salutary effects of tirilazad may be mediated by a reduction in LP within the subarachnoid blood and the underlying vasculature. They also imply that, at least in the context of SAH, tirilazad is able to penetrate the CSF and the subarachnoid blood clot.

VII. TIRILAZAD CLINICAL NEUROPROTECTION TRIALS

Currently, tirilazad is being actively investigated in Phase III clinical trials in head and spinal cord injury, ischemic stroke, and subarachnoid hemorrhage (SAH). The first completed Phase III trial of the efficacy of tirilazad has been a multinational European/Australian/New Zealand trial in aneurysmal SAH that included 1023 patients (86). The patients were randomly assigned to either 0.6, 2, or 6 mg/kg/day i.v. in four divided doses, or placebo. All patients received the calcium channel blocker nimodipine based on the prior regulatory approval of this drug for aneurysmal SAH. Treatment was begun within 48 h after SAH and continued for 8 to 10 days.

In the 6 mg/kg dose group (compared to vehicle), there was a 43% reduction in 3-month mortality ($p < 0.01$), a 21% improvement in the incidence of "good" recovery (Glasgow Outcome Scale) ($p < 0.012$), a 28% decrease in the incidence of clinical vasospasm ($p < 0.047$), a 46% lesser use of therapeutic "triple H" hypertensive/hypervolemic/hemodilution therapy ($p < 0.006$) and, most impressively, a 48% increase in the number of patients who could return to full-time work (86). However, the benefits were predominantly shown in males, not females. The reason for this gender difference in efficacy may be due to pharmacokinetic differences wherein the compound is more rapidly cleared in female patients. Ongoing current studies are exploring the use of higher doses of tirilazad in females. However, the available Phase III results show that this antioxidant neuroprotective nonglucocorticoid 21-aminosteroid possesses significant neuroprotective efficacy in man.

VIII.   2-METHYLAMINOCHROMANS

Additional discovery efforts were aimed at the possibility of further enhancing the cerebral antioxidant activity of the lazaroids by replacing the steroid functionality, which possesses only weak antioxidant activity without the complex amino substitution, with a known antioxidant. A series of compounds was synthesized in which the steroid of tirilazad was replaced by the antioxidant ring structure (i.e., chromanol) of α-tocopherol (vitamin E). One of these compounds, U-78517F (Fig. 1), has been demonstrated to have predictably more potent in vitro lipid antioxidant and in vivo cerebroprotective activity in models of injury and global cerebral ischemia (87,88). However, U-78517F does not possess significantly better BBB penetration than tirilazad (21).

IX.   PYRROLOPYRIMIDINES

As detailed above, the 21-aminosteroid lazaroid tirilazad mesylate has been demonstrated to be a potent inhibitor of lipid peroxidation and to reduce traumatic and ischemic damage in a number of experimental models. Tirilazad acts, in large part, to protect the microvascular endothelium and, consequently, to maintain normal BBB permeability and cerebral blood flow autoregulatory mechanisms. However, due to its limited penetration into brain parenchyma, tirilazad has generally failed to affect delayed neuronal damage to the selectively vulnerable hippocampal CA_1 and striatal regions (Sec. V.A). Recently, we discovered a new group of antioxidant compounds, the pyrrolopyrimidines (89,90), which possess significantly improved ability to penetrate the BBB and gain direct access to neural tissue. Several compounds in the series, such as U-101033E (Fig. 1), have demonstrated greater ability to protect the CA_1 region in the gerbil transient forebrain ischemia model with a postischemic therapeutic window of at least 4 h. In addition, U-101033E has been found to reduce infarct size in the mouse permanent middle cerebral artery occlusion model in contrast to tirilazad which is minimally effective. These results suggest that antioxidant compounds with improved brain parenchymal penetration are better able to limit certain types of ischemic brain damage compared to those which are localized in the cerebral microvasculature. On the other hand, microvascularly localized agents like tirilazad appear to have better ability to limit BBB damage.

REFERENCES

1. Kontos HA, Povlishock JT. Oxygen radicals in brain injury. CNS Trauma 1986; 3:257–263.

2. Braughler JM, Hall ED. Central nervous system trauma and stroke: I. Biochemical considerations for oxygen radical formation and lipid peroxidation. Free Rad Biol Med 1989; 6:289–301.

3. Hall ED, Braughler JM. Central nervous system trauma and stroke: II. Physiological and pharmacological evidence for the involvement of oxygen radicals and lipid peroxidation. Free Rad Biol Med 1989; 6:303–313.

4. Hall ED, Braughler JM. Free radicals in CNS injury. In Waxman SG, ed. Molecular and Cellular Approaches to the Treatment of Neurological Disease. New York, Raven Press: 81–105.

5. Siesjo BK, Agardh C-D, Bengtsson F. Free radicals and brain damage. Cerebrovasc Brain Metab Rev 1989; 1:165–211.

6. Schmidley JW. Free radicals in central nervous system ischemia. Stroke 1990; 21:1086–1090.

7. Traystman RJ, Kirsch JR, Koehler RC. Oxygen radical mechanisms of brain injury following ischemia and reperfusion. J Appl Physiol 71:1185–1195.

8. Sano K, Asano T, Tanishima T, Sasaki T. Lipid peroxidation as a cause of cerebral vasospasm. Neurol Res 1980; 2,253–272.

9. Asano T, Matsui T, Takuwa Y. Lipid peroxidation, protein kinase C and cerebral vasospasm. Crit Rev Neurosurg 1991; 1,361–371.

10. Hall ED. Neuroprotective pharmacology of methylprednisolone: a review. J Neurosurg 1992; 76,13–22.

11. Bracken MB, Shepard MJ, Collins WF, Holford TR, Baskin DS, Eisenberg HM, Flamm ES, Leo-Summers L, Maroon JC, Marshall LF, Perot PL, Piepmeier J, Sonntag VKH, Wagner FC, Wilberger JL, Winn HR, Young W. A randomized controlled trial of methylprednisolone or naloxone in the treatment of acute spinal cord injury. New Engl J Med 1990; 322; 1405–1411.

12. Bracken MB, Shepard MJ, Collins WF, Holford TR, Baskin DS, Eisenberg HM, Flamm ES, Leo-Summers L, Maroon JC, Marshall LF, Perot PL, Piepmeier J, Sonntag VKH, Wagner FC, Wilberger JL, Winn HR, Young W. Methylprednisolone or naloxone treatment after acute spinal cord injury: 1 year follow-up data. J Neurosurg 1992; 76:23–31.

13. Hall ED, McCall JM, Yonkers PA, Chase RL, Braughler JM. A nonglucocorticoid analog of methylprednisolone duplicates its high dose pharmacology in models of CNS trauma and neuronal membrane damage. J Pharmacol Exp Ther 1987; 242:137–142.

14. Braughler JM, Pregenzer JF, Chase RL, Duncan LA, Jacobsen EJ, McCall JM. Novel 21-aminosteroids as potent inhibitors of iron-dependent lipid peroxidation. J Biol Chem 1987; 262:10438–10440.

15. Hall ED, McCall JM, Braughler JM. New pharmacological treatments for spinal cord trauma. J Neurotrauma 1988; 5:81–89.

16. McCall JM, Hall ED, Braughler JM. A new class of 21-aminosteroids that are useful for stroke and trauma. In: Capildeo R, ed. Steroids and CNS Diseases. Chichester: Wiley, 1989:69–80.

17. Mathews WR, Marschke CK Jr, McKenna R. Tirilazad mesylate protects endothelium from damage by reactive oxygen. J Mol Cell Cardiol 1992; 24(suppl. III):517.

18. Rohn TT, Hinds TR, Vincenzi FF. Ion transport ATPases as targets for free radical damage: protection by an aminosteroid of Ca^{2+} pump ATPase and Na^+K^+ pump ATPase of human red blood cell membranes. Biochem Pharmacol 1993; 46:525–534.

19. Hall ED, Pazara KE, Braughler JM. Effect of tirilazad mesylate on post-ischemic brain lipid peroxidation and recovery of extracellular calcium in gerbils. Stroke 1991; 22:361–366.

20. Hall ED, Pazara KE, Braughler JM. 21-Aminosteroid lipid peroxidation inhibitor U-74006F protects against cerebral ischemia in gerbils. Stroke 1988; 19:997–1002.

21. Raub TJ, Barsuhn CL, Williams LR, Decker DE, Sawada GA, Ho NFH. Use of a biophysical-kinetic model to understand the roles of protein binding and membrane partitioning on passive diffusion of highly lipophilic molecules across cellular barriers. J Drug Targeting 1993; 1:269–286.

22. Eversole RR, Smith SL, Beuving LJ, Hall ED. Protective effect of the 21-aminosteroid lipid peroxidation inhibitor tirilazad mesylate (U-74006F) on hepatic endothelium in experimental hemorrhagic shock. Circ Shock 1993; 40:125–131.

23. Hall ED, Yonkers PA, Andrus PK, Cox JW, Anderson KD. Biochemistry and pharmacology of lipid antioxidants in acute brain and spinal cord injury. J Neurotrauma 1992; 9 (Suppl 2):425–442.

24. Anderson DK, Braughler JM, Hall ED, Waters TR, McCall JM, Means ED. Effects of treatment with U-74006F on neurological recovery following experimental spinal cord injury. J Neurosurg 1988; 69:562–567.

25. Holtz A, Gerdin B. Blocking weight-induced spinal cord injury in rats: therapeutic effect of the 21-aminosteroid U-74006F. J Neurotrauma 1991; 8:239–245.

26. Holtz A, Gerdin B. Efficacy of the 21-aminosteroid U-74006F in improving neurological recovery after spinal cord injury in rats. Neurol Res 1992; 14:49–52.

27. Anderson DK, Hall ED, Braughler JM, McCall JM, Means ED. Effect of delayed administration of U-74006F (tirilazad mesylate) on recovery of locomotor function following experimental spinal cord injury. J Neurotrauma 1991; 8:187–192.

28. Fowl RJ, Patterson RB, Gewirtz RJ, Anderson DK. Protection against postischemic spinal cord injury using a new 21-aminosteroid. J Surg Res 1990; 48:299–303.

29. Hall ED, Yonkers PA, Horan KL, Braughler JM. Correlation between attenuation of post-traumatic spinal cord ischemia and preservation of vitamin E by the 21-aminosteroid U-74006F: evidence for an in vivo antioxidant action. J Neurotrauma 1989; 6:169–176.

30. Hall ED. Effects of the 21-aminosteroid U-74006F on post-traumatic spinal cord ischemia in cats. J Neurosurg 1988; 68:462–465.

31. Hall ED, Yonkers PA, McCall JM, Braughler JM. Effects of the 21-aminosteroid U-74006F on experimental head injury in mice. J Neurosurg 1988; 68:456–461.

32. Sanada T, Nakamura T, Nishmura MC, Isayama K, Pitts LH. Effect of U-74006F on neurological function and brain edema after fluid percussion injury in rats. J Neurotrauma 1993; 10:65–71.

33. McIntosh TK, Thomas M, Smith D. The novel 21-aminosteroid U-74006F attenuates cerebral edema and improves survival after brain injury in the rat. J Neurotrauma 1992; 9:33–46.

34. Dimlich RVW, Tornheim PA, Kindel RM, Hall ED, Braughler JM, McCall JM. Effects of a 21-aminosteroid (U-74006F) on cerebral metabolites and edema after severe experimental head trauma. In: Long DA, ed. Advances in Neurology. Vol. 52. New York: Raven Press, 1990:365–375.

35. Smith SL, Andrus PK, Zhang JR, Hall ED. Direct measurement of hydroxyl radicals, lipid peroxidation and blood-brain barrier disruption following unilateral head injury in the rat. J Neurotrauma 1994; 11:393–404.

36. Hall ED, Andrus PK, Yonkers PA. Brain hydroxyl radical generation in acute experimental head injury. J Neurochem 1993; 60:588–594.

37. Sato PH, Hall ED. Tirilazad mesylate protects vitamins E and C in brain ischemia-reperfusion injury. J Neurochem 1992; 58:2263–2268.

38. Oostveen JA, Williams LR. Effects of the cytoprotective agent tirilazad mesylate (U-74006F) on the time course of neutrophil infiltration in cerebral ischemia. Neurosci Abs 1991; 17:1092.

39. Andrus PK, Hall ED, Taylor BM, Sam LM, Sun FF. Effects of the 21-aminosteroid tirilazad mesylate (U-74006F) on gerbil brain eicosanoid levels following ischemia and reperfusion. Brain Res 1994; 659:126–132.

40. Silvia RC, Piercey MF, Hoffmann WE, Chase RL, Tang AH, Braughler JM. U-74006F, an inhibitor of lipid peroxidation, protects against lesion development following experimental stroke in the cat: histological and metabolic analysis. Neurosci Abs 1987; 13:1499.

41. Takeshima R, Kirsch JR, Koehler RC, Traystman RJ. Tirilazad treatment does not decrease early brain injury after transient focal ischemia in cats. Stroke 1994; 25:670–676.

42. Gelb AW, Henderson SM, Zhang C. U-74,006F, a 21-aminosteroid, does not ameliorate feline focal cerebral ischemia. J Neurosurg Anesthesiol 1990; 2:240.

43. Boisvert DP, Hall ED. Effectiveness of post-ischemic administration of the 21-aminosteroid tirilazadmesylate (U-74006F) in preventing reperfusion brain edema after temporary focal ischemia in monkeys. Can J Neurol Sci 1996; 23:46–52.

44. Xue D, Bruederlin B, Heinecke E, Li H, Slivka A, Buchan AM. U-74006F reduces neocortical infarction, but does not attenuate selective hippocampal CA_1 necrosis. Stroke 1990; 21:178.

45. Xue D, Slivka A, Buchan AM. Tirilazad reduces cortical infarction after transient, but not permanent focal cerebral ischemia in rats. Stroke 1992; 23:894–899.

46. Seidman MD, Quirk WS. The protective effects of tirilazad mesylate (U-74006F) on ischemia and reperfusion-induced cochlear damage. Otolaryngol Head Neck Surg 1991; 105:511–516.

47. Mori E, Ember J, Copeland BR, Thomas WS, Koziol JA, DelZoppo GT. Effect of tirilazad mesylate on middle cerebral artery occlusion/reperfusion in non-human primates. Cerebrovasc Dis 1995; 5:342–349.

48. Young W, Wojak JC, DeCrescito V. 21-Aminosteroid reduces ion shifts and edema in the rat middle cerebral artery occlusion model of regional ischemia. Stroke 1988; 19:1013–1019.

49. Lythgoe DJ, Little RA, O'Shaughnessy CT, Steward MC. Effect of U-74006F on oedema and infarct volumes following permanent occlusion of the middle cerebral artery in the rat. Br J Pharmacol 1990; 100:454P.

50. Beck T, Bielenberg GW. The effects of two 21-aminosteroids on overt infarct size 48 hours after middle cerebral artery occlusion in the rat. Brain Res 1991; 560:159–162.

51. Park CK, Hall ED. Dose-response analysis of the 21-aminosteroid tirilazad mesylate (U-74006F) upon neurological outcome and ischemic brain damage in permanent focal cerebral ischemia. Brain Res 1994; 645:157–163.

52. Karki A, Westergren I, Widmer H, Johansson B. Tirilazad reduces brain edema after middle cerebral artery occlusion in hypertensive rats. Acta Neurochir 1994; (Suppl)60:310–313.

53. Bagenholm R, Andine P, Hagberg H, Kjellmer I. Effects of the 21-aminosteroid U-74006F on brain damage and edema following perinatal hypoxia-ischemia in the rat. J Cereb Blood Flow Metab 1991;11(Suppl 2):S134.

54. Myers RE, Kleinholz M, Wagner KR, deCourten-Myers, GM. Effects of experimental aminosteroid on outcome of cerebrovascular occlusion in cats. Stroke 1990; 21:179.

55. Umemura K, Wada R, Mizuno A, Nakashima M. Effect of the 21-aminosteroid lipid peroxidation inhibitor U-74006F in a rat middle cerebral artery occlusion model using resonance imaging. Eur J Pharmacol 1994; 251:69–74.

56. Wilson JT, Bednar MM, McAuliffe TL, Raymond S, Gross CE. The effect of the 21-aminosteroid U-74006F in a rabbit model of thromboembolic stroke. Neurosurgery 1992; 31:929–934.

57. Clark WM, Hotan T, Lauten J, Coull BM. Therapeutic efficacy of tirilazad in experimental multiple cerebral emboli. Stroke 1993; 24:175.

58. Hall ED, Braughler JM, McCall JM. Role of oxygen radicals in stroke: effects of the 21-aminosteroids (lazaroids), a novel class of antioxidants. In: Meldrum B, Williams M, eds. Current and Future Trends in Anticonvulsant, Anxiety, and Stroke Therapy. Baltimore: Williams & Wilkins, 1990:351–362.

59. Althaus JS, Andrus PK, Williams CM, VonVoigtlander PF, Cazers AR, Hall ED. The use of salicylate hydroxylation to detect hydroxyl radical generation in ischemic and traumatic brain injury: reversal by tirilazad mesylate (U-74006F). Molec Chem Neuropathol 1993; 20:147–162.

60. Lesiuk HJ, Sutherland GR, Peeling J, Wilkins D, McTavish J, Saunders JK. Effect of U-74006F on forebrain ischemia in rats. Stroke 1991; 22:896–901.

61. Sutherland G, Haas N, Peeling J. Ischemic neocortical protection with U74006F— a dose-response curve. Neurosci Lett 1993; 149:123–125.

62. Hoffman WE, Baughman VL, Polek W, Thomas C. The 21-aminosteroid U-74006F does not markedly improve outcome from incomplete ischemia in the rat. J Neurosurg Anesthesiol 1991; 3:96–102.

63. Beck T, Bielenberg GW. Failure of the lipid peroxidation inhibitor U-74006F to

improve neurologic outcome after transient forebrain ischemia in the rat. Brain Res 1990; 532:336–338.

64. Pahlmark K, Smith ML, Siesjo BK. Failure of U-74006F to ameliorate neuronal damage due to transient ischemia or hypoglycemia. J Cereb Blood Flow Metab 1991; 11(Suppl 2):S138.

65. Buchan AM, Bruederlin B, Heinicke E, Li H. Failure of the lipid peroxidation inhibitor U-74006F to prevent postischemic selective neuronal injury. J Cereb Blood Flow Metab 1992; 12:250–256.

66. Hall ED, Andrus PK, Althaus JS, VonVoigtlander PF. Hydroxyl radical production and lipid peroxidation parallels selective post-ischemic vulnerability in gerbil brain. J Neurosci Res 1993; 34:107–112.

67. Natale JE, Schott RJ, Hall ED, Braughler JM, D'Alecy LG. Effect of the 21-aminosteroid U-74006F after cardiopulmonary arrest in dogs. Stroke 1988; 19:1371–1378.

68. Zwemer CF, Whitesall SE, D'Alecy LG. Cardiopulmonary-cerebral resuscitation with 100% oxygen exacerbates neurological dysfunction following nine minutes of normothermic cardiac arrest in dogs. Resuscitation 1994; 27:159–170.

69. Perkins WJ, Milde LN, Milde JH, Michenfelder JD. Pretreatment with U-74006F improves neurologic outcome following complete cerebral ischemia in dogs. Stroke 1991; 22:902–909.

70. Helfaer MA, Kirsch JR, Hurn PD, Blizzard KK, Koehler RC, Traystman RJ. Tirilazad mesylate does not improve early cerebral metabolic recovery following compression ischemia in dogs. Stroke 1992; 23:1479–1486.

71. Maruki Y, Koehler RC, Kirsch JR, Blizzard KK, Traystman RJ. Effect of the 21-aminosteroid tirilazad on cerebral pH and somatosensory evoked potentials after incomplete ischemia. Stroke 1993; 24:724–730.

72. Haraldseth O, Gronas T, Unsgard G. Quicker metabolic recovery after forebrain ischemia in rats treated with the antioxidant U-74006F. Stroke 1991; 22:1188–1192.

73. Vande Linde AMQ, Chopp M, Lee SA, Schultz LR, Welch KMA. Post-ischemic brain tissue alkalosis suppressed by U-74006F. J Neurol Sci 1993; 114:36–39.

74. Hall ED, Yonkers PA. Attenuation of postischemic cerebral hypoperfusion by the 21-aminosteroid U-74006F. Stroke 1988; 19:340–344.

75. Sterz F, Safar P, Johnson DW, Oku K-I, Tisherman SA. Effects of U-74006F on multifocal cerebral blood flow and metabolism after cardiac arrest in dogs. Stroke 1991; 22:889–895.

76. Hall ED, Travis MA. Effects of the non-glucocorticoid 21-aminosteroid U-74006F on acute cerebral hypoperfusion following experimental subarachnoid hemorrhage. Exp Neurol 1988; 102:244–248.

77. Zuccarello M, Anderson DK. Protective effect of a 21-aminosteroid on the blood-brain barrier following subarachnoid hemorrhage in rats. Stroke 1989; 20:367–371.

78. Smith SL, Scherch HM, Hall ED. Protective effects of tirilazad mesylate and metabolite U-89678 against blood-brain barrier damage after subarachnoid hemorrhage and lipid peroxidation neuronal injury. J Neurosurg 1996; 84:229–233.

79. Hall ED, Travis MA. Inhibition of arachidonic acid-induced vasogenic brain edema

by the non-glucocorticoid 21-aminosteroid U-74006F. Brain Res 1988; 451:350–352.

80. Vollmer DG, Kassell NF, Hongo K, Ogawa H, Tsukahara T. Effect of the non-glucocorticoid 21-aminosteroid U-74006F on experimental cerebral vasospasm. Surg Neurol 1989; 31:190–194.

81. Zuccarello M, Marsch JT, Schmitt G, Woodward J, Anderson DK. Effect of the 21-aminosteroid U-74006F on cerebral vasospasm following subarachnoid hemorrhage. J Neurosurg 1989; 71:98–104.

82. Steinke DE, Weir BKA, Findlay JM, Tanabe T, Grace M, Kruschelnycky BW. A trial of the 21-aminosteroid U-74006F in a primate model of chronic cerebral vasospasm. Neurosurgery 1989; 24:179–186.

83. Kanamaru K, Weir BKA, Findlay JM, Grace M, MacDonald RL. A dosage study of the effect of the 21-aminosteroid U-74006F on chronic cerebral vasospasm in a primate model. Neurosurgery 1990; 27:29–38.

84. Matsui T, Asano T. Effects of new 21-aminosteroid tirilazad mesylate (U74006F) on chronic cerebral vasospasm in a "two-hemorrhage" model of beagle dogs. Neurosurgery 1994; 34:1035–1039.

85. Kanamaru K, Weir BKA, Simpson I, Witbeck T, Grace M. Effect of 21-aminosteroid U-74006F on lipid peroxidation in subarachnoid clot. J Neurosurg 1991; 74:454–459.

86. Kassell NF, Haley EC, Apperson-Hansen C, Alves WM. A randomized double-blind, vehicle-controlled trial of tirilazad mesylate in patients with aneurysmal subarachnoid hemorrhage: a cooperative study in Europe/Australia/New Zealand. J Neurosurg 1996; 84:221–228.

87. Hall ED, Braughler JM, Yonkers PA, Smith SL, Linseman KL, Means ED, Scherch HM, Jacobsen EJ, Lahti RA. U-78517F: a potent inhibitor of lipid peroxidation with activity in experimental brain injury and ischemia. J Pharmacol Exp Ther 1991; 258:688–694.

88. Hall ED, Pazara KE, Braughler JM, Linseman KL, Jacobsen EJ. The non-steroidal lazaroid U-78517F in models of focal and global ischemia. Stroke 1990; 21(suppl III):83–87.

89. Hall ED, Andrus PK, Smith SL, Oostveen JA, Scherch HM, Lutzke BS, Raub TJ, Sawada GA, Palmer JR, Banitt LS, Tustin JM, Belonga KL, Ayer DE, Bundy GL. Neuroprotective efficacy of microvascularly-localized versus brain-penetrating antioxidants. Acta Neurochir (Suppl) 1995; 66:107–113.

90. Hall ED, Smith SL, Andrus PK, Scherch HM, Lutzke BS, Raub TJ, VonVoigtlander PF, Fici GJ, Althaus JS, Palmer JR, Bundy GL. Pyrrolopyrimidines: a novel series of orally bioavailable brain-penetrating neuroprotective antioxidants. J Neurotrauma 1995; 12:967.

9

Antioxidant Properties of Ebselen

Noriko Noguchi and Etsuo Niki
University of Tokyo, Tokyo, Japan

I. INTRODUCTION

Several lines of evidence have been reported that the active oxygen-species-induced and free-radical-mediated oxidation of biological molecules, membranes and tissues occurring in vivo closely related to a variety of pathological events (1). Accordingly, the antioxidants have received much attention recently, and the function, action, and synthesis of natural and synthetic antioxidant drugs have been the subject of extensive studies. Hydroperoxides that are produced in aerobically living cells, either by oxidative reactions or by specific enzymes, can be harmful to biological structures. The most studied examples are the formation of hydroxyl (2) or alkoxyl (3) radicals from hydrogen peroxide or lipid hydroperoxide by a Fenton-type reaction between hydrogen peroxide or the hydroperoxide and Fe^{2+}. Hydrogen peroxide or organic hydroperoxides can be removed by Se-independent peroxidase such as catalase or glutathione S-transferase (4), respectively, as well as glutathione peroxidase (GSH Px). There are at least three different glutathione peroxidases, such as the classical GSH Px, phospholipid hydroperoxide GSH peroxidase (PHGSH Px), and plasma GSH Px. The catalytic mechanism at the selenocysteine moiety in the reaction center of these different enzyme proteins seems to be similar.

In 1984, the glutathione peroxidase-like activity of ebselen, 2-phenyl-1,2-benzisoselenazol-3(2H)-one, a synthetic selenium-containing heterocycle (called

PZ 51), was described (5,6). Ebselen has been found to have an antioxidant activity in model systems which is independent of GSH (5,7,8). Sies has reviewed ebselen as a glutathione peroxidase mimic (9). In this chapter, the action and role of ebselen as an antioxidant in the presence or absence of GSH will be reviewed.

II. CHEMICAL STRUCTURE AND PHYSICOCHEMICAL PROPERTIES

The chemical structure and physicochemical properties of ebselen are summarized in Table 1.

III. INHIBITION OF OXIDATION IN VITRO BY EBSELEN

A. Activity as a Radical-Scavenging Antioxidant

Potent radical-scavenging antioxidants react with stable radicals such as galvinoxyl and 2,2'-diphenyl-1-picrylhydrazyl (DPPH) rapidly. It has been found that ebselen does not react with DPPH at an appreciable rate, suggesting that it does not act as a potent radical scavenger (7).

Table 1   Chemical Structure and Physicochemical Properties of Ebselen

Chemical name :	2-phenyl-1,2-benzisoselenazol-3(2H)-one
Chemical structure :	
	$C_{13}H_9NOSe$
Molecular weight :	274.18
Appearance :	light yellow crystal powder
Taste, Smell :	none
Solubility :	DMF, DMSO > CHCl$_3$, CH$_2$Cl$_2$ >
	Methanol , Ethanol >> H$_2$O (<4.5 µg/ml)
Stability :	Stable for 2 years at room temperature

The activity of ebselen as a radical scavenger has been measured in the oxidation of methyl linoleate emulsions in aqueous dispersions induced either by a hydrophilic (AAPH) or lipophilic (AMVN) radical initiator. The addition of either AAPH or AMVN induces oxidation without any induction period and the oxygen consumption is observed at a constant rate. Ebselen has little effect on the rate of oxygen uptake (Fig. 1a,b), while α-tocopherol or pentamethyl

(a)

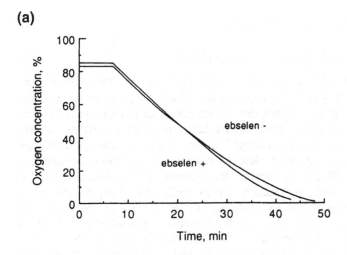

(b)

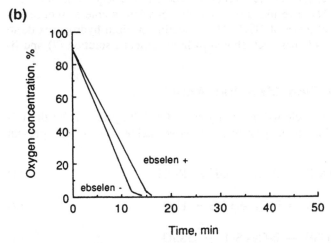

Figure 1 Rates of oxygen consumption in the oxidation of methyl linoleate (74 mM) emulsions in aqueous dispersions induced by (a) AAPH (2.0 mM) or (b) AMVN (2.0 mM) in the absence or presence of ebselen (50 μM) at 37°C in air.

chromanol (PMC) suppresses the oxidation very efficiently. The results show that the reactivity of ebselen as a peroxyl radical scavenger is very small.

B. Reduction of Hydroperoxide

The addition of ferrous ion also induces the oxidation of methyl linoleate emulsions (7). Probably, ferrous ion decomposes methyl linoleate hydroperoxide contained initially in methyl linoleate to give alkoxyl radical, which initiates the chain oxidation.

$$LOOH + Fe^{2+} \rightarrow LO\cdot + {}^-OH + Fe^{3+} \tag{1}$$

$$LO\cdot + LH \rightarrow LOH + L\cdot \rightarrow \text{chain oxidation} \tag{2}$$

Ebselen suppresses the ferrous-ion-induced oxidation of methyl linoleate emulsions dose-dependently (Fig. 2), implying that ebselen suppresses the oxidation by decomposing lipid hydroperoxides, which act as an oxygen radical precursor or by sequestrating iron ion. The oxidation of methyl linoleate micelles containing different amounts of methyl linoleate hydroperoxides clarifies these points. Thirty micromoles of ebselen can suppress the oxidation of methyl linoleate containing 19 μM hydroperoxides induced by the addition of 10 μM $FeSO_4$. However, when the oxidation of the micelles containing 63 μM methyl linoleate hydroperoxides initially is induced by 10 μM $FeSO_4$, the same concentration of ebselen cannot suppress the oxidation as markedly (Fig. 3a,b). These results suggest that the major role of ebselen is to reduce hydroperoxides and destroy the oxygen radical precursor, and that the sequestration of iron is minimal, if any. These results also show that ebselen is able to reduce hydroperoxides in the absence of GSH. The reduction of lipid hydroperoxide to corresponding hydroxide has been shown in homogeneous solution (7) and lipoprotein (10,11).

C. Glutathione Peroxidase-like Activity

The accepted reaction cycle for the catalysis of GSH Px proceeds by the following steps, involving the enzyme-bound selenocysteine, E-Cys-SeH, present as the selenol (12):

$$E\text{-Cys-SeH} + ROOH \rightarrow E\text{-Cys-SeOH} + ROH \tag{1}$$

$$E\text{-Cys-SeOH} + GSH \rightarrow E\text{-Cys-Se-SG} + H_2O \tag{2}$$

$$E\text{-Cys-Se-SG} + GSH \rightarrow E\text{-Cys-SeH} + GSSG \tag{3}$$

$$ROOH + 2GSH \rightarrow ROH + H_2O + GSSG \tag{4}$$

Equation (4) shows the overall reaction of GSH Px.

(a)

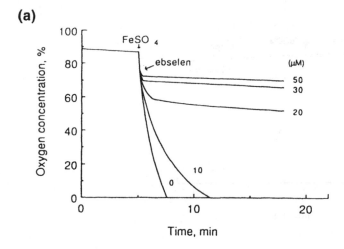

(b)

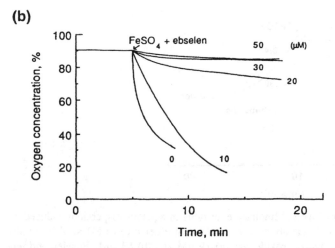

Figure 2   Rates of oxygen consumption in the oxidation of methyl linoleate (74 mM) emulsions in aqueous dispersions induced by ferrous ion. (a) Ferrous sulfate (10 μM) and ebselen were added at the point indicated by an arrow. (b) Ferrous sulfate (10 μM) and ebselen were mixed for 2 min before adding into aqueous dispersions at the point indicated by an arrow.

It has been shown that ebselen catalyzes GSSG formation from GSH in the presence of hydroperoxides (6), which suggests that ebselen exhibits GSH Px–like activity. Many other reports have also shown GSH Px–like activity of ebselen in the different experimental systems, which are summarized in Table 2. For example, ebselen inhibits the copper-induced oxidation of LDL com-

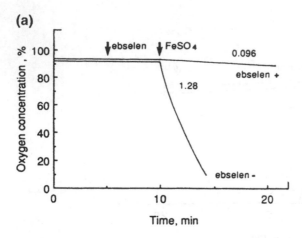

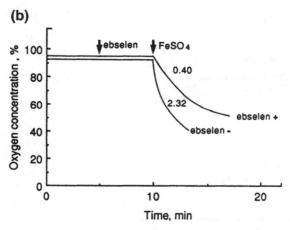

Figure 3 Oxidation of methyl linoleate emulsions in aqueous dispersions induced by ferrous sulfate (10 μM) in the absence or presence of ebselen (30 μM) at 37°C in air. Initial methyl linoleate hydroperoxide was (a) 19 μM and (b) 63 μM. Ebselen and ferrous sulfate were added 5 and 10 min after incubation, respectively, as indicated with an arrow. The numbers are the rate of oxygen uptake in μM/s.

pletely by reducing the hydroperoxides in LDL, since the initiation of copper-dependent oxidation requires the presence of a trace amount of hydroperoxides in LDL (10,11,13,14).

C. Mechanisms of Reduction of Hydroperoxides

The fact that a sulfur analog of ebselen is almost devoid of GSH Px activity (5) suggests that selenium is an essential structural component of ebselen as is

Table 2 Activities of Ebselen in Biological Model Systems

Biological systems	References	
Ethanol-induced liver cell damage	Cadenas	38
Lipid peroxidation in liver microsomes	Muller et al.	5
Inflammatory and liver metabolism	Wendel et al.	6
Secretory activities of macrophages	Parnham and Kindt	39
Oxidative damage to hepatocytes	Muller et al.	21
Inflammatory responses	Parnham et al.	40
Neutrophil lipoxygenase	Safayhi et al.	41
Lipid peroxidation in liver microsomes	Hayashi and Slater	42
Galactosamine/endotoxin-induced hepatitis	Wendel and Tiegs	43
Inhibition of cytochrome P450 reductase	Wendel et al.	44
Leukotriene B4 formation in leukocytes	Kuhl et al.	45,46
Exudative diathesis in chicken	Mercurio and Combs	47
Experimental allergic neuritis	Hartung et al.	48
Periarticular inflammation	Schalkwijk et al.	49
Gingivitis	Van Dyke et al.	50
Superoxide production in leukocytes	Ishikawa et al.	51
Diquat toxicity in hepatocytes	Cotgreave et al.	22
Pulmonary toxicity	Cotgreave and Moldeus	52
Alveolitis and bronchiolitis	Cotgreave et al.	53
Granulocyte oxidative burst	Cotgreave et al.	54
Cerebral ischemic edema	Tanaka and Yamada	55
Arachidonate metabolism in ocular tissue	Hurst et al.	56
Interaction with cytochrome P450	Kuhn-Velten and Sies	57
Microsomal drug biotransformation	Laguna et al.	58
Microsomal electron transport	Nagi et al.	59
Malarial toxicity	Huther et al.	60
Superoxide production in macrophage	Leurs et al.	61
Gastric injury	Kurebayashi et al.	62
Inflammatory and gastric injury	Leyck and Parnham	63
Gastric acid secretion	Beil et al.	64
Contractile responses in lung strips	Leurs et al.	65
Acute gastric mucosal injury	Ueda et al.	66
Endotoxin-induced fetal resorption	Gower et al.	67
Interferon and cytokine induction	Inglot et al.	68
Platinum(II) nephrotoxicity	Baldew et al.	69
Oxidative damage to mitochondria	Narayanaswami and Sies	70
Experimental diabetes	Flechner et al.	71
Ischemic brain edema	Johshita et al.	72
Superoxide prodution in leukocytes	Wakamura et al.	73
Carcinogenesis in kidney	Roy and Liehr	74
Carcinogenesis in mamma	Ip and Ganther	75
Bleomycin toxicity in melanoma cell	Kappus and Reinhold	76

(continued)

Table 2 (Cont.)

Biological systems	References	
Acute pancreatitis	Niederau et al.	77
Lymphocyte proliferation	Hunt et al.	78
Mitogenic activity in blood	Czyrski and Inglot	79
Mitomycin C toxicity	Gustafson and Pritsos	80
Calcium homeostasis in platelets	Brune et al.	81
IP3 receptor binding	Dimmeler et al.	82
Low-density lipoprotein oxidation	Thomas and Jackson	13
Renal preservation in ischemia	Gower et al.	83
Reoxygenation injury in Kupffer cells	Wang et al.	27
Nitric oxide synthesis in Kupffer cells	Wang et al.	27
Endothelial cell injury	Ochi et al.	84
Low-density lipoprotein oxidation	Maiorino et al.	14
Hydroperoxide-induced damage in leukemia cell	Geiger et al.	85
Endothelial cell damage	Thomas et al.	86
Gastric secretion and ulceration	Tabuchi and Kurebayashi	87
Leukocyte and lymphocyte migration	Gao and Issekutz	88
Polymorphonuclear leukocyte migration	Gao and Issekutz	89
Endothelial nitric oxide synthesis	Zembowic et al.	90
Plasma phosphatidylcholine hydroperoxide level	Miyazawa et al.	91
High- and low-density lipoprotein oxidation	Sattler et al.	10
Lipid peroxidation in microsome	Andersson et al.	92
Gastric $H^+,K(+)$-ATPase	Tabuchi et al.	93
Activity of glutathione S-transferase and papain	Nikawa et al.	37
Metal-induced cytotoxicity to epidermal keratinocyte	Kappus and Reinhold	94
Inactivation of $Na^+, K(+)$-ATPase in leukemia cell	Lin and Girotti	95
LTB4-mediated migration of neutrophil	Patrick et al	96
Low-density lipoprotein oxidation	Noguchi et al.	11
Mammalian lipoxygenase activity	Schewe et al.	8
Endothelial nitric oxide synthesis	Hatchett et al.	97
T-lymphocyte migration to arthritic joints	Gao and Issekutz	98
Kainic acid–induced seizures and neurotoxicity	Baran et al.	99
Inducible nitric oxide synthase	Hattori et al.	28
Alcohol-induced damage in perfused liver	Oshita et al.	100
Paracetamol-induced toxicity in hepatocytes	Li et al.	101
Low-density lipoprotein oxidation	Christison et al.	102
Ischemia-reperfusion injury in myocardial infarction	Hoshida et al.	103
Ethanol-induced gastric mucosal injury	Tabuchi et al.	104
Aminoguanidine-induced oxidation	Philis-Tsimikas et al.	105
Brain damage in ischemia	Dawson et al.	106

GSH Px (15,16). Several proposed mechanisms of hydroperoxide reduction have been reported with a kinetic study (17) or a product study using ^{1}H and [77]Se nuclear magnetic resonance (NMR) (18) or mass spectrometry (19). The modified proposed mechanism of hydroperoxide reduction by ebselen in the presence or absence of GSH is shown in Scheme 1. Cycle A in Scheme 1 is dependent on a low (or no) GSH concentration and a high hydroperoxide concentration. As shown in Fig. 4, α-tocopherol is not consumed during the reduction of linoleic acid hydroperoxide by ebselen in the absence of GSH, suggesting that the free radical does not mediate the reaction in cycle A. The mechanism in the presence of GSH is more complicated. Kinetic study of the catalysis of the GSH Px reaction by ebselen reveals (17) that the mechanism appeared identical to that of the enzyme reaction, a tert-uni ping pong, which is shown in cycle B. Cycle C is also presented by detection of diselenide and selenenic acid anhydride during the reaction. The second-order rate constants for hydrogen peroxide with ebselen and the intermediates which are formed by

Scheme 1

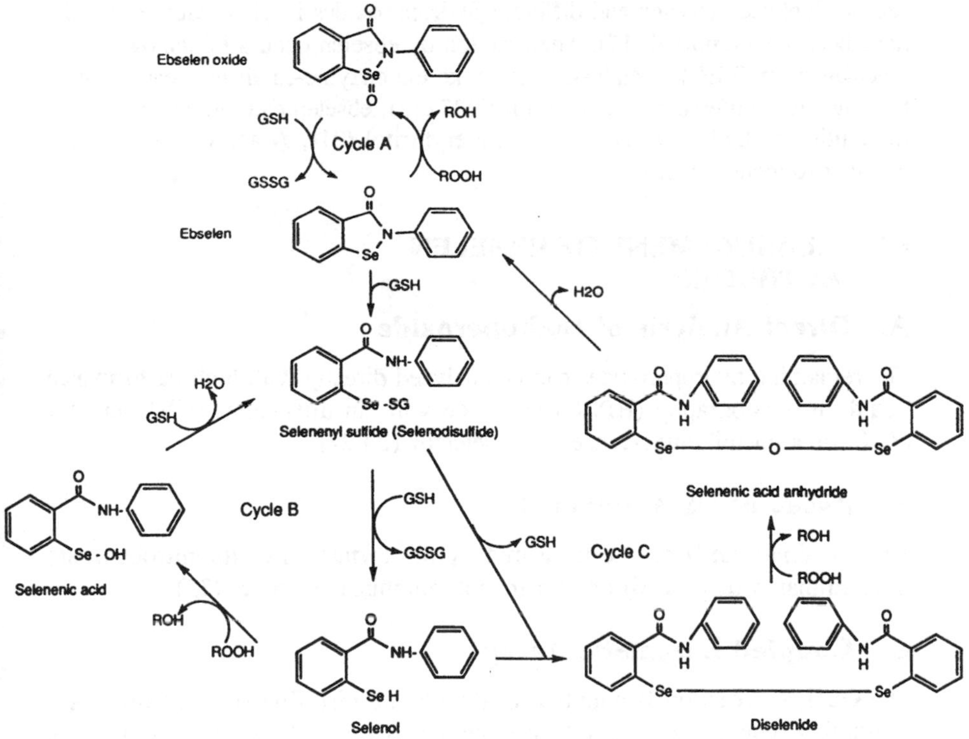

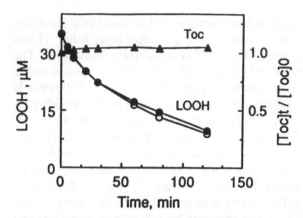

Figure 4   Reduction of linoleic acid hydroperoxide by ebselen (30 μM) in the absence (O) or presence (●) of α-tocopherol (50 μM) and consumption of α-tocopherol (▲).

the reaction of ebselen with glutathione reveal that selenol is particularly active in this respect (20). The apparent second-order rate constants (k_i) for the reaction between ebselen and different hydroperoxides in the presence of GSH have been also reported (17). The reaction by ebselen occurs in analogy to the mechanism of GSH Px catalysis, but unlike the enzyme-catalyzed reaction with binding sites conferring specificity for GSH (12), ebselen can utilize other thiols in addition to GSH (for example, dithioerythritol (21), N-acetylcysteine (22), or dihydrolipoate (19).

IV. MEASUREMENT OF EBSELEN ACTIVITIES

A. Direct Analysis of Hydroperoxide

The remaining hydroperoxides can be analyzed directly with high-performance liquid chromatography (HPLC) equipped with an ultraviolet (UV) detector (7,10) or a chemiluminescence (CL) detector (23,24).

B. Assay of GSH Removal

The remaining GSH can be measured by the formation of thionitrobenzoate from Ellman's reagent (6) or of a monobromobimane adduct (25).

C. Coupled Enzymatic Assay

The GSH Px reaction is coupled to glutathione disulfide (GSSG) reductase consuming reduced nicotinamide adenine dinucleotide phosphate (NADPH) as

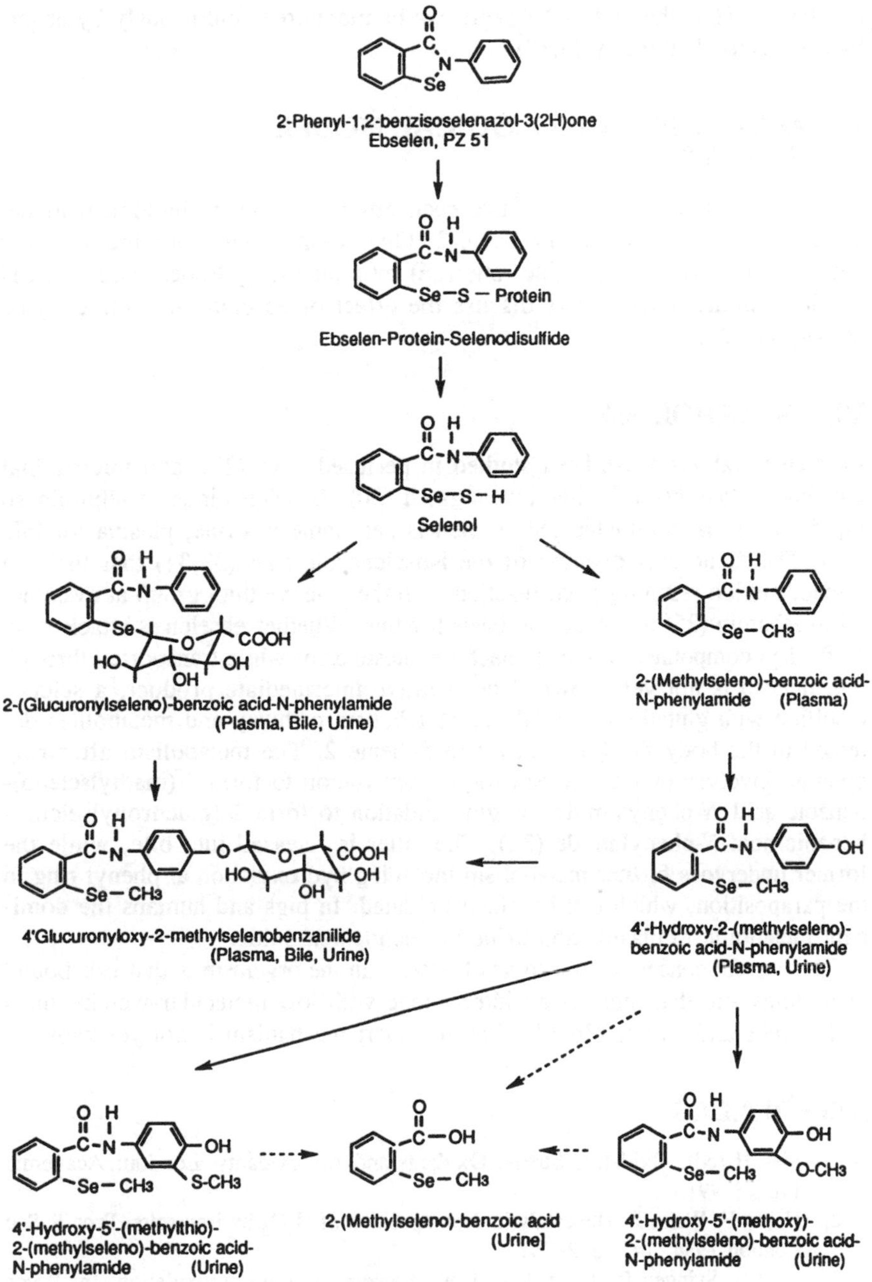

Scheme 2

a cofactor. Then the loss of NADPH can be monitored continuously by absorbance spectrophotometry (5,17).

V. ACTIVITIES IN BIOLOGICAL MODEL SYSTEMS

The diverse effects of ebselen have been observed in many biological model systems, which are shown in Table 2. One common basis for them is that ebselen can lower the peroxide tone (26) by reducing hydroperoxide as GSH Px does. Some other papers discuss the effect of ebselen on some enzyme activity (27,28).

VI. METABOLISM

Ebselen metabolism has been studied in perfused liver (29), and microsomal metabolism has been further investigated (30). Ebselen binds to albumin so rapidly (31) that no unchanged ebselen is detectable in urine, plasma, or bile (32). The facile ring opening of the isoselenazole ring (33,34) that forms a selenodisulfide resulting from reaction with the reactive thiol group at cysteine-34 in albumin (35) is a probable basis for this. Whether ebselen is attacked by sulfhydryl compounds in the stomach or intestines or while transported through the mucosa is not yet known. The putative intermediate product, a selenodisulfide with glutathione, is labile. Metabolism pathway and metabolites detected in the body fluid are shown in Scheme 2. The metabolism after ring opening involves two major pathways: methylation to form 2-(methylseleno)-benzoic acid N-phenylamide, or gluconidation to form 2-(glucuronylseleno)-benzoic acid-N-phenylamide (32). The latter is released into bile, while the former undergoes further metabolism including hydroxylation at phenyl ring in the paraposition, which can be glucuronidated. In pigs and humans the dominant metabolite in plasma and urine is selenoglucuronide.

The current concept of transport of ebselen in the organism is that it is bound to proteins and that there is an interchange with low-molecular-weight thiols within cells and tissues (36,37). The transport mechanism is not yet known.

REFERENCES

1. Sies H (ed). Oxidative Stress: Oxidants and Antioxidants. London: Academic Press, 1991.
2. Haber F, Weiss J. The catalytic decomposition of H_2O_2 by iron salts. Proc R Soc London 1934; 147: 332–351.
3. Aust S, Svingen B. The role of iron in enzymatic lipid peroxidation. In: Pryor W, ed. Free Radicals in Biology, Vol. 5. New York: Academic Press, 1982: 1–28.

4. Prohaska JR, Oh SH, Hoekstra WG, Ganter HE. Glutathione peroxidase: inhibition by cyanide and release of selenium. Biochem Biophys Res Commun 1977; 74: 64–71.

5. Muller A, Cadenas E, Graf P, Sies H. A novel biologically active seleno-organic compound. I. Glutathione peroxidase-like activity in vitro and antioxidant capacity of PZ 51 (ebselen). Biochem Pharmacol 1984: 33: 3235–3239.

6. Wendel A, Fausel M, Safayhi H, Tiegs G, Otter R. A novel biologically active seleno-organic compound. II. Activity PZ 51 in relation to glutathione peroxidase. Biochem Pharmacol 1984; 33: 3241–3245.

7. Noguchi N, Yoshida Y, Kaneda H, Yamamoto Y, Niki E. Action of ebselen as an antioxidant against lipid peroxidation. Biochem Pharmacol 1992; 44: 39–44.

8. Schewe C, Schewe T, Wendel A. Strong inhibition of mammalian lipoxygenases by antiinflammatory seleno-organic compound ebselen in the absence of glutathione. Biochem Pharmacol 1994; 48: 65–74.

9. Sies E. Ebselen, a selenoorganic compound as glutathione peroxidase mimic. Free Rad Biol Med 1993; 14: 313–323.

10. Sattler W, Maiorino M, Stocker R. Reduction of HDL- and LDL-associated cholesterylester and phospholipid hydroperoxides by phospholipid hydroperoxide glutathione peroxidase and ebselen (PZ 51). Arch Biochem Biophys 1994; 309: 214–221.

11. Noguchi N, Gotoh N, Niki E. Effects of ebselen and probucol on oxidative modifications of lipid and protein of low density lipoprotein induced by free radicals. Biochim Biophys Acta 1994; 1213: 176–182.

12. Flohe L. The selenoprotein glutathione peroxidase. In: Dolphin D et al, eds. Glutathione. New York: Wiley, 1989: 643–731.

13. Thomas C, Jackson R. Lipid hydroperoxide involvement in copper-dependent and independent oxidation of low density lipoproteins. J Pharmacol Exp Therap 1991; 256: 1182–1188.

14. Maiorino M, Roveri A, Ursini F. Antioxidant effect of ebselen (PZ 51): peroxidase mimetic activity on phospholipid and cholesterol hydroperoxides vs free radical scavenger activity. Arch Biochem Biophys 1992; 295: 404–409.

15. Rotruck JT, Pope AL, Ganther HE, Swanson AB, Hafeman DG, Hoekstra WG. Selenium: Biological role as a component of glutathione peroxidase. Science 1973; 179: 588.

16. Flohe L, Gunzler WA, Schock HH. Glutathione peroxidase: a selenoenzyme 1973; FEBS Lett 32: 132–134.

17. Maiorino M, Roveri A, Coassin M, Ursini E. Kinetic mechanism and substrate specificity of glutathione peroxidase activity of ebselen (PZ 51). Biochem Pharmacol 1988; 37: 2267–2271.

18. Fischer H, Dereu N. Mechanism of the Catalytic reduction of hydroperoxides by ebselen: A selenium-77 NMR study. Bull Soc Chim Belg 1987; 96: 757–768.

19. Haenen GR MM, de Rooij BM, Vermeulen NPE, Bast A. Mechanism of the reaction of ebselen with endogenous thiols: Dihydrolipoate is a better cofactor than glutathione in the peroxidase activity of ebselen. Mol Pharmacol 1990; 37: 412–422.

20. Morgenstern R, Cotgreave L, Engman L. Determination of the relative contri-

butions of the diselenide and selenol forms of ebselen in the mechanism of its glutathione peroxidase-like activity. Chem Biol Interact 1992; 84: 77–84.

21. Muller A, Gabriel H, Sies H. A novel biologically active selenoorganic compound. IV. Protective glutathione-dependent effect of PZ 51 (ebselen) against ADP-Fe induced lipid peroxidation in isolated hepatocytes. Biochem Pharmacol 1985; 34: 1185–1189.

22. Cotgreave IA, Sandy MS, Berggren M, Moldeus PM, Smith MT. N-acetylcysteine and glutathione dependent protective effect of PZ 51 (ebselen) against diquat induced cytotoxicity in isolated hepatocytes. Biochem Pharmacol 1987; 36: 2899–2904.

23. Yamamoto Y. Chemiluminescence-based high-performance liquid chromatography assay of lipid hydroperoxides. In: Packer L, ed. Methods in Enzymology. Vol. 233. London: Academic Press, 1994: 319–324.

24. Miyazawa T, Fujimoto K, Suzuki T, Yasuda K. Determination of phospholipid hydroperoxides using luminol chemiluminescence-high-performance liquid chromatography. In: Packer L, ed. Methods in Enzymology. London: Vol. 233. Academic Press, 1994: 324–332.

25. Cotgreave IA, Moldeus PM, Brattsand R, Hallberg A, Andersson CM, Engman L. Alpha-(phenylselenenyl)-aceto-phenone derivatives with glutathione peroxidase-like activity. Biochem Pharmacol 1992; 43: 793–802.

26. Hemler ME, Cook HW, Lands WEM. Prostaglandin biosynthesis can be triggered by lipid peroxides. Arch Biochem Biophys 1979; 193: 340–345.

27. Wang J-F, Komarov P, Sies H, de Groot H. Inhibition of superoxide and nitric oxide release and protection from reoxygenation injury by ebselen in rat Kupffer cells. Hepatology 1992; 15: 1112–1116.

28. Hattori R, Inoue R, Sase K, Eizawa H, Kosuga K, Aoyama T, Masayasu H, Kawai C, Sasayama S, Yui Y. Preferential inhibition of inducible nitric oxide synthase by ebselen. Eur J Pharmacol 1994; 267: R1–R2.

29. Muller A, Gabriel H, Sies H, Terlinden R, Fischer H, Romer A. A novel biologically active selenoorganic compound. VII. Biotransformation of ebselen in perfused rat liver. Biochem Pharmacol 1988; 37: 1103–1109.

30. John NJ, Terlinden R, Fischer H, Evers M, Sies H. Microsomal metabolism of 2-methylselenobenzanilide with thiols. Chem Res Toxicol 1990; 3: 199–203.

31. Nomura H, Hakusui H, Takegoshi T. Binding of ebselen to plasma protein. In: Wendel A, ed. Selenium in Biology and Medicine. Heidelberg: Springer-Verlag, 1989: 189–193.

32. Fischer H, Terlinden R, Lohr JP, Romer A. A novel biologically active selenoorganic compound. VIII. Biotransformation of ebselen. Xenobiotica 1988; 18: 1347–1359.

33. van Caneghem P. Influences comparatives de differentes substances seleniees et soufrees sur la fragilite des lysosomes et des mitochondries in vitro. Biochem Pharmacol 1974; 23: 3491–3500.

34. Kamigata N, Takata M, Matsuyama H, Kobayashi M. Novel ring opening reaction of 2-aryl-1,2-benziselenazol-3(2H)-one with thiols. Heterocycles 1986; 24: 3027–3030.

35. Peters T Jr. Serum albumin. Adv Prot Chem 1985; 37: 161–245.
36. Wagner G, Schuch G, Akerboom TP, and Sies H. Transport of ebselen in plasma and its transfer to binding sites in the hepatocyte. Biochem Pharmacol 1994; 48: 1137–1144.
37. Nikawa T, Schuch G, Wagner G, Sies H. Interaction of albumin-bound ebselen with rat liver glutathione S-transferase and microsomal proteins. Biochem Mol Biol Int 1994; 32: 291–298.
38. Cadenas E, Wefers H, Muller A, Brigelius R, Sies H. Active oxygen metabolites and their action in the hepatocyte studies on chemiluminescence responses and alkane production. Agents Actions 1982; 11: 203–216.
39. Parnham MJ, Kindt SA. A novel biologically active seleno-organic compound-III. Effects of PZ51 (ebselen) on glutathione peroxidase and secretory activities of mouse macrophages. Biochem Pharmacol 1984; 33: 3247–3250.
40. Parnham MJ, Leyck S, Dereu N, Winkelmann J, Graf E. Ebselen (PZ51)—a GSH-peroxidase-like organoselenium compound with antiinflammatory activity. Adv Inflam Res 1985; 10: 397–400.
41. Safayhi H, Tiegs G, Wendel A. A novel biologically active seleno-organic compound-V. Inhibition of ebselen (PZ51) of rat peritoneal neutrophil lipoxygenase. Biochem Pharmacol 1985; 34: 2691–2694.
42. Hayashi M, Slater TF. Inhibitory effects of ebselen on lipid peroxidation in rat liver microsomes. Free Radic Res Comm 1986; 2: 179–185.
43. Wendel A, Tiegs G. A novel biologically active seleno-organic compound-VI. Protection by ebselen (PZ 51) against galactosamine/endotoxin-induced hepatitis in mice. Biochem Pharmacol 1986; 35: 2115–2118.
44. Wendel A, Otter R, Tiegs G. Inhibition by ebselen of microsomal NADPH-cytochrome P 450-reductase in vitro but not in vivo. Biochem Pharmacol 1986; 35: 2995–2997.
45. Kuhl P, Borbe HO, Romer A, Fischer H, Parnharm MJ. Selective inhibition of leukotriene B4 formation by ebselen: A novel approach to antiinflammatory therapy. Agents and Actions 1985; 17: 366–367.
46. Kuhl P, Borbe HO, Fischer H, Roemer A, Safayhi H. Ebselen reduces the formation of LTB4 in human and porcine leukocytes by isomerisation to its 5S, 12R-6-trans-isomer. Prostaglandins 1986; 31: 1029–1048.
47. Mercurio SD, Combs GF Jr. Synthetic seleno-organic compound with glutathione peroxidase-like activity in the chick. Biochem Pharmacol 1986; 35: 4505–4509.
48. Hartung HP, Schaefer B, Heininger K, Toyka KV. Interference with arachidonic acid metabolism suppresses exprimental allergic neuritis. Ann Neurol 1986; 20: 168.
49. Schalkwijk J, van den Berg WB, van de Putte LBA, Joosten LAB. An experimental model for hydrogen peroxide-induced tissue damage. Effects of a single inflammatory mediator on (peri)articular tissues. Arthritis Rheum 1986; 29: 532–538.
50. van Dyke TE, Braswell L, Offenbacher S. Inhibition of gingivitis by topical application of ebselen and rosmarnic acid. Agents Actions 1986; 19: 376–377.

51. Ishikawa S, Omura K, Katayama T, Okamura N, Ohtsuka T, Ishibashi S, Masayasu H. Inhibition of superoxide anion production in guinea pig polymorphonuclear leukocytes by a seleno-organic compound, ebselen. J Pharmaco-bio-Dyn 1987; 10: 595–597.

52. Cotgreave IA, Moldeus P. Lung protection by thiol-containing antioxidants. Bull Eur Physicopathol Respir 1987; 23: 275–277.

53. Cotgreave IA, Johansson U, Westergren G, Moldeus PW, Brattsand R. The anti-inflammatory activity of ebselen but not thiols in experimental alveolitis and bronchiolisis. Agents Actions 1988; 24: 313–319.

54. Cotgreave IA, Duddy SK, Kass GEN, Thompson D, Moldeus P. Studies on the anti-inflammatory activity of ebselen. Ebselen interferes with granulocyte oxidative burst by dual inhibition of NADPH oxidase and protein kinase C? Biochem Pharmacol 1989; 38: 649–656.

55. Tanaka J, Yamada F. Ebselen (PZ51) inhibits the formation of ischemic brain edema. In: Wendel A, ed. Selenium in Biology and Medicine. Heidelberg: Springer-Verlag, 1989: 173–176.

56. Hurst JS, Paterson CA, Bhattacherjee P, Pierce WM. Effects of ebselen on arachidonate metabolism by ocular and non-ocular tissues. Biochem Pharmacol 1989; 38: 3357–3363.

57. Kuhn-Velten N, Sies H. Optical spectral studies of ebselen interaction with cytochrome P-450 of rat liver microsomes. Biochem Pharmacol 1989; 38: 619–625.

58. Laguna JC, Nagi MN, Cook L, Cinti D. Action of ebselen on rat hepatic microsomal enzyme-catalyzed fatty acid chain elongation, desaturation, and drug biotransformation. Arch Biochem Biophys 1989; 269: 272–283.

59. Nagi MN, Laguna JC, Cook L, Cinti DL. Disruption of rat hepatic microsomal electron transport chains by the selenium-containing anti-inflammatory agent ebselen. Arch Biochem Biophys 1989; 269: 264–271.

60. Huether AM, Zhang Y, Sauer A, Parnham MJ. Antimalarial properties of ebselen. Parasitol Res 1989; 75: 353–360.

61. Leurs R, Timmerman H, Bast A. Inhibition of superoxide anion radical production by ebselen (PZ51) and its sulfur analogue (PZ25) in guinea pig alveolar macrophages. Biochem Int 1989; 18: 295–299.

62. Kurebayashi Y, Tabuchi Y, Akasaki M. Gastric cytoprotection by ebselen against the injury induced by necrotizing agents in rats. Arzneimittelforschung 1989; 39: 250–253.

63. Leyck S, Parnham MJ. Acute inflammatory and gastric effects of the seleno-organic compound ebselen. Agents Actions 1990; 30: 425–431.

64. Beil W, Staar U, Sewing KF. Interaction of the anti-inflammatory seleno-organic compound ebselen with acid secretion in isolated parietal cells and gastric H^+/K^+-ATPase. Biochem Pharmacol 1990; 40: 1997–2003.

65. Leurs R, Bast A, Timmermann H. Ebselen inhibits constractile responses of guinea-pig parenchymal lung strips. Eur J Pharmacol 1990; 179: 193–199.

66. Ueda S, Yoshikawa T, Takahashi S, Naito T, Oyamada H, Takemura T, Morita Y, Tanigawa T, Sugino S, Kondo M. Protection by seleno-organic compound ebselen, against acute gastric mucosal injury induced by ischemia-reperfusion in rats. Adv Exp Med Biol 1990; 264: 187–190.

67. Gower JD, Baldock RJ, Sullivan AM, Dore CJ, Coid CR, Green CJ. Protection against endotoxin induced foetal resorption in mice by desferrioxamine and ebselen, Int J Exp Pathol 1990; 71: 433–440.

68. Inglot AD, Zielinska-Jenczylik J, Piasecki E, Syper L, Mlochowski J. Organoselenides as potential immunostimulants and inducers of interferon gamma and other cytokines in human peripheral blood leukocytes. Experientia 1990; 46: 308–311.

69. Baldew GS, McVie JG, van der Valk, MA, Los G, de Goeij JJ, Vermeulen NP. Selective reduction of cis-diammine-dichloroplatinum (II) nephrotoxicity by ebselen. Cancer Res 1990; 50: 7031–7036.

70. Narayanaswami V, Sies H. Oxidative damage to mitochondria and protection by ebselen and other antioxidants. Biochem Pharmacol 1990; 40: 1623–1629.

71. Flechner I, Maruta K, Burkart V, Kawai K, Kolb H, Kiesel U. Effects of radical scavengers on the development of experimental diabetes. Diabetes Res 1990; 13: 67–73.

72. Johshita H, Sasaki T, Matsui T, Hanamura T, Masayasu H, Asano T. Effects of ebselen (PZ51) on ischemic brain edema after focal ischemia in cats. Acta Neurochir Suppl 1990; 51: 239–241.

73. Wakamura K, Ohtsuka T, Okamura N, Ishibashi S, Masayasu H. Mechanism for the inhibitory effect of a seleno-organic compound, ebselen, and its analogues on superoxide anion production in guinea pig polymorphonuclear leukocytes. J Pharmacobiodyn 1990; 13: 421–425.

74. Roy D, Liehr JG. Inhibition of estrogen-induced kidney carcinogenesis in Syrian hamsters by modulators of estrogen metablism. Carcinogenesis 1990; 11: 567–570.

75. Ip C, Ganther HE. Activity of methylated forms of selenium in cancer prevention. Cancer Res 1990; 50: 1206–1211.

76. Kappus H, Reinhold C. Inhibition of bleomycin-induced toxic effects by antioxidants in human malignant melanoma cells. Adv Exp Med Biol 1990; 264: 345–348.

77. Niederau C, Ude K, Niederau M, Luthen R, Strohmeyer G, Ferrell LD, Grendell JH. Effects of the seleno-organic substance ebselen in two different models of acute pancreatitis. Pancreas 1991; 6: 282–290.

78. Hunt NH, Cook EP, Fragonas JC. Interference with oxidative processes inhibits proliferation of human peripheral blood lymphocytes and murine B-lymphocytes. Int J Immunopharmacol 1991; 13: 1019–1026.

79. Czyrski JA, Inglot AD. Mitogenic activity of selenoorganic compounds in human peripheral blood. Experientia 1991; 47: 95–97.

80. Gustafson DL, Pritsos CA. Inhibition of mitomycin C's aerobic toxicity by the seleno-organic antioxidant PZ 51. Cancer Chemother Pharmacol 1991; 28: 228–230.

81. Brune B, Diewald B, Ullrich V. Ebselen affects calcium homeostasis in human platelets. Biochem Pharmacol 1991; 41: 1805–1811.

82. Dimmeler S, Brune B, Ullrich V. Ebselen prevents inositol (1,4,5)-triphosphate binding to its receptor. Biochem Pharmacol 1991; 42: 1151–1153.

83. Gower JD, Lane NJ, Goddard G, Manek S, Ambrose IJ, Green CJ. Ebselen.

Antioxidant capacity in renal preservation. Biochem Pharmacol 1992; 43: 2341–2348.

84. Ochi H, Morita I, Murota S. Roles of glutathione and glutathione peroxidase in the protection against endothelial cell injury induced by 15-hydroperoxyeicosatetraenoic acid. Arch Biochem Biophys 1992; 294: 407–411.

85. Geiger PG, Lin F, Girotti AW. Selenoperoxidase-mediated cytoprotection against the damaging effects of tert-butyl hydroperoxide on leukemia cells. Free Radic Biol Med 1993; 14: 251–266.

86. Thomas JP, Geiger PG, Girotti AW. Lethal damage to endothelial cells by oxidized low density lipoprotein: role of selenoperoxidases in cytoprotection against lipid hydroperoxide- and iron-mediated reactions. J Lipid Res 1993; 34: 479–490.

87. Tabuchi Y, Kurebayashi Y. Antisecretory and antiulcer effects of ebselen, a seleno-organic compound, in rats. Jpn J Pharmacol 1993; 61: 255–257.

88. Gao JX, Issekutz AC. The effect of ebselen on polymorphonuclear leukocyte and lymphocyte migration to inflammatory reactions in rats. Immunopharmacology 1993; 25: 239–251.

89. Gao JX, Issekutz AC. The effect of ebselen on polymorphonuclear leukocyte migration to joints in rats with adjuvant arthritis. Int J Immunopharmacol 1993; 15: 793–802.

90. Zembowicz A, Hatchett RJ, Radziszewski W, Gryglewski RJ. Inhibition of endothelial nitric oxide synthase by ebselen. Prevention by thiols suggests the inactivation by ebselen of a critical thiol essential for the catalytic activity of nitric oxide synthase. J Pharmacol Exp Ther 1993; 267: 1112–1118.

91. Miyazawa T, Suzuki T, Fujimoto K, Kinoshita M. Elimination of plasma phosphatidylcholine hydroperoxide by a seleno-organic compound, Ebselen. J Biochem 1993; 114: 588–591.

92. Andersson CM. Hallberg A, Linden M, Brattsand R, Moldeus P, Cotgreave I. Antioxidant activity of some diarylselenides in biological systems. Free Radic Biol Med 1994; 16: 17–28.

93. Tabuchi Y, Ogasawara T, Furuhama K. Mechanism of the inhibition of the hog gastric H+,K(+)-ATPase by the seleno-organic compound ebselen. Arzneimittelforschung 1994; 44: 51–54.

94. Kappus H, Reinhold C. Heavy metal-induced cytotoxicity to cultured human epidermal keratinocyte and effects of antioxidants. Toxicol Lett 1994; 71: 105–109.

95. Lin F, Girotti AW. Cytoprotection against merocyanine 540-sensitized photoinactivation of the Na+,K(+)-adenosine triphosphatase in leukemia cells: glutathione and selenoperoxidase involvement. Photochem Photobiol 1994; 59: 320–327.

96. Patrick RA, Peters PA, Issekutz AC. Ebselen is a specific of LTB4-mediated migration of human neutrophils. Agents Actions 1993; 40: 186–190.

97. Hatchett RJ, Gryglewski RJ, Mochowski J, Zembowicz A, Radziszewski W. Carboxyebselen a potent and selective inhibitor of endothelial nitric oxide synthase. J Physiol Pharmacol 1994; 45: 55–67.

98. Gao JX, Issekutz AC. The effect of ebselen on T-lymphocyte migration to ar-

thritic joints and dermal inflammatory reactions in the rat. Int J Immunopharmacol 1994; 16: 279-287.

99. Baran H, Vass K, Lassmann H, Hornykiewicz O. The cyclooxygenase and lipoxygenase inhibitor BW755C protects rats against kainic acid-induced seizures and neurotoxicity Brain Res 1994; 646: 201-206.

100. Oshita M, Takei Y, Kawano S, Fusamoto H, Kamada T. Protective effect of ebselen on constrictive hepatic vasculature: prevention of alcohol-induced effects on portal pressure in perfused livers. J Pharmacol Exp Ther 1994; 27: 20-24.

101. Li QJ, Bessems JG, Commandeur JN, Adams B, Vermeulen NP. Mechanism of protection of ebselen against paracetamol-induced toxicity in rat hepatocytes. Biochem Pharmacol 1994; 48: 1631-1640.

102. Christison J, Sies H, Stocker R. Human blood cells support the reduction of low-density-lipoprotein-associated cholesteryl ester hydroperoxides by albumin-bound ebselen. Biochem J 1994; 304: 341-345.

103. Hoshida S, Kuzuya T, Nishida M, Yamashita N, Hori M, Kamada T, Tada M. Ebselen protects against ischemia-reperfusion injury in a canine model of myocardial infarction. Am J Physiol 1994; 267: H2342-2347.

104. Tabuchi Y, Sugiyama N, Horiuchi T, Furusawa M, Furuhama K. Ebselen, a seleno-organic compound, protects against ethanol-induced murine gastric mucosal injury in both in vivo and in vitro systems. Eur J Pharmacol 1995; 272: 195-201.

105. Philis-Tsimikas A, Parthasarathy S, Picard S, Palinski W, Witztum JL. Aminoguanidine has both pro-oxidant and antioxidant activity toward LDL. Arterioscler Thromb Biol 1995; 15: 367-376.

106. Dawson DA, Masayasu H, Graham DI, Macrae IM. The neuroprotective efficacy of ebselen (a glutathione peroxidase mimic) on brain damage induced by transient focal cerebral ischemia in the rat. Neurosci Lett 1995; 185: 65-69.

10

The Development of Diaryl Chalcogenides and α-(Phenylselenenyl) Ketones with Antioxidant and Glutathione Peroxidase-Mimetic Properties

Ian A. Cotgreave
Karolinska Institute, Stockholm, Sweden

Lars Engman
Uppsala University, Uppsala, Sweden

I. INTRODUCTION

Oxidative stress, generated by dynamic imbalance between the availability and activity of antioxidants and the flux of reactive oxygen metabolites, such as H_2O_2 and the free radicals $\cdot O_2^-$ and $\cdot OH$, in biological systems has been proported to constitute a common denominator in the development of many human disease states (1). The realization of this has spawned the search for xenobiotic structures which can support the function of endogenous antioxidant principles for application as therapeutic agents. Often, the search for such structures has been conducted in a retrospective manner, in which established compounds are shown to possess redox activity, which may then be implicated in the therapeutic mechanism of the compound. On the other hand, rational search, based on sound theoretical principles such as the appreciation of redox chemistry, physicochemical factors such as lipid-aqueous partitioning, structure-activity relationships and, importantly, the mechanism of activity of the compounds in the biological setting, has often been lacking.

The elements in group XVI possess unique but similar chemical properties which facilitate comparative studies of their reactions. Thus, increasing ease of electron donation from sulfur, through selenium to tellurium facilitates redox reactions, particularly direct chemical reduction of reactive oxygen metabolites and other oxygen-centered reactive intermediates, resulting in the eventual

305

production of oxides of the respective elements (2). The close homology in the redox chemistries of these atoms also facilitate chemical processes such as those of thiol/selenol-disulfide interchange (3,4). It is, therefore, not surprising that biological systems themselves have utilized this chemical homology in the evolution of antioxidant structures such as the thiol-containing tripeptide glutathione (GSH) and the selenocysteine-containing protein GSH peroxidase.

From our experience with the organoselenium antioxidant and glutathione (GSH) peroxidase mimetic ebselen (5–7) and other data obtained with this interesting molecule (detailed in Chap. 9), we embarked on studies to produce other chalcogen-containing structures which would delineate the apparent antioxidant and GSH peroxidase-mimetic properties and optimize their biological activities. During these studies we have attempted to apply a systematic approach based on the rational chemical/biochemical principles detailed above. This chapter details the development of two classes of compounds based on diaryl selenides/tellurides and α-(phenylselenenyl) ketones (Fig. 1). It is hoped that the reader will not only gain an appreciation of the individual properties of these compounds, but also be prompted to utilize some of the reasoning employed in future work in this area.

II. ANTIOXIDANTS

Classically, within the biological setting, an antioxidant is a compound which, at low concentration in relation to their biological molecules, will react with free-radical species before these react with other molecules. Much of this definition has been used to describe compounds which act as "chain-breaking" species in the process of lipid peroxidation. At the outset of our search for effective biological antioxidants, we were aware that the ultimate biological activity of the molecules would represent a balance between a suitably low redox potential of the electron-donating heteroatom and the physicochemical characteristics of the carrier molecules, such as their aqueous-lipid partitioning behavior.

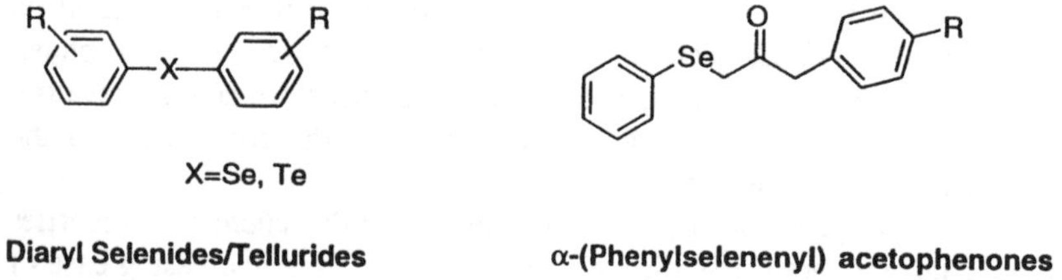

X=Se, Te

Diaryl Selenides/Tellurides   α-(Phenylselenenyl) acetophenones

Figure 1 Structures of the chalcogen-containing antioxidants and glutathione peroxidase-mimetics detailed in this chapter.

In our search for antioxidant species based on the redox chemistry of selenium, we centered around the possibility of manipulating the electron-donating capacity of the heteroatom by appropriate chemical derivatization. The diaryl selenide structure shown in Fig. 1 provided such a possibility in which the stability of the selenium-centered radical cation, arising through electron transfer, could be altered by substitution of the phenyl rings with electron-withdrawing (-I effect) or electron-donating (+I effect) substituents.

It can be seen from Table 1 that alteration of the substituents of the aromatic rings has profound effects on the antioxidant efficacy of the parent diphenyl selenide. Thus, symmetrical para-substitution of the rings with amine groups provides a supply of electrons through donation of the nitrogen lone pair to the aromatic system. This greatly potentiates the efficacy of the carrier structure. On the other hand, symmetrical para-substitution of the rings with electron-withdrawing cyano or nitro groups reflects in greatly diminished activity of the structure. These changes are also reflected in the oxidation potentials of the molecules. For instance, the oxidation potential of the parent molecule, as determined by cyclic voltammetry, is 1.38 V, while that of the symmetrical para-substituted amine derivative is 0.80 V, reflecting substantially increased oxidizability at the selenium heteroatom (8).

These studies, and others using a series of dibenzo(1,4)dichalcogenines, in which the oxidation potential of the molecules was altered by chalcogen substitution, without altering the lipid solubility (9), revealed that an optimal oxidation potential for biological activity lies between 0.65 and 0.85 V. Compounds with a lower oxidation potential, although chemically more reactive as electron donors, are probably too unstable in the biological setting, resulting

Table 1 Inhibition of Fe^{2+}/ADP/ Ascorbate-Induced Lipid Peroxidation in Rat Liver Microsomes by Symmetrical Diaryl Selenides

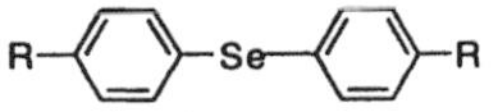

$$R-\langle\bigcirc\rangle-Se-\langle\bigcirc\rangle-R$$

Substitution	IC_{50} (μM)	% inhibition at 50 μM
R = H	>50	19
R = NH$_2$	3	100
R = Ph	8	65
R = CH$_3$	25	58
R = CN	>50	12

in diminished antioxidant activity (9). Here it should be mentioned that the oxidation potential of the naturally occurring antioxidant vitamin E is 0.84 V.

Having established an appropriate oxidation potential which provides considerable antioxidant effect in a biological system, we then attempted to manipulate the aqueous-lipid partitioning behavior of bis(4-aminophenyl) selenide by various N-substitutions. It can be seen from Table 2 that N-ethylation of one or both of the amino groups of the parent antioxidant decreased the IC_{50} in peroxidizing microsomes from 3 to 1 μM. This potentiation was postulated to be due to increased lipid-phase partitioning of the structures as the oxidation potential of the selenium was not expected to be considerably altered (8). Additionally, symmetrical N-isobutylation potentiated the antioxidant activity of the molecule, while substitution of one or both of the amine groups with bulky aliphatic alkane chains considerably diminished the efficacy of the carrier molecule. Again, little change in the oxidation potential of the compounds was expected (8). These structure-activity studies thus indicated that consideration of the accessibility of the carrier structure to the intramembranar sites of lipid peroxidation is as important as an appropriate oxidation potential when considering the design of biologically active antioxidants. In the case of the N-substituted bis(4-aminophenyl) selenides tested, this entailed that attempts to increase the lipophilicity of the molecule by derivatization with bulky aliphatic chains, reminiscent of the phytyl chains of the natural antioxidant α-tocopherol (vitamin E), actually restricted the access of the molecule into the membranes of the experimental systems used and impeded biological activity. Indeed, vitamin E itself is a poor antioxidant in such systems (8). Thus, through appro-

Table 2 The Effect of Various N-Substituents on the Antioxidant Activity of bis(4-aminophenyl) Selenide in Rat Liver Microsomes Undergoing Fe^{2+}/ADP/Ascorbate-Induced Lipid Peroxidation

$$RNH - \bigcirc - Se - \bigcirc - NHR_1$$

Substitution	IC_{50} (μM)	% inhibition at 50 μM
R= R_1 = H	3	100
R=R_1 = CH$_2$CH$_3$	1	100
R=R_1 = CH$_2$CH(CH$_3$)$_2$	1	100
R = CH$_2$CH$_3$, R_1 = H	1	100
R=R_1 = C$_{16}$H$_{33}$	>50	34
R = C$_{16}$H$_{33}$, R_1= H	>50	25

Experimental conditions and other analogues detailed in Ref. 8.

priate structure-activity studies, it is important to define a balance between the aqueous-lipid partitioning behavior of the structure and its physical size and shape in order to ensure optimal bioavailability of the molecule to sites of free-radical generation in membranes.

When the selenium heteroatom of the diphenyl selenide structure was ex-changed for tellurium, a remarkable property of these structures was revealed. Regardless of the para-substitution of the benzene rings, all diaryl tellurides tested were much more potent antioxidants in biological systems than their re-spective selenium analogs (10). Thus, even though the oxidation potential of the parent diaryl telluride was above that defined as optimal for biological an-tioxidant activity, according to the results obtained with the corresponding se-lenium analogs, and assuming that the aqueous-lipid partitioning of the molecule is not too different from the parent selenium analog, the molecule is a surpris-ingly potent antioxidant exhibiting an IC_{50} in microsomes of 180 nM as opposed to >50 μM for the selenium analog. This was generally the case for all ana-logs tested, with substitution of selenium by tellurium resulting in between 100- and 250-fold potentiation (Table 2 versus Table 3), without generally affecting the oxidation potential. Additionally, it will be noted that the potency of the diaryl tellurides seems to be systematically related to their oxidation potentials.

From these studies, it was suggested that the increased potency of the tel-luride molecules was due to the ability of GSH or ascorbate to reduce tellu-rium(IV) dihydroxides (hydrated telluroxides), formed by further oxidation/disproportionation and hydrolysis of the tellurium radical cation in the presence of water. This would establish a catalytic cycle in which the antioxidant would be regenerated at the expense of the reducing agent, thus allowing for effec-

Table 3 The Antioxidant Activity of Various Diaryl Tellurides in Fe^{2+}/ADP/ Ascorbate-Induced Lipid Peroxidation in Rat Liver Microsomes

Substitution	IC$_{50}$ (nM)	oxidation potential
R = H	85	0.95
R = NH$_2$	60	0.56
R = Ph	80	0.84
R = CH$_3$	100	0.89
R = NO$_2$	315	1.14

Experimental details and other analogues detailed in Ref. 10.

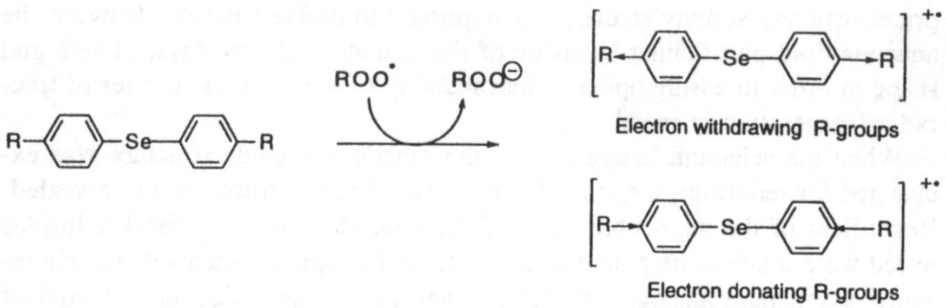

Figure 2 The formation of a selenium-centered radical cation during the reaction of a diaryl selenide with a peroxyl radical.

tive chain-breaking activity by much fewer molecules present in the membranes (Fig. 3, cycle I).

The evidence for such a catalytic antioxidant cycle in biological systems is compelling (11,12). Recent experiments in a simplified two-phase chemical system comprising water and chlorobenzene clearly demonstrate the ability of diaryl tellurides to react directly with free-radical species in a lipid-like phase, with the formation of tellurium (IV) dihydroxide intermediates, which are then rapidly re-reduced by thiols, such as N-acetylcysteine, present in the aqueous

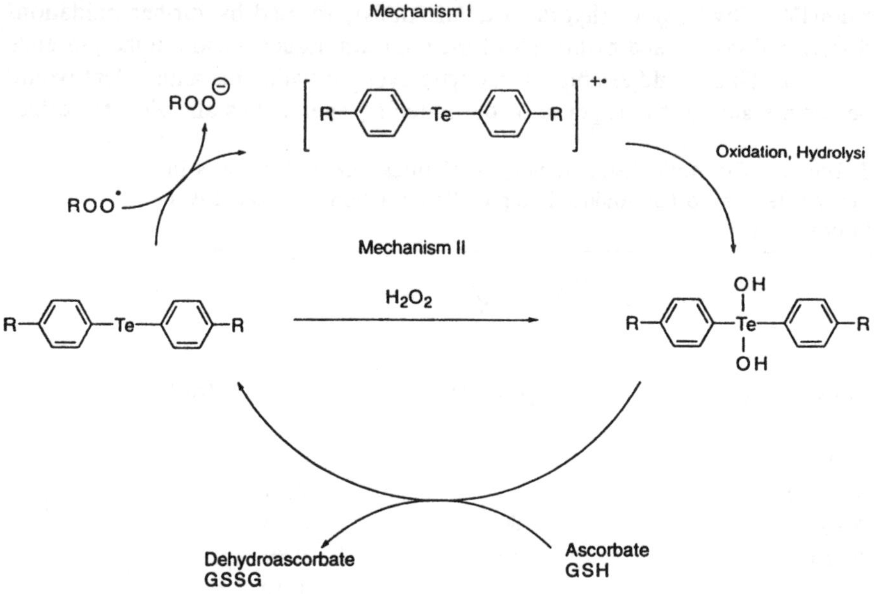

Figure 3 The proposed catalytic cycles for the antioxidant (mechanism I) and gluta-thione peroxidase-like (mechanism II) action of diaryl tellurides.

phase of the model. Further, diaryl telluroxides possess similar potency to their parent tellurides in this regenerating system (11). Here, it is important to state that the more water-soluble nature of these oxidized intermediates actually facilitates the access to reducing agents.

It is interesting to note that ebselen, which is reported to be a chain-breaking antioxidant (13,14), possessed poor activity in chemical models of lipid peroxidation. This in accordance with Niki et al., though, who have suggested it to be a poor antioxidant in biological systems (15). In fact, this is not surprising in view of its high oxidation potential (1.52 V) (12). However, vitamin E was shown to possess greater intrinsic antioxidant capacity than the tellurides in this system. However, in the presence of excess thiol in the aqueous phase, on a molecule for molecule basis, the overall antioxidant effect of the organotellurium compounds surpassed that of the naturally occurring compound.

The appreciation of the subtle nature of the recycling antioxidant mechanism of diaryl tellurides illustrates a dimension of the rational search for effective xenobiotic antioxidants which is not generally considered. Thus, appreciation of redox events after the initial donation of an electron can lead to the development of catalytic compounds, lowering the effective biological dose at sites of free-radical generation. Indeed, this is how nature has solved the problem of controlling lipid peroxidation events by the cyclic interaction of vitamin E radical in the membrane milieux with aqueous reducing agents such as ascorbate and GSH.

III. GLUTATHIONE PEROXIDASE MIMETICS

The reduction of hydroperoxides by GSH represents one of the most important antioxidative mechanisms in biological systems. Indeed, evolution has equipped both the intra- and extracellular milieux with enzymes, GSH peroxidases, which catalyze this two-electron transfer process. A number of GSH peroxidases utilize the redox chemistry of selenium in the form of selenocysteine in the active site, to perform the electron transfer. The catalytic cycle of the selenium-dependent GSH peroxidase contains a number of short-lived, oxidized selenium intermediates and it is considered that hydrogen peroxide is introduced via a reaction with the free selenol of the enzyme (16).

The search for xenobiotic molecules capable of catalyzing GSH peroxidase-like reactions was spurred by the discovery of the activity of the benzisoselenazolone ebselen (13). Despite considerable research into the chemical properties and biological effects of ebselen (17,18, and Chap. 9 of this volume), its mechanism of catalysis remained uncertain until the definitive identification of a highly reactive selenol intermediate produced during ebselen's reaction with thiols such as GSH (19). This indicated similarity in the catalytic mechanism between the native enzyme and this low-molecular-weight catalyst. Careful kinetic studies on the reactivities of various proported intermediates in the catalytic cycle of ebselen revealed the complexity of ebselen's interaction with thiols and peroxides (Fig. 4).

Several important facts are evident from the above cycle. Thus, ebselen itself possesses no reactivity with peroxides and initial reactions with excess thiol are necessary in order to generate the reactive selenol. Additionally, both ebselen selenol and ebselen diselenide possess reactivity with peroxide, but the rate of reaction of H_2O_2 with the former is far superior ($K = 2.8$ mM^{-1} min^{-1} versus 0.32 mM^{-1} min^{-1}). This, coupled to an appropriate steady-state concentration of the selenol in the presence of excess thiol indicates that the selenol form of ebselen carries the majority of the catalytic activity of the compound, despite it being present at lower concentrations than the diselenide (20).

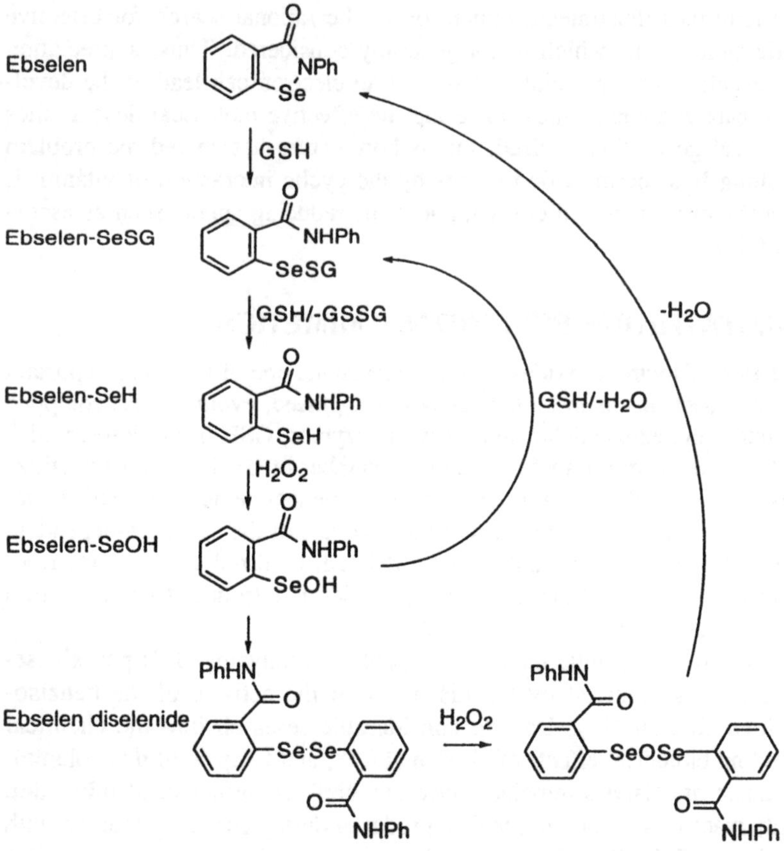

Figure 4 The mechanism of the glutathione peroxidase-mimetic action of ebselen.

The central role of a reactive selenol in the mechanism of action of both the native enzyme and ebselen stimulated research to produce other structures utilizing selenol-dependent catalytic mechanisms which might have improved catalytic efficiency over ebselen. Indeed, ebselen itself, although a catalyst, exhibits poor catalytic efficiency per mole selenium atom in comparison with the native enzyme. At the same time, it was envisaged that the intrinsic instability and reactivity of organic selenols may present several pharmaceutical, pharmacological, and toxicological problems during the development of drug candidates. Thus, substances catalyzing GSH peroxidase-like reactions in the absence of involvement of selenol intermediates were desirable. Here, we will detail one example of each of these classes of catalysts.

Table 4 indicates that α-(phenylselenenyl) acetophenone is a more efficient GSH peroxidase-like catalyst than is ebselen. Additionally, para-substitution at the acetophenone ring had considerable effect on the catalytic efficiency of the parent molecule. Thus, substitution with an electron-withdrawing nitro group greatly enhanced the activity of the molecule. This molecule now exhibits nearly three times the catalytic efficiency of ebselen. On the other hand, para-substitution with an electron-donating methoxy group almost abolished the activity of the compound down to the basal reaction rate (21).

These and other chemical studies in which scission products from the reaction of the parent molecule with thiols such as GSH were determined (22), have

Table 4 The Glutathione Peroxidase-Mimetic Activity of Substituted α-(Phenylselenenyl) Acetophenones

Substituent/ compound	% increase in the basal reaction between H_2O_2 and GSH [*]
R = H	790
R = NO$_2$	1207
R = OMe	192
Ebselen	538

[*]Assayed in the presence of GSSG reductase and NADPH as the rate of decrease in absorbance at 340 nm over the initial 1 min of the linear phase. The basal reaction of H_2O_2 with GSH was set at 100. Other experimental details and other derivatives are detailed in Ref. 21.

indicated that the mechanism of reaction of these α-(phenylselenenyl) ketones is similar to that of ebselen in that a central role for the catalysis is played by benzeneselenolate, which is generated from the parent molecule by thiol-dependent release of acetophenone, with the commensurate formation of S-(phenylselenenyl) glutathione. This then reacts with excess GSH to yield the selenol (Fig. 5). Thus, the α-(phenylselenenyl) acetophenones tested actually represent a group of procatalysts which release the active catalyst, a reactive selenol, upon reaction with thiol. The redox cycle then emanating from this selenol subsequently unifies the catalytic mechanism of these procatalysts with that of diphenyl diselenide(s), which is (are) also known to possess GSH peroxidase-mimetic activity (23,24).

From this mechanism it is also clear that the initial cleavage event induced by GSH is rate-limiting for the activity of these catalysts. Indeed, kinetic studies of the generation of benzeneselenolate from a variety of different α-(phenyl-selenenyl) ketones indicated that all catalysts reach similar catalytic performance with varying times of incubation of procatalyst with GSH. Thus, the electron-withdrawing (and enolate stabilizing) effect of the nitro substituent on the

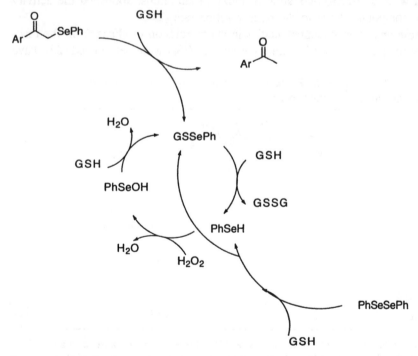

Figure 5   The common role of benzeneselenol in the mechanism of the glutathione peroxidase mimetic activity of α-(phenylselenenyl) ketones and diphenyl diselenide.

acetophenone (shown in Table 4) promotes nucleophilic attack by the thiol and subsequent bond cleavage, while the electron-donating methoxy group hinders the reactions, delaying the release of the active selenol (22).

In our efforts to study the mechanism of the catalytic antioxidant function of diaryl tellurides, we discovered a remarkable property of tellurium in these molecules. In addition to allowing one electron transfer from the tellurium, we demonstrated that certain of the diaryl tellurides tested actually allowed for direct reduction of peroxide by the transfer of two electrons (Table 5) (25,26). The resultant oxidized tellurium species, a hydrated telluroxide (dihydroxy telluride), then undergoes rapid reduction with thiols such as GSH to regenerate the parent molecule. This cycle (Fig. 3, cycle II) thus represents a catalytic GSH peroxidase-like mechanism which does not involve a free tellurol intermediate (27). The mechanism is very similar to the one proposed for the catalytic antioxidant action of these tellurides (Fig. 3, cycle I) except that the dihydroxy tellurium intermediate is reached directly upon reaction of the telluride with peroxide. Support for this simple mechanism comes from the demonstration that the telluroxide metabolites of some of the tellurides studied have very similar catalytic efficiencies to their parent molecules (Table 5). In analogy with

Table 5 The Glutathione Peroxidase-Mimetic Activities of Diaryl Tellurides (I). Diaryl Telluroxides (II), and bis(4-aminophenyl) Selenide (III)

I II III

Compound class /substituent	% increase in the basal reaction between H_2O_2 and GSH [*]
I	
R = H	190
R = CF$_3$	126
R = OMe	319
R = OH	1400
R = NH$_2$	2003
I I	
R = OMe	394
R = CF$_3$	100
I I I	
parent	100

[*]All details as in Table 4.

the antioxidant capacity of these structures, electron-donating para-substituents on the aromatic rings also promotes the GSH peroxidase mimetic activity of the tellurides, by facilitating two-electron oxidation of the heteroatom. Additionally, it will be noted that diaryl selenides, exemplified by bis(4-aminophenyl) selenide, are totally devoid of such GSH peroxidase-mimetic activity, indicating that the selenium heteroatom reacts only very sluggishly with hydrogen peroxide, even in the presence of electron-donating substituents (Table 5).

Thus, the diaryl telluride compounds described above represent a unique series of compounds which are able to interfere with lipid peroxidative process both by acting as a catalytic, chain-breaking antioxidant and as a catalytic GSH peroxidase-mimetic. In this latter case it will be noted that the catalytic efficiency of the diaryl tellurides increased with increasing lipophilicity of the peroxide substrate. Thus, with cumene and linoleic acid hydroperoxides, bis(4-aminophenyl) telluride catalyzed the reduction of the peroxides with GSH 82- and 75-fold, respectively. This was shown to be nearly one order of magnitude higher than the corresponding catalysis demonstrated by ebselen (25).

IV. SUMMARY AND CONCLUDING REMARKS

The redox-active compounds described in this chapter represent attempts to "captivate" the intrinsic chemical reactivities of selenium and tellurium and to direct these against the destruction of free radicals and peroxides generated in biological systems, particularly during lipid peroxidation. During their development, the emphasis has been on the use of sound chemical reasoning, extensive structure-activity studies, and attempts to define the mechanism of action of the compounds in the biological setting. This has resulted in the development of antioxidants which, as well as the well-established GSH peroxidase-mimetic action of the agents, also undergo cyclic interactions with endogenous reducing agents. Indeed, the unique chemical properties of tellurium impart both catalytic properties to the same molecule. It is hoped that the pharmacodynamic properties of these compounds will pave the way to the development of more effective medicinal agents for the treatment of human pathologies associated with the development and persistence of oxidative stress. Our extensive knowledge of the behavior of these compounds in the biological setting is already, however, providing unique opportunities for their use as experimental tools. For example, it was recently shown that the GSH peroxidase mimetics ebselen, α-(phenylselenenyl) acetophenone, and bis(4-aminophenyl) telluride, as well as several antioxidant diaryl selenides, all inhibit phorbol ester–induced down-regulation of gap junctional intercellular communication between epithelial cells in vitro. It has long been proported that this in vitro tumor promotive effect of phorbol esters relies on the induction of oxidative stress in the cells. The use of these compounds clearly demonstrates that both peroxides and free radicals are generated in the cells in response to phorbol esters and that both partake in the mechanism of the cellular response to phorbol esters (28).

ACKNOWLEDGMENTS

The authors acknowledge financial support from the Swedish Medical (MFR) and Natural Sciences (NFR) Research Councils.

REFERENCES

1. Gutteridge JMC. Free radicals and disease processes: A compilation of cause and consequence. Free Rad Res Comm 1993; 19: 141–158.
2. Engman L, Lind J, Merényi G. Redox properties of diaryl chalcogenides and their oxides. J Phys Chem 1994; 98: 3174–3182.
3. Cotgreave IA, Weiss M, Atzori L, Moldéus P. Glutathione and protein function. In: Vina J, ed. Glutathione: Metabolism and Physiological Functions. Boca Raton, FL: CRC Press, 1990: 155–175.
4. Schuppe-Koistinen I, Gerdes R, Moldéus P, Cotgreave IA. Studies on the reversibility of S-thiolation in human endothelial cells. Arch Biochem Biophys 1994; 315: 226–234.
5. Cotgreave IA, Sandy MS, Berggren M, Moldéus P, Smith MT. N-Acetylcysteine and glutathione-dependent protective effects of PZ 51 (Ebselen) against diquat-induced cytotoxicity in isolated hepatocytes. Biochem Pharmacol 1987; 36: 2899–2904.
6. Cotgreave IA, Johansson U, Westergren G, Moldéus P, Brattsand R. The antiinflammatory activity of ebselen but not thiols in experimental alveolitis and bronchiolitis. Agents Actions 1988; 27: 313–319.
7. Cotgreave IA, Duddy S, Kass GEN, Thompson D, Moldéus P. Studies on the antiinflammatory activity of ebselen. Ebselen interferes with granulocyte oxidative burst by dual inhibition of NADPH oxidase and protein kinase C. Biochem Pharmacol 1989; 38: 649–656.
8. Andersson CM, Hallberg A, Linden M, Brattsand R, Moldéus P, Cotgreave IA. Antioxidant activity of some diarylselenides in biological systems. Free Rad Biol Med 1994; 16: 17–28.
9. Cotgreave IA, Moldéus P, Engman L, Hallberg A. The correlation of the oxidation potentials of structurally-related dibenzo(1,4)dichalcogenines to their antioxidance capacity in biological systems undergoing free radical-induced lipid peroxidation. Biochem Pharmacol 1991; 42: 1481–1485.
10. Andersson CM, Brattsand R, Hallberg A, Engman L, Persson J, Moldéus P, Cotgreave IA. Diaryl tellurides as inhibitors of lipid peroxidation in biological and chemical systems. Free Rad Res 1994; 20: 401–410.
11. Vessman K, Ekström M, Berglund M, Andersson CM, Engman L. Catalytic antioxidant activity of diaryl tellurides in a two-phase lipid peroxidation model. J Org Chem 1995; 60: 4461–4467.

12. Engman L, Persson J, Vessman K, Ekström M, Berglund M, Andersson CM. Organotellurium compounds as efficient retarders of lipid peroxidation in methanol. Free Rad Biol Med 1995; 19: 441–452.

13. Müller A, Cadenas E, Graf P, Sies H. A novel biologically active seleno-organic compound I. Glutathione peroxidase-like activity in vitro and antioxidant capacity of PZ 51 (Ebselen). Biochem Pharmacol 1984; 33: 3235–3239.

14. Hayashi M, Slater T. Inhibitory effets of ebselen on lipid peroxidation in isolated rat liver microsomes. Free Rad Res Comm 1986; 2: 179–185.

15. Noguchi N, Yoshida Y, Kaneda H, Yamamoto Y, Niki E. Biochem Pharmacol 1992; 44: 39–44.

16. Flohé L, Loschen G, Günzler WA, Eichele E. Glutathione peroxidase V. The kinetic mechanism. Z Physiol Chem 1972; 353: 987–992.

17. Fischer H, Dereu N. Mechanism of the catalytic reduction of hydroperoxides by Ebselen: a selenium-77 NMR study. Bull Soc Chim Belg 1987; 96: 757–768.

18. Haenen GRMM, De Rooij BM, Vermeulen NPE, Bast A. Mechanism of the reaction of Ebselen with endogenous thiols: Dihydrolipoate is a better cofactor than glutathione in the peroxidase activity of Ebselen. Mol Pharmacol 1990; 37: 412–422.

19. Cotgreave IA, Morgenstern R, Engman L, Ahokas J. Characterisation and quantitation of a selenol intermediate in the reaction of ebselen with thiols. Chem Biol Interact 84: 69–76.

20. Morgenstern R, Cotgreave IA, Engman L. Determination of the relative contributions of the diselenide and selenol forms of ebselen in the mechanism of its glutathione peroxidase-like activity. Chem Biol Interact 1992; 84: 77–84.

21. Cotgreave IA, Moldéus P, Brattsand R, Hallberg A, Andersson CM, Engman L. α-(Phenylselenenyl) acetophenone derivatives with glutathione peroxidase-like activity. Biochem Pharmacol 1992; 43: 793–802.

22. Engman L, Andersson C, Morgenstern R, Cotgreave IA, Andersson CM, Hallberg A. Evidence for a common selenolate intermediate in the glutathione peroxidase-like catalysis of α-(phenylselenenyl) ketones and diphenyl selenides. Tetrahedron 1994; 50: 2929–2938.

23. Wilson SR, Zucker PA, Huang R-RC, Spector A. Development of synthetic compounds with glutathione peroxidase activity. J Am Chem Soc 1989; 111: 5936–5939.

24. Iwaoka M, Tomoda S. A model study on the effect of an amino group on the antioxidant activity of glutathione peroxidase. J Am Chem Soc 1994; 116: 2557–2561.

25. Andersson CM, Hallberg A, Brattsand R, Cotgreave IA, Engman L, Persson J. Glutathione peroxidase-like activity of diaryl tellurides. Bioorg Med Chem Lett 1994; 3: 2553–2558.

26. Engman L, Stern D, Pelcman M, Andersson CM. Thiol peroxidase-activity of diorganyl tellurides. J Org Chem 1994; 59: 1973.

27. Engman L, Stern D, Cotgreave IA, Andersson CM. Thiol peroxidase activity of diaryltellurides as determined by a ^{1}H NMR method. J Am Chem Soc 1992; 114: 9737–9743.

28. Hu J, Engman L, Cotgreave IA. Redox-active chalcogen-containing glutathione peroxidase-mimetics and antioxidants inhibit tumour promoter-induced downregulation of gap junctional intercellular communication between WB-F344 liver epithelial cells. Carcinogenesis 1995; 16: 1815–1824.

11

Modulation of Glutathione

Mary E. Anderson
Cornell University Medical College, New York, New York

I. INTRODUCTION

Cells are continuously assaulted by a variety of endogenous and exogenous toxic compounds and various types of oxidative stress, including free radicals. Many diverse pathological conditions have been associated with oxidative stress, including aging, atherosclerosis, cancer, arthritis, cataract formation, stroke, myocardial infarction, and viral infections, such as AIDS and hepatitis (1–14). Cells have an arsenal of defense mechanisms, including antioxidants that protect against such stresses (Fig. 1). α-Tocopherol, ascorbic acid, and glutathione (L-γ-glutamyl-L-cysteinylglycine; GSH) are several of the natural cellular antioxidants, with glutathione (Fig. 2) being present in the highest concentration. As the major cellular antioxidant, glutathione's metabolism and modulation is important in the protection against oxidative stress. This chapter discusses the various methods for modulating cellular glutathione levels.

A general overview of glutathione metabolism is given below (2, 15–17).The γ-glutamyl cycle describes the five enzymatic reactions that are involved in the synthesis and degradation of glutathione (2,15–17). Glutathione is a cofactor

---

This chapter is dedicated to Alton Meister (1921–1955), a pioneer in glutathione, amino acid, and enzyme biochemistry.

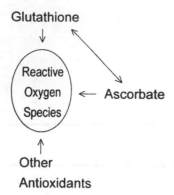

Figure 1 Overview of glutathione.

for several enzymes, for example, glyoxylase, maleylacetoacetate isomerase, formaldehyde dehydrogenase, prostaglandin endoperoxidase isomerase. Glutathione S transferases catalyze the formation of glutathione S-conjugates that, generally, lead to the detoxication of a variety of toxic compounds via the mercapturic acid pathway that forms *N*-acetyl cysteine conjugates that are excreted. Glutathione is oxidized to glutathione disulfide during the detoxication of hydrogen and organic peroxides (glutathione–dependent peroxidases). Glutathione disulfide is also produced by thiol transferases that are involved in the reduction of dehydroascorbate to ascorbate, the formation of deoxyribonucleotides and by protein disulfide isomerase.

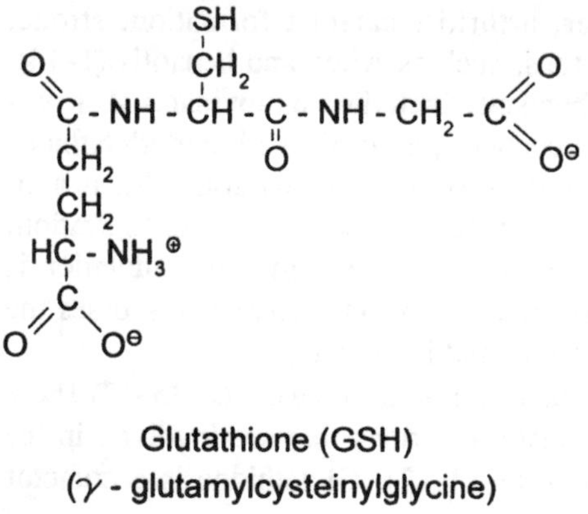

Glutathione (GSH)
(γ - glutamylcysteinylglycine)

Figure 2 Structure of glutathione.

II. BIOSYNTHESIS AND DEGRADATION OF GLUTATHIONE

Glutathione is synthesized intracellularly and exported from cells (2,15-17). Cellular glutathione is normally in the "reduced" form (>90%). Under conditions of severe stress, glutathione disulfide is formed and is also exported. Normally, glutathione disulfide is reduced to glutathione by glutathione disulfide reductase. The degradation of glutathione, glutathione disulfide, and glutathione S-conjugates is catalyzed by the externally facing membrane bound γ-glutamyl transpeptidase, the only enzyme known to catalyze the hydrolysis of the γ-glutamyl bond. One product, cysteinylglycine, may be cleaved by dipeptidase to cystine and glycine, which may be taken up by cells. The resulting γ-glutamyl enzyme may accept water, leading to the release of glutamate (hydrolysis). When the acceptors are amino acids, a γ-glutamyl amino acid is formed. Certain cells, such as those of the kidney, transport γ-glutamyl amino acids. Many amino acids serve as acceptors of the γ-glutamyl group, with cystine and methionine being among the best. After transport, the γ-glutamyl amino acid is hydrolyzed to amino acid and 5-oxoproline, a cyclic form of glutamate. When cystine is the acceptor, γ-glutamylcystine is formed and may be transported into certain cells. It may be reduced to γ-glutamylcysteine and be used directly for glutathione biosynthesis, thus bypassing the feedback-inhibited γ-glutamylcysteine synthetase (17-18). This pathway has been termed the alternative or salvage pathway of glutathione biosynthesis.

Glutathione is synthesized intracellularly by the consecutive actions of γ-glutamylcysteine (1) and glutathione (2) synthetases. Both reactions consume ATP and proceed through enzyme-bound acyl phosphate intermediates.

$$glu + cys + ATP \Leftrightarrow \gamma\text{-glu-cys} + ADP + P_i \qquad (1)$$

$$\gamma\text{-glu-cys} + gly + ATP \Leftrightarrow GSH + ADP + P_i \qquad (2)$$

The product, glutathione, feedback inhibits the first enzyme, γ-glutamylcysteine synthetase (K_i) ~ 1.5 mM) (19-21). Thus, intracellularly, this enzyme is not likely to function at a maximum rate. The rate of glutathione synthesis is also affected by the availability of substrates and the levels of the enzymes.

III. DECREASED LEVELS OF GLUTATHIONE

A. Inborn Metabolic Deficiencies of Glutathione-Related Enzymes

Several inborn deficiencies of enzymes involved in glutathione metabolism (22) have been described, such glutathione disulfide reductase deficiency which is associated with increased susceptibility to erythrocyte hemolysis. Less frequently observed is a deficiency of γ-glutamyl transpeptidase which is characterized by glutathione and γ-glutamylcyst(e)ine in both urine and plasma; similar findings were also observed in rodents treated with inhibitors of transpeptidase (23). Deficiency of γ-glutamylcysteine synthetase has not been frequently seen. Patients with the deficiency ($<3\%$ of normal enzyme activity) have hemolytic anemia, various neurological symptoms and amino aciduria (22). There are two types of glutathione synthetase deficiency, mild and severe, also called 5-oxo-prolinuria. In the severe form, glutathione synthetase and glutathione levels are markedly decreased in erythrocytes and fibroblasts (22,24). These patients excrete gram quantities of 5-oxoproline because of a futile cycle. γ-Gluta-mylcysteine synthetase is not feedback inhibited in the absence of glutathione; the excess formation of γ-glutamylcysteine, which is a substrate for γ-gluta-mylcyclotransferase, leads to the overproduction of 5-oxoproline. These patients are acidotic, have increased rates of hemolysis, have central nervous system symptoms, and have decreased formation of cysteinyl leukotrienes (22,25). Interestingly, treatment of patients with antioxidants seems to be beneficial (26), perhaps because of their "sparing" effect on glutathione (see below). The milder form of glutathione synthetase deficiency (22) seems to produce symptoms only in erythrocytes and is not characterized by 5-oxoprolinuria. In the milder form, the enzyme seems to be less stable, but this is apparently compensated for in cells with nuclei by new enzyme synthesis.

B. Experimental Glutathione Deficiency

To study glutathione metabolism and its role in protection against oxidative stress, it was desirable to study the effects of glutathione depletion. Glutathione levels may be reduced by treatment with oxidizing compounds, radiation, peroxides, or compounds that form glutathione-S-conjugates. The use of such nonspecific oxidizing agents, for example diamide, hydrogen peroxide, or t-butyl hydroperoxide, also leads to increased levels of glutathione disulfide and probably other disulfides (2). Compounds that form S-conjugates, such as diethylmaleate and phorone, also have other metabolic effects. These nonspecific agents, besides depleting glutathione, probably also cause the depletion and/or oxidation of other thiols.

Specific inhibition of glutathione biosynthesis was achieved with amino acid sulfoximines. (Inhibition of glutathione synthetase is not desirable for the propose of depleting cellular glutathione levels because, as discussed above, this would produce 5-oxoprolinuria.) Methionine sulfoximine (S-n-methyl homocysteine sulfoximine) administration to rodents produces convulsions; it inhibits both γ-glutamylcysteine synthetase and glutamine synthetase. These enzymes catalyze similar reactions via γ-glutamyl phosphate intermediates, but they have different acceptors (ammonia and cysteine, respectively). Selective inhibition of γ-glutamylcysteine synthetase was achieved by increasing the alkyl chain from methyl to propyl and butyl, prothionine sulfoximine, and buthionine sulfoximine (BSO), respectively (27,28). BSO resembles the presumed tetrahedral intermediate, γ-glutamyl phosphate. BSO is phosphorylated and binds tightly to the enzyme. When BSO is administered to rodents or applied to cultured cells, glutathione levels decline substantially because glutathione transport out of cells continues in the absence of synthesis.

Treatment with BSO leads to increase the toxicity of cadmium ions (29), mercury ions (30), cisplatin (31), radiation, and other compounds, and decreases lymphocyte activation (32). BSO also reverses certain types of tumor resistance to anticancer agents (33,91), and it is in clinical trial for combination anticancer therapy (34,35).

In a series of elegant studies (36–39), Meister and colleagues showed that experimental glutathione deficiency produced by long-term BSO treatment of rodents leads to severe mitochondrial damage in adult rodent muscle, lung, intestine, and brain. The finding that treatment of pregnant rats with BSO led to cataracts in offspring (40) led Meister to develop a model of endogenously produced oxidative stress in newborn rodents and guinea pigs (37,38). BSO does not cross the blood brain barrier, and thus its administration cannot produce cataracts in adult rodents. In these studies, BSO-induced glutathione deficiency also led to decreased tissue levels of ascorbate. Adult rodents can synthesize ascorbate, but newborn rodents and guinea pigs, like humans, are not able to synthesize it. BSO administration to animals that cannot synthesize ascorbate leads to severe mitochondrial damage and death. Treatment of guinea pigs with ascorbate prevented the lethal effects of glutathione deficiency, a "sparing" effect (37–41). Deficiency of ascorbate leads to scurvy and eventually death in guinea pigs; such effects are prevented by administration of glutathione monoester, which increases cellular glutathione levels (see below). Glutathione monoester or ascorbate, but not glutathione, administration prevents the formation of cataracts and other cellular effects of glutathione deficiency. This series of studies showed that, while glutathione and ascorbate have different functions, their metabolism is intimately connected.

IV. INCREASING CELLULAR GLUTATHIONE LEVELS

Increasing cellular glutathione levels may be useful in the treatment of various diseases. Several approaches, such as the administration of substrates, especially of the essential amino acid cysteine, may increase cellular glutathione levels. This approach may be particularly useful for patients on total parental nutrition whose transsulfuration pathway may not be fully functional, which may occur in neonates and certain surgical patients. It may also be useful for patients with high levels of oxidative stress where the use of glutathione may exceed the capacity for its synthesis.

Initially it might be thought that administration of glutathione might be effective in increasing cellular glutathione. However, while glutathione is exported from most cells, little, if any, intact glutathione is taken up by cells. Several reviews (2,15,17) have discussed this in detail. Briefly, when glutathione is administered to rodents after glutathione synthesis is inhibited by BSO, there is little or no change in cellular glutathione levels, even though plasma glutathione levels are extremely high (about 20 mM; normal rodent plasma is about 20 μM). Also, glutathione does not protect against the endogenous oxidative stress or cataract formation, as described above. That glutathione levels increase after administration without BSO treatment is attributed to its degradation and its resynthesis. Thus, glutathione administration may be thought of as a method to deliver cysteine.

A. Administration of Substrates

Cellular levels of glutamate and glycine are not normally limiting in glutathione biosynthesis. However, cellular levels of cysteine are substantially lower, about one-tenth that of glutathione. Thus, cysteine may be the limiting substrate for glutathione biosynthesis especially when there is oxidative stress. Administration of cysteine is not practical because it is readily oxidized. It is also reported to be toxic when given to rodents (42–44); however, the mechanism of toxicity is unclear. N-acetyl cysteine treatment may lead to increased cysteine levels, and thus to increased glutathione levels after it is deacetylated; it is effective in protecting against acetaminophen toxicity. Cysteine prodrugs, such as thiazolidine-4-carboxylate and 2-methyl thiazolidine-4-carboxylate, which lack a free thiol have been described (45–46); however, while forming cysteine, the former is reported toxic and the latter also yields the undesirable side product, acetaldehyde.

Another thiazolidine, L-2-oxothiazolidine-4-carboxylate (OTC), was developed from substrate specificity studies on 5-oxoprolinase (47–50). D-OTC is neither a substrate nor an inhibitor of 5-oxoprolinase. Studies using [^{35}S]OTC, showed that it is readily transported into most tissues (51). It is converted intracellularly by 5-oxoprolinase into cysteine (Fig. 3). Most tissues, except the lens and erythrocytes have 5-oxoprolinase activity. Administration of OTC raises cellular cysteine and/or glutathione levels in most tissues, and it is apparently nontoxic in rodents and humans. OTC has been shown to be more effective than N-acetyl cysteine in protecting against acetaminophen toxicity (49). OTC also promotes rat growth and increases organ glutathione levels in rats fed a sulfur amino acid–deficient diet (59). When OTC was administered

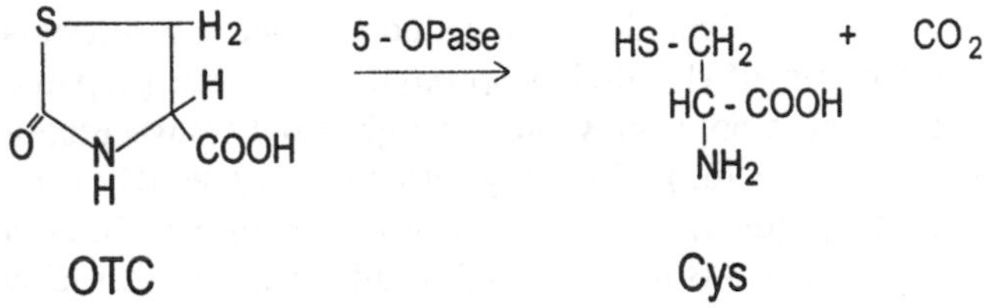

Figure 3 Metabolism of L-2-oxothiazolidine-4-carboxylic acid.

to healthy volunteers, lymphocyte glutathione and cysteine levels increased (60). OTC may be useful for peritoneal dialysis (52). It has been reported to protect Chinese hamster ovary cells from oxygen free-radical stress (53) and to protect against ischemic damage in isolated, perfused rat hearts (54). Preliminary reports (55,56) suggest that OTC administration decreases AZT bone marrow hypoplasia in mice and increases the antiviral activity of AZT in cultured lymphocytes. OTC treatment of peripheral blood mononuclear cells is reported to inhibit HIV expression (57). A phase I/II (58) study in asymptomatic HIV patients showed that whole blood glutathione levels increased and β_2-microglobin levels decreased. Further studies on the therapeutic usefulness of OTC are suggested by the available data.

B. γ-Glutamylcyst(e)ine

Raising glutathione levels by use of cysteine prodrugs is limited because of the feedback inhibition by glutathione on the first enzyme, γ-glutamylcysteine synthetase. Thus, the next compounds examined were substrates for the second enzyme, glutathione synthetase, which is not subject to feedback inhibition. Administration of γ-glutamylcysteine, γ-glutamylcysteine disulfide or of γ-glutamylcystine to mice led to increased kidney glutathione levels (18). Recent studies showed that intercerebroventricular administration of γ-glutamylcysteine raised rat brain glutathione levels (61). Any cell with the ability to transport γ-glutamyl amino acids would be expected to transport γ-glutamylcyst(e)ine and thus increase cellular glutathione levels. γ-Glutamylcysteine ester also is effective in preventing ischemia reperfusion injury (62,63). Given the nephrotoxicity of some anticancer drugs such as cisplatin, treatment with γ-glutamylcyst(e)ine might decrease toxicity (18). Further studies on the effectiveness of these compounds in raising glutathione levels of other tissues is indicated.

C. Glutathione Monoesters

Since glutathione administration is not a useful method of increasing cellular glutathione levels, derivatives of glutathione were prepared that were transported and converted intracellularly into glutathione. Glutathione mono(glycyl)esters (Fig. 4), where the ester was methyl, ethyl, isopropyl, etc., were prepared (64,65). Glutathione monoester is not a substrate for γ-glutamyl transpeptidase or glutathione disulfide reductase (65). The ethyl ester is very widely used because the other hydrolysis product, ethanol, is relatively nontoxic. Studies (66) in which [^{35}S]monoester that was administered to mice showed that label was found in many tissues. More label was found in tissues after the administration of [^{35}S]monoester than with [^{35}S]glutathione. Glutathione monoester is transported into kidney, liver, spleen, pancreas, heart, skeletal muscle, and lung. In neonates, glutathione monoester is also transported into the lens and brain (2,37,38,40). Because of the blood brain barrier in adults, little glutathione monoester is transported into the brain, but it has been found in their cerebrospinal fluid (70). The monoesters are effective in raising glutathione levels when administered intraperitoneally or orally. Glutathione monoesters also are transported into erythrocyte, lymphoid cells, fibroblasts, and ovarian tumor cells (67–69).

Glutathione monoester administration, but not glutathione, protects against the toxicity of acetaminophen, cyclophosphamide, monocrotaline, BCNU (1,3-bis-(2-chloroethyl)-1-nitrosourea), cadmium ions, and mercury ions (2,42, and references therein). Glutathione monoester treatment, and to a lesser extent glutathione treatment, protects against cisplatin toxicity (31). That glutathione protects suggests that cisplatin exhibits both intracellular and extracellular toxicity. It has also been reported that in a fibrosarcoma model (91), the monoester protected normal tissues but not tumor from chemotherapeutic agents. The monoester also protects when added before, and partially protects when added after, radiation, suggesting that glutathione participates in repair (67). Glutathione monoester administration, but not glutathione, protects against endog-

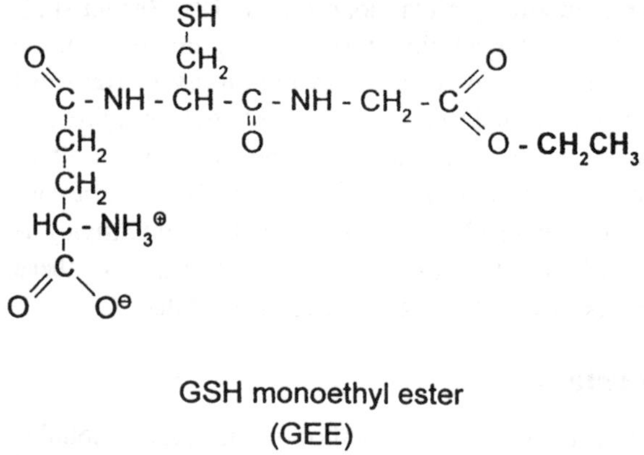

GSH monoethyl ester
(GEE)

Figure 4   Structure of glutathione monoester.

enous oxidative stress using the BSO model described above. Thus, treatment of neonates with glutathione monoester protected against BSO-induced cataract formation. Glutathione mono isopropyl ester is reported to protect against ischemic damage in rat brain (71). In studies on chronically infected monocytes, glutathione, N-acetyl cysteine, and glutathione monoester have been reported to decrease HIV replication (68).

Although monoesters are relatively easy to prepare, there have been occasional reports of toxicity (72–74), which we have attributed to various impurities especially metal ions (75,76). At extremely high doses, some toxicity due to alcohol intoxication might be observed.

D. Glutathione Diesters

We initially attributed occasional toxicity of glutathione monoester hydrochlorides to the slight contamination with glutathione diester (1–15%) (75–77); however, we found that preparations with more contaminating diester consistently gave slightly higher thiol levels in kidney and in erythrocyte. [The toxicity was later shown to by caused by trace metal ion contamination (75–76).] We synthesized pure diester, where both the α-glutamyl and the glycyl carboxyls of glutathione are esterified (Fig. 5), and found that it was not toxic and that it was effectively transported into erythrocytes, lymphocytes, and ovarian tumor cells. When erythrocytes were treated with glutathione diester, there were rapid increases in diester and monoester levels. When erythrocytes were loaded with diester, washed, and the medium examined, the diester was the major species found. These experiments suggest (Fig. 6) that the diester is rapidly taken up and rapidly exported. However, once inside a cell, it is rapidly converted into monoester which is not exported as rapidly as the diester. The monoester is then more slowly converted into glutathione. Glutathione diester

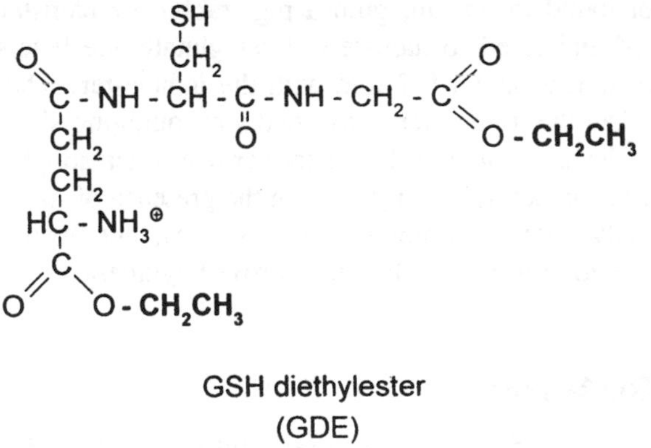

GSH diethylester
(GDE)

Figure 5 Structure of glutathione diester.

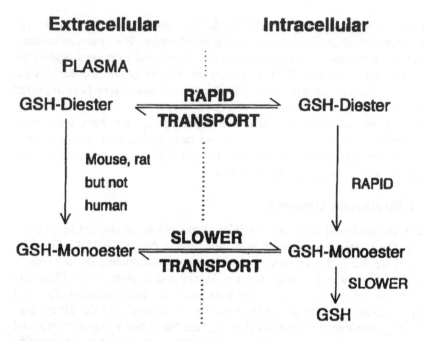

Figure 6   GSH monoester and diester transport.

is thus a "slow release" glutathione delivery agent or a "monoester delivery" agent.

Our initial studies (75) in which the diester was administered to mice showed the diester to be about as effective as the monoester. The plasma of mice and rats contains a diesterase that splits the diester into the monoester. Fortunately, this diesterase activity is not found in human, guinea pig, rabbit, or hamster plasma. When diester was administered to hamsters, liver glutathione levels increased about 3-fold, compared to about 0.7-fold with the monoester. That the diester is more effective than the monoester is especially encouraging since the diester releases two molecules of alcohol to the monoester one. Interestingly, the diester, unlike the monoester, is not apparently toxic in the presence of metal ions. We have seen no toxicity with the limited in vivo studies, but further studies are needed and will be greatly helped by an improved synthetic route to the diester.

V.   ENZYMATIC APPROACHES

It has long been known that some radiation or drug-resistant tumors have increased levels of glutathione. However, the levels of glutathione do not always correlate with the level of resistance. In studies using *E. coli* that contained expression vectors for the bacterial genes for both γ-glutamylcysteine and glutathione synthetases, we found this mutant had higher glutathione levels and was more radiation resistant than the wild-type strain (78). When the mutant was treated with BSO, glutathione levels remained the same, but now the mutant cells were sensitized to radiation. This study suggested that the capacity to synthesize glutathione is more important than the absolute levels of glutathione. Studies (33) with a series of human ovarian tumor cells with varying degrees of resistance to cisplatin showed that while the resistant cells had increased glutathione levels, the increase in glutathione levels were not as great as the degree of resistance to cisplatin. Northern blots showed that the levels of mRNA for the heavy subunit of γ-glutamylcysteine synthetase were greatly increased in the resistant cells as compared to the sensitive cells. Thus, the capacity of cells to synthesize glutathione is important for cell survival in response to stress. The use of molecular approaches may provide longer-lasting and selective methods for modulating cellular glutathione levels.

Both the *E. coli* and rat kidney enzyme catalyze the same reaction, and they are both inhibited by BSO, but only the mammalian enzyme is inhibited by γ-methylene glutamate, S-sulfocysteine and cystamine (79). The bacterial enzyme consists of a single chain (Mr, ~100,000). The cDNA for while the rat kidney enzyme is a heterodimer (Mr, ~100,000). The cDNA for rat kidney (80) and human liver (81) have been cloned and sequenced for the heavy subunit of γ-glutamylcysteine synthetase, and they have a high degree of homology. A putative antioxidant response element has been reported in the 5'-flanking region of the human heavy subunit gene (84). The human gene has been assigned to chromosome 6 (82). The heavy subunit (Mr, 72614) has catalytic activity, while the light subunit (Mr, 30548) appears to be regulatory (21). After overexpression of the heavy subunit in *E. coli* and purification, the heavy subunit has much lower activity than the holoenzyme (20). Overexpression of both subunits, separately or together, leads to fully active enzyme (21). Both the recombinant and isolated heavy subunit have about the same affinity for cysteine and the nonthiol cysteine analog, α-aminobutyrate, as the holoenzyme, but they have a much lower affinity for glutamate, 18 versus 1.8 mM. The heavy subunit is much more sensitive to feedback inhibition by glutathione than is the holoenzyme, and unlike the holoenzyme it is inhibited by the nonthiol glutathione analog, ophthalmic acid (γ-glutamyl-α-aminobutyrylglycine). As expected based on the *E. coli* studies, overexpression of both the heavy and the light subunits in human cells resulted in increased γ-glutamylcysteine synthetase activity, glutathione levels, and resistance to melphalan (83), and supports the view that glutathione plays a role in certain types of resistance of tumor cells to chemotherapy or radiation therapy, and such an approach may be useful for combating oxidative stress.

Normally the activity of glutathione synthetase is not limiting because, little γ-glutamylcysteine is found in cells; however, as discussed above, there are inborn errors associated with this enzyme, and in overexpression systems it is likely that overexpression of both γ-glutamylcysteine and glutathione synthetase will be needed. We recently cloned the rat kidney enzyme (85); the human brain has also been cloned (86). The rat kidney enzyme is a homodimer (subunit Mr, 52,344). Although there is no significant homology between the mammalian and the bacterial enzymes, there is some similarity with the yeast (87) and putative frog sequences (88). There is a conserved region of 13 amino acids among the species, which in the bacterial enzyme is a loop structure that may be important for substrate binding (89–91). The availability of the sequences will be useful for prenatal diagnosis and possibly therapy involving the modulation of glutathione.

VI. DISCUSSION

There is much interest in the role of antioxidants, and especially for glutathione, in diseases associated with oxidative stress. Basic research in this area is greatly facilitated by the availability of methods for modulating glutathione levels, such as BSO, OTC, glutathione esters, and the sequences of the cDNAs for the enzymes involved in its synthesis. These tools should also be useful in therapy.

VII. ABBREVIATIONS

GSH Glutathione
GSSG Glutathione disulfide
GEE Glutathione monoethylester
GDE Glutathione Diethylester
BSO L-Buthionine-SR-sulfoximine
ROS Reactive oxygen species

ACKNOWLEDGMENT

This work was supported in part by a grant from the PHS (NIH grant AI 31804).

REFERENCES*

1. Packer L, Cadenas E, eds. Biothiols in Health and Disease. New York: Marcel Dekker, 1995.
2. Meister A. Glutathione deficiency produced by inhibition of its synthesis, and its

---

*Due to space limitations and the large number of papers on GSH and GSH delivery compounds, all references in the literature could not be included.

reversal; applications in research and therapy. Pharmacol Therapeut 1991; 51: 155-194.

3. Cutler RG, Packer L, Bertram J, Mori A, eds. Oxidative Stress and Aging. Boston: Birkhauser Verlag, 1995.

4. Fields JA, Keshavarzian A, Eiznhamer D, Frommel T, Winship D, Holmes EW. Low levels of blood and colonic glutathione in ulcerative colitis. Gastroenterology 1994; 106: A680.

5. Ginter E. Marginal vitamin C deficiency, lipid metabolism, and atherogenesis. Adv Lipid Res 1978; 16: 167-220.

6. Eck H-P, Gmuender H, Hartmann M, Petzoldt D, Daniel V, Droege W. Low concentrations of acid soluble thiol (cysteine) in the blood plasma of HIV-1 ingfected patients. Biol chem Hoppe Seyler 1989; 370: 101-108.

7. Martinez-Cayuela M. Oxygen free radicals and human disease. Biochimie 1995; 77: 147-161.

8. Pace GW, Leaf CD. The role of oxidative stress in HIV disease. Free Rad Biol Med 1995; 19: 523-528.

9. Pacht ER, Timerman AP, Lykens MG, Merola AJ. Deficiency of alveolar fluid glutathione in patients with sepsis and the adult respiratory distress syndrome. Chest 1991; 100: 1397-1403.

10. Sabeh F, Baxter CR, Norton SJ. Skin burn injury and oxidative stress in liver and lung tissues of rabbit models. Eur J Clin Chem Clin Biochem 1995; 33: 323-328.

11. Suarez M, Beloqui O, Ferrer JV, Fil B, Qian C, Garcia N, Civeira P, Prieto J. Glutathione depletion in chronic hepatitis C. Int Hepatol Commun 1993; 1: 215-221.

12. Willis GC. An experimental study of the intimal ground substance in atherosclerosis. Can Med Assoc J (1953); 69: 17-22.

13. Buhl R, Holroyd KJ, Mastrangeli A, Cantin AM, Jaffe HA, Wells FB, Saltini C, Crystal RG. Systemic glutathione deficiency in symptom-free HIV-seropositive individuals. Lancet 1989; ii: 1294-1298.

14. Cantin AM, Hubbard RC, Crystal RG. Glutathione deficiency in the epithelial lining fluid of the lower respiratory tract in idiopathic pulmonary fibrosis. Am Rev. Respir Dis 1989; 139: 370-372.

15. Meister A. (1989). Metabolism and funtion of glutathione. In: Dolphin D, Poulson R, Avramovic O, eds. Glutathione: Chemical, Biochemical and Medical Aspects. New York: Wiley, 1989: 367-474.

16. Meister A. Strategies for increasing cellular glutathione. In: Packer L, Cadenas E, eds. Biothiols in Health and Disease. New York: Marcel Dekker, 1995; 165-188.

17. Meister A. (1989). Metabolism and function of glutathione. In: Dolphin D. Poulson

18. Anderson ME, Meister A. Transport and direct utilization of γ-glutamylcyst(e)ine for glutathione synthesis. Proc Natl Acad Sci USA (1983); 80: 707-711.

19. Richman PG, Meister A. Regulation of γ-glutamylcysteine synthetase by nonallosteric feedback inhibition of glutathione. J Biol Chem (1973); 250: 1422-1426.

20. Huang C-S, Chang L-S, Anderson ME, Meister A. Catalytic and regulatory properties of the heavy subunit of rat kidney γ-glutamylcysteine synthetase. J Biol Chem 1993; 19675-19689.

21. Huang C-S, Anderson ME, Meister A. Amino acid sequence and function of the light subunit of rat kidney γ-glutamylcysteine synthetase. J Biol Chem 1993; 268: 20578–20583.

22. Meister A, Larsson A. Glutathione synthetase deficiency and other disorders of the γ-glutamyl cycle. In: Scriver CR, Beaudet AL, Sly WS, Valle D, eds. The Metabolic Basis of Inherited Disease. 7th ed. New York: McGraw-Hill, 1994; 1461–1477.

23. Griffith OW, Meister A. Translocation of intracellular glutathione to membrane-bound γ-glutamyl transpeptidase as a discrete step in the γ-glutamyl cycle; glutathionuria after inhibition of transpeptidase. Proc Natl Acad Sci USA 1979; 76: 268–272.

24. Wellner VP, Sekura R, Meister A, Larsson A. Glutathione synthetase deficiency. An inborn error of metabolism involving the γ-glutamyl cycle in patients with 5-oxoprolinuria I(pyroglutamic aciduria). Proc Natl Acad Sci USA 1974; 71: 2969–2972.

25. Mayatepek E, Becker K, Carlsson B, Larsson A, Hoffmann GF. Deficient synthesis of cysteinyl leukotrienes in glutathione synthetase deficiency. Int J Tissue Reactions 1993; 15: 245–252.

26. Jain A, Buist NR, Kennaway NG, Powell BR, Auld PA, Martensson J. Effect of ascorbate or N-acetylcysteine treatment in a patient with hereditary glutathione synthetase deficiency. J Pediat 1994; 124: 229–233.

27. Griffith OW, Anderson ME, Meister A. Inhibition of glutathione biosynthesis by prothionine (S-n-propyl homocysteine sulfoximine), a selective inhibitor of γ-glutamylcysteine synthetase. J Biol Chem 1979; 254: 1205–1210.

28. Griffith, OW, Meister A. Potent and specific inhibition of glutathione synthesis by buthionine sulfoximine (S-n-butyl homocysteine sulfoximine). J Biol Chem 1979; 254: 7558–7560.

29. Singhal RK, Anderson ME, Meister A. Glutathione, a first line of defense against cadmium toxicity. FASEB J 1987; 1: 220–223.

30. Naganuma A, Anderson ME, Meister A. Cellular glutathione is a determinant of sensitivity to mercuric chloride toxicity; prevention of toxicity by giving glutathione monoester. Biochem Pharmacol 1990; 40: 693–697.

31. Anderson ME, Naganuma A, Meister A. Protection against cisplatin toxicity by administration of glutathione ester. FASEB J 1990; 4: 3251–3255.

32. Suthanthiran M, Anderson ME, Sharma VK, Meister A. Glutathione regulates activation-dependent DNA synthesis in highly purified T lymphocytes stimulated via CD2 and CD3 antigens. Proc Natl Acad Sci USA 1990; 87: 3343–3347.

33. Godwin AK, Meister A, O'Dwyer PJ, Huang C-S, Hamilton TC, Anderson ME. High resistance to cisplatin in human ovarian cancer cell lines is associated with marked increase of glutathione synthesis. Proc Natl Acad Sci USA 1992; 89: 3070–3074.

34. Bailey HH, Mulcahy RT, Tutsch KD, Arzoomanian RZ, Alberti D, Tombes MB, Wilding G, Pomplun M, Spriggs DR. Phase I clinical trial of intravenous L-buthionine sulfoximine and melphalan: An attempt at modulation of glutathione. J Clin Oncol 1994; 12: 194–205.

35. Yao K, Godwin AK, Ozols RF, Hamilton TC, O'Dwyer PJ. Variable baseline γ-glutamylcysteine synthetase messinger RNA expression in peripheral mononuclear cells of cancer patients, and its induction by buthionine sulfoximine treatment. Cancer Res 1993; 53: 3662–3666.

36. Martensson J, Meister A. Mitochondrial damage in muscle occurs after marked depletion of glutathione and is prevented by giving glutathione monoester. Proc Natl Acad Sci USA 1989; 86: 471–475.

37. Martensson J, Jain A, Stole E, Frayer W, Auld PAM, Meister A. Inhibition of glutathione synthesis in the newborn rat: a model for endogenously produced oxidative stress. Proc Natl Acad Sci USA 1991a; 88: 9360–9364.

38. Martensson J, Meister A. Glutathione deficiency decreases tissue ascorbate levels in newborn rats: ascorbate spares glutathione and protects. Proc Natl Acad Sci USA 1991b; 88: 4656–4660.

39. Martensson J, Han J, Griffith OW, Meister A. Glutathione ester delays the onset of scurvy in ascorbate-deficient guinea pigs. Proc Natl Acad Sci USA 1993; 90: 317–321.

40. Calvin HL, Medvedovsky C, Worgul BV. Near-total glutathione depletion and age-specific cataracts induced by buthionine sulfoximine in mice. Science 1986; 233: 553–555.

41. Meister A. Glutathione-ascorbate acid antioxidant system in animals. J Biol Chem 1994; 269: 9397–9400.

42. Anderson ME. Glutathione and glutathione delivery compounds. Adv Pharmcol 1996; 38: 65–77.

43. Nishiuch Y, Sasaki M, Nakayasu M, Oikawa A. Cytotoxicity of cysteine in culture media. In Vitro 1976; 12: 635.

44. Olney JW, Ho O-L, Rhee V. Cytotoxic effect of acid and sulphur containing amino acids on the infant mouse central nervous system. Brain Res 1971; 14: 61–76.

45. Nagasawa HT, Goon DJW, Zerat RT, Yuzon DI. Prodrugs of L-cysteine as liver protective agents. 2(RS)-methylthiazolidine-4(R)-carboxylic acid, a latent cysteine. J Med Chem 1982; 25: 489–491.

46. Nagasawa HT, Goon DJW, Muldoon WP, Zera RT. 2-Substituted thiazolidine-4(R)-carboxylic acids as prodrugs of L-cysteine. Protection of mice against acetaminophen hepatoxicity. J Med Chem 1984; 27: 591–596.

47. Williamson JM, Meister A. Stimulation of hepatic glutathione formation by administration of L-2-oxothiazolidine-4-carboxylate, a 5-oxo-L-prolinase substrate. Proc Natl Acad Sci USA 1981; 78: 936–939.

48. Williamson JM, Meister A. New substrates of 5-oxo-L-prolinase. J Biol Chem 1982; 257: 12039–12042.

49. Williamson JM, Boettcher B, Meister A. Intracellular cysteine delivery system that protects against toxicity by promoting glutathione synthesis. Proc Natl Acad Sci USA 1982; 79: 6246–6249.

50. Anderson ME, Meister A. Intracellular delivery of cysteine. Meth Enzymol 1987; 143: 313–325.

51. Meister A, Anderson ME, Hwang O. Intracellular delivery of cysteine and glutathione delivery systems. J Am Coll Nutr 1986; 5: 137–151.

52. Breborowicz A, Witowski J, Martis L, Oreopoulos DG. Enhancement of viability of human peritoneal mesothelial cells with glutathione precursor: L-2-oxothiazolidine-4-carboxylate. Adv Peritoneal Dialysis 1993; 9: 21–24.

53. Weitberg AB. The effect of L-2-oxothiazolidine on glutathione levels in cultured mammalian cells. Mutation Res 1987; 191: 189–191.

54. Shug AL, Madsen DC. Protection of the ischemic rat heart by procysteine and amino acids. J Nutr Biochem 1994; 5: 3356–359.

55. Wilson DM, White RD, Webb LE, Bender JG, Pippin LI, Goldberg DI. Amelioration of AZT-induced bone marrow hypoplasia in mice cotreated with glutathione prodrug, Procysteine™. IXth Int. Conf. on AIDS June 7–11, 1993, Berlin, abstract.

56. Josephs S, Asuncion C, Sun C, Jacobson P, Webb L. Procysteine (L-2-oxothiazolidine-4-carboyxlic-acid) reduces the toxicity of AZT (zidovudine) and enhances the antiviral activity of AZT in cultured peripheral blood mononuclear cells (PBMC). IXth Intl. Conf. on AIDS (June 6–11, 1993 Berlin), 1, abstract 230.

57. Lederman MM, Georger D, Dando S, Schmelzer R, Averill L, Goldberg D. L-2-oxothiazolidine-4-carboxylic acid (Procysteine) inhibits expression of the human immunodeficiency virus and expression of the interleukin-2 receptor alpha chain. J Acq Immune Def Syndromes 1995; 8: 107–115.

58. Kalayjian RC, Skowron G, Emgushov R-T, Chance M, Spell SA, Borum PR, Webb LS, Mayer KH, Jackson JB, Yen-Liberman B, Story KO, Rowe WB, Thompson K, Goldberg D, Trimbo S, Lederman MM. A phase I/II trial of intravenous L-2-oxothiazolidine-4-carboxylic acid (Procysteine) in asymptomatic HIV infected subjects. J Acq Immune Def Syndromes 1994; 7: 369–374.

59. Jain A, Madsen DC, Auld PAM, Frayer WW, Schwartz MK, Meister A, Martensson J. L-2-oxothiazolidine-4-carboxylate, a cysteine precursor, stimulates growth and normalizes tissue glutathione concentrations in rats fed a sulfur amino acid-deficient diet. J Nutr 1995; 125: 851–6.

60. Porta P, Aebi S, Summer K, Lauterburg BH. L-2-oxothiazolidine-4-carboxylic acid, a cysteine prodrug: pharmacokinetics and effects on thiols in plasma and lymphocytes in human. J Pharmacol Exp Ther 1991; 257: 331–334.

61. Pileblad E, Magnusson T. Increase in rat brain glutathione following intracerebroventricular administration of γ-glutamylcysteine. Biochem Pharmacol 1992; 44: 895–903.

62. Hoshida S, Kuzuya T, Yamshita N, Nishida M, Kitahara S, Hori M, Kamada T, and Tada M. γ-Glutamylcysteine ethyl ester for myocardial protection in dogs during ischemia and reperfusion. J Am Coll Cardiol 1994; 24: 1391–1397.

63. Ozaki M, Ozasa H, Fuchinoue S, Teraoka S, Ota K. Protective effects of glycine and esterified γ-glutamylcysteine on ischemia/reoxygenation injury or the rat liver. Transplantation 1994; 58: 753–755.

64. Puri RN, Meister A. Transport of glutathione, as γ-glutamylcysteinylglycyl ester, into liver and kidney. Proc Natl Acad Sci USA 1983; 80: 5258–5260.

65. Anderson ME, Meister A. Glutathione monoesters. Anal Biochem 1989b; 183: 16–20.

66. Anderson ME, Powrie F, Puri RN, Meister A. Glutathione monoethyl ester: preparation, uptake by tissues, and conversion to glutathione. Arch Biochem Biophys 1985; 239: 538–548.

67. Wellner VP, Anderson ME, Puri RN, Jensen GL, Meister A. Radioprotection by glutathione ester: transport of glutathione ester into human lymphoid cells and fibroblasts. Proc Natl Acad Sci USA 1984; 81: 4732–4735.

68. Kalebic T, Kinter A, Poli G, Anderson ME, Meister A, Fauci AS. Suppression of HIV expression in chronically infected monocytic cells by glutathione, glutathione ester, and N-acetyl cysteine, Proc Natl Acad Sci USA 1991; 88: 896–990.

69. Anderson ME, Levy EJ, Meister A. Preparation and use of glutathione monoesters. Meth Enzymol 1994; 234: 492–499.

70. Anderson ME, Underwood M, Bridges RJ, Meister A. Glutathione metabolism at the blood-cerebrospinal fluid barrier. FASEB J 1989; 3: 2527–2531.

71. Gotoh O, Yamamoto M, Tamura A, Sano K. Effect of YM737, a new glutathione analog, on ischemic brain edema. Acta Neurochir 1994; 60 Suppl: 318–320.

72. Tsan M-F, White JE, Rosano CL. Modulation of endothelial GSH concentrations: effect of exogenous GSH and GSH monoethyl ester. J Appl Physiol 1989; 66: 1029–1034.

73. Vos O, Roos-Verhey WSD. Endogenous versus exogenous thiols in radioprotection. Pharmacol Ther 1988; 39: 169–177.

74. Scadutto RC Jr, Gattone VH II, Grotyohann LW, Wertz J, Martin LF. Effect of an altered glutathione content on renal ischemic injury. Am J Physiol 1988; 255: F911–F921.

75. Levy EJ, Anderson ME, Meister A. Transport of glutathione diethyl ester into human cells. Proc Natl Acad Sci USA 1993; 90: 9171–9175.

76. Levy EJ, Anderson ME, Meister A. Preparation and properties of glutathione diethyl ester and related derivatives. Meth Enzymol 1994; 234: 499–505.

77. Minhas H, Thornalley PJ. Comparison of the delivery of reduced glutathione into P388D$_1$ cells by reduced glutathione and its mono- and diethyl ester derivatives. Biochem Pharmacol 1995; 49: 1475–1482.

78. Moore WB, Anderson ME, Meister A, Murata K, Kimura A. Increased capacity for glutathione synthesis enhances resistance to radiation in *Escherichia coli*: A possible model for mammalian cell protection. Proc Natl Acad Sci USA 1989, 86: 1461–1464.

79. Huang C-S, Moore W, Meister A. On the active site thiol of γ-glutamylcysteine synthetase: relationships to catalysis, inhibition, and regulation. Proc Natl Acad Sci USA 1988; 85: 2464–2468.

80. Yan N, Meister A. Amino acid sequence of rat kidney γ-glutamylcysteine synthetase. J Biol Chem 1990; 265: 1588–1593.

81. Gipp JH, Chang C, Mulcahy RT. Cloning and nucleotide sequence of full-length cDNA for human liver γ-glutamylcysteine synthetase. Biochem Biophys Res Commun 1992; 185: 29–35.

82. Sierra-Rivera E, Summar ML, Dasouki M, Krishnamani MR, Phillips JA, Freeman ML. Assignment of the gene (GLCLC) encodes the heavy subunit of γ-glutamylcysteine synthetase to human chromosome 6. Cytogene Cell Gene 1995; 70: 278–279.

83. Mulcahy RT, Bailey HH, Gipp JJ. Transfection of cDNAs for the heavy and light subunits of human γ-glutamylcysteine synthetase results in an elevation of intracellular glutathione and resistance to melphalan. Cancer Res 1995; 55: 4771–4775.

84. Mulcahy RT, Gipp JJ. Identification of a putative antioxidant response element in the 5'-flanking region of the human γ-glutamylcysteine synthetase heavy subunit gene. Biochem Biophys Res Commun 1995; 209: 2270233.

85. Huang C-S, He W, Meister A, Anderson ME. Amino acid sequence of rat kidney glutathione synthetase. Proc Natl Acad Sci USA 1995; 92: 1232–1236.

86. Gali RR, Board PG. Sequencing and expression of a cDNA for human glutathione synthetase. Biochem J 1995; 310: 353–358.

87. Mutoh N, Nakagawa CW, Hayashi Y. Molecular cloning and nucleotide sequencing of the γ-glutamylcysteine synthetase gene of the fission yeast *Schizosaccharomyces pombe*. J Biochem 1995; 117: 283–288.

88. Habenicht A, Hille S, Knochel W. Molecular cloning of the large subunit of glutathione synthetase from *Xenopus laevis* embryos. Biochim Biophys Acta 1993; 1174: 295–298.

89. Yamaguchi H, Kato H, Hata Y, Nishioka T, Kimura A, Oda J, Katsube Y. J Three-dimensional structure of the glutathione synthetase from Escherichia coli B at 2.0Å resolution. Molecular Biol 1993; 229: 1083–1100.

90. Hibi T, Nishioka T, Kato H, Tanizawa K, Fukui T, Katsube Y, Oda J. Structure of the multifunctional loops in the nonclassical ATP-binding fold of glutathione synthetase. Nature Structural Biol 1996; 3: 16–18.

91. Teicher BA, Crawford JM, Holden SA, Lin Y, Cathcart KNS, Luchette CA, Flatow J. Glutathione monoethyl ester can selectively protect from high dose BCN or cyclophosphamide. Cancer 1988; 62: 1275–1281.

12

Nitrone Radical Traps as Protectors of Oxidative Damage in the Central Nervous System

Robert A. Floyd, Guang-Jun Liu, and Peter K. Wong
Oklahoma Medical Research Foundation, Oklahoma City, Oklahoma

I. INTRODUCTION AND BASIC IMPORTANT CONCEPTS

Several basic facts have been established that are important to understanding the protective role of nitrone radical traps (NRTs) in brain oxidative damage. These are listed in Table 1. These concepts have been reviewed by ourselves (1,2) as well as others and will be treated only in a summary fashion here.

The concept that aerobic biological systems experience an oxidative stress at all times simply because they are utilizing oxygen is becoming widely accepted. The notion that very small levels of reactive oxygen species (ROS) exists continuously in aerobic biological systems is valid but due to technical problems difficult to prove. Yet evidence of their existence as revealed by the presence of oxidized biological components have been documented extensively. Another approach to exploring the presence of ROS is to add exogenous traps which react with them to yield unique quantifiable products. It is in this quest for proof of endogenous biological ROS production (as well as the characterization of other endogenous free radicals) which initially prompted the use of NRTs (also known as spin traps) in biochemical and biological systems. It was later that their protective (pharmacological) action was realized.

Oxidative stress can be viewed as the dynamic balance between the oxidative damage potential as opposed to the antioxidant defense capacity. Oxidative

Table 1 Important Concepts in Brain Oxidative Damage

I.	Oxidative stress in biological systems involves the dynamic balance of the oxidative damage potential as opposed by the antioxidative defense capacity.
II.	Oxidative damage results when the antioxidant defense capacity is not capable of inhibiting the oxidative damage potential.
III.	Brain is very susceptible to oxidative damage.
IV.	An ischemia/reperfusion insult to brain causes oxidative damage.
V.	Older brain is more prone to oxidative damage than younger brain.

damage is the result of oxidative stress. It is rational to question the definition of oxidative damage, especially since, as noted, there is normally at all times a constant low level of ROS and consequently a low level of oxidatively damaged molecules present. With reference to the discussion here, we would consider significant oxidative damage as being that when there is an observable (quantifiable) biological effect and/or that where the concentration of an oxidized biological component (or exogenous trap added) under question is increased significantly higher than that observed under normal (control) conditions. It is clear that oxidative damage potential as well as the antioxidant defense capacity may vary dramatically between tissues, under various physiological conditions, and upon the nutritive status of the organism, which also may vary with time. Oxidative damage may be either severe or nonconsequential depending on the extent of the damage that has occurred and the biological consequences involved.

Brain is one of the most sensitive of tissues to oxidative damage (2). This apparently is because of its high level of easily peroxidizable (20:4, 22:6) membrane fatty acids, its high level of oxygen consumption per unit weight, its relatively high content of iron in certain regions, its relatively low level of protective antioxidants (enzymes as well as vitamin E), and the fact that neurons are postmitotic cells. Brain homogenate readily peroxidizes when it is exposed to catalytic amounts of Fe in combination with ascorbate at levels normally found in brain. Not only is isolated brain easily oxidized, but the organ in the intact animal is also sensitive to oxidative damage. We showed (3,4) that the brain undergoes oxidative damage when it experiences an ischemia/reperfusion insult (IRI). The facts demonstrating that brain experiencing an IRI undergoes oxidative damage are outlined in Table 2. Much of the observations are based upon the use of salicylate hydroxylation as an index of hydroxyl free-radical flux or upon protein oxidation as well as loss of glutamate synthetase (GS) activity. The activity of GS is very sensitive to oxidative damage.

It is clear the data convincingly demonstrate that brain undergoing an IRI experiences a large amount of oxidative damage. The validity of these obser-

Table 2   Ischemia/Reperfusion Insult (IRI) to Mongolian Gerbil Brain Causes Oxidative Damage

I.   Significant salicylate hydroxylation was observed only in brains where oxygen was allowed to reenter after ischemia (3).

II. The extent of salicylate hydroxylation was directly related to the extent of injury caused by the IRI (3,5).

III. The extent of salicylate hydroxylation was brain-region-specific and was in proportion to the degree of sensitivity of the region to an IRI (i.e., hippocampus was more extensive than cortex); no increase was observed in brain stem or cerebellum which is not subjected to an IRI by common carotid occlusion (5).

IV. Protein oxidation was proportional to the degree of IRI-mediated injury (5).

V.   Glutamine synthetase activity decreased in brain regions where IRI-mediated injury occurred (4).

VI. IRI in older gerbils is more lethal than in younger gerbils (1).

vations is underscored by the fact that the parameters used for assessment of oxidative damage (salicylate hydroxylation, protein oxidation, and glutamine synthetase activity loss) are changed only in those regions of the brain which experiences an ischemia/reperfusion insult, whereas this does not occur in the unaffected brain regions in the same animal.

Another important fact is that lethality caused by a brain IRI is significantly enhanced in older animals as compared to younger animals. The reason for this is not known, but, as briefly discussed later, we consider this due to the lack of ability of the brain mitochondria to meet the enhanced metabolic demands caused by an IRI (6).

II.   FREE-RADICAL TRAPPING BY NRTS

NRTs have been used in analytical chemistry to characterize free radicals since the mid-1960s (7). The use of NRTs to characterize trapped free radicals ideally is based on the electron paramagnetic spectrum of the product formed, which is then compared to synthesized authentic trapped products. The reaction of α-phenyl-t-butyl nitrone (PBN) with a free radical is shown in the following equation:

PBN	Free radical	NRT-radical adduct
	(reactive)	(stable)

Thus, the reaction of PBN with a *reactive* free radical to form a *stable* adduct which is still a free radical, but not very reactive, prevents the possible further reactions of the original free radical.

In general, free-radical reactions are characterized by a low or nearly zero activation energy; thus, they readily react with many different chemical species leading to a chain of reactions. Many free radicals in biological systems cause chain reactions leading to oxidative damage such as that occurring in lipid peroxide reactions where oxygen adds to the lipid radicals formed, thus leading to further lipid radicals, etc., in a chain propagation series of reactions as denoted in the following reactions:

$$\dot{R} + L_0 \Rightarrow \dot{L} + O_2 \Rightarrow L_0O\dot{O}$$

$$L_0O\dot{O} + L_1H \Rightarrow L_0OOH + \dot{L}_1 + O_2 \Rightarrow L_1O\dot{O}$$

$$L_1O\dot{O} + L_2H \Rightarrow L_1OOH + \dot{L}_2 + O_2 \Rightarrow L_2O\dot{O}$$

Thus if PBN were to react with $\dot{L}_0$ to form a stable adduct, for instance, then the sequential series of reactions would be prevented (Fig. 1). It is possible that NRTs are then preventing oxidative damage in brain by some sort of quenching reaction as that alluded to here. This remains to be proven.

III. NRTS PROTECT IN EXPERIMENTAL MODELS OF NEURODEGENERATION

We first showed that PBN prevented brain IRI-mediated lethality in gerbils. As Table 3 shows, this has been followed up with several observations from many different laboratories, clearly demonstrating that NRTs do provide protection in several experimental neurodegenerative models. The most extensive research has been reported in experimental models of stroke, where PBN has been shown to be protective if given before as well as after brain ischemia/reperfusion. This has been shown to be true in both the Mongolian gerbil brain (global stroke model) and the rat middle cerebral occlusion artery (MCAO) model, either the permanent occlusion (12) or the transient occlusion model (13). It is not known how NRTs protect in the brain in the stroke model, but the fact that glutamine synthetase activity is decreased implies that glutamate, which is a neurotoxin, levels may increase, causing excitotoxic neurotoxicity. Glutamate levels do significantly increase in IRI-lesioned brain (17), reinforcing this notion. Thus in the global stroke model, NRTs may protect by inhibiting the oxidative damage that would normally occur, resulting in decreased GS activity (Fig. 2).

It is important to note that older gerbils which have had chronic PBN administration for 2 weeks become significantly less susceptible to an IRI. It is

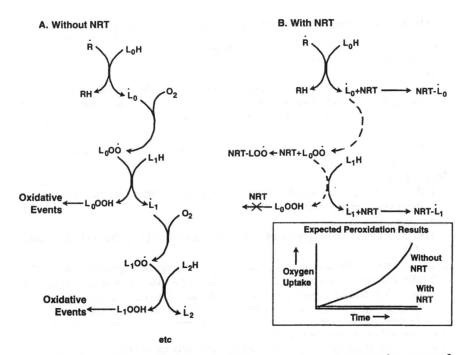

Figure 1 Schematic showing lipid peroxidation events in the absence and presence of enough NRT to trap lipid peroxidation free radicals.

remarkable that the decreased susceptibility to an IRI remains from 3 to 5 days after ceasing PBN administration. It should be noted that the PBN half-life in rats is 134 min (18). Thus, 3 to 5 days after ceasing PBN administration is time enough such that very little, if any, PBN remains in the animal. These observations clearly imply that NRTs are influencing processes and/or enzymatic systems that are altering the sensitivity of brain to oxidative damage. It is highly likely that functional processes associated with mitochondria are involved, but this has yet to be proven. One aspect of this notion is the idea that NRTs are influencing mitochondrial processes. This will be examined briefly and new data will be presented. It should be noted that reinforcing these ideas is the fact that NRTs have been shown to be protective in experimental models of mitochondrial dysfunction (16).

IV. NRT ACTION ON MITOCHONDRIA: EXPERIMENTAL FINDINGS

The question that arises from the striking demonstrations that NRTs protect against oxidative damage in brain is, do the NRTs protect by direct interaction

Table 3 Experimental Neurodegenerative Models Where NRTs Are Protective

Model	Observations

Brain stroke
   Lethality prevented if PBN given before (1) or shortly after stroke (8,9)[a]
   Administration of PBN prevents CA1 neuron loss if given before or after stroke (10)[a]
   PBN administration chronically to older gerbils prevents stroke-lethality long after ceasing PBN dosing (11)[a]
   PBN, before or after stroke, protects from brain necrosis (12,13), edema, and loss of behavioral scores (12)[b]
Brain aging
   PBN administered chronically decreased age-related brain protein oxidation and improved memory (14)[a]
Brain concussion
   PBN administered before prevented concussion-induced hydroxyl free-radical formation (15).
Brain excitotoxity, lowered energy production, and dopamine depletion models (16).
   2-sulfo-PBN suppressed neuronal death by NMDA (N-methyl-D-aspartate), AMPA (α-amino-3-hydroxy-5-methyl-4-isoxasole-propionate), KA (kianic acid), 3-AP (3-acetyl pyridine), by malonate or by MPP+ (1-methyl-4-phenyl pyridinium).

[a]Experiments done in Mongolian gerbils.
[b]Permanent (12) or transient (13) middle cerebral artery occlusion in rats.

with mitochondria? The answer is not known, but the available facts indicate that the NRT protective action is probably not explained simply by a direct interaction of these compounds with mitochondria. Beal's group (16) showed that in the malonate-mediated damage to brain, that even though 2-sulfo-PBN

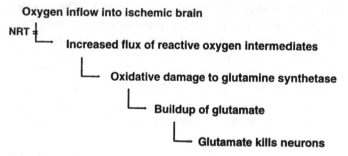

Figure 2 Scheme illustrating that oxidative damage to brain leads to glutamate buildup through a sequence of events removed from the initial free-radical formation. NRTs are considered to reduce oxidative damage in stroke in part by inhibiting the propagation of initial free radicals which cause oxidative damage to glutamine synthetase.

exerted considerable protective effect on brain necrosis, it had no action on the malonate-mediated lactate buildup or ATP loss. Lactate is oxidized to pyruvate which is then metabolized in the mitochondria, thus producing reducing equivalents to the electron transport chain where ATP is generated. If 2-sulfo-PBN protects by its direct interaction with mitochondria, it is expected that indices of mitochondrial action would vary collinearly with its protectiveness.

We have found that PBN at higher levels (4.7 mM) caused a slight decrease (15–24%) in malate/glutamate driven respiration in isolated liver mitochondria (19). In addition, it has been shown that NRTs trap free radicals produced by isolated brain and heart mitochrondia (20,21), but the amount of NRTs used (typically 10–50 mM) for these in vitro experiments and allowing for the fact that free-radical trapping efficacy may not be as high as 50% and that in vivo NRT concentrations may not be nearly this large, it is hard to rationalize that their protective action is due to direct interaction with mitochondria.

We have examined the action of NRTs on hydroxyl free radicals produced in isolated mitochondria experiencing an anoxia/reoxygenation challenge. We utilized an anoxia/reoxygenation challenge in vitro to replicate an IRI. Hydroxyl free-radical flux was assessed by salicylate hydroxylation. The data obtained are shown in Table 4 and Fig. 3. It is clear that anoxia/reoxygenation enhances the amount of $\dot{O}H$ formed by isolated mitochondria, and, surprisingly, the presence of PBN enhances salicylate hydroxylation up to about 7 mM PBN at which point it then starts to cause a decrease in the amount of salicylate hydroxylated. We do not understand these data, but it is possible that the observed changes with increasing PBN concentration is due to two counteracting processes. That

Table 4 Anoxia/Reoxygenation Effect on Salicylate Hydroxylation in Isolated Mitochrondria

Exp	Condition	2,3-DHBA[a] SA	Treatment[c] Ratio	Tot. DHBA[b] SA	Treatment[c] Ratio
A	Anoxia/oxygenation	0.110		0.164	
B	Control	0.092	1.20	0.129	1.27
C	Anoxia/oxygenation	0.083		0.146	
D	Control	0.073	1.14	0.112	1.30

[a]Refers to 2,3-dihydroxybenzoic acid to salicylate ratio times 1000.
[b]Refers to 2,3-DHBA plus 2,5-DHBA to salicylate ratio times 1000.
[c]Refers to results obtained when pig heart mitochondria in state 4 with malate/glutamate added as substrate was incubated without oxygen (i.e., 95% N_2 + 5% CO_2) for 20 min then with oxygen (95% O_2 to 5% O_2) for 40 min.

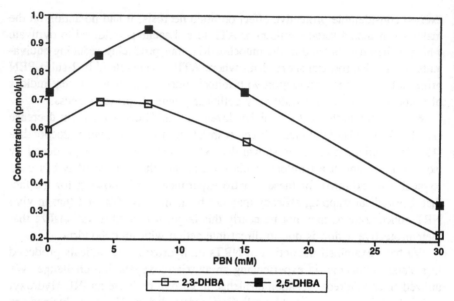

Figure 3 Formation of 2,3- and 2,5-dihydroxylbenzoic acid from salicylate in anoxia (20 min)/reoxygenation (40 min)-treated isolated pig heart mitochondria. (Unpublished data of G.-J. Lui, P.K. Wong, and R.A. Floyd.)

is, the presence of PBN may protect the mitochondria from oxidative damage caused by anoxia/reoxygenation and thus the enhanced amount of respiration in the protected mitochondria produces more ROS; but this enhanced ROS becomes more readily trapped by PBN as increasing amounts of this NRT are present.

Anoxia/reoxygenation of isolated mitochondria caused protein oxidation damage as shown in Table 5. We have not determined the effect of NRTs on anoxia/reoxygenation mediated protein oxidation in isolated mitochondria.

V. DO NRTs ACT AS ANTIOXIDANTS?

This is a difficult question to answer simply because the term "antioxidant" evokes concepts of the action of vitamin E on lipid peroxidation processes. It is clear that if simple lipid peroxidation processes could account for oxidative damage in the brain then NRTs may exert some of their action by this classical antioxidant mechanism (Fig. 1). Yet the fact that NRTs protect brain from IRI-mediated oxidative damage if given an hour or more after the start of

Figure 5 Anoxia/Reoxygenation Effect on Protein Oxidation in Isolated Mitochondria

Exp	Condition[a]	DNPH incorporated (nmole) Protein (mg)	Treatment ratio
1	A/R	11.30	
	C	8.76	1.29
2	A/R	9.06	
	C	6.24	1.45
3	A/R	28.30	
	C	21.80	1.30
4	A/R	12.80	
	C	8.74	1.47
5	A/R	10.10	
	C	5.75	1.76
6	A/R	14.10	
	C	9.92	1.42

[a]A/R refers to 20 min anoxia (95% N_2 + 5% CO_2) followed by 40 min of oxygen (95% O_2 + 5% CO_2) and C refers to oxygen for 60 min. The pig heart mitochondria were in state 4 respiration with malate/glutamate as substrate.

reperfusion clearly implicates that they act in a much more specific fashion such as in the prevention of the loss of glutamine synthetase activity (Fig. 2) or in the prevention of specific gene induction for instance. Therefore if the definition of antioxidant can be viewed in an expanded understanding, encompassing amplified action such as described above, the NRTs may fit in this domain. We consider that if a severe oxidative stress event causes an inducible gene product (X_i) to arise and the combined action of X_i plus the enhanced level of ROS leads to tissue injury, then administration of an NRT either continuously throughout the event cycle or after the severe oxidative stress insult when X_i is being induced will prevent the tissue injury (Fig. 4). Later NRT administration times (C in Fig. 4) would be less effective. The degree of NRT effectiveness would then coincide with its application when the maximum X_i induction time occurs after the severe oxidative stress insult has taken place. An extrapolation of this idea to the aged animal situation is possible if it is assumed that X_i as well as ROS (or an oxidation product) is enhanced in the aged condition (Fig. 5) and that application of NRT in a chronic fashion would depress not only X_i but ROS to younger levels, thus making the older animal less susceptible to a severe oxidative stress insult. Following this logic, then, cessation of

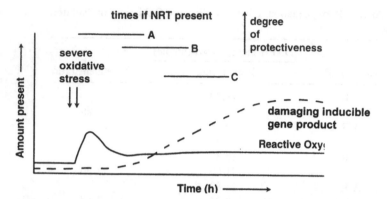

Figure 4 Scheme illustrating that time after a severe oxidative stress leads to at first a burst of ROS and then remains elevated for some time afterwards. The enhanced oxidative damage causes the induction of a gene leading to a damaging inducible gene product (X_i). Treatment with NRT for certain periods of time, i.e., during the severe oxidative stress and continuing to slightly after (A), beginning slightly after (B), or beginning much longer after the initial severe oxidant stress (C) leads to less and less protectiveness.

the chronic NRT dosing would then require some time to elapse before the treated older animal again regains its original susceptibility to the severe oxidative stress insult of an IRI. This model would explain our observations (Table 3).

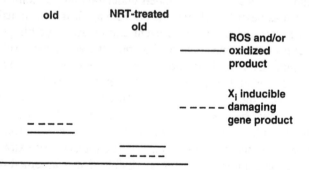

Figure 5 Illustration that the combination of ROS (and/or oxidized product) and X_i (inducible damaging gene product) is elevated in old brain and that NRT treament reduces these components back to younger levels, thus making the NRT-treated animals less susceptible to an IRI. With time after ceasing NRT treatment, these two components rise back to original levels.

ACKNOWLEDGMENTS

This work was supported in part by NIH grants NS 23307 and AG 09690. We thank Mr. Quentin Pye for help in mitochondria isolation.

REFERENCES

1. Floyd RA. Role of oxygen free radicals in carcinogenesis and brain ischemia. FASEB J 1990; 4: 2587–2597.
2. Floyd RA, Carney JM. Free radical damage to protein and DNA: Mechanisms involved and relevant observations on brain undergoing oxidative stress. Ann Neurol 1992; 32: S22–S27.
3. Cao W, Carney JM, Duchon A, Floyd RA, Chevion M. Oxygen free radical involvement in ischemia and reperfusion injury to brain. Neurosci Lett 1988; 88: 233–238.
4. Oliver CN, Starke-Reed PE, Stadtman ER, Liu GJ, Carney JM, Floyd RA. Oxidative damage to brain proteins, loss of glutamine synthetase activity, and production of free radicals during ischemia/reperfusion-induced injury to gerbil brain. Proc Natl Acad Sci USA 1990; 87: 5144–5147.
5. Carney JM, Tatsuno T, Floyd RA. The role of oxygen radicals in ischemic brain damage: Free radical production, protein oxidation and tissue dysfunction. In: Krieglstein J, Oberpichler-Schwenk H, eds. Pharmacology of Cerebral Ischemia Stuttgart: Wissenschaftliche Verlagsgesellschaft 1992: 321–331.
6. Floyd RA, Carney JM. Age influence on oxidative events during brain ischemia/reperfusion. Arch Gerontol Geriatr 1991; 12: 155–177.
7. Janzen EG. Spin trapping. Acc Chem Res 1991; 4: 31–40.
8. Phillis JW, Clough-Helfman C. Protection from cerebral ischemic injury in gerbils with the spin trap agent N-tert-butyl-α-phenylnitrone (PBN). Neurosci Lett 1990; 116: 315–319.
9. Phillis JW, Clough-Helfman C. Free Radicals and ischaemic brain injury: Protection by the spin trap agent PBN. Med Sci Res 1990; 18: 403–404.
10. Clough-Helfman C, Phillis JW. The free radical trapping agent N-tert-butyl-α-phenylnitrone (PBN) attenuates cerebral ischaemic injury in gerbils. Free Radic Res Commun 1991; 15: 177–186.
11. Floyd RA, Carney JM. Nitrone radicals traps (NRTs) protect in experimental neurodegenerative diseases. In: Chapman CA, Olanow CW, Jenner P, Youssim M, eds. Neuroproctetive Approaches to the Treatment of Parkinson's Disease and Other Neurodegenerative Disorders London: Academic Press, 1993: 70–90.
12. Cao X, Phillis JW. α-Phenyl-tert-butyl-nitrone reduces cortical infarct and edema in rats subjected to focal ischemia. Brain Res 1994; 644: 267–272.
13. Zhao Q, Pahlmark K, Smith M-I, Siesjo BK. Delayed treatment with the spin trap α-phenyl-N-tert-butyl nitrone (PBN) reduces infarct size following transient middle cerebral artery occlusion in rats. Acta Physiol Scand 1994; 152: 349–350.
14. Carney JM, Starke-Reed PE, Oliver CN, Landrum RW, Chen MS, Wu JF, Floyd

RA. Reversal of age-related increase in brain protein oxidation, decrease in enzyme activity, and loss in temporal and spacial memory by chronic administration of the spin-trapping compound N-tert-butyl-α-phenylnitrone. Proc Natl Acad Sci USA 1991; 88: 3633-3636.

15. Sen S, Goldman H, Morehead M, Murphy S, Phillis JW. α-Phenyl-tert-butylnitrone inhibits free radical release in brain concussion. Free Radic Biol Med 1994; 16: 685-691.

16. Schulz JB, Henshaw DR, Siwek D, Jenkins BG, Ferrante RJ, Cipolloni PB, Kowall NW, Rosen BR, Beal MF. Involvement of free radicals in excitotoxicity in vivo. J Neurochem 1995; 64: 2239-2247.

17. Delbarre G, Delbarre B, Calinon F, Loiret C, Ferger A. Accumulation of glutamate, asparate and GABA in the hippocampus of gerbil after transient cerebral ischemia. Soc Neurosci Abst 1989; 15: 1174 (abstr).

18. Chen G, Bray TM, Janzen EG, McCay PB. Excretion, metabolism and tissue distribution of a spin trapping agent, α-phenyl-N-tert-butyl-nitrone (PBN) in rats. Free Radic Res Commun 1990; 9: 317-323.

19. Wong PK, Poyer JL, DuBose CM, Floyd RA. Hydralazine-dependent carbon dioxide free radical formation by metabolizing mitochondria. J Biol Chem 1988; 263: 11296-11301.

20. Nohl H, Jordan W, Hegner D. Identification of free hydroxyl radicals in respiring rat heart mitochondria by spin trapping with the nitrone DMPO. FEBS Lett 1981; 123: 241-244.

21. Dykens JA. Isolated cerebral and cerebellar mitochondria produce free radicals when exposed to elevated Ca^{2+} and Na^+: Implications for neurodegeneration. J Neurochem 1994; 63: 584-591.

13

Antioxidant Properties of Nitroxides and Nitroxide SOD Mimics

Amram Samuni
Hebrew University Medical School, Jerusalem, Israel

Murali C. Krishna
National Cancer Institute, National Institutes of Health, Bethesda, Maryland

I. INTRODUCTION

A general definition of antioxidant includes enzymes such as SOD and SOD mimics which catalyze O_2^- dismutation; catalases, peroxidases, and mimics of such enzymes, which remove H_2O_2; chelators which bind redox-active transition metals; and low-molecular-weight antioxidants (LMWA) which rapidly react with ·OH and other oxygen, nitrogen- and carbon-centered radicals, reduce oxidized sites, and inhibit radical chain reactions. The less-sweeping definition generally relates to the latter class only. Natural or synthetic, hydrophilic or lipophilic LMWA include thiols such as glutathione, cysteine, n-acetyl-cysteine or cysteamine; ascorbate, urate, trolox, vitamin E and other tocopherols, β-carotene and other carotenoids, butylated hydroxytoluene (BHT), carnosine, spin traps such as phenyl-t-butyl nitrone, phenols and polyphenols such as hydroxybenzoic and caffeic acids, flavonoids, etc. Such LMWA are generally reducing agents, which act in a stoichiometric fashion. Therefore, a high flux of reactive oxygen-derived species (ROS) can rapidly deplete their level in the tissue. Another limitation of LMWA might result when at higher O_2 concentrations they lose their antioxidant activity and exert autocatalytic, prooxidative effect.

Nitroxides, sometimes denoted nitroxyls or aminoxyls, constitute a unique and unusual class of antioxidants. Their use offers a complementary, and sometimes alternative, powerful strategy of combating oxidative damage.

II. NITROXIDE ANTIOXIDANT ACTIVITY

A. General

The distribution, metabolism, and persistence of nitroxides in tissues along with their use as biophysical probes or contrast agents for nuclear magnetic resonance imaging have been extensively investigated and reviewed in the past (1–3). This chapter centers on their antioxidative activity. The biological activity of nitroxides, observed almost 3 decades ago (4) and recently reviewed (5), has been investigated using diverse means of insult, various nitroxides, and a variety of experimental models.

B. Diverse Insults

Nitroxide's protective activity has been studied using physical, chemical, biochemical, and cellular means of insult which initiate injurious processes mediated by ROS. Most studies employed H_2O_2 (6–8), organic peroxides, such as t-butyl hydroperoxide (t-BuOOH) or cumene hydroperoxide (6,9–11), and O_2^- generated enzymatically by hypoxanthine/xanthine oxidase (HX/XO) (6,12). Other studies, however, employed different insults such as ionizing radiation (13–16), ADP-Fe^{II} (17,18), tumor necrosis factor (19), 2,2'-azobis(2-amidinopropane) dihydrochloride (AAPH) and 2,2'-azobis(2,4-dimethylvaleronitrile) (AMVN) (20), postischemic reperfusion (21–23), xenobiotics such as streptonigrin (24), bleomycin (25), mitomycin (26), juglone (27), paraquat (28), primaquine (29), ethanol (30), nonsteroidal antiinflammatory drugs (30), gastric and colonic irritants such as TNB (31), mechanical trauma (32), hyperbaric oxygen, (33) or hyperthermia to 40°C (34).

C. Nitroxides Tested

Stable cyclic nitroxides, which have been tested for their biological activity, include several ring types such as (a) five-membered oxazolidine ring or doxyl, (b) five-membered saturated ring pyrrolidine or proxyl, (c) five-membered unsaturated ring pyrroline, and (d) six-membered saturated piperidine ring (Scheme [1]) as well as fused-ring systems.

For a given ring system, four oxidation states of the aminoxyl moiety are possible, namely (a) amine, (b) hydroxylamine, (c) nitroxide, and (d) oxoammonium (Scheme [2]), where the oxidation states are each separated by one electron.

[1]

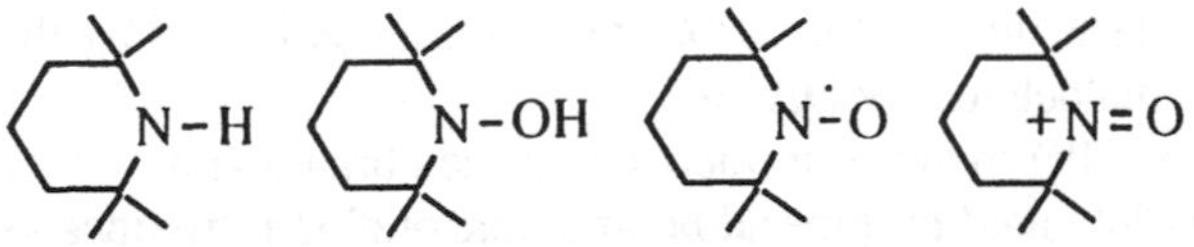

a   b   c   d

[2]

a   b   c   d

As nitroxides differ in size, charge, and lipophilicity, efforts were made to select or synthesize specific derivatives in order to achieve desired localization inside cells (35) and optimal activity. Most studies were performed using commercially available nitroxides; particularly TEMPO and TEMPOL. Additionally, especially tailored nitroxide derivatives have been synthesized in order to improve their biological activity (10,11,36). Alternative research direction centered on conjugating nitroxides with active moieties having specific biological or therapeutic activity. Nitroxide-labeled analogs of various drugs have been synthesized and tested for biological activity (37–40). Another approach has made use of tagging them to high-molecular-weight proteins, such as albumin (41,42) in order to increase their persistence in the blood circulation (42).

D. Experimental Model Systems

The experimental models tested varied by their complexities and ranged from isolated biomolecular preparations, through cells and organs up to laboratory animals.

1. Molecular and Subcellular Level

Isolated preparations of biomolecules subjected to oxidative stress served to study the effect of nitroxides in the molecular level. The activity of nitroxides and their respective hydroxylamines is reflected by their inhibitory effect on oxidation of proteins (20,34,43), lipids (4,17,34,36,44), and DNA (24,27).

Protein Tanfani et al. (10) have examined the effect of 1,2-dihydro-2,2-diphenyl-4-ethoxy-quinoline-1-*N*-oxyl on bovine serum albumin (BSA) oxidation induced by *t*-BuOOH in the presence of PbO$_2$. They showed that the elevation in carbonyl groups, reflecting protein oxidation, could be fully prevented by 400 μM nitroxide. This protective effect was attributed to the scavenging of peroxyl radicals by the nitroxide (10).

Lipids   Lipid peroxidation (LPO) induced by AAPH and AMVN in soybean and 18:2 PC liposomes was inhibited by *N*-oxyl-4,4'-dimethyloxazolidine derivatives of stearic acid (20). Several studies showed that both the nitroxide as well as its one-electron-reduced form, namely the hydroxylamine inhibit LPO (17,45,46). Inhibition by TEMPOLH of LPO has been observed for preparations of liver endoplasmic reticulum and in stored microsomes as judged by the accumulation of malonic dialdehyde (MDA) (45).

Nilsson et al. found that LPO induced in microsomal membranes and in pure phospholipid vesicles was inhibited by piperidine and oxazolidine nitroxides as well as by their reduced forms (17). The protective activities of hydroxylamines were about threefold higher than those of the respective nitroxides as judged by the concentrations required for 50% inhibition (17). The nitroxides' protective activity increased with their lipophilicity when LPO was induced by *t*-BuOOH, though not when ADP-Fe^{II} mediated the damage (17). Miura et al. studied LPO in microsomal preparations from rat liver supplemented with NADPH and examined the inhibitory effects of doxylstearate and piperidine derivatives (43). To evaluate the nitroxide inhibitory effect they measured the decrease in LPO as reflected by the accumulation of thiobarbituric acid reactive species, conjugated dienes, and lipid peroxides. They showed that hydroxylamines act like the respective nitroxides and that lipophilic derivatives such as doxylstearates were about 100-fold more effective than TEMPOL or 4-trimethylamino-2,2,6,6-tetramethylpiperidine-1-*N*-oxyl (CAT1) (43).

DNA   The inhibitory effect of nitroxide on cleavage of covalently closed circular DNA isolated from pUC19 plasmid has been demonstrated by measuring the extent of single- and double-strand breaks. TEMPAMINE at 1 mM fully prevented DNA scission inflicted by 100 μM streptonigrin or HX/XO as measured the by the electrophoretic mobilities of treated and untreated DNA (24). Similar results were obtained with the same experimental system where 2 mM TEMPAMINE prevented DNA scission exerted by 400 μM juglone together with 1 μM Fe^{II} and 2 mM GSH (27).

2.   *Cellular Level*

The biological activity of nitroxides has been demonstrated for bacterial and mammalian cells.

Bacterial Cells   Studies demonstrating the effect of nitroxides have used specific mutants such as *xthA* (7) and *recA* (47) cells of *E. coli*, which are particularly sensitive toward oxidative damage. The nitroxides' protective effect was reflected by preserving the colony-forming ability of the cell and preventing DNA degradation. Nitroxides have been found to protect bacterial cells exposed to H_2O_2, HX/XO (7), cytotoxic drugs such as streptonigrin (24), as well as the naphthoquinones juglone, plumbagin, and menadione (27). Such protection has been found also for hypoxic cells where injury involved neither O_2^- nor H_2O_2 (26,27), suggesting that the antioxidant activity of the nitroxides extends beyond combating ROS.

Mammalian Cells A significant protective activity of 5 mM TEMPOL, TEMPO, and TEMPAMINE was observed for CHO and V79 cells subjected to either H_2O_2 or organic peroxides and to ionizing radiation (6,48). These nitroxides at concentrations >40 µM protected, in a dose-dependent manner, cultured cardiomyocytes of newborn rats incubated with HX/XO, glucose/glucose oxidase (49), or H_2O_2 (50). The nitroxide did not prevent the decrease in ATP level (50), but fully inhibited leakage of cytoplasmic lactate dehydrogenase (LDH) from cardiomyocytes and preserved their autonomous beating (49).

TEMPOL which partially protected aerobic V79 cells from mitomycin c provided complete protection under hypoxic conditions (26). TEMPOL was also found to protect aerated human red blood cells (RBC) exposed to the antimalarial reagent primaquine. TEMPOL facilitated oxidation of RBC hemoglobin to methemoglobin but inhibited cell hemolysis (29).

In radiation studies, TEMPOL, though not its hydroxylamine (TEMPOLH), provided dose-dependent protection against aerobic-radiation-induced cytotoxicity at all concentrations tested. Under hypoxia, TEMPOL did not protect against radiation but rather caused a modest radiosensitization. In another study, where the structural determinants of modification of aerobic radioresponse were studied, it was found that the charge on the substituents rather than the ring type or size is important in providing aerobic radioprotection. Positively charged substituents such as amino groups were effective, whereas neutral or negatively charged substituents were minimally effective (15).

TEMPOL has been also shown to protect cultured epithelial cells of rabbit lens against H_2O_2 which inhibited cell growth, caused blebbing of cell membrane, decrease in NAD^+ and in the formation of single-strand DNA breaks (51).

3. *Organ Level*

The effect of nitroxides on isolated rat heart perfused in the Langendorff configuration has been studied (52) following postischemic regional (21) and global (53) ischemia. Gelvan et al. demonstrated that 0.4–1 mM TEMPO diminished the postischemic release of LDH and reduced the duration of reperfusion arrhythmia, which reflect the reperfusion damage (21).

More recent work indicated that most of the heart function, impaired upon 20 min global ischemia, as demonstrated by hemodynamic parameters, could be restored following reperfusion. Copper[II] was found to potentiate the heart injury as the work index was restored only to 17% of the preischemic value. The inclusion of 100 µM TEMPOL during reperfusion abolished the copper-induced sensitization, but it provided no protection in the absence of copper (53). The hemodynamic effects of TEMPOL have been studied also using perfused isolated rat hearts under normoxia by monitoring the left ventricular pressure, coronary flow rate, and heart rate (52). TEMPOL up to a concentration of 10–30 mM increased the coronary flow rate up to 51% but did not affect the cGMP values, whereas TEMPOL concentration higher than 30 mM decreased the coronary flow. These results led to the conclusion that the vasodilatory activity of nitroxide was not attributable to its effect on nitric oxide (52). Conversely, in another study where the spin trap 5,5-dimethyl-1-pyrroline-N-oxide (DMPO) was found to protect an isolated reperfused rat heart the protective effect has been attributed to the persistent nitroxide spin-adduct resulting from DMPO rather than to the spin trap itself (23).

4. Whole-Body Level

The protective activity of nitroxides has been studied in several experimental models of laboratory animals.

Gastric Mucosal Injury Nitroxides protect rats against inflammatory bowel disease (IBD), such as gastric mucosal damage induced by several irritants such as ethanol 96%, aspirin, indomethacin, 25% NaCl, or 0.6N HCl (30). Both TEMPO and TEMPOL prevented formation of mucosal lesions and attenuated the increase in LTB_4 and LTC_4 generation induced by ethanol. Doses of 100 mg TEMPOL/kg.b.w. given i.g., i.v. or i.p. 5 min before induction of damage fully prevented lesion formation (30). Although nitroxides rapidly disappear through reduction to hydroxylamines and via clearance from the body, TEMPOL retained its protective effect even when given 1 h prior to induction of damage. Lower doses, however, provided partial protection only.

Experimental Colitis Protection by TEMPOL has been demonstrated also in models of experimental colitis induced in rats with trinitrobenzene sulfonic acid (TNB) or acetic acid (31). The hemorrhagic and ulcerative damage to the distal colon, induced by intracolonic administration of TNB/ethanol is manifested by colonic lesions, increase in colon weight, elevation of lipooxygenase products and of myeloperoxidase (MPO) activity. Intragastric administration of 0.5 g TEMPOL/kg.b.w., immediately after intracolonic administration of 30 mg TNB in 0.25 ml 50% ethanol, and once daily thereafter, significantly decreased colonic mucosal lesion area assessed after 1, 3, and 7 days, the wet weight of the 10 cm distal colonic segment (day 3), myeloperoxidase (MPO) activity (day 7), and mucosal LTB_4 generation (day 3). Similar results have been observed when acetic acid was also used to induce ulcerative colitis (31).

Mechanical Trauma The neuroprotective properties of TEMPOL and TEMPO were investigated in a rat model of mechanical trauma (32). Closed head injury (CHI) was induced to ether anesthetized rats by a weight-drop device, and recovery was followed for up to 24 h. The severity of injury was determined by evaluating the motor function, edema formation, and impairment of brain blood barrier. The clinical status of the rats was evaluated according to a set of criteria defined as neurological severity score (NSS) which test reflexes and motor functions by giving points on the basis of a deficit in these functions. The clinical status spontaneously improves during the recovery period and the NSS measured at 24 h post-CHI is lower than that at 1 h post-CHI. The difference between the NSS at 1 and 24 h (ΔNSS), which reflects the degree of recovery, increases by neuroprotective treatments. Vasogenic and cytotoxic edema, which follows CHI, was assessed by measurement of water content of 24 h. The integrity of the blood brain barrier (BBB), impaired by CHI, was investigated using Evans Blue extravasation. The neuroprotective activity of 50 mg TEMPOL/kg.b.w. was reflected by facilitated clinical recovery, reduced edema formation, and amelioration of BBB disruption. Preliminary studies also demonstrated the therapeutic window of TEMPOL to be in the range of 4 h after CHI and that TEMPO has a greater neuroprotective effect than TEMPOL (32).

Radiation Injury   Previous studies indicated a radiosensitizing effect when *E. coli* B/r cells were exposed to 0.1 mM TEMPO under anoxia (13,54). More recent studies, however, showed that TEMPOL exerts a significant radioprotection to mice subjected to whole body irradiation (16). C3H mice were injected i.p. with the maximally tolerated doses of TEMPOL (275 mg/kg) which resulted in maximal blood levels between 5 and 10 min after injection. Mice treated with saline controls or with TEMPOL at maximally tolerated doses were γ-irradiated and the survival was assessed for 30 days, as a measure of bone-marrow toxicity. TEMPOL provided whole body radioprotection when given prior to radiation treatment but failed to protect when given 24 h after radiation treatment [16].

Hyperoxia   The effect of TEMPOL and TEMPO on the development of hyperoxic-induced seizures has been studied using rats as an experimental model (33). Rats implanted with chronic cortical electrodes for continuous EEG monitoring were injected i.p. with 1–80 mg nitroxide/kg.b.w. or vehicle alone 10 min before exposure to 0.5 mPa oxygen. The duration of the latent period until the appearance of electrical discharges in the EEG was used as an index of CNS oxygen toxicity. Complete protection has been provided by 2.5 mg TEMPO/kg.b.w. TEMPOL was effective only at doses above 60 mg/kg. These results provide additional support for the role of reactive oxygen species in the development of hyperoxia-induced seizures (33).

Anticancer Effect   It has also been reported that nitroxides increase the tolerance of laboratory animals to cytostatic reagents (55). When pyrroline and piperidine nitroxides were co-injected with lethal doses of antitumor cytotoxic agents such as 6-mercaptopurine, cyclophosphimide, a decrease in lethality was observed. In this study the nitroxide effect has been attributed to an inhibition of enzymes such as the cytochromes P450 (55).

Alopecia Topical application of TEMPOL was evaluated for possible protective effects against radiation-induced alopecia using guinea pig skin as a model. For single acute x-ray doses up to 30 Gy, TEMPOL, when topically applied 15 min prior to irradiation, increased the rate and extent of new hair recovery when compared to untreated skin (56). Animals were treated topically with TEMPOL 15 min prior to x-ray treatment. With this mode of application, no systemic levels of TEMPOL were detected when blood samples were withdrawn for EPR studies. Following a single acute dose of 30 Gy at a dose rate of 3.75 Gy min^{-1}, hair loss was followed for 10 weeks. While hair loss was similar for 7 weeks from TEMPOL treated vs untreated animals, protective effects of TEMPOL treatment were observed beyond 7 weeks.

Hypotension The hypotensive activity of nitroxides has also been investigated (57). Some, though low, hypotensive effect has been found for cats treated with 25 mg TEMPO/kg.b.w. (57).

Apoptosis The potential effect of TEMPOL in the development of apoptosis has also been examined (58). Apoptosis has been induced in rat thymocytes by dexamethasone and etoposide and the effect of various antioxidants, including TEMPOL, was studied. TEMPOL at 1 mM and 5 mM partially and fully, respectively, prevented the apoptosis (58).

Ischemic Shock The effect of thiol-containing derivatives of reduced nitroxides on the survival of rats subjected to ischemic shock has been previously investigated (44) by following a long-term tourniquetting of both rats' hind limbs, causing more than 90% mortality within 5 days. Dose of 35 mg 4,4-dithiol-TEMPOH/kg.b.w. injected 1 h prior to tourniquet removal increased survival by 45% (44).

III. THE ACTIVE SPECIES

By undergoing one-electron transfer reactions, nitroxides are readily reduced to hydroxylamines or oxidized to oxo-ammonium cations. Consequently, all three forms can be present in the tissue as shown for TEMPO in Scheme [3]. In the presence, however, of two-electron reductant such as NADPH the oxo-ammonium cation can be reduced to hydroxylamine (59). Inside the body the nitroxide and hydroxylamine predominantly coexist. The metabolism of nitroxides in biological systems has been thoroughly studied and reviewed by Swartz and co-workers (1). Nitroxides in tissues are primarily, though not exclusively (60), reduced through enzyme-associated mechanisms (1). This bioreduction, which predominantly occurs intracellularly in the mitochondria, yielding almost

$$+N=O \quad \text{Oxo-ammonium cation}$$

$$\text{Nitroxide radical} \quad \overset{\cdot}{N}\text{-O}$$

$$e \qquad 2e \qquad [3]$$

$$e$$

$$N\text{-OH} \quad \text{Hydroxylamine}$$

exclusively the respective hydroxylamines (1, and references therein), is very rapid and can lead to a 99% reduction within several minutes (1).

The oxidation of hydroxylamines to nitroxides can also occur at very high rates (61–63) and with some nitroxides can be comparable to the rate of nitroxide reduction. Consequently a distribution of nitroxide/hydroxylamine is soon achieved, regardless which of the two forms is administered. Therefore, although both the nitroxide and its reduced form are effective antioxidants, this rapid metabolic exchange limits the distinction between the protective activities of the two forms.

IV. ADVERSE EFFECTS

The need to evaluate any potential adverse activity of nitroxides stems from their potential applicability as antioxidants or contrast agents for MRI (3). Because hydroxylamines are less toxic than their respective nitroxides (36), the toxicity tests focused primarily on the activity of nitroxides themselves.

A. Mutagenicity

Contrasting evidence exists regarding mutagenic effect of nitroxides (64,65). Nitroxides at concentration >50 mM reportedly exert a mutagenic effect in *Salmonella typhimurium* (66). Sies and Mehlhorn using *Salmonella typhimurium* strain TA 104, which is sensitive to oxidative damage, demonstrated a mutagenic activity of 8 mM TEMPOL or 3-hydroxymethyl-PROXYL a well as potentiation of the mutagenic effect of H_2O_2 (64). Another study showed that TEMPON incubated with monolayered cultured fibroblasts from male Chinese hamsters caused chromosome aberrations (67,68). A more recent study examined the effect of TEMPOL on radiation-induced chromosomal aberrations in peripheral human blood (69). In this study, TEMPOL at 10 mM and 50 mM did not cause chromosomal aberrations. However, treatment of lymphocytes with TEMPOL before irradiation with x-rays resulted in an inhibition of chromosome aberration induction.

In another study, when TEMPOL was tested using the XPRT forward mutation assay in hypermutable CHO AS52 cells, no mutagenic activity was detected (25,65). Moreover, TEMPOL demonstrated antimutagenic activity by preventing the mutagenic effect of bleomycin (25) or HX/XO (65), though not of ethylmethane sulfonate (65).

B. Cytotoxicity

The potential cytotoxic effect of nitroxides has been examined in several test systems. Early tests of the TEMPONE effect on survival of laboratory animals showed that mice tolerated up to 0.5 g TEMPONE/kg.b.w. given i.p. (13). In later studies, with the exception of 4-maleimido-TEMPO, neither hydrophilic nor lipid-soluble nitroxides at 1 mM exhibited cytotoxic effects in rats and in cell cultures (70,71). Similarly, no decrease by 14.5 mM TEMPOL in the viability of human embryo lung fibroblasts has been found (9). The survival of monolayered CHO cells was unaffected by piperidine (TEMPOL and TEMPAMINE), pyrroline 3-carbamoyl-2,2,5,5-tetramethyl-3-pyrroline-1-yloxy, (CTPO), 3-carbamoyl-2,2,5,5 tetramethyl-3-pyrroline-1-oxyl-3-carboxylic acid (CTPC) and doxyl (5-doxylstearate, 12-doxylstearate, 16-doxylstearate) derivatives at concentration as high as 1 mM (71). Similar results were obtained when no effect of 10 mM TEMPOL on survival of cultured CHO AS52 cells could be detected (25,65), nor by 10 μM–10 mM of TEMPO incubated for 2–48 h on the survival of MCF-7 human breast cancer cells (72).

Exposure of epithelial cells cultured from rabbit lens to 5 mM TEMPOL showed no effect on cell growth (51). In another study no effect on the viability of isolated rat hepatocytes incubated with 1 mM 2-ethyl-2,5,5-trimethyl-3-oxazolidine-1-N-oxyl (OXANO) or 5 mM 1-hydroxy-2-ethyl-2,5,5-trimethyl-3-oxazolidine (OXANOH) was observed, as judged by dye exclusion, despite a significant decrease in GSH level (18).

On the other hand, acting as SOD mimics, nitroxides are anticipated to facilitate O_2^- conversion to H_2O_2. Indeed, the incubation of cells with TEMPO has been found to increase the production of H_2O_2 (72). This result led to the assumption that TEMPO increases H_2O_2 production by enhancing the activity of P450 peroxidase, and consequently "exerts a prooxidant effect by increasing the cellular hydrogen peroxide concentration" (72). It was also found that nitroxide at concentration >5 mM potentiates juglone-induced cell injury of *xthA* mutant of *E. coli* (27). More recent study showed that nitroxides exert a bactericidal effect on *E. coli recA* cells and potentiate H_2O_2 cytotoxicity in this cell line (47). This sensitizing effect by nitroxides appeared to stem from their effect on DNA synthesis and repair and is manifested only in cells which lack a functional RecA protein.

C. Putative Mechanism(s)

Several hypotheses have been considered to account for the potential adverse effects of nitroxides: The mutagenic effect was found to increase with the increase in the nitroxide susceptibility to reduction. The distinction between a direct mutagenic effect of the nitroxide and its promutagenic activity has been previously considered (66), though not resolved. Mutagenicity has been ascribed to reactive species derived upon oxidation of GSH by the nitroxide or to injurious species produced in the presence of $O_2^{.-}$ (64,66). It was also assumed that nitroxides exert a pro-oxidative effect by increasing the cellular H_2O_2 concentration (72) or that a direct interaction of nitroxide with DNA, and/or generation of secondary reactive species from nitroxide after being metabolized within cells (72).

According to an alternative assumption nitroxides promote the formation of oxidized thiols, which might exert destructive effects in biological environments (64). However, the particular susceptibility of *E. coli recA* cells toward nitroxides could not be accounted for by the above-mentioned explanations, since there was no apparent decrease in GSH cellular concentration. Since the toxicity of hydroxylamines, which have no oxidant properties, is lower than that of nitroxides it appears that the adverse effect result from the oxidation of certain, yet undefined, critical cellular targets.

V. MECHANISMS UNDERLYING ANTIOXIDATIVE ACTIVITY

A. SOD Mimic Activity

The SOD-mimic activity of nitroxides has been previously established both by direct and indirect assays. Two mechanisms have been identified by which nitroxides catalyze the dismutation of $O_2^{.-}$, yielding O_2 and H_2O_2.

1. *Oxidative Mode*

Exposure of piperidine derivatives, such as TEMPO, to a continuous flux of superoxide results with a rapid dismutation of $O_2^{.-}$ radicals. At the same time the nitroxide concentration does not decay but rather remains time invariant. The mechanism that underlies this SOD-mimic activity involves flip-flopping of the catalyst between two oxidation states, namely nitroxide and oxo-ammonium cation (59).

$$\text{(N-O} + \cdot O_2^- + 2H^+ \rightleftarrows {}^+N=O + H_2O_2) \qquad \{1\}$$

$$({}^+N=O + \cdot O_2^- \rightleftarrows \text{N-O} + O_2) \qquad \{2\}$$

Preliminary independent experiments could demonstrate the reverse reaction {-1} by reacting 1 M H_2O_2 with oxo-ammonium cation, independently formed in electrochemical generator. Therefore, as neither the nitroxide nor its oxo-ammonium cation reacts with oxygen, the apparent catalytic rate constant

$$k_{app} = \frac{2k_1 k_2}{k_1 + k_2 + k_{-1}[H_2O_2]/[O_2^-]}$$

measured in the presence of catalase (or even under physiological levels of H_2O_2) k_{app} approaches its limiting value (k_{cat}):

$$k_{cat} = \frac{2k_1 k_2}{k_1 + k_2}$$

The failure to detect any decrease in nitroxide concentration under high flux of O_2^- suggests that $k_2 \gg k_1$ and, consequently, $k_{cat} \sim 2k_1$.

2. Reductive Mode

Oxazolidine derivatives such as OXANO can catalyze O_2^- dismutation via a reductive mode, flip-flopping between the nitroxide and its reduced form (48,73):

$$\text{N-O} + \cdot O_2^- + H^+ \rightleftarrows \text{N-OH} + O_2 \qquad \{3\}$$

$$\text{N-OH} + \cdot O_2^- + H^+ \rightleftarrows \text{N-O} + H_2O_2 \qquad \{4\}$$

The reverse reaction {-3} has been independently studied (74), but no appreciable reaction of the oxazolidine derivative with H_2O_2 was detected.

As was previously discussed (75) for other SOD mimics, the actually observed value (k_{app}) of the catalytic rate constant differs from its limiting value (k_{cat}):

$$k_{app} = \frac{2k_3 k_4}{k_3 + k_4 + k_{-3}[O_2]/[O_2^-]}$$

$$k_{cat} = \frac{2k_3 k_4}{k_3 + k_4}$$

Due to the oxidation (reaction −3) of the reduced catalyst by oxygen k_{app} is always smaller than k_{cat}. However, considering the values (73) of k_4, k_3, and k_{-3} then under high O_2^- flux, while

$$\frac{k_{-3}[O_2]}{[O_2^-]} \ll (k_3 + k_4)$$

k_{app} approaches k_{cat}.

Although nitroxides always inhibit damage inflicted in the presence of superoxide radicals, the immediate relevance of the SOD-mimic activity of nitroxides to their antioxidant activity has not been conclusively established. This is because deleterious species, other than O_2^-, also operate in most cases, hence the distinction between their roles and the role played by O_2^- per se is not straightforward.

B. Oxidation of Semiquinone Radicals

The reaction of nitroxide with semiquinone radicals generated upon bioactivation underlies its protective effect against cytotoxicity of various quinone-based xenobiotics. Such protection has been demonstrated for several quinone-based xenobiotics, including streptonigrin (24), mitomycin c (26), juglone (27), and paraquat (28). In all cases, a direct oxidation of the semiquinone by nitroxides, such as TEMPO and TEMPOL, has been demonstrated. The reaction with the nitroxide inhibited further reaction of deleterious semiquinones and prevented or decreased their cytotoxicity, as schematically presented in Scheme [4] for streptonigrin. Thus, the chain of reactions initiated through cell-induced reduction of the drug, yielding site-specifically formed ·OH radicals, is blocked when nitroxides detoxify the streptonigrin semiquinone (24).

[4]

C. Oxidation of Reduced Metal Ions

Complex formation between nitroxide and copper or iron ions has been studied previously (76), yet there is no evidence that such complexes are involved in its protective activity. The oxidation of reduced transition metals can preempt the Fenton reaction and consequently its injurious consequences. Nitroxides acting as mild oxidants can oxidize Fe^{II} and Cu^{I} and thus prevent oxidative damage (7,17). This nitroxide-induced oxidation of metal ions appears to be the most important mechanism underlying the nitroxides' protective effect.

$$\begin{array}{c} \ce{N-O} + Fe^{II} + H^+ \quad \xrightleftharpoons{\qquad\qquad} \quad \ce{N-OH} + Fe^{III} \end{array} \qquad \{5\}$$

Preliminary kinetic studies of TEMPOL reaction with Fe^{II} in MOPS buffer pH7, in the presence of BSA indicate that k_5, which increases with [OH⁻], equals 0.044 M^{-1} s^{-1}, and it increases by orders of magnitude in the presence

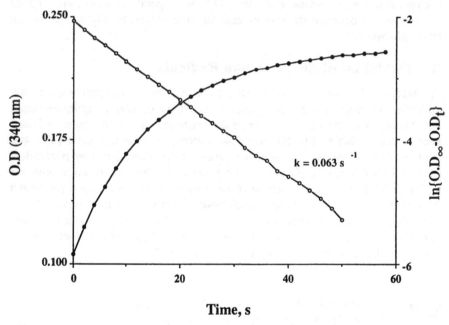

$$k = 0.063\ s^{-1}$$

Figure 1 Reaction between TEMPOL and Fe^{II}-BSA. 1.5 mM TEMPOL was mixed with 100 μM $FeSO_4$, 25 mg/ml BSA in 50 mM MOPS buffer pH7 at 30°C. Fe^{II} oxidation was determined by monitoring spectrophotometrically the O.D change at 340 nm (closed symbols). The pseudo first order was evaluated from time dependence of ln $\{O.D\infty - O.D_t\}$ (open symbols).

of phosphate (77). Evidently, practically all Fe^{II} ions in the tissue would be oxidized within a couple of minutes upon administration of nitroxide. Consequently, by maintaining the metal ions at their higher oxidation state they become less labile and cannot participate in the Fenton reaction.

D. Procatalatic-Mimic Activity and Detoxification of Hypervalent Metals

Ferryl-myoglobin (ferryl-Mb) and ferryl-hemoglobin (ferryl-Hb) as well as the globin radical have been previously implicated as capable of causing peroxidative damage to membranes (78–80). Therefore, by detoxifying these species, nitroxides are anticipated to provide protection particularly in cardiac injury. Mehlhorn and Swanson have studied the nitroxides' effect on the pseudocatalase activity of heme proteins (81). They show that nitroxides interact with heme proteins treated with H_2O_2 and stimulate in a catalytic fashion their catalase-like conversion of H_2O_2 to O_2 (catalatic activity). Unlike common antioxidants which react in a stoichiometric fashion, nitroxides have been shown to detoxify hypervalent metals as well as H_2O_2 while their concentrations do not change. The nitroxide-induced procatalatic activity can be inhibited by a two-electron reductant which reduces the oxo-ammonium cation to its respective hydroxylamine (82).

E. Interruption of Radical Chain Reactions

The nitroxide's inhibitory effect of lipid peroxidation has been recognized long ago (4). This led to the suggestion that nitroxides compete with oxygen for intermediate lipid free radicals and, in doing so, inhibit the chain reaction lead-

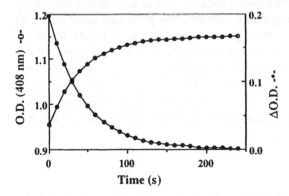

Figure 2 The reduction of ferryl-Mb by TEMPOL: Time dependence of the optical density at 408 nm upon mixing of 10 µM ferryl-Mb with 400 µM TEMPOL in 50 mM phosphate buffer, pH 6.5 at room temperature. OD (O), ΔOD (●).

ing to lipid peroxidation (4,83). This hypothesis was based on earlier findings demonstrating the inhibition by nitroxides of polymerization reaction (84). Nilsson et al. attribute the nitroxide inhibitory effect on LPO to blocking of primary and secondary LPO initiation, and scavenging of lipid radicals (17). Nitroxides have been shown to react with carbon-centered as well as oxygen-centered radicals (84,85). Miura et al. (43) suggest that hydrophilic nitroxides interfere as "preventive antioxidants" with the generation of ROS, whereas more lipophilic derivatives, as well as their reduced forms, acting as "chain-breaking antioxidants" inhibit the formation of lipid alkyl radicals (43).

F. Modulation of NO Level

No direct reaction of nitroxides with NO has been observed, except for nitronyl nitroxide (52); however, they are capable of indirectly increasing the steady-state concentration of nitric oxide. Because $O_2^{\cdot-}$ radicals readily react with NO yielding peroxynitrite, nitroxides, like any reagent which removes superoxide would increase the concentration of NO. This indirect effect, which might underlie the hypotensive activity of nitroxides (52,57), has been previously considered but could not be supported by results obtained with isolated perfused heart (52).

VI. POTENTIAL APPLICATIONS

Almost any model system or insult employed, so far, for testing the biological effect of nitroxides demonstrated their antioxidant activity. The only exception observed, adriamycin (86), lends further support to the conclusion that nitroxides are extremely effective modulators of processes mediated by paramagnetic species (radicals and transition metals). This feature makes the nitroxides a useful research tool for probing reactions and processes associated with free radicals. The second important potential application of nitroxides is the treatment or prevention of pathological conditions. The diverse experimental conditions manageable by nitroxides, as listed above, show the potential therapeutic value of nitroxides. Their use might offer a novel and promising treatment as antiulcerative agents and protectants against IBD in general, neuroprotective, antihemolytic, anti-inflammatory, cardioprotective, hypotensive, and radioprotective drugs.

VII. SUMMARY

The multimodal activity of nitroxides and the diverse mechanisms underlying their activities make nitroxides an unusual and unique type of antioxidants. This activity results from their capability to dismutate $O_2^{\cdot-}$, detoxify semiquinones, oxidize reduce metals, reduce hypervalent metals, detoxify lipid radicals, react with peroxyl, alkoxyl, and carbon-centered radicals. The features characterizing the biological activity of nitroxides can be summarized as follows:

1. Nitroxides enter cells readily and therefore do not require any preincubation period in order to exert their intracellular antioxidant activity.

2. Primarily nitroxides are neither toxic nor immunogenic, with the exception of *recA* mutants of *E. coli*, as long as their concentrations do not exceed the mM range.

3. Nitroxides, unlike other antioxidants, react neither with molecular O_2 nor with H_2O_2, yet they inhibit the toxic effects resulting from hyperoxia and H_2O_2.

4. Unlike common antioxidants, nitroxides preempt oxidative damage while acting as mild oxidants.

5. The nitroxides act, through oxidative or reductive mode, as genuine catalysts such that their concentrations remain time-invariant, while intra- and extracellular $O_2^{\cdot-}$ radicals are continuously removed.

6. Unlike common antioxidants which act in a suicidal mode, the rapid metabolic exchange between nitroxides and their reduced forms replenishes the nitroxide level.

7. In certain cases the antioxidative activity of hydroxylamines resembles, and even exceeds, that of their respective nitroxides.

8. In contrast with common antioxidants, which upon scavenging free radicals always yield secondary paramagnetic intermediate species, nitroxides terminate free-radical chain reactions forming diamagnetic end products.

ACKNOWLEDGMENT

This research was supported by a Grant # 309/931 from the Israel Science Foundation of the Israel Academy of Science, Jerusalem, Israel.

SYMBOLS

AAPH	2,2'-azobis(2-amidinopropane) dihydrochloride
AMVN	2,2'-azobis(2,4-dimethylvaleronitrile)
BBB	blood brain barrier
BHT	butylated hydroxy toluene
CAT-1	4-trimethylamino-2,2,6,6-tetramethylpiperidine-1-*N*-oxyl
CHDO	2-spirocyclohexane doxyl(2-spirocyclohexane-5,5-dimethyl-3-oxazolidinoxyl)
CHI	closed head injury

CHO	Chinese hamster ovary
CTPC	3-carbamoyl-2,2,5,5 tetramethyl-3-pyrroline-1-oxyl-3-carboxylic acid
CTPO	3-carbamoyl-2,2,5,5-tetramethyl-3-pyrroline-1-yloxy
DMPO	5,5-dimethyl-1-pyrroline-N-oxide
n-doxylstearate	n-(N-oxyl-4,4'-dimethyloxazolidin)-stearic acid
EPR	electron paramagnetic resonance
Hb	hemoglobin
HX/XO	hypoxanthine/xanthine oxidase
IBD	inflammatory bowel disease
i.p.	intraperitoneally
LMWA	low-molecular-weight antioxidants
LPO	lipid peroxidation
LTB_4 and LTC_4	leukotrienes B_4 and C_4
Mb	myoglobin
MDA	malonic dialdehyde
metHb	methemoglobin
MPO	myeloperoxidase
MRI	nuclear magnetic resonance imaging
NSS	neurological severity score
ΔNSS	differences of NSS
OP	1,10-phenanthroline
LDH	lactate dehydrogenase
OXANO	2-ethyl-2,5,5-trimethyl-3-oxazolidine-1-N-oxyl
OXANOH	1-hydroxy-2-ethyl-2,5,5-trimethyl-3-oxazolidine
PBN	phenyl-butyl nitrone
PROXYL	2,2,5,5-tetramethylpyrroline-1-N-oxyl
SOD	superoxide dismutase
t-BuOOH	t-butyl hydroperoxide
TNB	trinitrobenzene sulfonic acid
TEMPO	2,2,6,6,tetramethylpiperidine-1-N-oxyl
GSH	glutathione
TEMPOH	1-hydroxy-2,2,6,6,tetramethylpiperidine
TEMPOL	4-hydroxy-2,2,6,6-tetramethylpiperidine-1-N-oxyl
TEMPOLH	1,4-dihydroxy-2,2,6,6-tetramethylpiperidine
TEMPONE	4-oxo-2,2,6,6-tetramethyl-piperidine-1-N-oxyl
TEMPAMINE	4-amino-2,2,6,6-tetramethylpiperidine-1-N-oxyl

REFERENCES

1. Swartz HM. Principles of the metabolism of nitroxides and their implications for spin trapping. Free Radic Res Commun 1990; 9: 399–405.

2. Ionnone A, Tomasi A. Nitroxide radicals, their use as metabolic probes in biological model systems: an overview. Acta Pharm Yugosl 1991; 41: 277–297.

3. Brasch RC, London DA, Wesbey GE, Tozer TN, Nitecki DE, Williams RD, Doemeny J, Tuck LD, Lallemand DP. Work in progress: nuclear magnetic resonance study of a paramagnetic nitroxide contrast agent for enhancement of renal structures in experimental animals. Radiology 1983; 147: 773–779.

4. Weil TJ, Van der Veen J, Olcott HS. Stable nitroxides as lipid antioxidants. Nature 1968; 219: 168–169.

5. Zhdanov RI. Bioactive spin labels. New York: Springer-Verlag; 1992.

6. Mitchell JB, Samuni A, Krishna MC, DeGraff WG, Ahn MS, Samuni U, Russo A. Biologically active metal-independent superoxide dismutase mimics. Biochemistry 1990; 29: 2802–2807.

7. Samuni A, Godinger D, Aronovitch J, Russo A, Mitchell JB. Nitroxides block DNA scission and protect cells from oxidative damage. Biochemistry 1991; 30: 555–561.

8. Gelvan D, Moreno V, Clopton DA, Chen Q, Saltman P. Sites and mechanisms of low-level oxidative stress in cultured cells. Biochem Biophys Res Commun 1995; 206: 421–428.

9. Taylor L, Menconi MJ, Polgar P. The participation of hydroperoxides and oxygen radicals in the control of prostaglandin synthesis. J Biol Chem 1983; 258: 6855–6857.

10. Tanfani F, Carloni P, Damiani E, Greci L, Wozniak M, Kulawiak D, Jankowski K, Kaczor J, Matuszkiewics A. Quinolinic aminoxyl protects albumin against peroxyl radical mediated damage. Free Radic Res Commun 1994; 21: 309–315.

11. Antosiewicz J, Popinigis J, Wozniak M, Damiani E, Carloni P, Greci L. Effect of indolinonic and quinolinic aminoxyls on protein and lipid peroxidation of rat liver microsomes. Free Radic Biol Med 1995; 18: 913–917.

12. Gadzheva V, Ichimori K, Nakazawa H, Raikov Z. Superoxide scavenging activity of spin-labeled nitrosourea and triazene derivatives. Free Radic Res Commun 1994; 21: 177–186.

13. Emmerson PT. Enhancement of the sensitivity of anoxic *E. coli* B/r to x-rays by triacetoneamine-*N*-oxyl. Rad Res 1967; 30: 841–849.

14. Mitchell JB, DeGraff WG, Kaufman D, Krishna MC, Samuni A, Finkelstein E, Ahn MS, Hahn SM, Gamson J, Russo A. Inhibition of oxygen-dependent radiation-induced damage by the nitroxide superoxide dismutase mimic, tempol. Arch Biochem Biophys 1991; 289: 62–70.

15. Hahn SM, Wilson L, Krishna CM, Liebmann J, DeGraff W, Gamson J, Samuni A, Venzon D, Mitchell JB. Identification of nitroxide radioprotectors. Radiat Res 1992; 132: 87–93.

16. Hahn SM, Tochner Z, Krishna CM, Glass J, Wilson L, Samuni A, Sprague M, Venzon D, Glatstein E, Mitchell JB, et al. Tempol, a stable free radical, is a novel murine radiation protector. Cancer Res 1992; 52: 1750–1753.

17. Nilsson UA, Olsson LI, Carlin G, Bylund FA. Inhibition of lipid peroxidation by spin labels. Relationships between structure and function. J Biol Chem 1989; 264: 11131–11135.

18. Nilsson UA, Olsson LI, Thor H, Moldeus P, Bylund FA. Detection of oxygen

radicals during reperfusion of intestinal cells in vitro. Free Radic Biol Med 1989;
6: 251–259.

19. Pogrebniak H, Matthews W, Mitchell J, Russo A, Samuni A, Pass H. Spin trap
    protection from tumor necrosis factor cytotoxicity. J Surg Res 1991; 50: 469–474.
20. Takahashi M, Tsuchiya J, Niki E. Scavenging of radicals by vitamin E in the
    membrane as studied by spin label technique. J Am Chem Soc 1989; 111: 6350–
    6353.
21. Gelvan D, Saltman P, Powell SR. Cardiac reperfusion damage prevented by a
    nitroxide free radical. Proc Natl Acad Sci USA 1991; 88: 4680–4684.
22. Charloux C, Ishii K, Paul M, Loisance D, Astier A. Reduction production of the
    deleterious hydroxyl free radical during the final reperfusion of isolated rabbit heart
    with the use of an improved sodium lactobionate-based cardioplegic medium. J
    Heart Lung Transplant 1994; 13: 481–488.
23. Tosaki A, Haseloff RF, Hellegouarch A, Schoenheit K, Martin VV, Das DK,
    Blasig IE. Does the antiarrhythmic effect of DMPO originate from its oxygen
    radical trapping property or the structure of the molecule itself? Basic Res Cardiol
    1992; 87: 536–547.
24. Krishna MC, Halevy RF, Zhang R, Gutierrez PL, Samuni A. Modulation of strep-
    tonigrin cytotoxicity by nitroxide SOD mimics. Free Radic Biol Med 1994; 17:
    379–388.
25. An J, Hsie AW. Effects of an inhibitor and a mimic of superoxide dismutase on
    bleomycin mutagenesis in Chinese hamster ovary cells. Mutat Res 1992; 270: 167–
    175.
26. Krishna MC, DeGraff WG, Tamura S, Gonzalez FJ, Samuni A, Russo A, Mitchell
    JB. Mechanisms of hypoxic and aerobic cytotoxicity of mitomycin C in Chinese
    hamster V79 cells. Cancer Res 1991; 51: 6622–6628.
27. Zhang R, Hirsch O, Mohsen M, Samuni A. Effects of nitroxide stable radicals on
    juglone cytotoxicity. Arch Biochem Biophys 1994; 312: 385–391.
28. Aronovitch J, Samuni A, Godinger D, Krishna MC, Mitchell JB. Personal com-
    munication, 1995.
29. Grinberg NL, Samuni A. Nitroxide stable radical prevents primaquine-induced lysis
    of red blood cells. Biochim Biophys Acta 1994; 1201: 284–288.
30. Rachmilewitz D, Karmeli F, Okon E, Samuni A. A novel antiulcerogenic stable
    radical prevents gastric mucosal lesions in rats. Gut 1994; 35: 1181–1188.
31. Karmeli F, Eliakim R, Okon E, Samuni A, Rachmilewitz D. A stable nitroxide
    radical effectively decreases mucosal damage in experimental colitis. Gut 1994; 35:
    1181–1188.
32. Beit-Yannai E, Renliang Zhang R, Trembovler V, Samuni A, Shohami E. Unpub-
    lished results, 1995.
33. Bitterman N, Samuni A. Nitroxide stable radicals protect aganst hyperoxic-induced
    eizures in rats. Undersea Hyperb Med 1995; 22 Suppl: 47–48.
34. Antosiewicz J, Bertoli E, Damiani E, Greci L, Popinigis J, Przybylski S, Tanfani
    F, Wozniak M. Indolinonic and quinolinic aminoxyls as protectants against oxi-
    dative stres. Free Radic Biol Med 1993; 15: 203–208.
35. Hu HP, Sosnovsky G, Li SW, Rao NU, Morse P2, Swartz HM. Development of
    nitroxides for selective localization inside cells. Biochim Biophys Acta 1989; 1014:
    211–218.

36. Zhdanov RI, Komarov PG. Sterically-hindered hydroxylamines as bioactive spin labels. Free Radic Res Commun 1990; 9: 367–377.
37. Sosnovsky G, Li SW. In the search for new anticancer drugs. X. $N,N;$ N',N'-bis(1,2-ethanediyl)-N''-(1-oxyl-2,2,6,6-tetramethyl-4-piperidinylaminocarbonyl) phosphoric triamide—a new potential anticancer drug of high activity and low toxicity. Life Sci 1985; 36: 1473–1477.
38. Sosnovsky G, Li SW. In the search for new anticancer drugs XI. Anticancer activity of nitroxyl labeled phosphoric $N,N;N',N';N,N$-tris[1,2-ethanediyl]triamide(TEPA) and phosphorothioic $N,N;N',N';N,N$-tris[1,2-ethanediyl]triamide(thio-TEPA) derivatives. Cancer Lett 1985; 25: 255–260.
39. Sosnovsky G. A critical evaluation of the present status of toxicity of aminoxyl radicals. J Pharm Sci 1992; 81: 496–499.
40. Sosnovsky G, Baysal M, Erciyas E. In the search for new anticancer drugs. 28. Synthesis and evaluation of highly active aminoxyl labeled amino acid derivatives containing the [N'-(2-chloroethyl)-N'-nitrosoamino]carbonyl group. J Pharm Sci 1994; 83: 999–1005.
41. Chan HC, Sun KQ, Magin RL, Swartz HM. Potential of albumin labeled with nitroxides as a contrast agent for magnetic resonance imaging and spectroscopy. Bioconjug Chem 1990; 1: 32–36.
42. Liebmann J, Bourg J, Krishna CM, Glass J, Cook JA, Mitchell JB. Pharmacokinetic properties of nitroxide-labeled albumin in mice. Life Sci 1994; 54: PL506–PL509.
43. Miura Y, Utsumi H, Hamada A. Antioxidant activity of nitroxide radicals in lipid peroxidation of rat liver microsomes. Arch Biochem Biophys 1993; 300: 148–156.
44. Komarov P, Morgunov A, Zhdanov R, Bilenko M. Anti-ischaemic action of N-hydroxy spin label: a novel antioxidant agent. J Mol Cell Cardiol 1988; 20 Suppl V: S26.
45. Komarov PG, Taskaeva ON, Zhdanov RI, N-Hydroxytetramethylpiperidines as bioantioxidants. Dokl Akad Nauk SSSR 1987; 297: 734–737.
46. Nilsson UA, Carlin G, Bylund FA. The hydroxylamine OXANOH and its reaction product, the nitroxide OXANO, act as complementary inhibitors of lipid peroxidation. Chem Biol Interact 1990; 74: 325–342.
47. Wang G, Godinger D, Aronovitch J, Samuni A. Opposing effects of nitroxide free radicals in *E. coli* mutants deficient of DNA repair. Unpublished results, 1995.
48. Samuni A, Krishna CM, Mitchell JB, Collins CR, Russo A. Superoxide reaction with nitroxides. Free Radic Res Commun 1990; 9: 241–249.
49. Mohsen M, Pinson A, Zhang R, Samuni A. Do nitroxides protect cardiomyocytes from hydrogen peroxide or superoxide? Mol Cell Biochem 1995; 145: 103–110.
50. Samuni A, Winkelsberg D, Pinson A, Hahn SM, Mitchell JB, Russo A. Nitroxide stable radicals protect beating cardiomyocytes against oxidative damage. J Clin Invest 1991; 87: 1526-1530.
51. Reddan JR, Sevilla MD, Giblin FJ, Padgaonkar V, Dziedzic DC, Leverenz V, Misra IC, Peters JL. The superoxide dismutase mimic TEMPOL protects cultured rabbit lens epithelial cells from hydrogen peroxide insult. Exp Eye Res 1993; 56: 543–554.

52. Konorev EA, Tarpey MM, Joseph J, Baker JE, Kalyanaraman B. Nitronyl nitroxides as probes to study the mechanism of vasodilatory action of nitrovasodilators, nitrone spin traps, and nitroxides: role of nitric oxide. Free Radic Biol Med 1995; 18: 169–77.

53. Zeltcer G, Berenshtein E, Samuni A, Chevion M. The effects of copper and stable radicals on postischemic reperfusion injury in isolated rat heart. Personal communication 1995.

54. Emmerson P, Howard-Flanders P. Sensitization of anoxic bacteria to x-rays by di-t-butyl nitroxide and analogues. Nature 1964; 204: 1005–1006.

55. Konovalova NP, Diatchkovskaya RF, Volkova LM, Varfolomeev VN. Nitroxyl radicals decrease toxicity of cytostatic agents. Anticancer Drugs 1991; 2: 591–595.

56. Goffman T, Cuscela D, Glass J, Hahn S, Krishna CM, Lupton G, Mitchell JB. Topical application of nitroxide protects radiation-induced alopecia in guinea pigs. Int J Radiat Oncol Biol Phys 1992; 22: 803–806.

57. Brethernick L, Lee G, Lund E, Wragg W, Edge N. Congegers of pempidine with high ganglion-blocking activity. Nature 1959; 184: 1707–1708.

58. Wolfe JE, Ross D, Cohen JM. A role for metals and free radicals in the induction of apoptosis in thymocytes. FEBS Lett 1994; 352: 58–62.

59. Krishna MC, Grahame DA, Samuni A, Mitchell JB, Russo A. Oxoammonium cation intermediate in the nitroxide-catalyzed dismutation of superoxide. Proc Natl Acad Sci USA 1992; 89: 5537–41.

60. Belkin S, Mehlhorn RJ, Hideg K, Hankovsky O, Packer L. Reduction and destruction rates of nitroxide spin probes. Arch Biochem Biophys 1987; 256: 232–243.

61. Chen K, Swartz HM. Oxidation of hydroxylamines to nitroxide spin labels in living cells. Biochim Biophys Acta 1988; 970: 270–277.

62. Chen K, Glockner JF, Morse P2, Swartz HM. Effects of oxygen on the metabolism of nitroxide spin labels in cells. Biochemistry 1989; 28: 2496–2501.

63. Chen K, Swartz HM. The products of the reduction of doxyl stearates in cells are hydroxylamines as shown by oxidation by 15N-perdeuterated Tempone. Biochim Biophys Acta 1989; 992: 131–133.

64. Sies H, Mehlhorn R. Mutagenicity of nitroxide-free radicals. Arch Biochem Biophys 1986; 251: 393–396.

65. DeGraff WG, Krishna MC, Russo A, Mitchell JB. Antimutagenicity of a low molecular weight superoxide dismutase mimic against oxidative mutagens. Environ Mol Mutagen 1992; 19: 21–26.

66. Gallez BC D, Debuyst R, Dejehet F, Dumont P. Mutagenicity of nitroxyl compounds: structure-activity relationships. Toxicol Lett 1992; 63: 35–45.

67. Dawkins B, Patterson RM. Antioxidant regulation of free radical activity in mammalian cells. Cytogenetic and electron spin resonance studies. J Cell Biol 1979; 83: 172A.

68. Lyn-Cook BD, Patterson RM. Inhibition of free radical induced chromosomal aberrations in vitro by alfa-tocopherol. Nutr Res 1984; 4: 989–993.

69. Johnstone PA, DeGraff WG, Mitchell JB. Protection from radiation-induced chromosomal aberrations by the nitroxide Tempol. Cancer 1995; 75: 2323–7.

70. Afzal V, Brasch RC, Nitecki DE, Wolff S. Nitroxyl spin label contrast enhanc-

ers for magnetic resonance imaging. Studies of acute toxicity and mutagenesis. Invest Radiol 1984; 19: 549–552.

71. Ankel EG, Lai CS, Hopwood LE, Zivkovic Z. Cytotoxicity of commonly used nitroxide radical spin probes. Life Sci 1987; 40: 495–498.

72. Voest EE, van FE, van AB, Neijt JP, Marx JJ. Increased hydrogen peroxide concentration in human tumor cells due to a nitroxide free radical. Biochim Biophys Acta 1992; 1136: 113–118.

73. Samuni A, Krishna CM, Riesz P, Finkelstein E, Russo A. A novel metal-free low molecular weight superoxide dismutase mimic. J Biolc Chem 1988; 263: 17921–17924.

74. Rosen GM, Finkelstein E, Rauckman EJ. A method for the detection of superoxide in biological systems. Arch Biochem Biophys 1982; 215: 367–378.

75. Czapski G, Goldstein S. The uniqueness of superoxide dismutase (SOD)-why cannot most copper compounds substitute SOD in vivo? Free Radic Res Commun 1988; 4: 225–229.

76. Easton SS, Eaton GR. Interaction of spin labels with transition metals. Coord Chem Rev 1978; 26: 207–262.

77. Bar-Or P, Mohsen M, Chevion M, Samuni A. Unpublished results, 1995.

78. Turner JJO, Rice-Evans CA, Davis MJ, Newman ESR. The formation of free radicals by cardiac myocytes under oxidative stress and the effects of electron-donating drugs. Biochem J 1991; 277: 833–837.

79. Osawa Y, Korzekwa K. Oxidative modification by low levels of HOOH can transform myoglobin to an oxidase. Proc Natl Acad Sci USA 1991; 88: 7081–7085.

80. Osawa Y, Darbyshire JF, Meyer CA, Alayash AI. Differential susceptibilities of the prosthetic heme of hemoglobin-based red cell substitutes. Implications in the design of safer agents. Biochem Pharmacol 1993; 46: 2299–305.

81. Mehlhorn RJ, Swanson CE. Nitroxide-stimulated H_2O_2 decomposition by peroxidases and pseudoperoxidases. Free Radic Res Commun 1992; 17: 157–175.

82. Mehlhorn RJ, Gomez J. Hydroxyl and alkoxyl radical production by oxidation products of metmyoglobin. Free Radic Res Commun 1993; 18: 29–41.

83. Rauckman EJ, Rosen JM, Griffith LK. Enzymatic reactions of spin labels. In: Holtzman X, ed. Spin Labeling in Pharmacology. New York: Academic Press, 1984: 175–190.

84. Brownlie IT, Ingold KU. Can J Chem 1967; 45: 2427–2432.

85. Bowery VW, Ingold KU. Kinetics of nitroxide radical trapping. 2. Structural effects. J Am Chem Soc 1992; 114: 4992–4997.

86. DeGraff WG, Hahn SM, Mitchell JB, Krishna MC. Free radical modes of cytotoxicity of adriamycin and streptonigrin. Biochem Pharmacol 1994; 48: 1427–1435.

14

Mimics of Superoxide Dismutase

Kevin M. Faulkner
Texas Wesleyan University School of Law, Irving, Texas

Irwin Fridovich
Duke University Medical Center, Durham, North Carolina

I. INTRODUCTION

Chemists have attempted to mimic such enzymes as P450s, catalases, hemoglobin (and myoglobin), photosynthetic reaction centers, and superoxide dismutases to name a few (1,2). An ideal mimic would catalyze the same reaction as the parent enzyme with specificity and efficiency matching that of the native enzyme. This remains an elusive goal, but definitely one worth striving toward.

Superoxide dismutases are metalloproteins that catalyze the dismutation of superoxide ion, $O_2^{\cdot-}$, to dioxygen and hydrogen peroxide [Eq. (1)].

$$O_2^{\cdot-} + O_2^{\cdot-} + 2H^+ \rightarrow H_2O_2 + O_2 \tag{1}$$

Uncatalyzed, this reaction occurs with a rate constant of approximately 2×10^5 M^{-1} s^{-1} at pH 7.4 (3). Superoxide dismutases catalyze the reaction with a rate constant of about 2×10^9 M^{-1} s^{-1} (4,5). The superoxide dismutase catalyzed rate of reaction (1) is among the fastest enzymatic rates known, and is diffusion-limited.

All SOD enzymes are metalloproteins, and these metal centers are essential for their function (6–8). There are three major classes of SODs: Mn-centered SODs (MnSOD), the Cu–Zn-centered SODs (CuZnSOD), and Fe-centered

SODs (FeSOD). The structures of all three types of SOD have been determined by x-ray crystallography (9–15).

During the catalytic process the metal center cycles between two oxidation states. For the MnSOD and the FeSOD, the cycle is between $3+/2+$, while in the CuZnSOD the Cu cycles between $2+/1+$. Since it is known that large inner-sphere reorganization can limit electron transfer processes in metal coordination complexes, the SOD enzyme structure must avoid the inner-coordination sphere changes during the redox change of the metal center (16). In the case of both Cu and Mn coordination complexes, it is known that there is a large change in the coordination geometry upon oxidation or reduction of the metal (17–19).

Cu(I) complexes tend to be 4-coordinate planar, the Cu(II) complexes tend to be tetrahedral or 6-coordinate distorted octahedral (9,17). In CuZnSODs, the Cu inner-coordination sphere is 5-coordinate distorted tetrahedral, between that of a planar and tetrahedral/distorted octahedral geometry(9). The structure of the inner-coordination sphere is shown in Fig. 1. The distorted tetrahedral geometry of CuZnSOD imparts to the Cu center a transition-state-like structure, thus lowering the barrier to electron transfer. Azurins and plastocyanins also have a similar arrangement of ligands around the metal center, lowering the activation barrier to electron transfer (16).

Manganese complexes also show a large inner-coordination sphere change in going from the Mn^{2+} to Mn^{3+} states (17,18). In unrestricted Mn complexes such as $Mn(H_2O)_6$, $Mn(urea)_6$, $Mn(bpyO_2)_3$ ($bpyO_2$ = 1,1'-dioxo-2,2'-bipyridine), there is as much as a 25% change in average bond distances during a redox cycle (compared to a 14% change in average bond distances in similar Fe complexes, and less than 5% change in Ru complexes) (18). This large change in metal to ligand bond lengths presents a large activation barrier to electron transfer (16–19). However, the activation barrier to the $Mn^{2+/3+}$ couple is overcome when more lipid ligands such as EDTA, CDTA, and 3,6,10, 13,16,19-hexaazabicyclo[6.6.6]eicosane (18).

The rigid protein environment of metalloproteins such as MnSOD act in much the same way: the protein offers structural rigidity at the metal center through the series of H-bonding, hydrophobic, and ionic interactions throughout a protein's tertiary structure. In the active site of the MnSOD, the geometry around the Mn(III) center is a trigonal bipyramid as shown in Fig. 2, with one axial position open for water coordination (12). The x-ray crystallographic structure for the MnSOD, where the Mn is in the $+2$ oxidation state, reveals little change in the inner-coordination sphere of the Mn center (15). As with other redox-active metalloproteins, not only does the inner-coordination sphere stabilize the metal center in oxidized and reduced states, the entire enzyme structure does so.

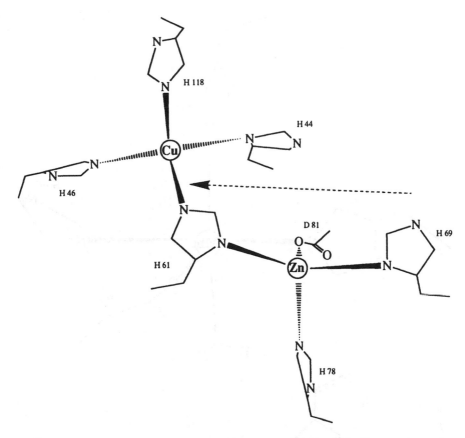

Figure 1 Inner-coordination sphere structure of CuZnSOD. This diagram shows the metal centers in the CuZnSOD, with the liganding amino acid residues. The arrow points to solvent/superoxide access channel. (Adapted from Refs. 9 and 10.)

The arrangement and identity of amino acid ligands around the metal centers has another important consequence. The inner-coordination sphere of the SODs metal center control the redox potentials (E^0) of the metal centers. In spite of the large variation in redox potentials for coordination compounds of Cu and Mn, both MnSOD and CuZnSODs have nearly the same E^0 value of +300 mV (20). This is not a coincidence, for it is this E^0 value that gives the dismutation reaction its highest driving force. Taken separately, the half-cell E^0 in Eq. a for the H_2O_2/O_2^- couple is +890 mV (versus NHE, at pH 7), while that for the O_2/O_2^- is –330 mV (versus NHE, at pH 7) (20). The E^0 value for

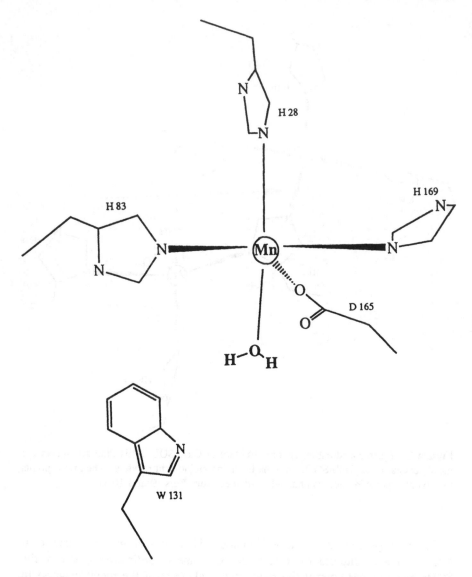

Figure 2 Inner-coordination sphere structure of MnSOD. This diagram shows the metal center in MnSOD, with the liganding amino acids and water molecule. Tryptophan 131 (from MnSOD of *Thermus termophilus* HB8) is about 3.5 Å from the active site, but not a ligand of the metal center. (Adapted from Refs. 11–15.)

SODs is between these E^0 values, giving a relative driving force of 600 mV for each half-reaction during the $Cu^{+/2+}$ and $Mn^{2+/3+}$ redox cycles in SODs.

Another stabilizing factor in metalloproteins is the hydrophobic nature of the "pocket" in which the metal resides. This hydrophobic region limits the interaction of the metal center for solvent molecules, and serves to tailor the metal center for its specific function (9,12). For both the CuZnSOD and MnSOD, the inner-coordination and outer-coordination spheres of the metal are arranged such that only one position on the metal is available for ligand exchange. This is presumably where the O_2^- molecule would interact with the metal center of the enzyme. Such is the case for MnSOD, where several conserved tryptophan residues around the metal center impart a hydrophobic environment, thus allowing the interaction of only one water molecule at the metal center (12).

In contrast to the stabilizing effect of the hydrophobic amino acid residues in the active site is the presence of positively charged amino acid residues near the open coordination site on the metal center (9,12). At physiological pH, the superoxide ion is an anion as depicted in Eq. a. Thus, the positively charged amino acid residues that face the solution will serve to attract the O_2^- species and hence increase the collision frequency between substrate and catalyst (21–23). Since the rate of a bimolecular reaction is limited by the frequency at which two reacting species collide, electrostatic attraction between SOD and O_2^- increases the rate of reaction. Recent studies show that when the native arginine residue at the active site of CuZnSOD is replaced by neutral or negatively charged residues, rates of reaction with O_2^-, or $Fe(II)(CN)_6^{4-}$ decrease (24).

Given the above facts, it is a daunting task for the chemist to set about trying to mimic nature in the case of SODs. Yet, much work has been directed to that goal. And, as our understanding of how oxidative processes contribute to irreversible biological tissue damage, the interest in a mimic of SOD is growing. We will now try to review attempts at mimicking SOD enzymes.

II. INORGANIC COORDINATION COMPLEXES AS SOD MIMICS

Several criteria determine which synthetic compounds will be the best mimics of SOD. These are (1) stability of the complex, or compound, in both oxidation states under the conditions in which it will be used, (2) an E^0 value at or near $+300$ mV versus NHE, and (3) the incorporation of cationic groups in the complex to enhance electrostatic attraction for O_2^-. Obviously, another consideration is the overall toxicity of the synthetic mimic in living systems. The intact mimic, and any dissociation or decomposition or metabolic products, should have minimal toxicity and be easily eliminated from the organism.

Our discussion will begin with metal-centered SOD mimics. The sheer number of different compounds tested as SOD mimics precludes a detailed discussion of each. Our hope is to give some historical context while discussing some studies that are representative of each class of compounds. The discussion will be subdivided into complexes of Mn, Cu, and Fe, and finally, nonmetallic organic compounds that mimic SOD.

A. Mn-Based SOD Mimics

Lactobacillus plantarum, which is aerotolerant but devoid of SOD, was seen to accumulate 25–30 mM Mn(II) as a functional replacement for SOD (25–28). Extracts of these cells contained a catalyst of the dismutation of O_2^- which was dialyzable and heat stable, yet appeared macromolecular, in that it would not pass through an ultrafilter. These cells also accumulate high-molecular-weight polyphosphate and the activity in the extracts appeared to be due to a Mn(II) polyphosphate. Since the Mn(II) polyphosphate acted as a true catalyst, it seemed likely that it was undergoing oxidation and then reduction during sequential interactions with O_2^-. Yet, studies in solutions lacking ligands suggested that O_2^- is incapable of oxidizing Mn(II) to Mn(III) (29), yielding MnO_2^+ or MnO_2H^{2+} instead.

This seeming discrepancy was resolved in a subsequent study (30), which demonstrated the formation of Mn(III) from Mn(II), when ligands were present which could stabilize Mn(III). Trivalent manganese is a potent oxidant and can be reduced by O_2^- or by other reductants. This can lead to puzzling results. Thus, in the classical xanthine oxidase/cytochrome c assay for SOD activity (6), done in phosphate buffer, Mn(II) inhibited the reduction of cytochrome c, but also caused the oxidation of ferrocytochrome c (30). The explanation lay in the following reactions:

$$\text{Mn(II) phosphate} + O_2^- + 2H^+ \rightarrow H_2O_2 + \text{Mn(III) phosphate} \qquad (2)$$

$$\text{Mn(III) phosphate} + \text{urate} \rightarrow \text{Mn(II) phosphate} + \text{urate radical}^+ \qquad (3)$$

$$\text{Urate radical}^+ + \text{ferrocyt c} \rightarrow \text{urate} + \text{ferricyt c} \qquad (4)$$

The inhibition of cytochrome c reduction by catalytic levels of Mn(II) was biphasic in a pyrophosphate-buffered $(P \sim P)$ medium; an indication that the oxidation of $Mn(II)P \sim P$ by O_2^- was more rapid than the subsequent reduction of $Mn(III)P \sim P$ by O_2^-. Since $Mn(III)P \sim P$ has a strong absorption band at 259 nm, conversion of the manganous to the manganic compound by O_2^- could be observed directly (30). Further evidence for the redox cycling of Mn(II) pyrophosphate by O_2^- was obtained. Thus $Mn(II)P \sim P$ was oxidized to $Mn(III)P \sim P$ by a flux of O_2^- and then could be converted to Mn(II)EDTA by the O_2^-. Clearly

EDTA could not displace P ~ P from the Mn(III) compound but could do so from the Mn(II) compound and O_2^- was the agent that reduced the former to the latter, since SOD halted the conversion (30). Phosphate could replace pyrophosphate in supporting these conversions, but higher concentrations were needed.

Mn(III), as the phosphate or the pyrophosphate, could also be reduced to Mn(II) by H_2O_2, yet there was no production of O_2^- during this reaction suggesting the overall reaction: 2 Mn(III) + H_2O_2 → $2H^+$ + O_2 + 2Mn(II). Mechanistically, one might suppose this reaction to occur via binuclear complexes of Mn(III). The effect of a variety of ligands in the O_2^- dismuting activity of manganese was reported (30).

These studies point to the importance of the stability and reactivity of both oxidation states of Mn in the presence of the ligand. The specific nature of the ligand determines if it will enhance the SOD activity of complexed Mn relative to free Mn^{2+}, or eliminate it. Thus, a complex such as MnEDTA is very stable, but inactive, while Mn^{2+}/polyphosphate mixtures at 1:4 ratios are active and stable (30). Since the stability constants are too high for EDTA complexes of Mn(II) and Mn(III) and it is known that MnEDTA is not a catalyst for superoxide dismutation, a screening test of any mimic of SOD might be to determine its persistence in the presence of EDTA.

1. Bidentate Ligand Complexes

Recently, di-Schiff base metal complexes have been studied as possible SOD mimics (31–36). These complexes, due to their bidentate nature, tend to form more stable complexes than their monodentate counterparts (37). Baudry et al. have synthesized a series of Mn-salen complexes as shown in Fig. 3 (34). These complexes have been studied as catalysts for expoxidation of alkenes, but it was recognized that they may have SOD activity as well. Of the six compounds tested, one (Fig. 3, (i)) showed high SOD activity in the 1–3 μM range in aqueous media (34). Of all the compounds tested in the study of Baudry, most were inhibited by EDTA accept compound (i) (34). They found that the nature of the aromatic ring substituents had a profound influence on the SOD activity. In particular, the bulkier of ring substituents the less it was inhibited by EDTA and the higher SOD activity it possessed. Steric hindrance may provide stability to the metal center in much the same way that the hydrophobic pocket of the active sites in SOD enzymes do, by preventing solvent molecules from complexing the metal.

More recently, Liu et al. have synthesized and tested a Mn-Schiff base complex as an SOD mimic (35). Figure 3a shows the ligand used in the work of Liu et al. In the xanthine-xanthine oxidase/cytochrome c assay (6), the Mn-salophen complex had an activity of 0.67 μM per SOD unit (35). The complex was found to behave in a catalytic manner in aqueous media, and was an im-

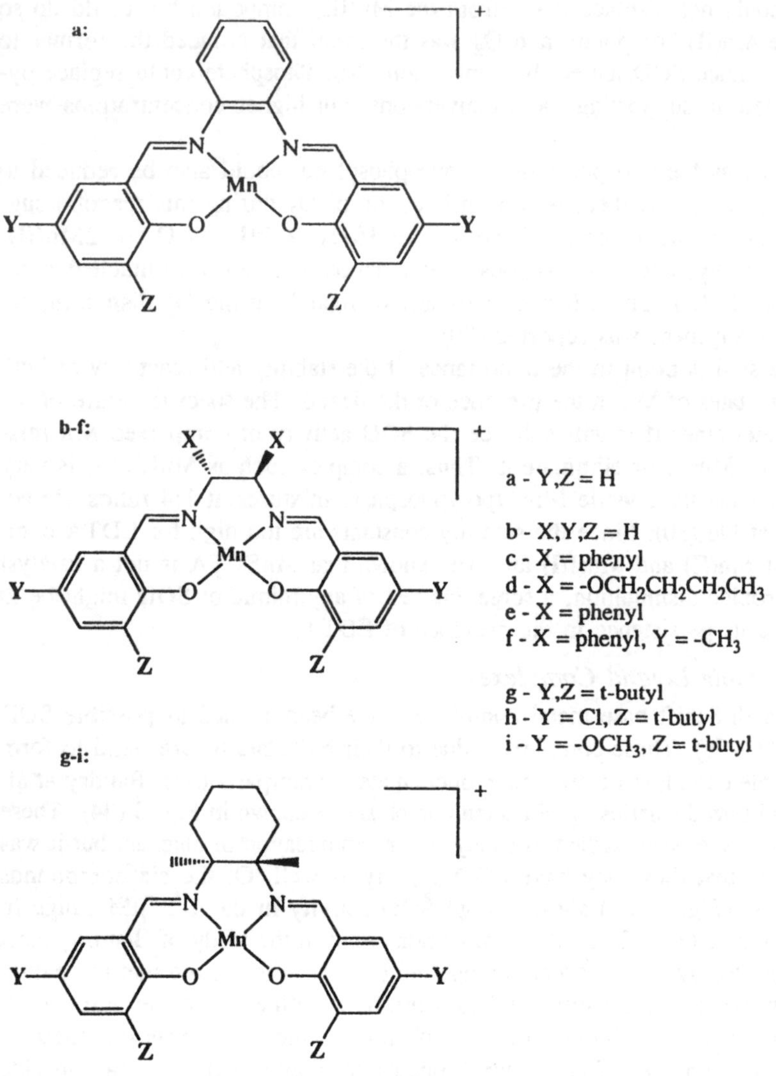

Figure 3 Mn-Schiff base complexes as SOD mimics. This diagram shows the structures, along with the charges in solution at pH 7.5, of the various Schiff base complexes of Mn from Baudry et al. (34) and Liu et al. (35).

provement over the previous Mn(picolinato)$_3$ complexes. However, in the presence of 1 mM EDTA the activity was inhibited by 50%. This is consistent with the findings of Boggess et al. that in aqueous solution the electrochemistry of these Mn complexes are quasi-reversible, hence indicating that there may be some ligand dissociation during the reduction cycle (31).

This complex also proved to be successful in vivo. As little as 5–10 μM of the Mn-salophen complex greatly enhanced the aerobic growth of *Escherichia coli* QC779 SOD-null cells over a 24-h period. However, 50 μM of the complex proved lethal to the *E. coli* QC779 cells, as did the ligand alone (35). In the same experiment, 50 μM Mn^{2+} was not lethal.

Recently, another Mn complex has been tested for its biological activity (36). In their study, synaptic transmissions in hippocampal cells were followed during anoxic episodes. Anoxia causes irreversible damage to synaptic transmission on CA1 pyramidal cells which was protected by 50 μM EUK-8 (36). It was also shown that the EUK-8 complex prevented lipid peroxidation.

2. Tridentate Ligand Complexes

SOD mimics based on trihydroxamic acid coordination compounds were used (38,39) as an extension of the highly stable Fe desferrioxamine B complex (40). Desferrioxamine B ($DFBH^+$; see Fig. 4) proved to form a stable complex with Mn (38,39,41). The tridentate nature of the $DFBH^+$ ligand is expected to have an advantage over monodentate ligands in coordinating ability. Also, the $Mn(III)DFBH^+$ complex will have a positive charge at pH 7–8, thus contributing favorably to its possible catalytic activity.

The reaction of $DFBH^+$ with MnO_2 resulted in a green solution with a broad metal to ligand charge transfer band at 610 nm (39). There was some argument as to the exact structure of this product (39,41). Faulkner et al. showed that the green complex could be formed with Mn^{2+}, $Mn(OH)_3$, or with MnO_2 (42). In the case of MnO_2, a reducing agent, either NH_2OH or the ligand itself, was needed, while with Mn^{2+} O_2 was needed as an oxidant (42). It was also shown by electrospray ionization mass spectroscopy that the green complex in aqueous solution is a hexadentate, 1:1 complex of Mn(III) and $DFBH^+$. The structure of the complex is shown in Fig. 4. Thus, the $Mn(III)DFBH^+$ complex is an analog of the $Fe(III)DFBH^+$ complex.

Initial in vitro studies of the $MnDFBH^+$ complex showed that 10 μM of the complex was equivalent to 1 unit of SOD activity (38,42). This activity was due to the $MnDFBH^+$ complex and not free Mn^{2+} (42). However, it was found that this activity was eliminated by EDTA. Thus, in spite of the tridentate nature of the $DFBH^+$ ligand which stabilizes Mn(III), it appears that during the redox cycle the complex falls apart when forced to compete with EDTA, which has a higher affinity for Mn^{2+}.

The function of an SOD mimic in vivo is complicated by the presence of many ligands, oxidants, and reductants that might interact with the complex. In vivo studies of the $MnDFBH^+$ complex demonstrated that it protected *Dunaliella salina* cells against light-dependent bleaching by paraquat (43). The $DFBH^+$ ligand alone was toxic to this alga, and Mn^{2+} did not protect, suggesting that the green $MnDFBH^+$ complex was responsible for the protective effects observed. In the absence of EDTA, the $MnDFBH^+$ was stable in O_2^- (42).

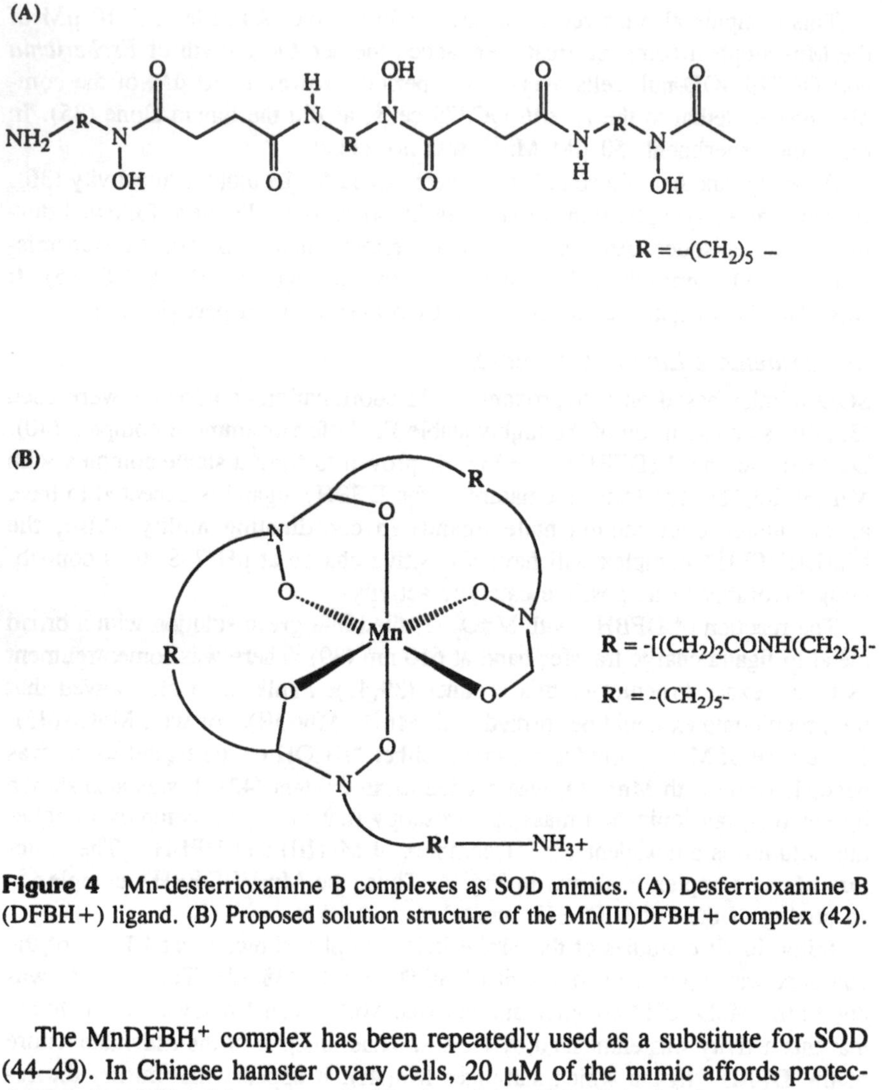

(A)

$R = -(CH_2)_5 -$

(B)

$R = -[(CH_2)_2CONH(CH_2)_5]-$

$R' = -(CH_2)_5-$

Figure 4 Mn-desferrioxamine B complexes as SOD mimics. (A) Desferrioxamine B (DFBH+) ligand. (B) Proposed solution structure of the Mn(III)DFBH+ complex (42).

The MnDFBH+ complex has been repeatedly used as a substitute for SOD (44–49). In Chinese hamster ovary cells, 20 µM of the mimic affords protection against 200 µM paraquat-induced toxicity (44). Selenite-dependent cytotoxicity in rat hepatocytes was also diminished by MnDFBH+ (45). Bhuyan et al. have shown that MnDFBH+ protects rabbit lenses against diquat-induced cataract formation (46).

Some in vivo studies suggest that the MnDFBH+ mimic may confer other modes of protection. Thus in V79 Chinese hamster cells (47,48), the MnDFBH+ complex was found to protect against t-butyl hydroperoxide damage. The MnDFBH+ mimic has also been found to protect human erythrocytes, human lymphocytes, and human bladder carcinoma cells against riboflavin-sensitized phototoxicity. Micromolar concentrations of the mimic have been shown to be sufficient to protect against photochemical toxicity.

3.   Cyclic Ligand Complexes

The cyclic analog of DFBH$^+$, desferrioxamine E (DFE) was reacted with MnCl$_2$ in an attempt to form a more stable complex (42). The DFE ligand formed a dark green complex with Mn(II), which is presumably a 1:1, 6-co-ordinate complex. However, this complex proved to be a very poor SOD mimic. The complex was stable toward ligand dissociation during O$_2^-$ flux.

Manganese macrocyclic amine complexes have been studied as SOD mimics (50–53). The most successful of the 20 complexes studied were the Mn(II) complexes of 1,4,7,10,13-pentaazacyclopentadecane (Mn([15]aneN$_5$)Cl$_2$) and 1,4,7,10,13-pentaazacyclohexadecane (Mn([16]aneN$_5$)Cl$_2$) (Fig. 5) (50). Using

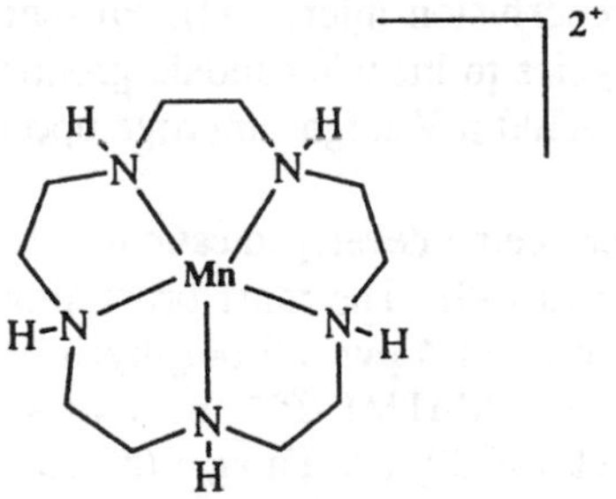

Mn(II)[15]aneN$_5$

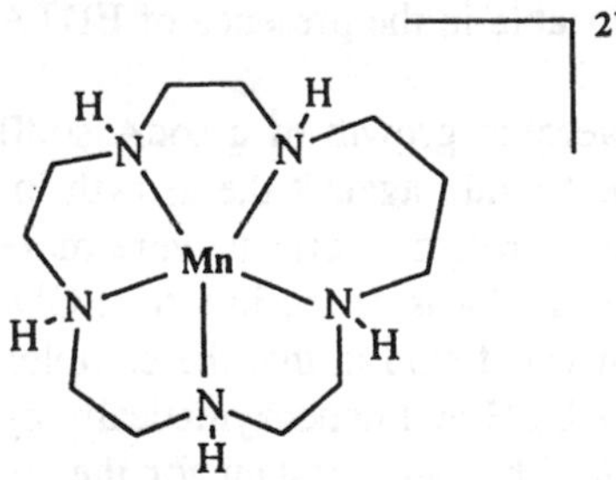

Mn(II)[16]aneN$_5$

Figure 5   Mn-macrocyclic complexes as SOD mimics. These are the proposed solution structures for Mn(II)(1,4,7,10,13-pentaazacyclopentadecane)$^{2+}$ (Mn([15]aneN$_5$)$^{2+}$) and Mn(II)(1,4,7,10,13-pentaazacyclohexadecane)$^{2+}$ (Mn([16]aneN$_5$)$^{2+}$) as adapted from Ref. 50.

stopped-flow analysis to measure the rates of reaction with KO_2, rates for the 15- and 16-member ring complexes are 2×10^7 M^{-1} s^{-1} and 1×10^6 M^{-1} s^{-1} in HEPES buffer at pH 8.2, respectively. Crystal structure data on the $Mn([15]aneN_5)Cl_2$ complex reveal a pentacoordinate complex with the ring nitrogens occupying distorted equatorial positions, and two axial Cl^- ions. The cyclic voltammetry of the complexes in nonaqueous media indicates a reversible couple centered at +770 mV (50). However, this couple was not reversible in aqueous media (54). This is also true of the $MnDFBH^+$ complex (42).

The $Mn([15]aneN_5)Cl_2$ (MnPAM) complex, and two other complexes, manganese(II)dichloro(*trans*-2,3-cyclohexano-1,3,7,10,13-pentaazacyclopentadecane) (MnTAM), manganese(II)dichloro(*cis*-2,3-cyclohexano-1,4,7,10,13-pentaazacyclopentadexane) (MnCAM), were found to protect endothelial cells against damage inflicted by neutrophils which had been activated by TNFα and complement component C5a (51).

Further in vivo studies using the Mn macrocyclic compounds suggest a protective effect against myocardial ischemia/reperfusion injury (52). Further refinement of the structure of these Mn macrocycles to include cationic groups and substituents to shift the E^0 value closer to +300 mV might improve upon these new compounds.

At the same time that the Mn macrocycles were being developed cationic Mn porphyrins were being examined as SOD mimics (55). The most promising of these mimics was $Mn(III)tetrakis(1-methyl-4-pyridyl)porphyrin^{5+}$ ($MnTMPyP^{5+}$), shown in Fig. 6 (55). The ability of $MnTMPyP^{5+}$ to act as an SOD scavenger was recognized first by Pasternack (56,57) and Faraggi (58,59). The rate of SOD catalysis by MnTMPyP was 4×10^7 M^{-1} s^{-1} consistent with the flash photolysis results of Faraggi for $MnTMPyP^{5+}$. The $MnTMPyP^{5+}$ showed reversible redox behavior at +60 mV in aqueous media, and both the reduced and oxidized states of the complex are stable in the presence of EDTA (55).

The $MnTMPyP^{5+}$ complex facilitated the aerobic growth of a sodA sodB strain of *E. coli* and protected a SOD-competent strain against the growth inhibition imposed by 30 μM paraquat (55). These protective effects were more impressive than might have been anticipated on the basis of the in vitro SOD-like activity of this complex. Some explanation was found in that the complex was reduced enzymatically at the expense of NADPH and nonenzymatically by GSH and was thus kept reduced within *E. coli*. The rate constant for the reaction of the reduced complex with O_2^- is 100-fold greater than that of the oxidized complex (60). It was thus anticipated that the complex acted in vivo not as a superoxide dismutase but as superoxide NADPH (or GSH) oxidoreductase (55). The ability of the complex to protect against paraquat was subsequently shown to be at least partially due to its ability to prevent the active uptake of paraquat by the *E. coli* (61).

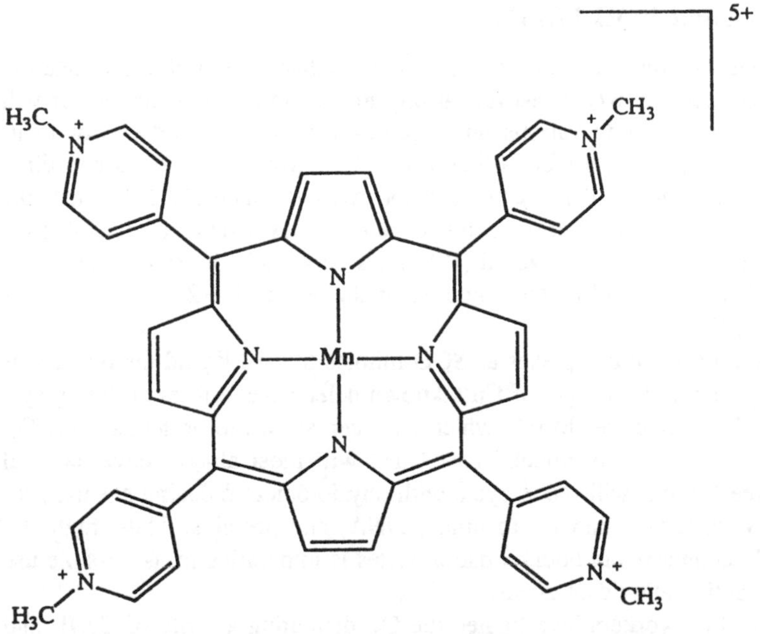

Figure 6 Mn-porphyrin complexes as SOD mimics. The proposed solution structure of Mn(III)(5,10,15,20-tetrakis(1-methyl-4-pyridyl)-porphyrin^{5+} (MnTMPyP^{5+}) as adapted from Refs. 55, 57, and 140.

Interestingly, an anionic Mn porphyrin, Mn(III) tetrakis(4-benzoic acid)-porphyrin^{3-} (MnTBAP^{3-}) also protected SOD-null *E. coli.* However, the in vitro rate of O$_2^-$ dismutation could not be assessed due to the inhibitory effect of the MnTBAP^{3-} complex on xanthine oxidase (which was not observed with the MnTMPyP^{5+} complex) (55).

Other Mn coordination complexes have been studied as mimics of SOD. Some of these are Mn saccharinate complexes studied by Apella et al. (62), some aminocarboxylate complexes of Mn studied by Koppenol et al. (63), and a recently synthesized pentacoordinate (benzoato)Mn(3,5-isopropylpyrazolyl borate) {Mn(OBz)(HB(3,5-iPr$_2$pz)$_3$)} complex by Kitajima et al. (64). The most promising of these is the latter, which has an IC$_{50}$ value of 0.75 μM (64). As the authors propose, the high in vitro activity of the Mn(OBz)(HB(3,5-iPr$_2$pz)$_3$) complex may be due to its similarity to the inner coordination sphere of the MnSOD enzyme. The x-ray crystal structure of Mn(OBz)(HB(3,5-iPr$_2$pz)$_3$) indicates a pentacoordinate complex with four pyrazine nitrogens and one benzoate oxygen coordinated to the Mn^{2+} site (an N$_4$O donor set), while MnSOD has three histidyl nitrogens, one aspartate oxygen, and one H$_2$O (N$_3$O$_2$ donor set).

B. Cu-Based SOD Mimics

It was found that free Cu^{2+} ions also catalyze the dismutation of superoxide ion in acidic medium (65–67). However, at physiological pH, the metal center will either hydrolyze or bind various nucleophiles within a cellular environment. And, unlike Mn, which prefers a 6-coordinate environment, the Cu coordination geometry is highly dependent upon its oxidation state (17,19). Also, the reduced form of Cu^+ complexes tends to be easily oxidizable at physiological pH, hence the Cu^{2+} state is favored (19). Thus, any study of possible Cu-based SOD mimics must stabilize the complex in the +1 and +2 oxidation states solution at pH 7.

The stability of Cu complexes as SOD mimics toward ligand dissociation is particularly important in light of Cu's known deleterious effects on living systems (68–73). Unlike free Mn^{2+}, which has been shown to protect against $O_2^{\cdot-}$ damage in some in vivo models, free Cu^{2+} will most always have harmful effects. Free Cu ions will redox cycle with physiological reducing agents, promoting oxygen radical damage to lipids, DNA, and proteins. Thus, high stability of Cu complexes in both oxidation states is imperative to its possible use as a therapeutic antioxidant agent.

Weser and co-workers investigated the $O_2^{\cdot-}$ dismuting activity of Cu(II) and of simple complexes of Cu(II) (66,67). The second-order rate constant of $Cu(lys)_2$ with $O_2^{\cdot-}$ was 0.6×10^9 M^{-1} s^{-1}, while free Cu^{2+} had a rate constant of 2.7×10^9 M^{-1} s^{-1}. However, bovine serum albumin quenched nearly all of the activity of the free Cu^{2+} (66), whereas $Cu(gly-his)_2$ complex was stable toward BSA. These workers concluded that some complexes of Cu are able to mimic SOD (67). Recently this same complex has been shown to exhibit cytotoxic effects toward human fibroblasts (73).

Younes and Weser (74) and others (75–78) reported that a complex of penicillamine with copper was stable and an SOD mimic. It was subsequently shown (79) that freshly prepared Cu-penicillamine, which had been freed of extraneous loosely bound Cu(II) by treatment with Chelex-100, was devoid of $O_2^{\cdot-}$ dismuting activity. This complex appeared to decompose spontaneously. Thus, incubation for 15 min at 35°C was sufficient to restore some activity; but this activity was EDTA-sensitive.

Sorenson and co-workers have also studied many complexes of Cu for SOD activity. A complex of Cu with EDTA was one of these (80). They reported that CuEDTA was an SOD. This was subsequently shown to have been in error (81) and actually due to excess of Cu(II) over EDTA.

The binary complex of Cu^{2+} with 3,5-diisopropylsalicylate $(Cu_2(DIP)_4)$ has been reported to be as effective as Cu_2Zn_2SOD in preventing ischemia-reperfusion injury (82), and has anti-inflammatory (83), anticonvulsant (84,85), anticarcinogenic (86–88), antimutagenic (89), antidiabetic (90), analgesic (91), and radiation protection and recovery activities (92,93). The solid-state structure and the proposed aqueous equilibrium structure of the $Cu_2(DIP)_4$ complex [$Cu(DIP)_2$] (94) are shown in Fig. 7. Sorenson and co-workers were interested in the state of the $Cu(DIP)_2$ complex in vivo (95). To this end they studied the complex in the presence of human serum albumin (HAS), which is known to greatly influence the transport of drugs to all tissues (96). Since it was known that $Cu(salicylate)_2$ complexes have the same stability constants as $Cu(histinate)_2$ (a possible complexation mode of Cu in vivo (96), $K = 10^{19}$ (97), it was of interest to determine if and how the $Cu(DIP)_2$ or $Cu_2(DIPS)_4$ complex binds to HSA (95).

Ultraviolet spectra indicates that the ternary $Cu(DIP)_2$ complex binds to HSA to form a 5-coordinate complex with SOD activity (95). This is consistent with the findings of other workers using $Cu_2(indomethacinate)_4$ (98), $Cu_2(aspirinate)_4$ (99), $Cu_2(lonazolac)_4$ (100), $Cu(salicylate)_2$, $Cu(PuPy_2)$, and $Cu(PuIm_2)$ (101). The IC_{50} values for the aqueous solution of the $Cu(DIPS)_2$ complex, the HSA bound complex, and the HSA bound and dialyzed complex are 1 μM, 1.4 μM, and 2.1 μM in H_2O at pH 7.5, respectively (95). However, it appears that the preparation had some Cu impurities or that the complex was at least partially unstable, as indicated by the 30% less of activity upon dialysis of the HSA-$Cu(DIPS)_2$ complex.

Many other Cu-based SOD mimics have been synthesized and studied, some of which have been reviewed by Sorenson (102). The vast majority are complexes of two bidentate nitrogen donors (2N) or one nitrogen and one oxygen (NO) or sulfur (NS). Some of the other ligands used in SOD mimics include pyrimine (2N) (103), bisthiosemicarbazone (2N2S) (104), nicotinic acid (NO) (105), substituted 1,10-phenanthrolines (2N) (106), coumarins (NO or 20), (107), triazine (N) (108), and cimetidine (NS) (109). While the in vitro SOD activities of all of these compounds are high, there are mixed results in vivo, as has been the case with many Cu-based mimics. For instance, the 1:0.5 complex of 1,10-phenanthroline and Cu^{2+} has high in vitro SOD activity, but in vivo acts to inhibit the growth of Ehrlich ascites tumor cells by generation of damaging oxygen radicals (110). Thus, as Sorenson and co-workers found (95), many Cu complexes often have dual activities as oxidant and pro-oxidant.

The oxidant/antioxidant nature of Cu complexes is highlighted in the case of a Cu-putrescine-pyridine (CuPuPy) SOD mimic studied by Nagele et al. (111). This complex has been shown to have high SOD activity, but at the same time it also sustains the production of H_2O_2 via redox cycling with glutathione. In small doses (0.05 mM CuPuPy), the complex was not toxic to Chinese hamster ovary cells. But at 0.1 to 0.5 mM CuPuPy, there was marked toxicity as glutathione became oxidized and depleted. On the other hand, CuPuPy

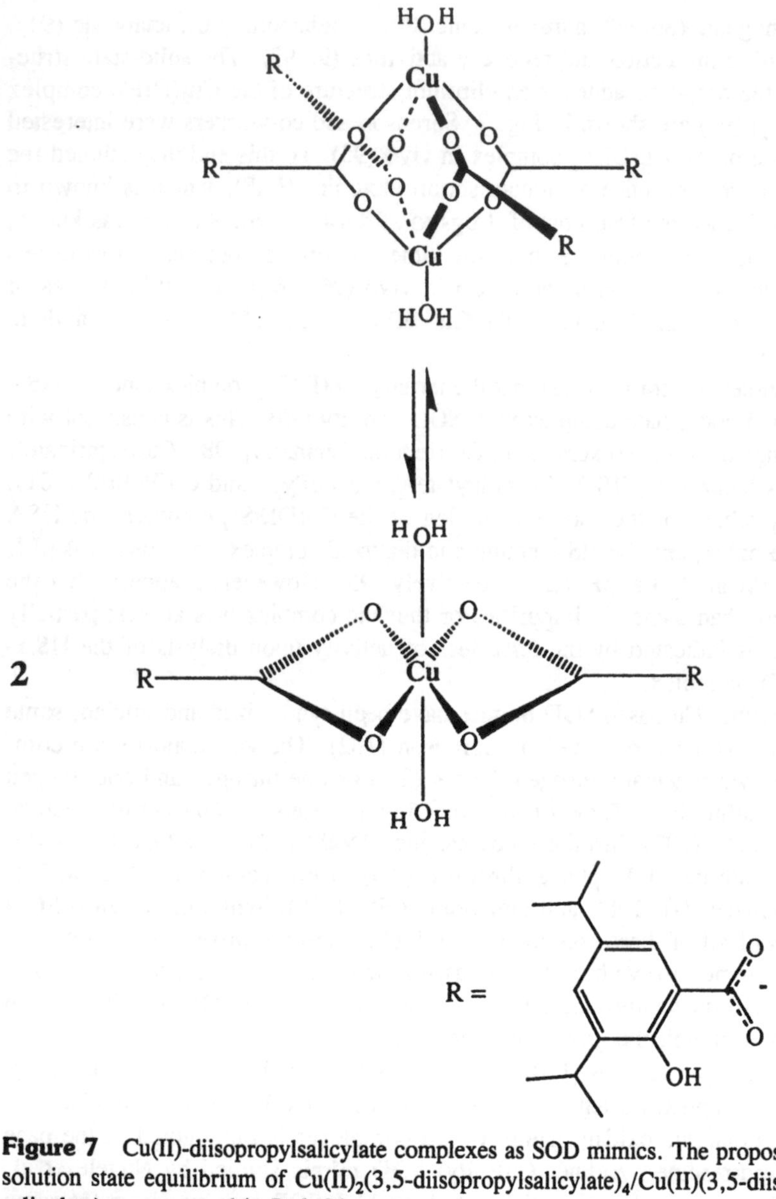

Figure 7 Cu(II)-diisopropylsalicylate complexes as SOD mimics. The proposed solid/solution state equilibrium of Cu(II)$_2$(3,5-diisopropylsalicylate)$_4$/Cu(II)(3,5-diisopropyl-salicylate)$_2$ as proposed in Ref. 94.

protected against H_2O_2 toxicity in the same system. Thus, it was concluded by Nagele et al. that this complex combines oxidant and antioxidant modes of action (111).

Figure 8 shows the reaction scheme which is proposed to be operative for CuPuPy and perhaps many metal-based mimics of SOD such as Cu(tyr)$_2$ and

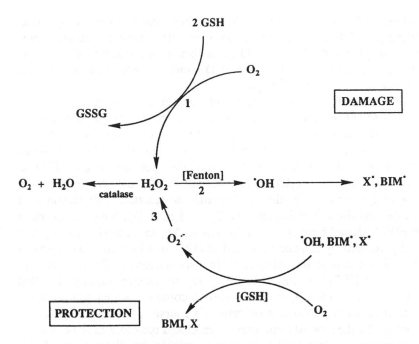

Figure 8 Oxidant and antioxidant mechanisms of SOD mimics. Possible pathways SOD mimics can take as oxidants or antioxidants. In path (1), the metal complex can be reduced by GSH and react with O_2 to produce H_2O_2. In path (2) the metal complex can act as a Fenton agent, catalyzing the production of damaging radicals ($X\cdot$, and biologically important molecules, $BIM\cdot$). Pathway (3) represents the antioxidant action of a metal complex as a mimic of SOD. (Adapted from Ref. 111.)

MnTMPyP. As shown in Fig. 8, mimics of SOD and other exogenous chemicals may influence the level of GSH directly through path 1 (111,112). The oxidation of GSH to GSSG is accompanied by the formation of H_2O_2. If the mimic is capable of catalyzing Fenton chemistry (path 2), cellular damage will ensue (111,113). Of course, the mimic will also dismutate O_2^- to form H_2O_2 (path 3). The key to the effectiveness of any mimic may be its ability to catalyze Fenton chemistry, since all pathways result in H_2O_2 production. Mn-based mimics are advantageous in this regard since they do not participate in Fenton chemistry.

C. Fe-Based SOD Mimics

The study of Fe complexes as mimics of SOD is complicated by the fact that Fe and Fe complexes, like Cu and Cu complexes, have been known for some time to catalyze Fenton chemistry, the end result of which is the production of OH· radicals (114–118). Iron and Fe complexes of EDTA, NTA, EGTA, ci-

trate, ADP, ATP, and pyrophosphate have all been shown to be strong inducers of damage to DNA, thought to be caused by OH· generation upon reaction
of the Fe complex with H_2O_2 (117,118). In light of the toxicity of Fe in living systems, it is vital that an Fe-based SOD mimic be very stable and that the
complex not act as a Fenton catalyst.

Iron-EDTA had been thought capable of catalyzing the dismutation of O_2^-,
but this proved not to be the case. This story is worth reviewing since it illustrates the potential artifacts which dog the footsteps of those seeking SOD
mimics. The first report of the SOD-like activity of iron-EDTA (119) was based
on the xanthine oxidase/cytochrome c(6) and on the xanthine oxidase/NBT (12)
assays of SOD activity. Fe(II)-EDTA, but not Fe(III)-EDTA, inhibited the
reduction of cytochrome c by the xanthine oxidase reaction. This should have
raised suspicions since, if acting catalytically to dismute O_2^-, both valence states
of iron-EDTA should have been equally active. Yet the catalytic activity was
supposedly verified by stopped-flow studies (121) in which the rate constants
of both half-reactions of the catalytic cycle were measured. These results indicated that Fe(III)EDTA was reduced by O_2^- reasonably rapidly, but that
Fe(II)EDTA reacted with O_2^- to give a peroxo complex, which was so stable
that it accumulated in the reaction mixtures. Of course, a sequence of reactions
is rate-limited by the slowest step. These data should therefore have been taken
as an indication that iron-EDTA could not catalyze the dismutation of O_2^-.
Inexplicably, these workers (121) came to the opposite conclusion.

The effects of iron-EDTA on the assays for SOD activity were subsequently
examined and explained (122). Fe(II)EDTA was found to reduce cytochrome
c very rapidly. Hence, addition of Fe(II)EDTA reduced a stoichiometric amount
of cytochrome c within the few seconds needed to mix it in and start the re-
corder of the spectrophotometer. This left less ferricytochrome c to compete,
with the spontaneous dismutation, for O_2^-. This was one factor which slowed
the observed reduction of cytochrome c. The other factors was the production
of HO· by the Fe(III)EDTA-catalyzed Haber-Weiss reaction (123). The HO·
would have the effect of reoxidizing the ferrocytochrome c and thus add to the
appearance of inhibition of the rate of cytochrome c reduction. The contribution of HO· was revealed by adding mannitol to intercept HO·, and that diminished the effect of Fe(II)EDTA. It was also shown by adding catalase to remove
H_2O_2 and thus prevent HO· production, and that too diminished the effect of
added Fe(III)EDTA (122).

The inhibition of NBT reduction by iron EDTA (119) required a different
explanation because both valence states were effective. It was demonstrated that
NBT augments the rate of reduction of Fe(III)EDTA by O_2^- (122). This led to
the following scheme:

$$NBT^{2+} + O_2^- \leftrightarrow NBT^{·+} + O_2 \tag{5}$$

$$NBT^{\cdot+} + Fe(III)EDTA \leftrightarrow NBT^{2+} + Fe(II) \, EDTA \tag{6}$$

$$Fe(II)EDTA + O_2 \leftrightarrow Fe(III)EDTA + O_2^- \tag{7}$$

$$O_2^- + O_2^- + 2H^+ \leftrightarrow H_2O_2 \tag{8}$$

The explanation for the inhibition of NBT reduction goes as follows: O_2^- reduced NBT to the tetraazoinyl radical, as in reaction (5). This reaction is known to proceed and its rate constant is $\sim 6 \times 10^4 \, M^{-1} \, s^{-1}$ (124). The tetraazoinyl radical was demonstrably able to reduce Fe(III)EDTA, as in reaction (6). Fe(II)EDTA is known to autoxidize (125) as in reaction (7) and the spontaneous dismutation of O_2^- goes at a rate of $\sim 10^5 \, M^{-1} \, s^{-1}$ in neutral aqueous solution (3). The net effect of these reactions is that addition of iron EDTA to the xanthine oxidase/NBT assay diminishes the rate of NBT reduction because it decreases the concentration of the tetraazoinyl radical, which is the precursor of the formazan being measured.

To finally demolish the possibility that iron-EDTA catalyzes the superoxide dismutation, recourse was had to an SOD assay which depends on the photochemical oxidation of the chromogen dianisidine (126). Iron-EDTA was completely inactive in this assay and did not interfere with the effect of SOD on the assay (122). Other workers subsequently reported that iron-EDTA lacks the ability to catalyze the dismutation of O_2^- (127,128). Interestingly, the workers who had originally failed to realize that one very slow step vitiates an entire pathway (121) refused to admit their error and published two additional obfuscating reports (129,130).

Fe(II)-tetrakis-N,N,N',N' (2-pyridyl methyl) ethylenediamine and related compounds have been reported to mimic SOD activity in the xanthine oxidase/cytochrome c assay and to protect $E.\ coli$ against the toxicity of paraquat (see Fig. 9) (131,132). Unfortunately, further study of these complexes indicated that they are unstable to H_2O_2 and are readily reduced by ascorbate and GSH and participate in Fenton chemistry (133).

Much work in determining the in vitro SOD activity of Fe porphyrins has been done, particularly by Pasternack and co-workers (56,57,134), Faraggi and co-workers (58,59,135,136), and Byrnes et al. (137). The first Fe porphyrin to be studied was a water-soluble Fe(III)tetrakis(1-methyl-4-pyridyl)porphyrin[5+] (FeTMPyP[5+]) porphyrin (see Fig. 6), which was found to catalyze the dismutation of O_2^- with a rate of $1-4 \times 10^7 \, M^{-1} \, s^{-1}$ (56). The mechanism of catalysis is proposed to be similar to that for the SOD enzyme, with the Fe center redox cycling between the Fe^{3+} and Fe^{2+} states (56,134). The high activity of the FeTMPyP[5+] complex is due to its favorable E^0 value of $+130$ mV versus NHE (138,139) and its high positive charge of $+5$ at pH 7 (140).

Faraggi et al. have studied a class of Fe porphyrins which have "picket-fence-like" groups attached to the porphyrin ring (135,136). These "picket-fence

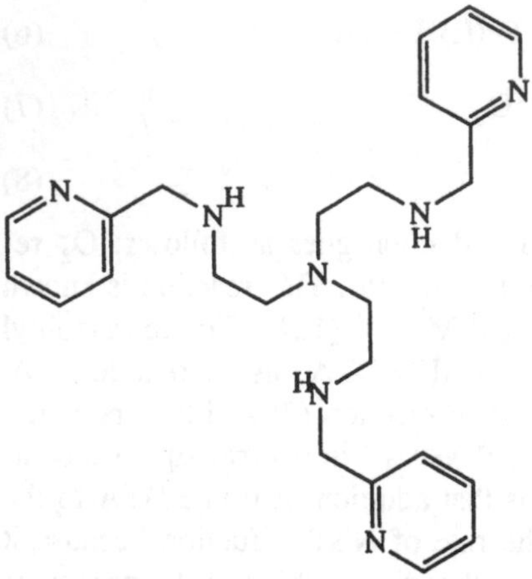

Figure 9 Iron complexes as SOD mimics. The ligands tetrakis-N,N,N',N'(2-pyridyl-methyl)ethylenediamine (TPEN) and tris[N-(2-pyridylmethyl)-2-aminoethyl]amine (TPAA) that complex Fe(III) to form mimics as described in Ref. 132.

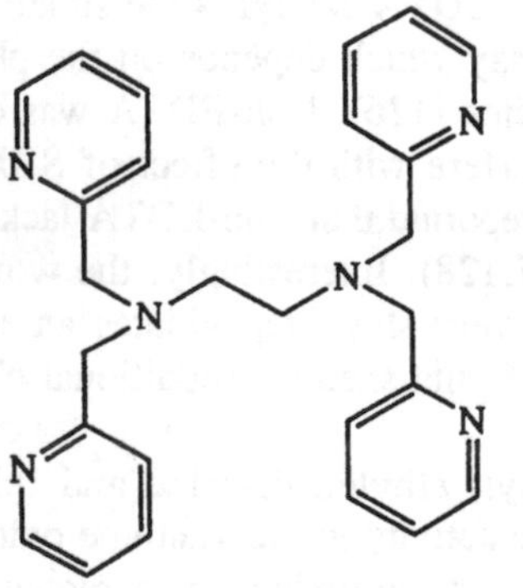

porphyrins" (FePFP) were first synthesized as possible mimics of myoglobin, since one axial face of the porphyrin is enclosed in four N-methyl-isonicotinamidophenyl groups which hinder aggregation of the porphyrins and binding of large axial ligands (134,135). The complex in which all groups faced the same direction (a^4-FePFP^{5+}) was the most efficient SOD mimic (134,135). A rate constant of 7.6×10^7 M^{-1} s^{-1} using pulse radiolysis was determined for the a^4-FePFP^{5+} reaction with O$_2^-$. The reaction of Fe(III)PFP^{5+} with O$_2^-$ is fast, with a rate constant of 8.1×10^8 M^{-1} s^{-1} as determined by pulse radiolysis.

Unfortunately, Byrnes et al. found that FeTMPyP^{5+} was an active DNA strand-cutting agent (137). Likewise, Faulkner et al. have found FeTMPyP^{5+}

to be toxic to *E. coli*, even if freed of excess Fe ions (55). The toxicity exhibited by Fe porphyrins is presumably due to the ability of the porphyrin to participate in Fenton chemistry, generating toxic OH· radicals. Hence, the in vivo efficacy of the FeTMPyP^{5+} complex is questionable and will need further study.

There is a large body of evidence against the usefulness of Fe complexes in vivo (114–118). This has been corroborated in studies that allude to the participation of Fe in Fenton chemistry (141–145). It is known that free Fe^{3+} participates in Fenton chemistry in acidic media where the Fe^{3+} species is highly soluble. At physiological conditions, Fe^{3+} is not highly soluble, forming insoluble hydrolysis products (19,37). Complexants of Fe^{3+}, such as EDTA, increase the solubility of Fe^{3+}, increasing its ability to potentiate OH· radical formation. Graf et al. studied Fe-catalyzed OH· radical formation and found that, at pH 7, complexes of Fe^{3+} that left a coordination site open were the best propagators of Fenton chemistry (141). In fact, complexes of Fe^{3+} with DFBH$^+$ which lack an open coordination site do not participate in Fenton chemistry. Hence, the DFBH$^+$ ligand has been found to be protective in vivo against Fe-mediated damage by complexing free or weakly complexed Fe (142,143,145).

III. ORGANIC SOD MIMICS

The best-studied organic SOD mimic is 4-hydroxy-2,2,6,6-tetramethylpiperidinyl-1-oxyl, or TEMPOL, the structure of which is shown in Fig. 10, along with its proposed mode of action (146).

TEMPOL and its derivatives are cyclic nitroxide compounds which can form stable one-electron radical compounds. Ironically, cyclic nitroxide compounds were widely used for at least 20 years before being recognized for their potential SOD activity. Their first uses were as "spin labels" which could be detected in vitro or in vivo using e.s.r. techniques to study such things as heme-protein coordination chemistry (147–149), water structure in muscle tissue (150), and other in vivo systems to study various tissues (151–154). The use of cyclic nitroxides was ideal due to the ability to prepare synthetic derivatives that have specific properties such as charge and hydrophobicity which can be utilized in various biophysical studies (155,156).

But it was Samuni et al. who first recognized that cyclic nitroxides were active as an SOD mimic (157–159). The first study to show this was with 2-ethyl-2,5,5-trimethyl-3-oxazolidinoxyl (OXANO), which was shown to catalyze $O_2^{\cdot-}$ dismutation in a catalytic, metal-independent manner (157). The catalytic rate dismutation is only 6.7×10^4 M^{-1} s^{-1}, using the xanthine-xanthine oxidase system to generate $O_2^{\cdot-}$ and e.s.r. to follow the reaction. However, Samuni et al. argue that since OXANOL is highly soluble in water and nontoxic, it can

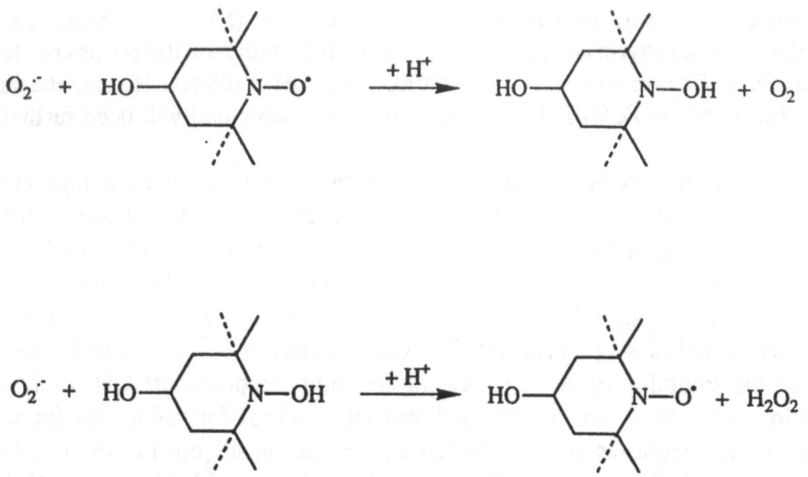

Figure 10 Cyclic nitroxides as mimics of SOD. The structure and mechanism of action of 4-hydroxy-2,2,6,6-tetramethylpiperidinyl-1-oxyl (TEMPOL) as it reacts with two molecules of O_2^- to form H_2O_2 and O_2. (Adapted from Refs. 146 and 160.)

be used in very high concentrations in vivo and thus has potential therapeutic uses as an SOD mimic.

The TEMPOL compound and other cyclic nitroxides were further characterized by cyclic voltammetry, UV-visible spectroscopy, and e.s.r (146,160). The results of these studies helped to elucidate the mode of activity as shown in Fig. 10, where the oxoammonium/nitroxide couple at +810 mV versus NHE (pH 7.4) corresponds to the O_2^- reactive couple.

Since TEMPOL and its derivatives have been found to be active in vitro as SOD mimics at high concentrations (IC_{50} = 2 mM), many in vivo studies have been carried out by Mitchell and co-workers (161–165), An et al. (166,167), and others (168). The toxicity and mutagenicity of these compounds has also been reviewed by Sosnovsky (169). TEMPOL and related compounds have been found to protect in a variety of model systems, including tumor necrosis factor-sensitive L929 cells (161), Chinese hamster cells exposed to H_2O_2 or O_2^- (162,166,167), x-ray-mediated cell damage (163,165), and in H_2O_2-induced rabbit lens damage (168). Due to the low toxicity of TEMPOL, its use as a therapeutic agent merits further study.

IV. CONCLUSIONS

It is clearly not easy to come to the efficiency and specificity of enzymes, which have been honed to perfection by a billion years of evolution. Yet, in the case of SOD, the task does not seem overwhelming because of the simplicity of the dismutation reaction and because under specified conditions, simple metal cations, such as Cu(II), can do the job. Moreover, the role of O_2^- in so many pathological processes, and in senescence, indicates that specific, stable, active, cell-permeable mimics would be very useful pharmaceuticals. At present, the goal is still in the future, but the search goes on and a good start has been made.

REFERENCES

1. Karlin KD. Metalloenzymes, structural motifs, and inorganic models. Science 1993; 261: 701–708.
2. Weighardt K. The active sites in manganese-containing metalloproteins and inorganic model complexes. Angew Int Ed 1989; 28: 1153–1172.
3. Bielski BHJ. Reevaluation of the spectral and kinetic properties of $HO_2\cdot$ and O_2^- free radicals. Photochem Photobiol 1978; 28: 645–649.
4. Argese E, Girotto R, Orsega EF. Comparative kinetic study between native and chemically-modified Cu,Zn superoxide dismutase. Biochem J 1993; 292: 451–455.
5. Rotilio G, Bray RC, Fielden EM. A pulse radiolysis study of superoxide dismutase. Biochim Biophys Acta 1972; 268: 605–609.
6. McCord JM, Fridovich I. Superoxide dismutase—an enzymic function for erythrocuprein (hemocuprein). J Biol Chem 1969; 244: 6049–6055.
7. Ose DE, Fridovich I. Superoxide dismutase. Reversible removal of manganese and its substitution by cobalt, nickel or zinc. J Biol Chem 1976; 251: 1217–1218.
8. Yamakura F, Suzuki S. Reconstitution of iron superoxide dismutase. Biochem Biophys Res Comun 1976; 72: 1108–1115.
9. Tainer JA, Getzoff ED, Beem KM, Richardson JS, Richardson DC. Determination and analysis of the 2 Å structure of copper, zinc superoxide dismutase. J Mol Biol 1982; 160: 181–217.
10. Tainer JA, Getzoff ED, Richardson JS, Richardson DC. Structure and mechanism of copper, zinc superoxide dismutase. Nature 1983; 306: 284–287.
11. Stallings WC, Pattridge KA, Strong RK, Ludwig ML. Manganese and iron superoxide dismutases are structural homologues. J Biol Chem 1984; 259: 10695–10699.
12. Stallings WC, Pattridge KA, Strong RK, Ludwig ML. The structure of manganese superoxide dismutase from *Thermus thermophilus* HB8 at 2.4 Å resolution. J Biol Chem 1985; 260: 16424–16432.
13. Stoddard BL, Howell PL, Ringe D, Petsko GA. The 2.1 Å Resolution structure of iron superoxide dismutase. Biochemistry 1990; 29: 8885–8893.
14. Ludwig ML, Metzger AL, Pattridge KA, Stallings WC. Manganese superoxide dismutase from *Thermus thermophilus*. A structural model refined at 1.8 Å resolution. J Mol Biol 1991; 219: 335–358.
15. Lah MS, Dixon MM, Pattridge KA, Stallings WC, Fee JA, Ludwig ML. Structure-function in *Escherichia coli* iron superoxide dismutase: comparisons with the manganese enzyme from *Thermus thermophilus*. Biochemistry 1995; 34: 1641–1660.

16. Marcus RA, Sutin N. Electron transfer in chemistry and biology. Biochim Biophys Acta 1985; 811: 265–322.

17. Glusker JP. Structural aspects of metal liganding to functional groups in proteins. Adv in Protein Chem 1991; 42: 1–76.

18. Macartney DH, Thompson DW. Electron-transfer reactions of manganese(II) and -(III) polyamino carboxylate complexes in aqueous media. Inorg Chem 1989; 28: 2195–2199.

19. Cotton FA, Wilkinson G. Advanced Inorganic Chemistry. A Comprehensive Text. 4th ed. New York: Wiley, 1980.

20. Koppenol WH, Butler J. Energetics of interconversion reactions of oxyradicals. Adv Free Radic Biol Med 1985; 1: 93–131.

21. Koppenol WH. The physiological role of the charge distribution of superoxide dismutase. In: Oxygen, Oxy-Radicals Chem Biol [Proc Int Conf 1980], 1981: 671–674.

22. Getzoff ED, Tainer JA, Weiner PK, Kollman PA, Richardson JS, Richardson DC. Electrostatic recognition between superoxide and superoxide dismutase. Nature 1983; 306: 284–290.

23. Getzoff ED, Cabelli DE, Fisher CL, Parge HE, Viezzoli MS, Bonii L, Hallewell RA. Faster superoxide dismutase mutants designed by enhancing electrostatic guidance. Nature 1992; 358: 347–351.

24. Bertini I, Hiromi K, Hirose J, Sola M, Viezzoli MS. Electron transfer between copper and zinc superoxide dismutase and hexacyanoferrate(II). Inorg Chem 1993; 32: 1106–1110.

25. Archibald FS, Fridovich I. Manganese and defences against oxygen toxicity in *Lactobacillus plantarum*. J Bacteriol 1981; 145: 442–451.

26. Archibald FS, Fridovich I. Manganese superoxide dismutase, and oxygen tolerance in some lactic acid bacteria. J Bacteriol 1981; 146: 928–930.

27. Archibald F. Manganese: its acquisition by and function in the lactic acid bacteria. Crit Rev Microbiol 1986; 13: 63–109.

28. Archibald FS, Fridovich I. Investigations of the state of manganese in *Lactobacillus plantarum*. Arch Biochem Biophys 1982; 215: 589–596.

29. Bielski BHJ, Chan PC. Products of reaction of superoxide and hydroxyl radicals with Mn(II) cation. J Am Chem Soc 1978; 100: 1920–1921.

30. Archibald FS, Fridovich I. The scavenging of superoxide radical by manganous complexes: in vitro. Arch Biochem Biophys 1982; 214: 452–463.

31. Boggess RK, Hughes JW, Coleman WM, Taylor LT. Preparation and electrochemical studies of tetradentate manganese (III) Schiff base complexes. Inorg Chim Acta 1980; 38: 183–189.

32. Matsushita T, Shono T. The preparation and characterization of dichloromanganese(VI) Schiff base complexes. Bull Chem Soc Jpn 1981; 54: 3743–3748.

33. Weser U, Miesel R, Linss M. Reactivity of active center analogues of Cu_2Zn_2 superoxide dismutase. Adv Exp Med Biol 1990; 264: 51–57.

34. Baudry M, Etienne S, Bruce A, Palucki M, Jacobsen E, Malfroy B. Salen-manganese complexes are superoxide dismutase mimics. Biochem Biophys Res Comm 1993; 192: 964–968.

35. Liu ZX, Robinson GB, Gregory EM. Preparation and characterization of Mn-salophen complex with superoxide scavenging activity. Arch Biochem Biophys 1994; 315: 74–81.

36. Musleh W, Bruce A, Malfroy B, Baudry M. Effects of EUK-8, a synthetic catalytic superoxide scavenger, on hypoxia- and acidosis-induced damage in hippocampal slices. Neuropharmacology 1994; 33: 929–934.

37. Hancock RD, Martell AE. Ligand design for selective complexation of metal ions in aqueous solution. Chem Rev 1989; 89: 1875–1914.

38. Darr D, Zarilla KA, Fridovich I. A mimic of superoxide dismutase activity based upon desferrioxamine B and manganese(IV). Arch Biochem Biophys 1987; 258: 351–355.

39. Beyer WR, Fridovich I. Characterization of a superoxide dismutase mimic prepared from desferrioxamine and MnO_2. Arch Biochem Biophys 1989; 271: 149–156.

40. Monzyk B, Crumbliss AL. Factors that influence siderophore mediated iron bioavailability: catalysis of interligand iron(III) transfer from ferrioxamine B to EDTA by hydroxamic acids. J Inorg Biochem 1983; 19: 19–39.

41. Rush JD, Maskos Z, Koppenol WH. The superoxide dismutase activities of two higher valent manganese complexes, Mn(VI) desferrioxamine and Mn(III) cyclam. Arch Biochem Biophys 1991; 289: 97–102.

42. Faulkner KM, Stevens RD, Fridovich I. Characterization of Mn(III) complexes of linear and cyclic desferrioxamines as mimics of superoxide dismutase activity. Arch Biochem Biophys 1994; 310: 341–346.

43. Rabinowitch HD, Privalle CT, Fridovich I. Effects of paraquat on the green alga *Dunaliella salina*: protection by the mimic of superoxide dismutase, desferal-Mn(IV). Free Radic Biol Med 1987; 3: 125–131.

44. Darr DJ, Yanni S, Pinnell SR. Protection by Chinese hamster ovary cells from paraquat-mediated cytotoxicity by a low molecular weight mimic of superoxide dismutase. Free Radical Biol Med 1988; 4: 357–363.

45. Kitahara J, Seko Y, Imura N. Possible involvement of active oxygen species in selenite toxicity in isolated rat hepatocytes. Arch Toxicol 1993; 67: 497–501.

46. Bhuyan KC, Bhuyan DK, Chiu W, Malik S, Fridovich I. Desferal-Mn(IV) in the therapy of diquat induced cataract in rabbit. Arch Biochem Biophys 1991; 288: 525–532.

47. Hahn MS, Krishna CM, Samuni A, Mitchell JB, Russo A. Mn(III)-desferioxamine superoxide dismutase mimic: alternative modes of action. Arch Biochem Biophys 1991; 288: 215–219.

48. Samuni A, Mitchell JB, DeGraff W, Krishna CM, Samuni U, Russo A. Nitroxide SOD-mimics: modes of action. Free Rad Res Comm 1991; 12–13: 187–194.

49. Ortel B, Gange RW, Hasan T. Investigations of a manganese-containing mimic of superoxide dismutase in riboflavin phototoxicity in human cells in vitro. J Photochem Photobiol B-Biology 1990; 7: 261–276.

50. Riley DP, Weiss RH. Manganese macrocyclic ligand complexes as mimics of superoxide dismutase. J Am Chem Soc 1994; 116: 387–388.

51. Hardy MM, Flickinger AG, Riley DP, Weiss RH, Ryan US. Superoxide dismutase mimetics inhibit neutrophil-mediated human aortic endothelial cell injury in vitro. J Biol Chem 1994; 269: 18535–18540.

52. Kilgore KS, Friedrichs GS, Johnson CR, Schasteen CS, Riley DP, Weiss RH, Ryan U, Lucchesi BR. Protective effects of the SOD-mimetic SC-52608 against ischemia/reperfusion damage in the rabbit isolated heart. J Mol Cell Cardiol 1994; 26: 995–1006.

53. Black SC Schasteen CS, Weiss RH, Riley DP, Driscoll EM, Lucchesi BR. Inhibition of in vivo myocardial ischemic and reperfusion injury by a synthetic manganese-based superoxide dismutase mimetic. J Pharm Exp Therapeut 1994; 270: 1208–1215.

54. Weiss RH, Riley DP. Personal communication, 1994.

55. Faulkner KF, Liochev SI, Fridovich I. Stable Mn(III) porphyrins mimic superoxide dismutase in vitro and substitute for it in vivo. J Biol Chem 1994; 269: 23471–23476.

56. Pasternack RF, Skowronek WR. Catalysis of the disproportionation of superoxide by metalloporphyrins. J Inorg Biochem 1979; 11: 261–268.

57. Pasternack RF, Banth A, Pasternack JM, Johnson CS. Catalysis of the disproportionation of superoxide by metalloporphyrins III. J Inorg Biochem 1981; 15: 261–267.

58. Peretz P, Solomon D, Weintraub D, Faraggi M. Chemical properties of water-soluble porphyrins 3. The reaction of superoxide radicals with some metalloporphyrins. J Rad Biol Relat Stud Phys Chem Med 1982; 42: 449–456.

59. Weinraub D, Levy P, Faraggi M. Chemical properties of water soluble porphyrins. 5. Reactions of some manganese (III) porphyrins with the superoxide and other reducing radicals. Intl J Rad Biol Relat Stud Phys Chem Med 1986; 50: 649–658.

60. Faraggi M. In: Bors W, Saran M, Tait D, eds. Oxygen Radicals in Chemistry and Biology. Berlin: Walter de Gruyter, 1984: 419–430.

61. Liochev SI, Fridovich I. A cationic porphyrin inhibits uptake of paraquat by *Escherichia coli*. Arch Biochem Biophys. In press.

62. Apella MC, Totaro R, Baran EJ. Determination of superoxide dismutase-like activity in some divalent metal saccharinates. Biol Trace Element Res 1993; 37: 293–299.

63. Koppenol WH, Levine F, Hatmaker TL, Epp J, Rush JD. Catalysis of superoxide dismutation by manganese aminopolycarboxylate complexes. Arch Biochem Biophys 1986; 251: 549–599.

64. Kitajima N, Osawa M, Tamura N, Moro-oka Y, Hirano T, Hirobe M, Nagano T. Monomeric(benzoato)manganese(II) complexes as manganese superoxide dismutase mimics. Inorg Chem 1993; 32: 1879–1880.

65. Klug Roth, Fridovich I, Rabani J. Pulse radiolytic investigation if superoxide catalyzed disproportionation. Mechanisms for bovine superoxide dismutase 1973. J Am Chem Soc 1973; 95: 2786–2790.

66. Joester KE, Jung G, Weber U, Weser U. Superoxide dismutase activity of Cu^{2+} amino acid chelates. FEBS Lett 1972; 25: 25–28.

67. Brigelius R, Spottl R, Bors W, Lengfelder E, Saran M, Weser U. Superoxide dismutase activity of low molecular weight Cu^{2+}-chelates studied by pulse radiolysis. FEBS Lett 1974; 47: 72–75.

68. Marva E, Chevion M, Golenser J. The effect of free radicals induced by paraquat and copper in the in vivo development of *Plasmodium falceparum*. Free Rad Res Comm 1991; 12–13: 137–146.
69. Kadiiska MB, Hanna PM, Mason RP. In vivo ESR spin trapping evidence for hydroxyl radical-mediated toxicity of paraquat and copper in rats. Tox Appl Pharm 1993; 123: 187–192.
70. Czapski G, Goldstein S. When do metal complexes protect the biological system from superoxide toxicity and when do they enhance it? Free Rad Res Comm 1986; 1: 157–161.
71. Aruoma OI, Kaur H, Halliwell B. Oxygen free radicals and human diseases. J R Soc Health 1991; 111: 172–177.
72. Chevion M. Protection against free radical-induced and transition metal-mediated damage: The use of "pull" and "push" mechanisms. Free Rad Res Comm 1991; 12–13: 691–696.
73. Arena G, Bindoni M, Cardile V, Maccarrone G, Riello MC, Rizzarelli E, Sciuto S. Cytotoxic and cytostatic activity of copper(II) complexes. Importance of the speciation for the correct interpretation of the in vitro biological results. J Inorg Biochem 1993; 50: 31–45.
74. Younes M, Weser U. Superoxide dismutase activity of copper-penicillamine: possible involvement of Cu(I) stabilized sulfur radical. Biochem Biophys Res Commun 1977; 78: 1247–1253.
75. Lengfelder E, Elstner EF. Determination of the superoxide dismutating activity of D-penicillamine copper. Hopper-Seyler's Z. Physiol Chem 1978; 359: 751–757.
76. Lengfelder E, Fuchs C, Younes M, Weser U. Functional aspects of the superoxide dismutative action of Cu-penicillamine. Biochim Biophys Acta 1979; 567: 492–502.
77. Youngman RJ, Dodge AD, Lengfelder E, Elstner EF. Inhibition of paraquat phytotoxicity by a novel copper chelate with superoxide dismutating activity. Experientia 1979; 35: 1295–1296.
78. Youngman RJ, Dodge AD. Mechanism of paraquat action: inhibition of the herbicidal effect by a copper chelate with superoxide dismuting activity. Z Naturforsch Sec C 34, 1032–1045.
79. Robertson P, Fridovich I. Does copper-D-penicillamine catalyze the dismutation of superoxide. Arch Biochem Biphys 1980; 203: 830–831.
80. Willingham WM, Sorenson JRJ. Copper(II) ethylenediamine-tetraacetate does disproportionate superoxide. Biochem Biophys Res Comm 1988; 150: 252–258.
81. Beyer WF, Fridovich I. Does copper(II) ethylenediaminetetraacetate disproportionate superoxide. Anal Biochem 1988; 173: 160–165.
82. Hernandez LA, Grisham MB, Granger DN. Effects of Cu-DIPS on ischemia-reperfusion injury. In: Sorenson JRJ, ed. Biology of Copper Complexes. Clifton, NJ: Humana Press, 1987: 201–214.
83. Sorenson JRJ. Copper chelates as possible active forms of the antiarthritic agents. J Med Chem 1976; 19: 135–148.
84. Sorenson JRJ, Rauls DO, Ramakrishna K, Stull RE, Voldeng AN. Anticonvulsant activity of some copper complexes. In: Hemphill DD, ed. Trace Substances in

Environmental Health. XIII. Columbia, MO: University of Missouri Press, 1979: 360–367.

85. Dollwet HHA, McNicholas JB, Pezeshk A, Sorenson JRJ. Superoxide dismutase-mimetic activity of antiepileptic drug copper complexes. Trace Element Med 1987; 4: 13–20.

86. Leuthauser SWC, Oberley LW, Oberley TD, Sorenson JRJ, Ramakrishna K. Antitumor effects of compounds with superoxide dismutase activity. J Natl Cancer Inst 1981; 66: 1077–1081.

87. Sorenson JRJ, Oberley LW, Crouch RK, Kensler TW, Kishore V, Leuthauser SWC, Oberley TD, Pezeshk A. Pharmacologic activities of copper compounds in chronic diseases. Biol Trace Element Res 1983; 5: 257–273.

88. Kensler TW, Bush DM, Kozumbo WJ. Inhibition of tumor promotion by a biomimetic superoxide dismutase. Science 1983; 221: 75–77.

89. Solanski V, Yotti L, Logani MK, Slaga TJ. The reduction of tumor initiating activity and cell mediated mutagenicity of dimethylbenz[a]anthracine by a copper coordination compound. Carcinogenesis 1984; 5: 129–131.

90. Gandy SE, Buse MG, Sorenson JRJ, Crouch RK. Attenuation of streptozoticin diabetes with superoxide dismutase-like copper(II)(3,5-diisopropylsalicylate)$_2$ in the rat. Diabetologia 1983; 24: 437–440.

91. Okuyama S, Hasimoto S, Aihara H, Willingham WM, Sorenson JRJ. Copper complexes of non-steroidal anti-inflammatory agents: analgesic activity and possible opioid receptor activation. Agents Actions 1986; 21: 130–140.

92. Sorenson JRJ. Bis(3,5-diisopropylsalicylato)copper(II), a potent radioprotectant with superoxide dismutase mimetic activity. J Med Chem 1984; 27: 1747–1749.

93. Sorenson JRJ, Soderberg LSF, Barnett JB, Baker ML, Salari H, Bond KB. Radiation protection with Cu(II)(3,5-DIPS)$_2$. Rec Trav Chim 1987; 106: 391.

94. Greenway FT, Norris LJ, Sorenson JRJ. Mononuclear and binuclear copper(II) complexes of 3,5-diisopropylsalicylic acid. Inorg Chim Acta 1988; 145: 279–284.

95. Shuff ST, Chowdhary P, Khan MF, Sorenson JRJ. Stable superoxide dismutase (SOD)-mimetic ternary human serum albumin-Cu(II)(3,5-diisopropylsalicylate)$_2$/Cu(II)2(3,5-diisopropylsalicylate)$_4$ complexes in tissue distribution of the binary complex. Biochem Pharm 1992; 43: 1601–1612.

96. McMenamy RH. Albumin binding sites. In: Rosenoer VM, Oratz M, Rothschild MA, eds. Albumin Structure, Function, and Uses. New York: Pergamon Press, 1977: 143–158.

97. Albert A. Design of chelating agents for selected biological activity. Fed Proc 20 1961; Suppl. 10: 137–146.

98. Weser U, Sellinger KH, Lengfelder E, Werner W, Strahle J. Structure of Cu$_2$(indomethacin)$_4$ and the reaction with superoxide in aprotic systems. Biochim Biophys Acta 1980; 631: 232–245.

99. Brown DH, Dunlop J, Smith WE, Teape J, Lewis AJ. Total serum copper and ceruloplasmin levels following administration of copper aspirinate to rats and guinea pigs. Agents Actions 1989; 10: 465–470.

100. Deuschle U, Weser U. Reactivity of Cu$_2$(lonozolac)$_4$, a lipophilic copper acetate derivative. Inorg Chim Acta 1984; 91: 237–242.

101. Miesel R, Hartman H-J, Li Y, Weser U. Reactivity of active center analogs of the Cu_2Zn_2 superoxide dismutase on activated polymorphonuclear leukocytes. Inflammation 1990; 14: 409–419.

102. Sorenson JRJ. Copper complexes offer a physiological approach to treatment of chronic diseases. Prog Med Chem 1989; 26: 437–568.

103. Itami C, Matsunaga H, Sawada T, Sakurai H, Kimura Y. Superoxide dismutase mimetic activities of metal complexes of L-2(2pyridyl)-1-pyrroline-5-carboxylic acid (pyrimine). Biochem Biophys Res Comm 1993; 197: 536–541.

104. Wada K, Fujibayashi Y, Yokoyama A. Copper(II)[2,3-butanedionebis(N^4-methylthiosemicarbazone)], a stable superoxide dismutase-like copper complex with high membrane penetrability. Arch Biochem Biophys 1994; 310: 1–5.

105. el-Saadani MA, Nassar AY, Abou el-Ela SH, Metwally TH, Nafady AM. The protective effect of copper complexes against gastric mucosal ulcer in rats. Biochem Pharm 1993; 46: 1011–1018.

106. Bijloo GJ, van der Goot H, Bast A, Timmerman H. Copper complexes of 1,10-phenanthroline and related compounds as superoxide dismutase mimetics. J Inorg Biochem 1990; 40: 237–244.

107. Vladimirov IA, Parfenov EA, Epanchintseva OM, Sharov VS, Dremina ES, Smirnov LD. [Antiradical activity of complex copper compounds (II) on coumarin ligand base]. Biull Eksp Biol Med 1992; 113: 479–481.

108. Tomas E, Popescu A, Titire A, Cajal N, Critescu C, Tomas S. Viral infection correlated with superoxide anion radicals production and natural and synthetic copper complexes. Virologie 1989; 40: 305–312.

109. Goldstein S, Czapski G. Determination of the superoxide dismutase-like activity of cimetidine-Cu(II) complexes. Free Rad Res Comm 1991; 12–13: 205–210.

110. Byrnes RW, Antholine WE, Petering DH. Interaction of 1,10-phenanthroline and its copper complex with Ehrlich cells. Free Rad Biol Med 1992; 12: 457–469.

111. Nagele A, Felix K, Lengfelder E. Induction of oxidative stress and protection against hydrogen peroxide-mediated cytotoxicity by the superoxide dismutase-mimetic complex copper-putrescine-pyridine. Biochem Pharm 1993; 47: 555–562.

112. Sies H, Brigelius R, Wefers H, Muller A, Cadenas E. Cellular redox changes and response to drugs and toxic agents. Fund Appl Tox 1983; 3: 200–208.

113. Maestre P, Lambs L, Thouvenot JP, Berthon G. Copper-ligand interactions and physiological free radical processes. pH-dependent influence of Cu^{2+} ions on Fe^{2+}-driven OH· generation. Free Rad Res Comm 1992; 15: 305–317.

114. Baker MS, Gebicki JM. The effects of pH on the conversion of superoxide to hydroxyl free radicals. Arch Biochem Biophys 1984; 234: 258–264.

115. Sutton HC, Winterbourn CC. On the participation of higher oxidation states of iron and copper in Fenton reactions. J Free Rad Biol Med 1989; 6: 53–60.

116. Aust SD, Morehouse LA, Thomas CE. Role of metals in oxygen radical reactions. J Free Rad Biol Med 1985; 1: 3–25.

117. Vile GF, Winterbourn CC, Sutton HC. Radical-driven Fenton reactions: studies with paraquat, adriamycin, and anthroquinone 6-sulfonate and citrate, ATP, ADP, and pyrophosphate iron chelates. Arch Biochem Biophys 1987; 259: 616–626.

118. Gutteridge JM. Superoxide-dependent formation of hydroxyl radicals from ferric-

complexes and hydrogen peroxide: an evaluation of fourteen iron chelators. Free Rad Res Comm 1990; 9: 119–25.

119. Halliwell B. The superoxide dismutase activity of iron complexes. FEBS Lett 1975; 56: 34–38.
120. Beauchamp CO, Fridovich I. Superoxide dismutase: improved assays and an assay applicable to polyacrylamide gels. Anal Biochem 1971; 44: 276–287 (1971).
121. McClune GJ, Fee JA, McCluskey GA, Groves JT. Catalysis of superoxide dimustase by iron-ethylenediaminetetraacetic acid complexes. Mechanism of the reaction and evidence for the direct formation of an iron(III) ethylenediaminetetraacetic acid peroxo complex from the reaction of superoxide with iron(II) ethylenediaminetetraacetic acid. J Am Chem Soc 1977; 99: 5220–5222.
122. DiGuiseppi J, Fridovich I. Putative superoxide dismutase activity of iron-EDTA: a reexamination. Arch Biochem Biophys 1980; 203: 145–150.
123. Beauchamp CO, Fridovich I. A mechanism for the production of ethylene from methional. The generation of the hydroxyl radical by xanthine oxidase. J Biol Chem 1970; 245: 4641–4646.
124. Bielski BHJ, Shive GG, Bajuk S. Reduction of nitro blue tetrazolium by CO_2^- and O_2^- radicals. J Phys Chem 1980; 84: 830–833.
125. Tadolini B. Iron oxidation in MOPS buffer. Effect of EDTA, Hydrogen peroxide and ferric chloride. Free Rad Res Comm 1987; 4: 173–182.
126. Misra HP, Fridovich I. Superoxide dismutase: a photochemical augmentation assay. Arch Biochem Biophys 1977; 181: 308–312.
127. Pasternack RF, Banth A, Pasternack JM, Johnson CS. Catalysis of the disproportionation of superoxide by metalloporphyrins. J Inorg Biochem 1981; 15: 261–267.
128. Butler J, Halliwell B. Reaction of iron-EDTA chelates with the superoxide radical. Arch Biochem Biophys 1982; 218: 174–178.
129. Bull C, Fee JA, O'Neill P, Fielden EM. Iron-ethylenediaminetetraacetic acid (EDTA)-catalyzed superoxide dismutase revisited: an explanation of why the dismutase activity of Fe-EDTA cannot be detected in the cytochrome c/xanthine oxidase assay system. Arch Biochem Biophys 1982; 215: 551–555.
130. Bull C, McClune GJ, Fee JA. The mechanism of Fe-EDTA catalyzed superoxide dismutase. J Am Chem Soc 1983; 105: 5290–5300.
131. Nagano T, Hirano T, Hirobe M. Superoxide dismutase mimics based on iron in vivo. J Biol Chem 1989; 264: 9243–9249.
132. Nagano T, Hirano T, Hirobe M. Novel iron complexes behave like superoxide dismutase in vivo. Free Rad Res Comm 1991; 12–13: 221–227.
133. Iuliano L, Pedersen JZ, Ghiselli A, Pratico D, Rotilio G, Violi F. Mechanism of reaction of a suggested superoxide dismutase mimic, Fe(II)-N,N,N',N'-tetrakis(2-pyridylmethyl)ethylenediamine. Arch Biochem Biophys 1992; 293: 153–157.
134. Halliwell B, Pasternack RF. Model compounds with superoxide dismutase activity: iron porphyrins and other iron complexes. Biochem Soc Trans 1978; 6: 1342–1343.
135. Faraggi M, Peretz P, Weintraub D. Chemical properties of water-soluble porphyrins. 4. The reaction of a 'picket-fence-like' iron(III) complex with the superoxide oxygen couple. Int J Rad Biol Relat Stud Phys Chem Med 1986; 49: 951–968.

136. McLaughlin VB, Faraggi M, Leussing DL. Reactions of Fe(III)(*meso-a,a,a*-tetrakis[*o*-(*N*-methylisonicotinamido)phenyl]-porphyrin)$^{5+}$ and Fe(III)(*meso*-tetrakis[*N*-methylpyridinium-4-yl]porphyrin)$^{5+}$ with CN$^-$, CO$_2^-$, O$_2^-$. Inorg Chem 1993; 32: 941–947.

137. Byrnes RW, Field RJ, Datta-Gupta N. DNA strand scission by iron complexes of *meso*-tetra(*N*-methylpyridyl)porphyrins. Chem Biol Interact 1988; 67: 225–241.

138. Forshey PA, Kuwana T. Electrochemical and spectral speciation of iron tetrakis (*N*-methyl-pyridyl)porphyrin in aqueous media. Inorg Chem 1981; 20: 693–700.

139. Barley MH, Rhodes MR, Meyer TJ. Electrocatalytic reduction of nitrite to nitrous oxide and ammonia based on the *N*-methylated cationic iron porphyrin complex Fe(III)(H$_2$O)(TMPyP)$^{5+}$. Inorg Chem 1987; 26: 1746–1750.

140. Harriman A, Porter G. Photochemistry of manganese porphyrins. 1. Characteristics of some water soluble complexes. J Chem Soc Faraday Trans 1979; 275: 1532–1542.

141. Graf E, Mahoney JR, Bryant RG, Eaton JW. Iron-catalyzed hydroxyl radical formation. J Biol Chem 1984; 259: 3620–3624.

142. Puntarulo S, Cederbaum AI. Comparison of the ability of ferric complexes to catalyze microsomal chemiluminescence, lipid peroxidation, and hydroxyl radical generation. Arch Biochem Biophys 1988; 264: 482–491.

143. Henriksson R, Grankvist K. Protective effect of iron chelators on epirubicin-induced fibroblast toxicity. Cancer Lett 1988; 43: 179–183.

144. Kukielka E, Cederbaum AI. NADH-dependent microsomal interaction with ferric complexes and production of reactive oxygen intermediates. Arch Biochem Biophys 1989; 275: 540–550.

145. Sotomatsu A, Nakano M, Hirai S. Phospholipid peroxidation induced by the catechol-Fe^{3+} (Cu^{2+}) complex: a possible mechanism of nigrostriatal cell damage. Arch Biochem Biophys 1990; 283: 334–341.

146. Krishna MC, Grahame DA, Samuni A, Mitchell JB, Russo A. Oxoammonium cation intermediate in the nitroxide-catalyzed dismutation of superoxide. Proc Natl Acad Aci USA 1992; 89: 5537–5541.

147. Azzi A, Tamburro AM, Farnia G, Gobbi E. Cytochrome c interaction with the mitochondrial membrane: a spin label study. Biochim Biophys Acta 1972; 256: 619–624.

148. Asakura T. Heme-spin-label studies on human hemoglobin. Ann NY Acad Sci 1973; 222: 68–85.

149. Giangrande M, Kim YW, Mizukami H. *N*-terminal spin label studies of hemoglobin, ligand, and pH dependence. Biochim Biophys Acta 1975; 412: 187–193.

150. Belagyi J. Water structure in striated muscle by spin labelling technique. Acta Biochim Biophys Acad Sci Hung 1975; 10: 63–70.

151. Feldman A, Wildman E, Bartolinini G, Piette LH. In vivo electron spin resonance in rats. Phys Med Biol 1975; 20: 602–612.

152. Maruyama K, Onishi S. A spin-label study of the photosynthetic bacterium, *Rhodospirillum rubrum*; reduction and regeneration of nitroxide spin-labels. J Biochem 1974; 75: 1153–1164.

153. Morrisett JD, Broomfield CA, Hackley BE. A new spin label specific for the active site of serine enzymes. J Biol Chem 1969; 244: 5758–5761.

154. Bartosz G. Aging of the erythrocyte. IV. Spin-label studies of membrane lipids, proteins, and permeability. Biochim Biophys Acta 1981; 664: 69–73.

155. Ross AH, McConnell HM. Permeation of a spin-label phosphate into the human erythrocyte. Biochem 1975; 14: 2793–2797.

156. Ankel EG, Lai CS, Hopwood LE, Zivkovic Z. Cytotoxicity of commonly used nitroxide radical spin probes. Life Sci 1987; 40: 495–498.

157. Samuni A, Krishna CM, Riesz P, Finkelstein E, Russo A. A novel metal-free low molecular weight superoxide dismutase mimic. J Biol Chem 1988; 263: 17921–17924.

158. Samuni A, Krishna CM, Mitchell JB, Collins CR, Russo A. Superoxide reaction with nitroxides. Free Rad Res Comm 1990; 9: 241–249.

159. Mitchell JB, DeGraff W, Kaufman D, Krishna MC, Samuni A, Finkelstein E, Ahn MS, Hahn SM, Gamson J, Russo A. Inhibition of oxygen-dependent radiation-induced damage by the nitroxide superoxide dismutase mimic. Tempol Arch Biochem Biophys 1991; 289: 62–72.

160. Charkoudian JC, Shuster L. Electrochemistry of norcocaine nitroxide and related compounds: implications for cocaine hepatotoxicity. Biochem Biophys Res Comm 1985; 130: 1044–1051.

161. Pogrebniak H, Matthews W, Mitchell J, Russo A, Samuni A, Pass H. Spin trap protection from tumor necrosis factor cytotoxicity. J Surgical Res 1991; 50: 469–474.

162. DeGraff WG, Krishna MC, Russo A, Mitchell JB. Antimutagenicity of a low molecular weight superoxide dismutase mimic against oxidative mutagens. Env Mol Mutagenesis 1992; 19: 21–26.

163. Goffman T, Cuscela D, Glass J, Hahn S, Krishna CM, Lupton G, Mitchell JB. Topical application of niroxide protects radiation-induced alopecia in guinea pigs. Int J Rad Oncol Biol Phys 1992; 22:803–806.

164. DeGraff WG, Krishna MC, Russo A, Mitchell JB. Antimutagenicity of a low molecular weight superoxide dismutase mimic against oxidative mutagens. Env Mol Mutagenesis 1992; 19: 21–26.

165. Liebmann J, DeLuca AM, Epstein A, Steinberg SM, Morstyn G, Mitchell JB. Protection from lethal irradiation by the combination of stem cell factor and TEMPOL. Rad Res 1994; 137: 400–404.

166. An J, Hsie AW. Effects of an inhibitor and a mimic of superoxide dismutase on bleomycin mutagenesis in Chinese hamster ovary cells. Mutation Res 1992; 270: 167–175.

167. An J, Hsie AW. Polymerase chain reaction-directed DNA sequencing of bleomycin-induced "nondeletion"-type, 6-thioguanine-resistant mutants in chinese hamster ovary cell derivative AS52: effects of an inhibitor and a mimic of super-oxide dismutase. Env Mol Mutagenesis 1994; 23: 101–109.

168. Reddan JR, Sevilla MD, Giblin FJ, Padgaonkar V, Dziedzic DC, Leverenz V,

Misra IC, Peters JL. The superoxide dismutase mimic TEMPOL protects cultured rabbit lens epithelial cells from hydrogen peroxide insult. Exp Eye Res 1992; 56: 543–554.

169. Sosnovsky G. A critical evaluation of the present status of toxicity of aminoxyl radicals. J Pharm Sci 1991; 81: 496–499.

15

Antioxidant Properties of Synthetic Iron Chelators

Clifford S. Collis and Catherine A. Rice-Evans
UMDS Guy's Hospital, London, England

I. INTRODUCTION

Metal chelators have the potential to exert a bifunctional role as antioxidants: on the one hand, as binders of transition metal ions, such as iron and copper, which otherwise may catalyze hydroxyl radical formation, and, on the other hand, as direct scavengers of free radicals. Concerning the latter function, there are several factors which determine the potency of antioxidants as free-radical scavengers in biological systems:

1. Redox potential of the antioxidant in relation to the radical/nonradical pair in question and the rate constant for the interaction
2. Accessibility of the antioxidant to the radical which depends on the localization of the antioxidant as well as the site of radical generation
3. Reactivity and fate of the subsequent antioxidant radical formed, its stabilization and decay
4. Partition coefficients of the antioxidants
5. Redox characteristics of the antioxidant and the implications for the interaction of the antioxidant radical with other antioxidants

In this chapter we are concerned with the mode of action of synthetic mono-hydroxamates as antioxidants, in contrast to their iron chelating properties.

II. HYDROXAMATES

A range of monohydroxamate compounds have been synthesized (1) and their efficacies as free-radical scavenging agents have been investigated compared to the naturally occurring hexadentate trihydroxamate desferrioxamine and the bidentate dihydroxamate rhodotorulic acid (Fig. 1). Their abilities to act as antioxidants are described, in reducing activated heme protein-derived radical species, in modulating deoxyribose degradation mediated by iron-induced hydroxyl radicals and as inhibitors of LDL oxidation, as well as their role in protecting against myocardial reperfusion injury in isolated hearts.

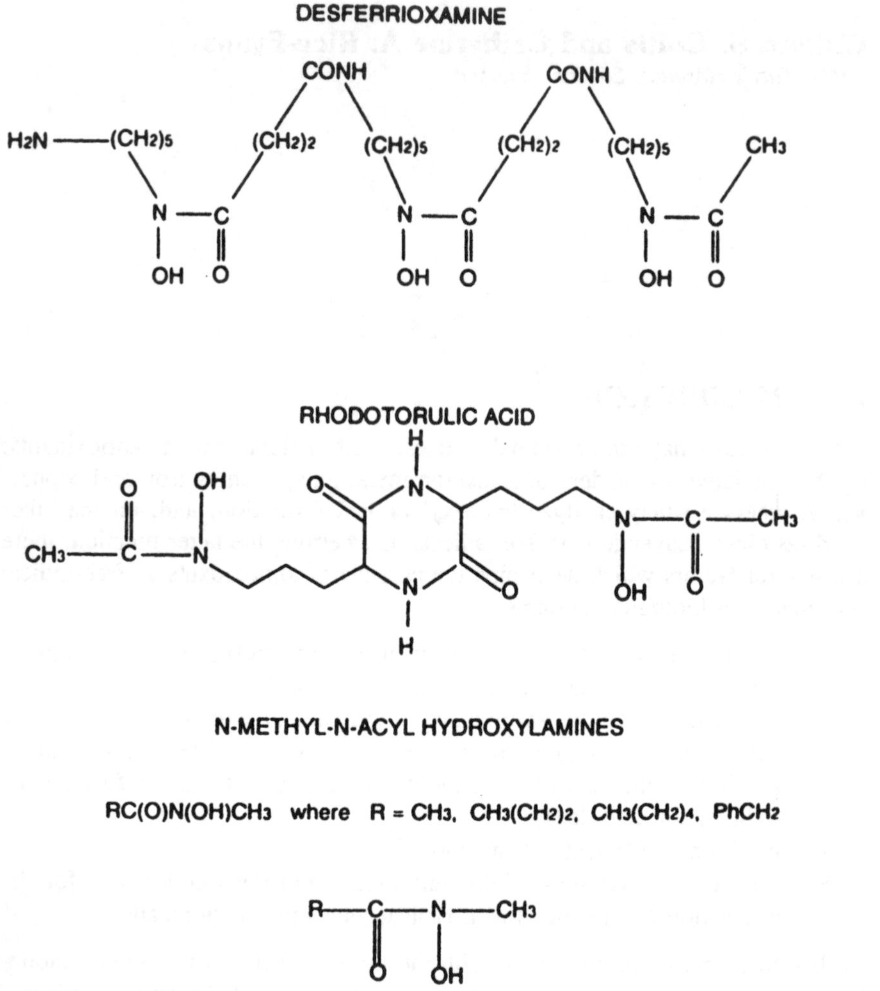

DESFERRIOXAMINE

RHODOTORULIC ACID

N-METHYL-N-ACYL HYDROXYLAMINES

RC(O)N(OH)CH₃   where   R = CH₃, CH₃(CH₂)₂, CH₃(CH₂)₄, PhCH₂

Figure 1   Structures of hydroxamate compounds.

Much attention over the last 20 years has focused on the development of hydroxamate compounds as oral synthetic iron chelators (2). Hydroxamic acids are known to have a plethora of biological activities including the inhibition of the activities of enzymes such as urease (3), ribonucleotide reductase (4), lipoxygenase (5,6), and cyclooxygenase (7). They have also been reported to function as direct-acting vasodilators (8). In light of these activities, a series of novel monohydroxamate derivatives has been synthesized [N-methylacetohydroxamic acid (NMAH), N-methylbutyrohydroxamic acid (NMBH), N-methylhexanoylhydroxamic acid (NMHH), and N-methylbenzoylhydroxamic acid (NMBzH)] (Fig. 1) and investigated for their efficacy as antioxidants and free-radical scavengers, in comparison with desferrioxamine and rhodotorulic acid. All these compounds are iron chelators, desferrioxamine, and rhodotorulic acid promoting iron excretion from iron-loaded rats in vivo (9–11). Of the monohydroxamates discussed here only N-methylacetohydroxamic acid has been evaluated as an iron chelator in in vivo animal models; NMAH, while having a high and selective affinity for iron ($k = 10^{28}$), shows only a low tendency for depleting liver iron (12).

Desferrioxamine is a trihydroxamate hexadentate iron chelator applied clinically for the treatment of iron overload (13). The hexadentate nature of the ligand (Fig. 2) allows chelation of iron III ions on an equimolar basis, with a binding constant of 10^{31} (14). This high binding constant together with the redox properties of the ferrioxamine complex make it a very poor catalyst for iron-catalyzed free-radical production (15).

Several studies have demonstrated that desferrioxamine has an activity other than as a metal chelator (16–25) through the activity of the trihydroxamate moiety as a hydrogen atom donor or electron donor to a variety of systems including activated horseradish peroxidase (17) and activated cytochromes (18), ferryl myoglobin (19,26) and ferryl hemoglobin, as well as the superoxide

Bidentate ligand 3 : 1   Hexadentate ligand 1 : 1

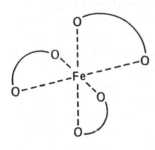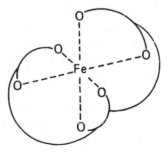

Figure 2 Bidentate ligand (sharing two oxygens with iron) and hexadentate ligands (sharing six oxygens). (From Ref. 13.)

radical (16). In addition, the ability of desferrioxamine to intercept the propagation phase of membrane and LDL peroxidation has also been demonstrated (19–22) with the formation of the desferrioxamine nitroxide radical (21). These findings will be reviewed here alongside the evidence for the action of the novel *mono*hydroxamates as hydrogen-donating antioxidants.

III. THE REDUCTION OF FERRYL SPECIES BY HYDROXAMATES

A. Ferrylmyoglobin

The peroxidase action of heme proteins has been recognized for many years and has been well reviewed from a range of standpoints (27,28). The interaction of metmyoglobin with peroxides involves activation to the ferryl species (29,30) which consists of an iron (IV)-oxo complex, in which the heme iron is one oxidizing equivalent above that of metmyoglobin, and one oxidizing equivalent is present on the globin moiety forming a protein radical on the surface of the protein (31–34). It is not clear which of these species is responsible for oxidative damage to biological substrates. It has been proposed that the radical on the surface of the heme protein is located on tyrosine-103, in equine myoglobin (35,34).

The protein radical has been characterized by EPR spectroscopy, using both stopped-flow and spin trapping with DMPO. It forms rapidly (in 10 s) and decays rapidly, the DMPO adduct being reduced to approximately 10% of its initial intensity in minutes. It is suggested that the first detectable product is tyrosine phenoxyl radical and that this can react with molecular oxygen to produce a peroxyl radical (34), although the latter reaction is disputed (36). Thus, the precise identification of the state of the initiating species is not yet established; recent studies have shown that it is the myoglobin-derived species, whatever their nature, on the surface of the heme protein which initiate oxidative damage to a variety of substrates including erythrocyte membranes (37), low-density lipoproteins (38), microsomes, and model lipid membranes (18).

The activities of monohydroxamate compounds NMAH, NMHH, NMBH, NMBzH have been compared to the trihydroxamate desferrioxamine and the dihydroxamate rhodotorulic acid, in terms of their suppressive effects on the oxidative activation of the heme protein, using two experimental approaches: first, investigating the effect of addition of the activating agent to the myoglobin after pretreatment with the hydroxamate, the latter thus being present during the formation of the ferryl myoglobin species; and, second, addition of the hydroxamate to preformed ferryl myoglobin under plateau conditions, i.e., 15 min after activation.

B. Optical Spectroscopic Studies

Activation of metmyoglobin by hydrogen peroxide, under the condition of a 1:1.25 molar ratio induces a rapid development of the spectral characteristics attributable to ferryl myoglobin (peaks at 515, 550, and 585 nm) with the concomitant loss of the well-defined metmyoglobin absorption at 630 nm. Application of the Whitburn algorithms (39), which allow the relative proportions of ferryl, met, and oxy to be calculated, indicates that the concentration of ferryl myoglobin, under these conditions, reaches a maximum value at 10 min, at which time it constitutes ca. 60% of the total heme (26) and remains at approximately this level for 60 min.

Figure 3 demonstrates and Table 1A summarizes the effects of the hydroxamates on time to 50% reduction of maximal ferryl myoglobin formation. The results show that NMHH is the most effective of the monohydroxamates and is very much more efficacious in scavenging ferryl myoglobin than desferrioxamine on a mole-for-mole basis, especially since desferrioxamine has three hydroxamate groups. Furthermore, the data show that in the case of NMBH, high concentrations (50 μM) initially inhibit and subsequently apparently enhance ferryl myoglobin formation. Thus, it seems that the monohydroxamates, depending on the varying nature of the group attached to the carbonyl function have differential rates of reaction with ferryl myoglobin radicals, different reactivities and abilities to stabilize the resulting radical species. Interestingly, in the presence of 10 μM concentrations of desferrioxamine, i.e., less than that of the initial metmyoglobin, an apparent stimulation of ferryl myoglobin formation was initially observed in the first few minutes after activation, followed by reduction by the hydroxamate. This enhancement of ferryl myoglobin formation also occurred with the other hydroxamates tested, rhodotorulic acid, NMAH and NMBzH showing more pronounced effects than desferrioxamine.

C. The Order of Reactivity of the Monohydroxamates

The rate constants for the reaction of ferryl myoglobin with the hydroxamate compounds, under the pretreatment conditions, have been determined. To model the data, the equation

$$N = \frac{\lambda_1}{\lambda_2 - \lambda_1}(e^{-\lambda_1 t} - e^{-\lambda_2 t})$$

has been adopted (24). In this context no physical significance is attached to the constant λ_1, but λ_2 is a fair description of the exponential decay of the efficacy of the drug. The equation has been fitted to the data with a least-squares

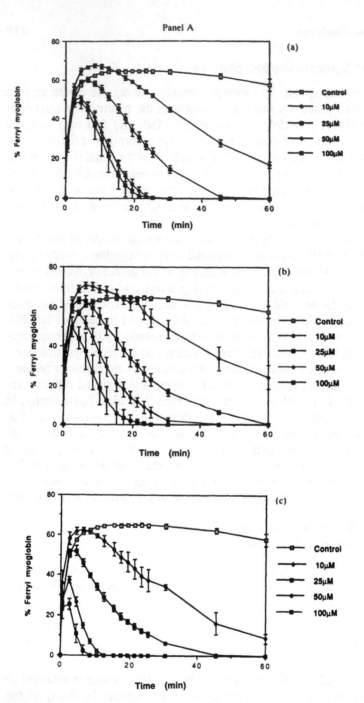

Figure 3 The effects of hydroxamate compounds on ferryl myoglobin formation. (Panel A) Hydroxamates added prior to activation. (Panel B) Hydroxamates added 15 min after ferrylmyoglobin formation. (a) desferrioxamine, (b) rhodotorulic acid, (c) NMHH, (d) NMBzH, (e) NMAH, (f) NMBH. [Metmyoglobin: hydrogen peroxide ratio 1:1.5.] (From Ref. 24, with permission.)

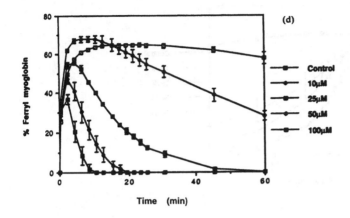

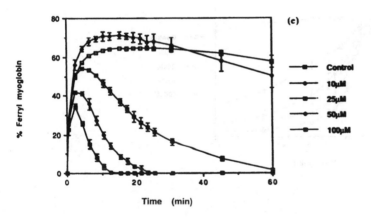

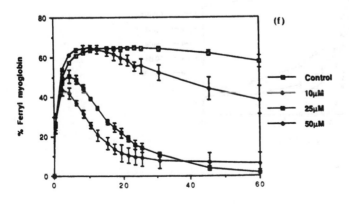

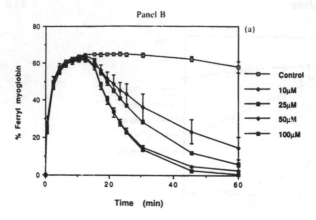

Panel B

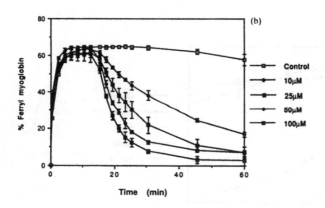

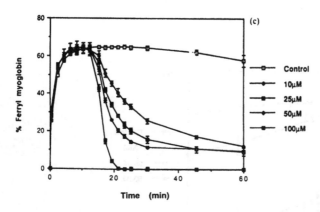

Figure 3 Continued

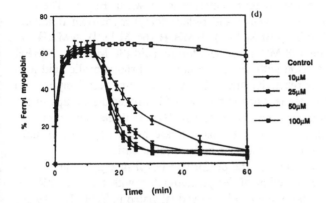

Table 1 Time (min) to 50% Reduction of Ferryl Myoglobin

A.   By the hydroxamate added at time of activation

Concentration (μM)	DFO	RDA	NMHH	NMBzH	NMAH	NMBH
10	45	55	37	58	≫60	>60
25	22	25	13	18	20	18
50	12	13	4	10	10	10
100	11	8	a	5	5	b

B.   Myoglobin after addition of hydroxamate[c]

Concentration (μM)	DFO	RDA	NMHH	NMBzH
10	24	26	14	12
25	15	12	8	6
50	7	7	5	5
100	7	5	2	4

Source: From Ref. 23.
[a]Extent of inhibition always >50% at all time points.
[b]NMBH at 100 μM showed enhanced ferrylmyoglobin formation.
[c]Hydroxamate added 15 min after myoglobin activation.

minimization program. A typical representation is shown in Fig. 4. The data show that the hydroxamates scavenge and reduce ferryl myoglobin relatively slowly, in the order NMHH (110 M^{-1} s^{-1}), NMBzH (66 M^{-1} s^{-1}), NMAH (50 M^{-1} s^{-1}), desferrioxamine (38 M^{-1} s^{-1}), rhodotorulic acid (30 M^{-1} s^{-1}), NMBH (27 M^{-1} s^{-1}). The comparison with desferrioxamine and rhodotorulic acid lead to the deduction that their reactivity and efficacy in reducing ferryl myoglobin to metmyoglobin are independent of the number of hydroxamate groups in the molecule. The data emphasize the notion that NMHH is three times more efficacious than desferrioxamine and that this increase in activity is even more significant if calculated on a per mole of hydroxamate basis.

Addition of the hydroxamates to the preformed ferryl myoglobin at $t = 15$ min induced progressive reduction of ferryl myoglobin with time to 50% reduction as a function of concentration (Fig. 3) and tabulated in Table 1B. Again, NMHH is the most effective of the monohydroxamates studied with total inhibition of ferryl myoglobin 5 min after adding the drug (100 μM).

D. EPR Studies

Examination of the above reaction mixtures by EPR spectroscopy reveal that on activation of metmyoglobin to ferryl under the above conditions (addition of hydrogen peroxide at 1.25 molar excess), spectral changes characteristic of

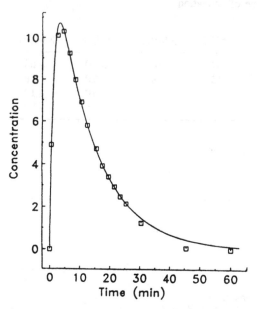

Figure 4   Typical least-squares fit for assessing rate constants for the reaction of ferryl myoglobin with NMHH. (From Ref. 24 with permission.)

the formation of a significant proportion of ferryl myoglobin are observed (26). Optical spectroscopy has shown that in the presence of the hydroxamates, ferryl myoglobin is reduced to the met form (23) with the hydroxamates present at either 50 or 100 μM, signals which could be assigned to the corresponding nitroxide radicals formed by oxidation of the N–OH group to N–O$^\cdot$ were detected as the ferryl myoglobin is reduced to the met form (Fig. 5) (23,25). NMAH, NMBH, NMHH, and NMBzH produced signals which were consistent with the presence of radicals with partial structure $CH_3N(O^\cdot)C(O)-$ (with parameters a_N 0.780, a_{3H} 0.886 mT), and rhodotorulic acid gave signals assignable to a radical with partial structure $CH_2N(O^\cdot)C(O)-$ (a_N 0.767, a_{2H} 0.661 mT) as would be expected for the nitroxide radicals produced from the parent compounds (24). Signals from the desferrioxamine nitroxide radical with hyperfine coupling constants have been determined previously (16).

In order to detect the signals associated with the reduction of the myoglobin protein radical by the hydroxamate and the formation of the nitroxide, radical low-temperature EPR spectroscopic studies were applied. Using frozen samples at liquid helium temperatures greater sensitivity is achieved, both intrinsically and because signal averaging is possible (as the signals are stable at low temperature); low-temperature experiments were therefore carried out using identical (20 μM) protein concentrations to those previously employed for optical spectroscopy (24). Figure 6 shows the $g = 2$ region of the EPR spectrum 5 s after peroxide treatment (25). In the absence of NMBH (Fig. 6a) two distinct peaks are seen at $g = 2.035$ and $g = 2.004$. Similar results have been reported by other workers (30,32,39,40) and attributed to two distinct radical species, with the signal at $g = 2.004$ being the major component. This is consistent with previous work, suggesting that there is more than one radical

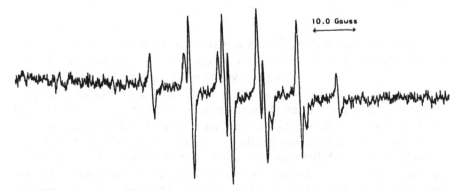

Figure 5 EPR spectra of nitroxide radicals derived from NMAH on interaction with ferryl myoglobin. For hyperfine coupling constants see text (10 Gauss = 1 mT).

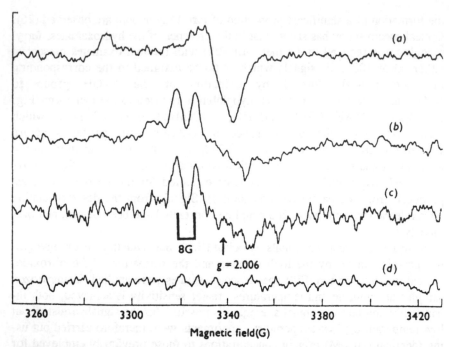

Figure 6 Free radicals formed from the scavenging of ferryl myoglobin by NMBH. (a) Metmyoglobin + hydrogen peroxide [1:1.25 mole ratio] frozen after 10 s (relative gain × 1). (b) As (a) + 100 μM NMBH, frozen after 10 s (relative gain × 4). (c) Ferricyanide + NMBH, frozen after 60 s (relative gain × 2). (d) Spectrum (b) minus spectrum (c). EPR conditions: temperature 100 K, microwave power 20 mW, microwave frequency 9.38 GHz, modulation frequency 100 kHz, modulation amplitude 2 Gauss, time constant 0.33 s, sweep time 5 Gauss/s, signal is average of 20 scans. (From Ref. 25 with permission.)

formed at room temperature (34). A consistent observation from many groups revealed that it has never been possible to detect a stoichiometric formation of the radical(s), relative to the initial level of metmyoglobin (30,32,39).

In the presence of NMBH (Fig. 6b) a much less intense free-radical signal is detected; it also has different characteristics. The radical is only observed when metmyoglobin, hydrogen peroxide, and NMBH are all present. This signal is attributed to the anisotropically broadened (i.e., broadened due to slow molecular motion) NMBH nitroxide radical; both the g-value (2.006) and the one resolved hyperfine splitting observable (8 Gauss) are consistent with this assignment (25). The low-temperature spectrum obtained in the presence of metmyoglobin, H_2O_2, and NMBH is identical to that seen following addition of excess ferricyanide to NMBH (Fig. 6c) (subtraction of the two spectra yield a flat baseline, Fig. 6d). Therefore, the radical seen in Fig. 6b must be the product of a one-electronc oxidation of NMBH. This confirms that NMBH decreases the yield of ferryl myoglobin by acting as an electron donor.

The desferrioxamine nitroxide radical has also been observed using low-temperature EPR spectroscopy (Fig. 7). Again immediately after peroxide addition the myoglobin protein radicals (a) are replaced by a nitroxide radical (b). This is confirmed by comparing this spectrum with that obtained via chemical oxidation of desferrioxamine with ferricyanide (c). In the absence of peroxide no EPR signals are detectable in the $g = 2$ region (d). All these findings are consistent with the hydrogen-donating antioxidant properties of the hydroxamates.

At high peroxide:heme ratios it is known that there is damage to the prosthetic group and release of iron (41). However, at the low ratios used in this study no evidence for this process was observed, as evidenced by a lack of high-spin non-heme iron EPR signals at $g = 4.3$. This is consistent with our previous optical studies (19), showing that there is no iron release at low

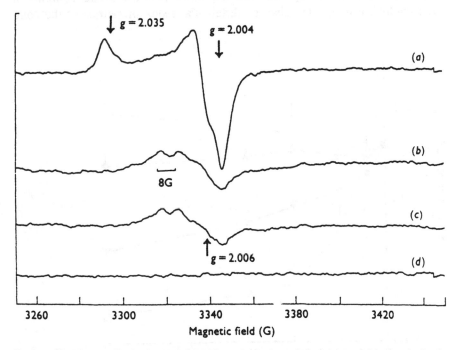

Figure 7 Free radicals formed from the scavenging of ferrylmyoglobin by desferrioxamine: (a) metmyoglobin + hydrogen peroxide (gain × 1); (b) ferryl myoglobin + 100 μM desferrioxamine (gain × 4); (c) desferrioxamine + ferricyanide (gain × 4); (d) metmyoglobin alone (gain × 4). Samples frozen 5 s after peroxide addition. EPR conditions as Fig. 6. (From Ref. 25 with permission.)

peroxide:heme ratios. Thus the effect of both desferrioxamine and NMBH in decreasing ferryl myoglobin–induced lipid peroxidation is via reduction of the ferryl heme iron and/or the protein-free radicals rather than via iron chelation.

IV. INHIBITION OF 2-DEOXYRIBOSE DEGRADATION BY HYDROXAMATES

The reactivity of the hydroxamate compounds as antioxidants against radicals capable of inducing the degradation of deoxyribose (24) is not consistent with what might be predicted from reaction rates (Fig. 8). Studies show that 50 μM NMAH and desferrioxamine are equally effective in inhibiting radical-induced degradation of deoxyribose, whereas at these concentrations NMHH and NMBH are ineffective. As the concentration increases, desferrioxamine was increasingly and proportionately more effective, whereas this was not the case with NMAH. These observed effects may relate not only to the radical-scavenging ability of the monohydroxamates but also the differential redox properties of the com-

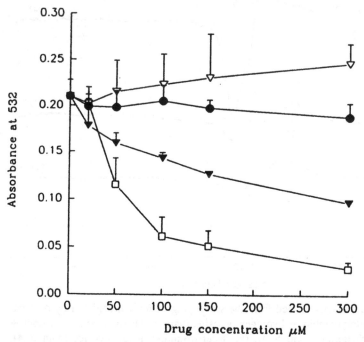

Figure 8 Deoxyribose degradation and the effects of NMBH (∇), NMHH (●), NMAH (▼), desferrioxamine (□). (From Ref. 24 with permission.)

plexes with iron. Thus, in systems containing "available" iron, the hydroxamates may exert a dual role in protection from free-radical damage.

V. HYDROXAMATES AS INHIBITORS OF LDL OXIDATION

A. Ferryl Myoglobin-Mediated LDL Oxidation

The antioxidant properties of the hydroxamates have been investigated further by studying their effect on the comparative mechanisms by which myoglobin and its activated counterpart, ferryl myoglobin, enhance LDL oxidation. A variety of heme-continuing systems have been shown to mediate LDL oxidation including oxy, met, and ferryl myoglobin (38), met and ferryl myoglobin (22,42), hemoglobin from erythrocytes (43), hemin (44), and the suppression of the oxidation by a variety of hydrogen-donating and electron-donating antioxidants (22,45–48).

Changes in the oxidation state as the heme protein is activated by hydrogen peroxide in the presence and absence of LDL are shown in Fig. 9. The spectra are analyzed for the presence of ferryl myoglobin applying the Whitburn algorithms. Activation of metmyoglobin in the absence of LDL generates ferryl myoglobin to the extent of 60% of the total heme, with very little change after 90 min interaction, as previously reported (23). In the presence of LDL, the maximal ferryl formation is attained more rapidly and the actual level is enhanced initially, 72% by 6.5 min interaction, sustained up to 14.5 min, but declines progressively as the reaction with LDL continues, becoming progressively reduced to metmyoglobin with only 39% ferryl myoglobin remaining after 90 min interaction.

Pretreatment of myoglobin with monohydroxamate and trihydroxamate hydrogen-donating compounds prior to activation with hydrogen peroxide (in the absence of LDL) substantially inhibits the development of the ferryl myoglobin (Fig. 9). In the absence of LDL, pretreatment with NMHH (100 µM) was more effective at reducing the ferryl myoglobin back to the met state than DFO when at concentrations in excess of the myoglobin but not at lower concentrations (10 µM). At hydroxamate concentrations of 10 µM, both DFO and NMHH inhibited the oxidation of LDL, as shown by the suppression of the altered electrophoretic mobility, the monohydroxamate being more effective than the trihydroxamate. At excess concentration of both drugs (100 µM) the change in surface charge is totally inhibited.

Inhibition of peroxidation of the polyunsaturated fatty acyl chains of the LDL by the drugs when present at the time of activation to ferryl myoglobin in the presence of the LDL is shown in Fig. 10. At 10 µM levels, the monohydroxamate was almost as effective as 100 µM concentrations, but 10 µM

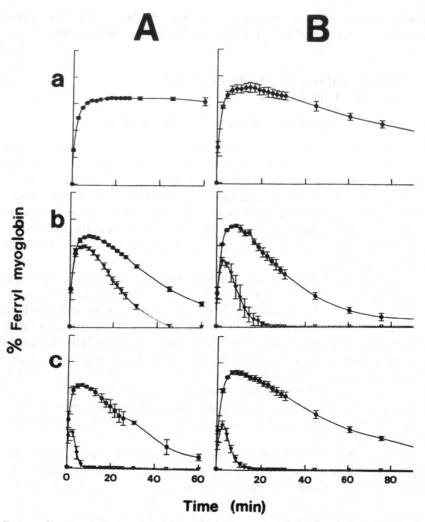

Figure 9 Changes in the oxidation state as metmyoglobin is activated to ferryl myoglobin in the absence (A) and presence (B) of LDL and the effects of the free-radical scavenging hydroxamates. (a) no hydroxamates; (b) desferrioxamine 10 μM (●), 100 μM (∇); (c) NMHH 10 μM (●), 100 μM (∇).

DFO only inhibited to the extent of 60%. While the mechanism of action of the hydroxamates here is clearly those of hydrogen-donating antioxidants (no chelatable iron being available), the relative contributions to reducing ferryl myoglobin and to scavenging peroxyl and alkoxyl radicals is by no means clear.

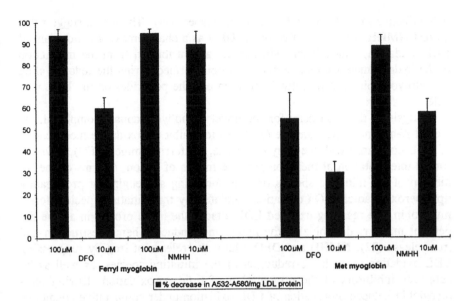

Figure 10 The percentage inhibition of myoglobin-mediated LDL oxidation by desferrioxamine and *N*-methyl-*N*-hexanoyl hydroxamate.

B. Metmyoglobin-Induced LDL Oxidation

The mechanism of LDL oxidation by metmyoglobin is hydroperoxide-dependent (42), i.e., preexisting LDL hydroperoxide is necessary for metmyoglobin to oxidatively and reductively catalyze the decomposition of peroxides, leading to the formation of alkoxyl and peroxyl radicals, capable of reinitiating further rounds of oxidation. The mechanism by which hydroxamates may inhibit such oxidation may be as hydrogen donors to LO·, alkoxyl, and LOO·, peroxyl radicals and as reductants for ferryl myoglobin.

During the interaction between LDL and metmyoglobin in the absence of hydrogen peroxide the LDL is oxidatively modified both in terms of lipid peroxidation and altered surface charge, evidence for metmyoglobin-mediated propagation of oxidation (Fig. 10). No changes in the optical spectrum of the heme protein are observed, suggesting that the oxidation is catalyzed through the redox-cycling of small amounts of metmyoglobin as it amplifies the oxidation of preexisting LDL hydroperoxides. That this cycling occurs through the intermediate oxidation to the ferryl state has been confirmed (42).

Metmyoglobin-induced propagation of LDL oxidation is inhibited extensively by 100 μM monohydroxamate, but, in this case, 100 μM and 10 μM DFO were only effective to 55% and 30% inhibition, respectively. These data suggest that 10 μM NMHH is twice as effective as DFO as a chain-breaking antioxidant of LDL oxidation, although this will include a contribution from the interaction of the hydroxamate with the ferryl myoglobin formed during the redox cycling of metmyoglobin as it reductively decomposes the peroxides on the LDL particle.

Analysis of the propensity for the novel monohydroxamate compound, *N*-methyl-*N*-hexanoyl hydroxamate (NMHH) to inhibit the oxidative modification of LDL compared with the trihydroxamate, desferrioxamine (DFO), have afforded interpretation of their comparative modes of action in terms of the inhibition of the initiating species or chain-breaking scavenging of propagating lipid peroxyl radicals. On oxidation mediated by the initiating species, ferryl myoglobin, the resulting oxidized LDL attacks the heme protein in a time-dependent manner, destabilizing the heme ring, inducing heme destruction and iron release (22). NMHH and DFO inhibit the release of iron by suppressing LDL oxidation through the reduction of the initiating species, as well as by acting as inhibitors of the propagation of lipid peroxidation. During metmyoglobin-induced propagation of LDL oxidation under comparative conditions of interaction, the LDL oxidation is less extensive and the heme protein remains intact.

VI. MONOHYDROXAMATES AS ATTENUATORS OF POSTISCHEMIC REPERFUSION DYSFUNCTION IN ISOLATED HEART MODELS

The development of ischemia-reperfusion injury in the heart is a multicomponent problem associated with a range of myocardial dysfunctions ranging from arrhythmias, transient mechanical dysfunction, and cell death (49,50) to which the generation of free radicals is believed to make a significant contribution (51–55). The generation of free radicals has been shown by EPR spectroscopy to occur within the first 3 min of reperfusion (54,56,57) and the deleterious effects of such production are thought to be further exacerbated by the cycling of redox-active iron ions (58,59). The possible sources of the latter are transferrin or ferritin (60), myoglobin (61), or hemoglobin (62). Prolonged ischemia increases the amount of iron available for these reactions (63) and induces a depletion of the heart's natural antioxidant defenses (64).

Desferrioxamine has been shown to inhibit electrical and contractile dysfunction in myocardial ischemia-reperfusion-induced injury (65–67). For example, Bolli et al. (68) have suggested as a result of experiments using open chest dogs undergoing ischemia and reperfusion that a substantial portion of the damage responsible for myocardial stunning is caused by iron-catalyzed free-radical reactions that developed in *the initial seconds* of reperfusion and that this could be attenuated with DFO. However, the effectiveness of desferrioxamine-treatment is controversial as the degree of success varies with different models and perfusion protocols. In Bolli's experiments the extent of myocardial stunning could be reduced by the administration of the iron chelator *just before* reflow but not *after* reperfusion had commenced. The possible reasons for limited protection may be related to the pharmacokinetic properties of this drug. In particular, desferrioxamine principally enters only those cells undergoing active pinocytosis (69), has a relatively short half-life of 5–7 min in the plasma and can exhibit pro-oxidant effects at high concentrations. The selective penetration of this drug could account for the requirement to administer it before reperfusion and for its subsequent lack of effect when applied on reperfusion. Furthermore, although desferrioxamine may be effective against an initial burst of free-radical production it would have little effect in cells not undergoing pinocytosis or on the chronic production of free radicals, e.g., by activated neutrophils once treatment was discontinued. This latter effect may account for the observations of Reddy et al. (70) where DFO provided a reduction in infarct size in a canine model after 4 h reperfusion but not after 24 h.

Studies with NMHH in an isolated rat heart model of reperfusion injury (71) demonstrated a significant improvement in the recovery of left ventricular developed pressure (LVDP) compared to control hearts when administered on reperfusion in contrast with DFO (Fig. 11). A contributing factor may be that NMHH is able to penetrate cells, from studies showing protection of endothelial cells from hydrogen peroxide-induced *intracellular* free-radical damage (72).

All the other monohydroxamate studies have also demonstrated an improved recovery of LVDP compared to controls (Fig. 11) (73), although only the recovery with NMBH-treated hearts was significantly different from control hearts. However, the results of these studies are consistent with the proposition that the administration of iron chelators and free-radical scavengers is effective in the attenuation of acute reperfusion-induced contractile dysfunction after ischemia.

The duration of ischemia is an important factor in determining the type and extent of damage, and has led to the development of different types of models for studying ischemia-reperfusion injury. Longer periods of ischemia (>20 min) have been associated with more permanent mechanical dysfunction described above (74), whereas short-duration transient ischemia (10–15 min) in experimental models consistently produces reperfusion arrhythmias indicating electrical dysfunction (75,76). Further, provided the arrhythmias induced are not lethal, e.g., irreversible ventricular fibrillation, contractile function usually is nearly completely recovered; however, arrhythmias provide some of the earliest evi-

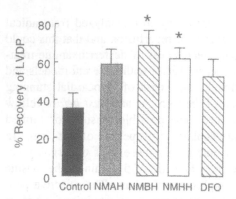

Figure 11 The effects of novel monohydroxamates and desferrioxamine (DFO) administered at reperfusion on the recovery of left ventricular developed pressure (LVDP) after 60-min reperfusion, following a 50-min ischemic insult in the isolated rat heart. Values are mean $\pm$ sem % of the preischemic absolute value, 100% = 90 ($\pm$5) mmHg; $n = 10$ for control hearts and 8 for drug-treated hearts; *$p < 0.05$ vs. control by Dunnett's test. (Redrawn from Refs. 71 and 73.)

dence for myocardial damage. It has been suggested that one possible target of free-radical damage is the ion pumps in the cellular and intracellular membranes (49,55); the thiol groups of such pumps in particular are vulnerable to attack (77). Damage to these pumps would lead to ionic imbalance and could account for the electrical dysfunction seen. Prolonged ischemia through depletion of the natural antioxidant defences, exacerbates free-radical damage on reperfusion. This could result in increased ionic imbalance, in particular calcium overload, and in addition damage to the membranes themselves contributing to irreversible contractile dysfunction and eventually cellular necrosis.

The effects of antioxidants on electrical dysfunction provides another mode of investigation of the efficacies of these compounds. DFO and the monohydroxamates have also been tested in an isolated heart model for reperfusion-induced arrhythmias (78). All the monohydroxamate drugs provided protection against reperfusion-induced arrhythmias as evidenced by a reduction in the incidence of sustained left ventricular fibrillation and an increase in the time spent in sinus rhythm postreperfusion (Fig. 12). NMAH was the more efficacious of the monohydroxamates and NMAH-treated hearts showed a reduced incidence of sustained ventricular fibrillation, a reduced duration of fibrillation, and an increased time spent in sinus rhythm. This difference in efficacy may be due to NMAH being the smallest of the compounds and therefore able to reach the sites of radical production more effectively in this type of acute insult.

A. Sustained Ventricular Fibrillation

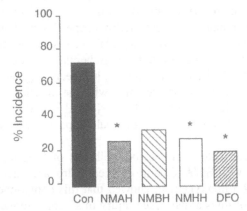

B. Sinus Rhythm

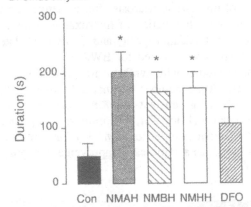

Figure 12 The effects of novel monohydroxamates and desferrioxamine (DFO) on (A) the % incidence of sustained ventricular fibrillation (duration > 3 min) and (B) the duration of sinus rhythm. (A) 100% (n) = 15 hearts; *$p < 0.05$ vs. control, Fisher exact test. (B) Values are mean ± sem; maximum possible duration 300 s; *$p < 0.05$ vs. control, Mann-Whitney-Wilcoxon Test. (Redrawn from Ref. 78.)

Interestingly, in previous studies, NMBH and NMHH were more effective than NMAH in producing an improvement in the recovery of contractile function after an ischemic insult (71,73). Whether this is due to differences in the biochemistry of the compounds or differences in the nature of the pathophysiology of arrhythmogenesis as opposed to contractile failure is a matter of speculation. For example, both NMAH and DFO were equally effective in inhibiting degradation of 2-deoxyribose, with NMHH and NMBH less so (24). It was suggested that this effect may be due to the longer alkyl chains on these compounds resulting in steric hindrance and/or different redox properties of their complexes with iron rather than their radical-scavenging abilities (as suggested by the above rate constants). However, the durations of ischemia and reperfusion in the two studies described are different and since it has been suggested that the amount of iron released on reperfusion increases with the duration of the ischemic period (59,63) one would have predicted that NMAH would have been more effective against contractile dysfunction than would appear to be the case.

There is evidence for some hydroxamic acids to be inhibitors of 5-lipoxygenase (8); however, experiments with BWA4C, a novel acetohydroxamic acid which was a selective inhibitor of 5-lipoxygenase, showed that this compound did not influence the progression of myocardial tissue injury in a canine model (79). These authors did not believe that the products of 5-lipoxygenase played a significant role in the propagation of myocardial necrosis after ischemia. This suggests that the principal modes of protective action of hydroxamic acids in the reperfusion model lies in both their antioxidant capacity and ability to chelate iron, although neither of these properties was quoted for BWA4C.

There are comparatively few other studies with synthetic iron chelating antioxidants. In a study by Katoh et al. (80), a number of other substances with iron chelating properties (with iron-binding constant log Km) were tested in models of myocardial ischemia, i.e., catechol (log Km, 43), mimosine (log Km, 36), kojic acid (log Km, 27), and desferrioxamine (log Km, 31) for comparison. All of these except kojic acid produced a significant improvement of postreperfusion LVDP.

VII. CONCLUDING COMMENTS

This work has described the effects of synthetic monohydroxamates with both antioxidant and iron chelating abilities in a range of biochemical systems and compared their efficacies with the dihydroxamate rhodotorulic acid and the trihydroxamate DFO. The reactivity of these compounds as hydrogen-donating antioxidants is independent of the number of hydroxamate groups in the molecule. Further, in a physiological system these monohydroxamates have been shown to ameliorate aspects of myocardial ischemia-reperfusion injury. Although this type of pathology is a multicomponent dysfunction, studies from many laboratories suggest that free radicals and iron-mediated radical reactions play a significant role in its development, and thus compounds which can inhibit these processes have significant potential in contributing to its amelioration. However, as yet, it is not possible to discern the relative importance of iron chelation compared to antioxidant mechanisms in the action of the mono-hydroxamates.

ABBREVIATIONS

DFO	Desferrioxamine
LVDP	Left ventricular developed pressure
NMAH	*N*-methyl acetohydroxamic acid
NMBH	*N*-methyl butyrohydroxamic acid
NMHH	*N*-methyl hexanoyl hydroxamic acid
NMBzH	*N*-methyl benzoyl hydroxamic acid

REFERENCES

1. Ulrich H, Sayigh AAR. Hydroxyamino derivatives from formaldehyde. Their reaction with acyl halides. J Chem Soc 1963; 1098–1101.
2. Grady RW, Graziano JH, White CP, Jacobs A, Cerami A. The development of new iron chelating drugs. J Pharmacol Exp Ther 1978; 205: 757–765.
3. Griffith DP. In: Kehl H, ed. Chemistry and Biology of Hydroxamic Acids. Basel: Karger, 1982: 186.
4. Elford HL, Van't Riet B. Inhibition of nucleoside diphosphate reductase by hydroxybenzohydroxamic acid derivatives [Review]. Pharmacol Therapeut 1985; 29: 239.
5. Cashman JR. Leukotriene biosynthesis inhibitors. Pharmac Res 1985; 6: 253–261.
6. Clapp CH, Bannerjee A, Rotenburg SA. Inhibition of soybean lipoxygenase I by *N*-alkyl hydroxylamines. Biochemistry 1985; 24: 1826–1830.
7. Kehl H, Fountain K, Early T. Structure-activity studies of hydroxamic acids as direct vasodilators. Arzneim Forsch 1978; 28: 2087.
8. Summer JB, Kim KH, Mazdiyasni H, Holmes JH, Ratajczyk JD, Stewart AO, Dyer RD, Carter GW. Hydroxamic acid inhibitors of 5-lipoxygenase. J Med Chem 1989; 33: 992–998.
9. Grady RW, Peterson CM, Jones JH. Rhodotorulic acid—investigation of its potential as an iron-chelating drug. J Pharmacol Exp Ther 1979; 209: 342–349.
10. Hershko C, Grady RW, Cerami A. Mechanism of iron chelation in the hypertransfused rat: definition of two alternative pathways of iron mobilisation. J Lab Clin Med 1978; 92: 144–151.
11. Baker E, Page H, Torrace J, Grady R. Effect of desferrioxamine, rhodotorulic acid and cholylhydroxamic acid on transferrin and iron exchange with hepatocytes in culture. Clin Physiol Biochem 1985; 3: 277–288.
12. Pitt CG, Gupta G, Estes WE, Rosenkrantz H, Metterville JJ, Crumbliss AL, Palmer RA, Nordquest KW, Sprinkle Hardy KA, Whitcomb DR, Byers RR, Arceneaux JEL, Garies CG, Sciortino CV. The selection and evaluation of new

chelating agents for the treatment of iron overload. J Pharmacol Exp Ther 1979; 208: 12–18.

13. Porter JB Huehns ER, Hider RC. The development of iron chelating drugs. In: Hershko C, ed. Bailliere's Clinical Haematology. London, Bailliere Tindall, Saunders, 1989: 257–292.

14. Keberle H. The biochemistry of desferrioxamine and its relationship to iron metabolism. Ann NY Acad Sci 1965; 119: 758–768.

15. Halliwell B. Use of desferrioxamine as a "probe" for iron-dependent formation of hydroxyl radicals. Evidence for a direct reaction between desferal and the superoxide radical. Biochem Pharmacol 1985; 34: 229–233.

16. Davies MJ, Donkor R, Dunster CA, Gee CA, Jonas S, Willson RL. Desferrioxamine (Desferal) and superoxide free radicals formation of an enzyme-damaging nitroxide. Biochem J 1987; 246: 725–729.

17. Morehouse KM, Flitter WD, Mason RP. The enzymatic oxidation of Desferal to a nitroxide free radical. FEBS Lett 1987; 222: 246–250.

18. Kanner J, Harel S. Desferrioxamine as an electron donor. Inhibition of membranal lipid peroxidation initiated by H_2O_2-activated metmyoglobin and other peroxidizing systems. Free Rad Res Commun 1987; 3: 309–317.

19. Rice-Evans C, Okunade G, Khan R. The suppression of iron release from activated myoglobin by physiological electron donors and desferrioxamine. Free Rad Res Comm 1989; 7: 45–54.

20. Darley-Usmar VM, Hersey A, Garland L, Leonard P, Wilson MT. Iron chelators and inhibition of lipid peroxidation. In: Rice-Evans C, ed. Free Radical Disease States and Antiradical Interventions London: Richelieu Press, 1989: 183–200.

21. Hartley A, Davies MJ, Rice-Evans C. Desferrioxamine as a lipid chain-breaking antioxidant in sickle erythrocyte membranes. FEBS Lett 1990; 264: 145–148.

22. Rice-Evans C, Green E, Paganga G, Cooper C, Wrigglesworth J. Oxidised low density lipoproteins induce iron release from networked myoglobin. FEBS Lett 1993; 326: 177–182.

23. Green ESR, Cooper CE, Davies MJ, Rice-Evans C. Antioxidant drugs and the inhibition of low-density lipoprotein oxidation. Biochem Soc Trans 1993a; 21: 362–366.

24. Green ESR, Evans H, Rice-Evans P, Davies MJ, Salah N, Rice-Evans C. The efficacy of monohydroxamates as free radical scavenging agents compared with di- and tri-hydroxamates. Biochem Pharmacol 1993b; 45: 357–366.

25. Cooper C, Green ESR, Rice-Evans C, Davies MJ, Wrigglesworth JM. A hydrogen-donating monohydroxamate scavengers ferryl myolgobin radicals. Free Rad Res 1994; 20: 219–227.

26. Turner JJO, Rice-Evans C, Davies MJ, Newman ESR. The formation of free radicals by cardiac myocytes under oxidative stress and the effects of electron donating drugs. Biochem J 1991; 277: 833–837.

27. Everse J, Everse KE, Grisham MB. Peroxidases in chemistry and biology. Boca Raton, FL: CRC Press, 1991.

28. Cadenas E. Oxygen activation and reactive oxygen species detoxification. In: Ahmad S, ed. Oxidative Stress and Antioxidant Defences in Biology. New York: Chapman and Hall, 1995: 1–61.

29. George P, Irvine DH. The reaction between metmyoglobin and hydrogen peroxide. Biochem J 1952; 52: 511–517.
30. Yonetani T, Schleyer H. Studies on cytochrome c peroxidase. J Biol Chem 1967; 242: 1974–1979.
31. Gibson JF, Ingram DJE, Nichols P. Free radical produced in the reaction of metmyoglobin with hydrogen peroxide. Nature 1958; 181: 1398–1399.
32. King WK, Looney FD, Winfield ME. Amino acid free radicals in oxidised metmyoglobin. Biochim Biophys Acta 1967; 133: 65–82.
33. Harada K, Yamazaki I. Electron spin resonance spectra of free radicals formed in the reaction of metmyoglobins with ethylhydroperoxide. J Biochem 1987; 101: 283–286.
34. Davies MJ. Identification of a globin free radical in equine myoglobin treated with peroxides. Biochim Biophys Acta 1991; 1077: 86–90.
35. Tew D, Ortiz de Montellano P. The myoglobin protein radical. Coupling of Tyr-103 to Tyr-151 in the H_2O_2-mediated cross-linking of sperm whale myoglobin. J Biol Chem 1988; 263: 17880–17886.
36. Kelman DJ, Mason RP. The myoglobin-derived radical formed on reaction of metmyoglobin with hydrogen peroxide is not a tyrosine peroxyl radical. Free Rad Res Commun 1992; 16: 27–33.
37. Newman ESR, Rice-Evans CA, Davies MJ. Identification of initiating agents in myoglobin-induced lipid peroxidation. Biochem Biophys Res Commun 1991; 179: 1414–1419.
38. Dee G, Rice-Evans C, Bruckdorfer KR, Obeyesekera S, Meriji S, Jacobs M. The modulation of ferryl myoglobin formation and its oxidative effects on LDL by nitric oxide. FEBS Lett 1991; 294: 38–42.
39. Whitburn KD, Shieh JJ, Sellers RM, Hoffman MZ, Taub IA. Redox transformations in ferrimyoglobin induced by radiation-generated free radicals in aqueous solution. J Biol Chem 1982; 257: 1860–1869.
40. Petersen RL, Symons MCR, Taiwo FA. Application of radiation and electron spin resonance spectroscopy to the study of ferryl myoglobin. J Chem Soc Far Trans 1989; 85: 2435–2443.
41. Catalano CE, Choe YS, Ortiz de Montellano P. Reactions of the protein radical in peroxide-treated myoglobin. J Biol Chem 1989; 264: 10534–10541.
42. Hogg N, Rice-Evans C, Darley-Usmar V, Wilson MT, Paganga G, Bourne L. The role of lipid hydroperoxides in the myoglobin-dependent oxidation of LDL. Arch Biochem Biophys 1994; 314: 39–44.
43. Paganga G, Rice-Evans C, Rule R, Leake D. The interaction between ruptured erythrocytes and low density lipoproteins. FEBS Lett 1992; 303: 154–158.
44. Balla G, Jacob HS, Eaton JW, Belcher JD, Vercellotti GM. Hemin: a possible physiological mediator of low density lipoprotein oxidation and endothelial injury. Arterioscler Thromb 1991; 11: 1700–1711.
45. Salah N, Miller NJ, Paganga G, Tijburg L, Bolwell GP, Rice-Evans C. Polyphenolic flavenols as scavengers of aqueous phase radicals and as chain-breaking antioxidants. Arch Biochem Biophys 1995; 322: 339–346.
46. Castelluccio C, Paganga G, Melikian N, Bolwell GP, Pridham J, Sampson J, Rice-

Evans C. Antioxidant potential of intermediates in phenylpropanoid metabolism in higher plants. FEBS Lett 1995; 368: 188–192.

47. Damiani E, Paganga G, Lucedio G, Rice-Evans C. Inhibition of copper-mediated low density lipoprotein peroxidation by quinoline and indolinone nitroxide radicals. Biochem Pharmacol 1994; 48: 1155–1161.

48. Rice-Evans C, Miller N, Paganga G. Structure-antioxidant activity relationships of flavonoids and phenolic aids. Free Rad Biol Med 1996; 20: 933–956.

49. Hearse DJ, Bolli R. Reperfusion induced injury: manifestations, mechanisms and clinical relevance. Cardiovasc Res 1992; 26: 101–108.

50. Bolli R. Myocardial ischaemia: metabolic disorders leading to cell death. Rev Port Cardiol 1994; 13: 649–653.

51. Garlick PB, Davies MJ, Hearse DJ, Slater TF. Direct detection of free radicals in the reperfused rat heart using electron spin resonance spectroscopy. Circ Res 1987; 61: 757–760.

52. Blasig IE, Ebert B, Hennig C, Pali T, Tosaki A. Inverse relationship between ESR spin trapping of oxy-radicals and degree of functional, recovery during myocardial reperfusion in isolated working rat heart. Cardiovas Res 1990; 24: 263–270.

53. Blasig IE, Shuter S, Garlick P, Slater T. Relative time-profiles for radical trapping, coronary flow, enzyme leakage arrhythmias and function during myocardial reperfusion. Free Rad Biol Med 1994; 16: 35–41.

54. Tosaki A, Bagchi D, Pali T, Cordia GA, Das DK. Comparison of ESR and HPLC methods for the detection of ·OH radicals in ischaemic/reperfused hearts. A relationship between the genesis of free radicals and reperfusion arrhythmias. Biochem Pharmacol 1993; 45: 961–969.

55. Collis CS, Rice-Evans C. Free radicals in hypoxia and reoxygenation. In: Haddad GG, Lister G, eds. Tissue Oxygen Deprivation: From Molecular to Integrated Function. New York: Marcel Dekker, 1996.

56. Bolli R, Jeroudi MO, Patel BS, Aruoma OI, Halliwell B, Lai EK, McCay PB. Marked reduction of free radical generation and contractile dysfunction by antioxidant therapy begun at the time of reperfusion. Cir Res 1989; 65: 607–22.

57. Bolli R, Patel BS, Jeroudi MO, Li X-Y, Triana JF, Lai EK, McCay PB. Iron-mediated radical reactions upon reperfusion contribute to myocardial "stunning." Am J Physiol 1990; 259: H1901–1911.

58. Krawatowska-Prokopczuk E, Czarnowska E, Beresewic A. Iron availability and free radical induced injury in the isolated ischemic/reperfused rat heart. Cardiovasc Res 1992; 26: 58–66.

59. Chevion M, Jiang Y, Har-El R, Berenshtein E, Uretzky G, Kitrossky N. Copper and iron are mobilized following myocardial ischaemia: possible predictive criteria for tissue injury. Proc Natl Acad Sci USA 1993; 90: 1102–1106.

60. Halliwell B, Gutteridge JMC. Oxygen toxicity, oxygen radicals, transition metals and disease. Biochem J 1984; 219: 1–14.

61. Drexel H, Durozak E, Kirchmair W, Miulz M, Puschendorf B, Dienstl F. Myoglobinaemia in the early phase of acute myocardial infarction. Am Heart J 1983; 105: 641–651.

62. Gutteridge JMC. Iron promoters of the Fenton reaction and lipid peroxidation can be released from haemoglobin by peroxides. FEBS Lett 1986; 201: 291–295.

63. Voogd A, Sluiter W, Koster JF. The increased susceptibility to hydrogen peroxide of the (post-)ischaemic rat heart is associated with the magnitude of the low molecular weight iron pool. Free Rad Biol Med 1994; 16: 453–458.

64. Porreca E, Del Boccio G, Lapenna D, Di Febbo C, Pennelli A, Cipollone F, Di Ilio C. Myocardial antioxidant defense mechanisms: time related changes after reperfusion of the ischemic rat heart. Free Rad Res 1994; 20: 171–179.

65. Bernier M, Hearse DJ, Manning AS. Reperfusion-induced arrhythmias and oxygen-derived free radicals: studies with anti-free radical interventions and a free radical generating system in the isolated perfused rat heart. Circ Res 1986; 58: 331–340.

66. Bernier M, Manning AS, Hearse DJ. Reperfusion arrhythmias: dose-related protection by anti-free radical interventions. Am J Physiol 1989; 256: H1344–1352.

67. Reddy BR, Kloner RA, Przyklenk K. Early pretreatment with desferrioxamine limits myocardial ischemia/reperfusion injury. Free Rad Biol Med 1989; 7: 45–52.

68. Bolli R, Patel BS, Zhu W-X, O'Neil PG, Hartley CJ, Charalat ML, Roberts R. The iron chelator desferrioxamine attenuated postischaemic ventricular dysfunction. Am J Physiol 1987; 253: H1372–80.

69. Lloyd JB, Cable H, Rice-Evans C. Evidence the desferrioxamine cannot enter cells by passive diffusion. Biochem Pharmacol 1991; 41: 1361–1363.

70. Reddy BR, Wynne J, Kloner RA, Przyklenk K. Pretreatment with the iron chelator desferrioxamine fails to provide sustained protection against myocardial ischemia-reperfusion injury. Cardiovasc Res 1991; 25: 711–719.

71. Collis CS, Davies MJ, Rice-Evans C. Comparison of N-methyl hexanoyl-hydroxamic acid, a novel antioxidant, with desferrioxamine and N-acetyl cysteine against reperfusion-induced dysfunctions in isolated rat heart. J Cardiovasc Pharmacol 1993a; 22: 336–342.

72. De Bono DP, Yang WD, Davies MJ, Collis CS, Rice-Evans CA. Effects of N-methyl hexanoylhydroxamic acid (NMHH) and myoglobin on endothelial damage by hydrogen peroxide. Cardiovasc Res 1994; 28: 1641-1646.

73. Collis CS, Davies MJ, Rice-Evans C. The effects of N-methyl butyrohydroxamic acid and other monohydroxamates on reperfusion-induced damage to contractile function in the isolated rat heart. Free Rad Res Commun 1993b; 18: 269–277.

74. Kramer JH, Misik V, Weglicki WB. Lipid peroxidation derived-free radical production and postischaemic myocardial reperfusion injury. Ann NY Acad Sci 1994; 723: 180–196.

75. Manning AS, Hearse DJ. Reperfusion induced arrhythmias: mechanisms and prevention. J Mol Cell Cardiol 1984; 16: 497–518.

76. Jeroudi MO, Hartley CJ, Bolli R. Myocardial reperfusion injury: role of oxygen radicals and potential therapy with antioxidants. Am J Cardiol 1994; 73: 2B-7B.

77. Curtis MJ, Pugsley MK, Walker MJA. Endogenous chemical mediators of ventricular arrhythmias in ischaemic heart disease. Cardiovasc Res 1993; 27: 703–719.

78. Collis CS, Rice-Evans C. Novel monohydroxamate drugs attenuate myocardial reperfusion-induced arrhythmias. Int J Biochem Cell Biol 1996; 28: 405–413.
79. Maxwell MP, Marston C, Hadley MR, Salmon JA, Garland LG. Selective 5-lipoxygenase inhibitor BWA4C does not influence progression of tissue injury in a canine model of regional myocardial ischaemia and reperfusion. J Cardiovasc Pharmacol 1991; 17: 539–545.
80. Katoh S, Toyama J, Kodama I, Kamiya K, Akita T, Abe T. Protective action of iron-chelating agents (catechol, mimosine, deferoxamine, and kojic acid) against ischemia-reperfusion injury of isolated neonatal rabbit hearts. Eur Surg Res 1992; 24: 349–355.

Index

About the Editors

LESTER PACKER is Professor of Molecular and Cell Biology, Division of Cell and Developmental Biology, University of California, Berkeley. Dr. Packer is the author of over 500 published articles and coeditor of the *Handbook of Antioxidants, Oxidative Stress in Dermatology, Vitamin E in Health and Disease, Retinoids: Progress in Research and Clinical Applications*, and *Biothiols in Health and Disease* (all titles, Marcel Dekker, Inc.). He is President of the Oxygen Club of California, President of the International Society of Free Radical Research, and a member of the Oxygen Society, the American Society of Biochemistry and Molecular Biology, and the American Institute of Nutrition, among others. Dr. Packer received the B.S. (1951) and M.S. (1952) degrees in biology and chemistry from Brooklyn College, Brooklyn, New York, and the Ph.D. degree (1956) in microbiology and biochemistry from Yale University, New Haven, Connecticut.

ENRIQUE CADENAS is a Professor in and Chairman of the Department of Molecular Pharmacology and Toxicology in the School of Pharmacy and a Professor of Biochemistry in the School of Medicine, University of Southern California, Los Angeles. The author of over 160 professional publications and coeditor of the *Handbook of Antioxidants* and *Biothiols in Health and Disease* (both titles, Marcel Dekker, Inc.), Dr. Cadenas is a member of the American

Chemical Society, the American Society of Photobiology, the Biochemical Society (United Kingdom), and the European Society for Photobiology, among others. Dr. Cadenas received the M.D. degree (1973) and the Ph.D. degree (1977) in biochemistry from the University of Buenos Aires School of Medicine, Argentina.

Milton Keynes UK
Ingram Content Group UK Ltd.
UKHW020010071024
449327UK00031B/2721